IN-SITU CHARACTERIZATION OF HETEROGENEOUS CATALYSTS

IN-SITU CHARACTERIZATION OF HETEROGENEOUS CATALYSTS

Edited by

JOSÉ A. RODRIGUEZ
Brookhaven National Laboratory
Upton, New York, USA

JONATHAN C. HANSON
Brookhaven National Laboratory
Upton, New York, USA

PETER J. CHUPAS
Argonne National Laboratory
Lemont, Illinois, USA

WILEY

Published by John Wiley & Sons, Inc., Hoboken, New Jersey
Published simultaneously in Canada

For general information on our other products and services or for technical support, please contact our Customer Care Department within the United States at (800) 762-2974, outside the United States at (317) 572-3993 or fax (317) 572-4002.

Wiley also publishes its books in a variety of electronic formats. Some content that appears in print may not be available in electronic formats. For more information about Wiley products, visit our web site at www.wiley.com.

Library of Congress Cataloging-in-Publication Data:
In-situ characterization of heterogeneous catalysts / edited by José A. Rodriguez, Brookhaven National Laboratory, Upton, New York, USA, Jonathan C. Hanson, Brookhaven National Laboratory, Upton, New York, USA, Peter J. Chupas, Argonne National Laboratory, Lemont, Illinois, USA.
pages cm
Includes bibliographical references and index.
ISBN 978-1-118-00016-8 (hardback)
1. Heterogeneous catalysis. I. Rodriguez, José A., editor of compilation. II. Hanson, Jonathan C., editor of compilation. III. Chupas, Peter J., editor of compilation.
QD505.I49 2013
541'.395–dc23

2012045180

ISBN: 9781118000168

10 9 8 7 6 5 4 3 2 1

CONTENTS

CONTRIBUTORS

Miguel A. Bañares, Catalytic Spectroscopy Laboratory, CSIC—Instituto de Catalisis y Petroleoquímica, Madrid, Spain

Andrew M. Beale, Department of Inorganic Chemistry and Catalysis, Debye Institute for NanoMaterials Science, Utrecht University, Utrecht, The Netherlands

Hendrik Bluhm, Chemical Sciences Division, Lawrence Berkeley National Laboratory, Berkeley, CA

Angelika Brückner, Leibniz-Institut für Katalyse e. V. an der Universität Rostock, Rostock, Germany

Karena W. Chapman, X-ray Science Division, Advanced Photon Source Argonne National Laboratory, Argonne, IL

Wei-Ren Chen, Chemical and Engineering Materials Division, Spallation Neutron Source, Oak Ridge, TN

Fang Cheng, Department of Chemistry and Biochemistry, University of Notre Dame, Notre Dame, IN

Peter J. Chupas, X-ray Science Division, Advanced Photon Source, Argonne National Laboratory, Argonne, IL

Andrew J. Dent, SERC Daresbury Laboratory, Warrington, UK

Marcos Fernández-García, CSIC—Instituto de Catalisis y Petroleoquímica, Madrid, Spain

Anatoly I. Frenkel, Physics Department, Yeshiva University, New York, NY

Jan-Dierk Grunwaldt, Karlsruher Institut für Technologie (KIT), Institute of Technical Chemistry and Polymer, Chemistry (ITCP), Campus South and Institute of Catalysis Research and Technology (IKFT), Campus North, Karlsruhe, Germany

Jonathan C. Hanson, Chemistry Department, Brookhaven National Laboratory, Upton, NY

Michael Hävecker, Division Solar Energy Research, Elektronenspeicherring BESSY II, Berlin, Germany

Ashfia Huq, Chemical and Engineering Materials Division, Spallation Neutron Source, Oak Ridge, TN

Syed Khalid, National Synchrotron Light Source, Brookhaven National Laboratory, Upton, NY

Axel Knop-Gericke, Department of Inorganic Chemistry, Fritz-Haber-Institut der Max-Planck-Gesellschaft, Berlin, Germany

Zhi Liu, Advanced Light Source, Lawrence Berkeley National Laboratory, Berkeley, CA

Gerhard Mestl, SÜD-CHEMIE AG, Bruckmühl, Germany

Marco Milanesio, Dipartimento di Scienze e Tecnologie Avanzate and Nano-SiSTeMI Interdisciplinary Centre, Università del Piemonte Orientale "A. Avogadro," Alessandria, Italy

Kumudu Mudiyanselage, Chemistry Department, Brookhaven National Laboratory, Upton, NY

Maarten Nachtegaal, Paul Scherrer Institute, Viiligen, Switzerland

Mark A. Newton, European Synchrotron Radiation Facility, Grenoble, France

Luan Nguyen, Department of Chemistry and Biochemistry, University of Notre Dame, Notre Dame, IN

Poul Norby, Technical University of Denmark, Department of Energy Conversion and Storage, Roskilde, Denmark

José A. Rodriguez, Chemistry Department, Brookhaven National Laboratory, Upton, NY

Javier Ruiz-Martinez, Department of Inorganic Chemistry and Catalysis, Debye Institute for NanoMaterials Science, Utrecht University, Utrecht, The Netherlands

Christian G. Schroer, Technische Universität Dresden, Institute for Structural Physics, Dresden, Germany

Darío J. Stacchiola, Chemistry Department, Brookhaven National Laboratory, Upton, NY

David E. Starr, Center for Functional Nanomaterials, Brookhaven National Laboratory, Upton, NY

Eli Stavitski, National Synchrotron Light Source, Brookhaven National Laboratory, Upton, NY

Franklin (Feng) Tao, Department of Chemistry and Biochemistry, University of Notre Dame, Notre Dame, IN

Atsushi Urakawa, Institute of Chemical Research of Catalonia, Tarragona, Spain

Wouter van Beek, Swiss-Norwegian Beamlines at ESRF, Grenoble Cedex, France

Lei Wang, Department of Chemistry and Biochemistry, University of Notre Dame, Notre Dame, IN

Bert M. Weckhuysen, Department of Inorganic Chemistry and Catalysis, Debye Institute for NanoMaterials Science, Utrecht University, Utrecht, The Netherlands

Yingchun Ye, Department of Chemistry and Biochemistry, University of Notre Dame, Notre Dame, IN

Shiran Zhang, Department of Chemistry and Biochemistry, University of Notre Dame, Notre Dame, IN

Yuan Zhu, Department of Chemistry and Biochemistry, University of Notre Dame, Notre Dame, IN

INTRODUCTION: GOALS AND CHALLENGES FOR THE *IN-SITU* CHARACTERIZATION OF HETEROGENEOUS CATALYSTS

José A. Rodriguez, Jonathan C. Hanson, and Peter J. Chupas

CATALYSIS AND THE NEED TO CHARACTERIZE ACTIVE SITES IN DIFFERENT TYPES OF MATERIALS AND CHEMICAL ENVIRONMENTS

Catalysis is central to the production of fuels and chemicals, including more than 70% of today's chemical products. About 20% of the value of all commercial products manufactured in the United States is derived from processes involving catalysis [1–3]. Thus, the understanding and optimization of heterogeneous catalysts is a critical need within the chemical industry [2, 3]. The most important considerations when designing a new catalyst or when optimizing the performance of an existing one are activity, selectivity, and durability [3]. The catalyst must be able to perform a reaction of interest at an acceptable rate under a practical set of temperatures and pressures. There is a commercial incentive to tune catalysts to operate at low temperatures and pressures [2, 3]. It is concurrently important that the rate of side reactions be minimal, and the catalyst must be able to sustain the desired reaction over long periods of time [3]. When designing or optimizing a catalyst, one must know the nature of its active sites and how they interact with the reactants, intermediates, and products of a catalytic process. The structural and electronic properties of the active

In-situ *Characterization of Heterogeneous Catalysts*, First Edition.
Edited by José A. Rodriguez, Jonathan C. Hanson, and Peter J. Chupas.

sites are usually affected by the chemical environment in which they operate. This book describes experimental techniques that have been developed to characterize catalysts and study reaction mechanisms *in situ* under academic and technical or industrial conditions.

Heterogeneous catalysts can be metals, oxides, carbides, sulfides, nitrides, practically any type of material [1–3]. In the nineteenth century it was already known that all the regions or atoms in the surface of a heterogeneous catalyst are not involved in the real catalytic process. The concept of the "active site" in catalysis has been used since its introduction in the 1920s by Taylor and Langmuir [4, 5]. However, despite considerable scientific effort over many years, our knowledge of what exactly the active sites are and how they really work is in most cases very primitive [6]. First, it is necessary to identify the active site and fully characterize its electronic and structural properties under reaction conditions. Since industrial catalysts and catalytic processes are extremely complex [2, 3], the development of techniques for the characterization of catalytic systems *in situ* as they evolve in time with a changing chemical environment is a very challenging task. Figure I.1 shows four different types of systems which are frequently used in catalytic studies. Figure I.1A displays an image of high-resolution transmission electron microscopy (HR-TEM) for a high surface area Pt-Ru/CeO_2 catalyst [7]. One sees particles of a Pt–Ru alloy (1.5–3 nm in size) on top of a ceria support that mainly exhibits the (111) face of the oxide. In the case of Figure I.1B, the HR-TEM image corresponds to a Pt/CeO_x/TiO_2 catalyst [8]. The loading of Pt is very low (0.5 wt%) and the small particles of the metal (0.4–0.5 nm) are difficult to detect with HR-TEM. The TiO_2 support nanoparticles are monocrystalline and present an average size of 10–15 nm. The bright spots as indicated by the arrows represent CeO_2 nanoparticles with an average diameter of about 4–5 nm. In the Pt/CeO_x/TiO_2 catalyst, the very low content of Pt points to the need for characterization techniques with a high sensitivity. Many catalytic processes are carried out on the type of metal/oxide powder catalyst shown in Figure I.1A,B, with the surface exhibiting a high degree of structural heterogeneity [1–3]. In contrast, when studying correlations between surface "structure" and reactivity, it may be advantageous to do experiments on well-defined single-crystal surfaces such as those displayed in Figure I.1C,D [9, 10]. Furthermore, the system in Figure I.1A has the typical metal/oxide configuration seen in many industrial catalysts [1–3]. However, to enhance the participation of ceria in catalytic reactions, one may adopt an inverse configuration in which nanoparticles of this oxide are deposited on top of the surface of a metal or another oxide [11]. Thus, the inverse CeO_2/CuO and CeO_2/CuO_x/Cu(111) catalysts shown in Figure I.2 exhibit a very high activity for the low-temperature oxidation of CO and the CO preferential oxidation (PROX) reactions [12, 13]. In practical terms, one needs to develop techniques for the *in-situ* characterization of all the systems shown in Figure I.1 and Figure I.2, and even much more complex materials in which there are multiple phases coexisting in a three-dimensional space [14]. When focusing on spatial resolution, characteristic length scales

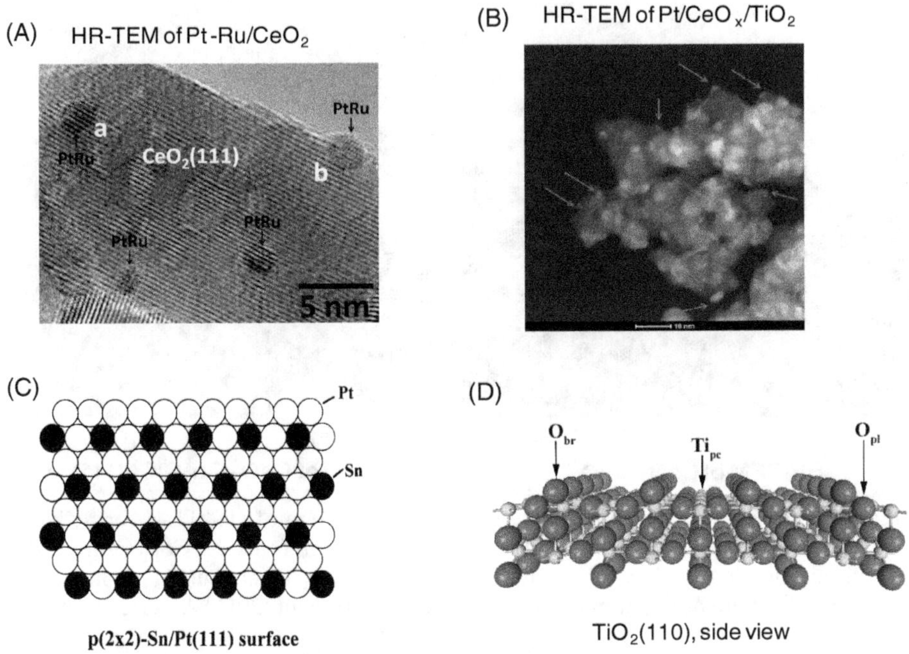

Figure I.1. Different types of systems investigated in studies in the area of catalysis. (A) Image of HR-TEM for a Pt–Ru/CeO2 catalyst. Reprinted with permission from Reference 7. Copyright (2012) Elsevier. (B) Image of HR-TEM for a Pt/CeO_x/TiO_2 catalyst. Reprinted with permission from Reference 8. Copyright (2012) American Chemical Society. (C) Top view of a Sn–Pt(111) model catalyst. (D) Side view of a TiO_2(110) surface.

relevant for catalytic materials range from a few millimeters (usual sample size) to nanometers (typical size of catalytic nanoparticles). If one is interested in the geometrical structure of molecules adsorbed on the surface of the catalyst as reactants or intermediates, then the significant length scale is in the order of angstroms or picometers.

Time is a very important parameter when studying a catalytic process. A phenomenological rate law gives the disappearance of reactants or appearance of products as a function of time for a particular set of reaction conditions (pressure, temperature, and chemical concentrations) [6]. At a microscopic level, molecules are transformed and the chemical environment around the active sites of a catalyst changes as a function of time [6, 10]. In order to obtain a full understanding of the key transformations associated with a catalytic process, one must be able to track the time evolution of the structural and electronic properties of the active site. The activation and breaking of a chemical bond inside a molecule occurs in the picosecond regime but the completion

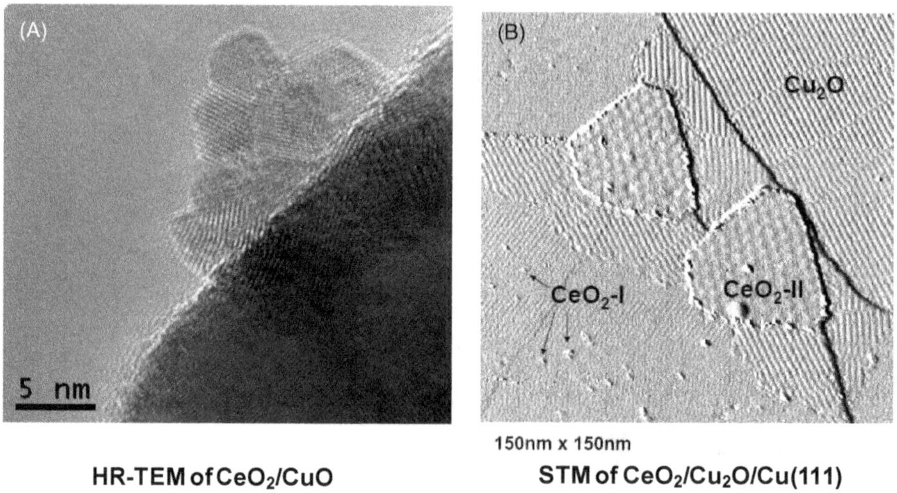

Figure I.2. Inverse ceria/copper oxide catalysts. (A) Image of HR-TEM for a CeO_2/CuO powder catalyst. Reprinted with permission from Reference 12. Copyright (2010) American Chemical Society. (B) Image of STM for a CeO_2/CuO_x/Cu(111) model catalyst. Reprinted with permission from Reference 13. Copyright (2011) American Chemical Society.

of a full reaction cycle on the surface of a catalyst usually takes longer times. Based on turnover rates for typical reactions (10^{-2}–10^3 molecules/site•second) [9, 10], one can estimate that the timescale for a catalytic transformation on the surface of a catalyst is usually in the range of minutes to milliseconds under typical reaction conditions. Ideally, one must be able to track transformations of the catalytic material in this time range. Very valuable information about catalytic processes can be obtained under steady-state conditions (i.e., no variations in temperature, pressure, and reaction rate), but to obtain a detailed understanding of reaction mechanism one frequently has to do experiments in a non-steady-state or transient mode in which the properties of the catalytic system are perturbed by changing the temperature (see Fig. I.3), or by pulses (fluctuations) in the pressures of the reactants [5, 6, 15]. In principle, the perturbations in the reaction conditions can affect the structural properties of the catalyst (Fig. I.3 and Fig. I.4) and the lifetime of surface intermediates that are produced during the reaction.

On the basis of the considerations discussed in the previous paragraphs, one finds that the ideal tool for a detailed study of typical catalytic processes should have high sensitivity and allow us the fast acquisition of data in a milliseconds timescale with a spatial resolution in the range of nanometers or smaller, monitoring simultaneously the properties of the catalyst active sites and the adsorbed reaction intermediates. This is a highly demanding set of requirements. At the present time, no single technique can accomplish all of these tasks. For a complete characterization of a catalytic process, one must

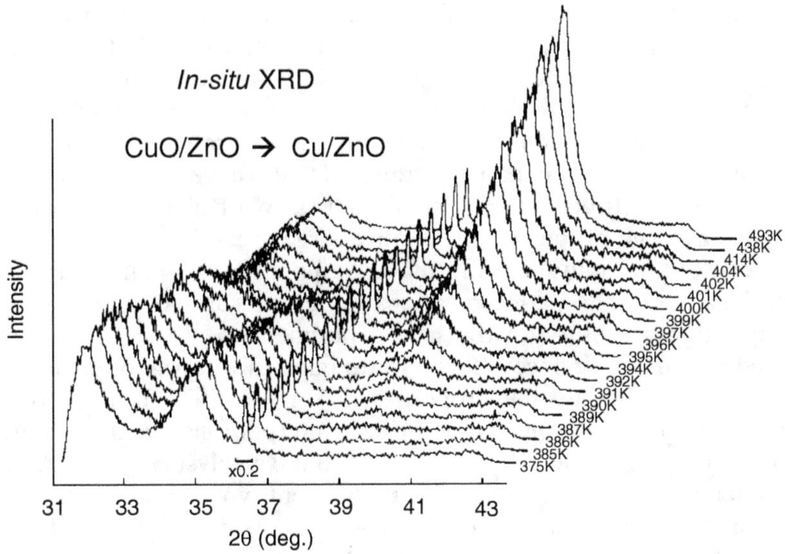

Figure I.3. Three-dimensional plot of *in-situ* XRD patterns collected during the reduction/activation of a powder CuO/ZnO catalyst in a mixture of 0.25% CO, 0.25% CO_2, and 4% H_2 in argon. The XRD patterns were collected at temperatures between 375 and 493 K. A CuO → Cu structural transformation was seen around 400 K. Reprinted with permission from Reference 5. Copyright (1991) Elsevier.

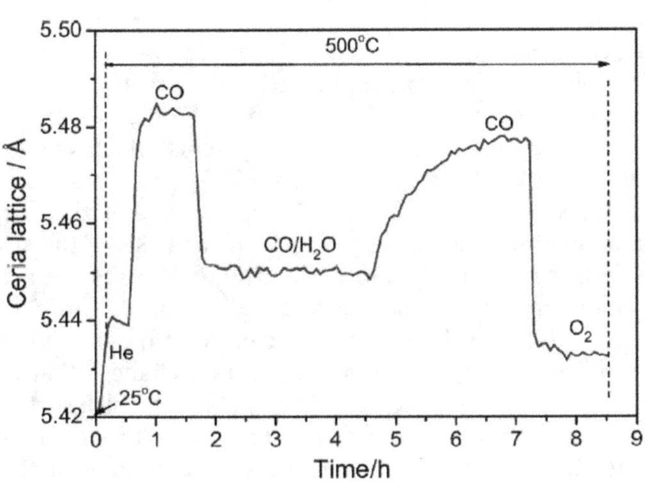

Figure I.4. Variation of the lattice constant of ceria after exposing a powder Au–CeO_2 catalyst to CO, the reaction mixture for the water–gas shift reaction (CO/H_2O), CO, and O_2. The reported values were obtained after analyzing results of time-resolved XRD by Rietveld refinement. Reprinted with permission from Reference 15. Copyright (2006) American Chemical Society.

combine different techniques. Many of the existing techniques focus only on examining the properties of the catalyst, while others are better suited for studying the surface chemistry associated with the reaction process. In practical terms, it is necessary to find a reasonable balance between temporal and spatial resolutions. Techniques which can provide morphological information at the nanometer or subnanometer range do not have the time resolution required for most transient or kinetic studies. Most characterization techniques give average properties for relative large areas (from micrometers to millimeters) of the catalyst sample, making emphasis on obtaining high sensitivity for dealing with low concentrations of elements and for allowing the fast data acquisition necessary for transient experiments.

Nowadays surface science offers a quite impressive array of experimental techniques to investigate the properties of surfaces [6, 10]. Many of these techniques are based on some type of excitation (photons, electrons, neutrons, ions, electromagnetic field, heat, etc.) to which the catalyst responds. A similar type of excitation can be associated with different types of phenomena yielding information about different properties. For example, X-ray diffraction (XRD), X-ray absorption spectroscopy (XAS), and X-ray photoelectron spectroscopy (XPS) all use photons for probing the sample and are among the most frequently employed techniques for catalyst characterization [5, 6]. XRD occurs in the elastic scattering of X-ray photons by atoms in a periodic lattice and can be used to obtain structural information. In contrast, irradiating a catalyst with X-ray photons can induce excitation of electrons from occupied core levels to empty valence levels (XAS) or generate photoelectrons (XPS), yielding information about the composition of the sample and the electronic properties of the elements present.

Many surface science techniques operate under vacuum or high-vacuum conditions [6, 10] and cannot be applied in the characterization of catalysts under normal reaction conditions. In order to overcome this problem, several laboratories have developed experimental systems which combine a high-pressure system with an ultra-high-vacuum (UHV) analysis chamber [9, 10, 16]. The high-pressure reactor allows the kinetics of catalytic reactions to be measured on a given surface (see Fig. I.5), while analysis of the structure and composition of the surface both before and after reaction can be accomplished in the UHV chamber. This approach has provided valuable insights into many catalytic processes [9, 10, 16], but it is now clear that the combination of *prenatal* and *postmortem* analysis can miss important changes that occur to the surface of a catalyst under reaction conditions [17–19]. Since surfaces are flexible entities which can reconstruct after interacting with adsorbed molecules (see Fig. I.6 and References [17–19]), one must perform an *in-situ* characterization of the catalyst. In recent years, notable advances in design and instrumentation have added moderate-pressure XPS (Fig. I.7) and high-pressure scanning tunneling microscopy (HP-STM) [17, 20, 21] to the arsenal of techniques that is available for the *in-situ* characterization of catalysts. Furthermore, there has been an extraordinary effort to develop or improve characterization

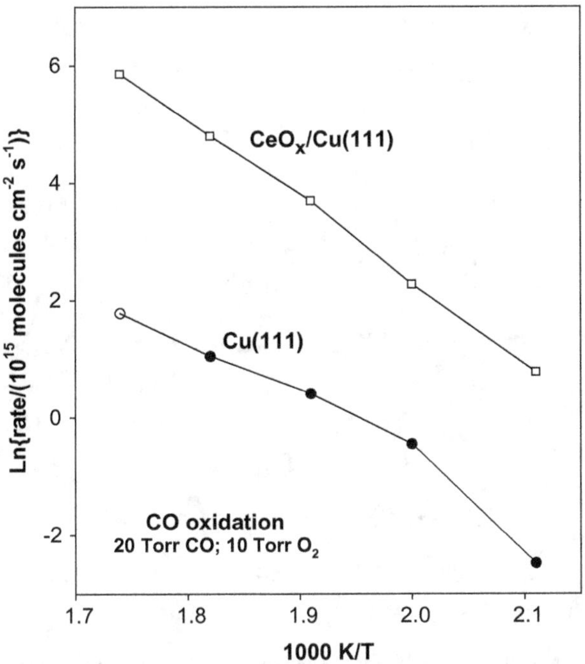

Figure I.5. Arrhenius plots for the $2CO + O_2 \rightarrow 2CO_2$ reaction on Cu(111) and on a surface in which ceria nanoparticles covered ~18% of the copper substrate. $P_{CO} = 20$ Torr, $P_{O2} = 10$ Torr. Reprinted with permission from Reference 13. Copyright (2011) American Chemical Society.

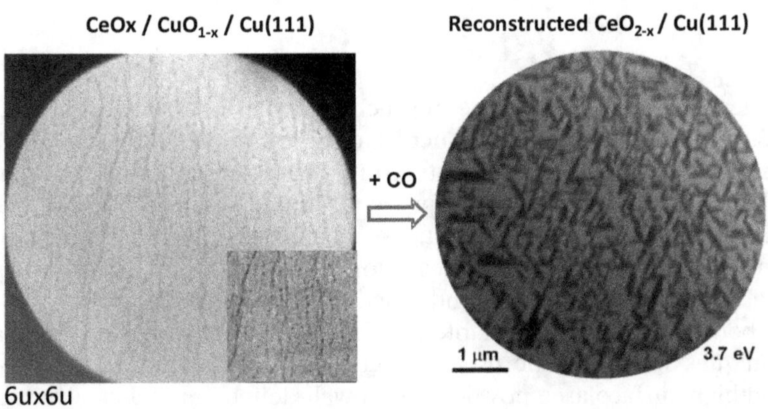

Figure I.6. Images of *in-situ* low-energy electron microscopy obtained before and during the exposure of a CeO$_x$/CuO$_x$/Cu(111) model catalyst to CO [19]. In the final step, the system consisted of ceria nanoparticles dispersed on a reconstructed Cu(111) surface. Reprinted with permission from Reference 19. Copyright (2012) American Chemical Society.

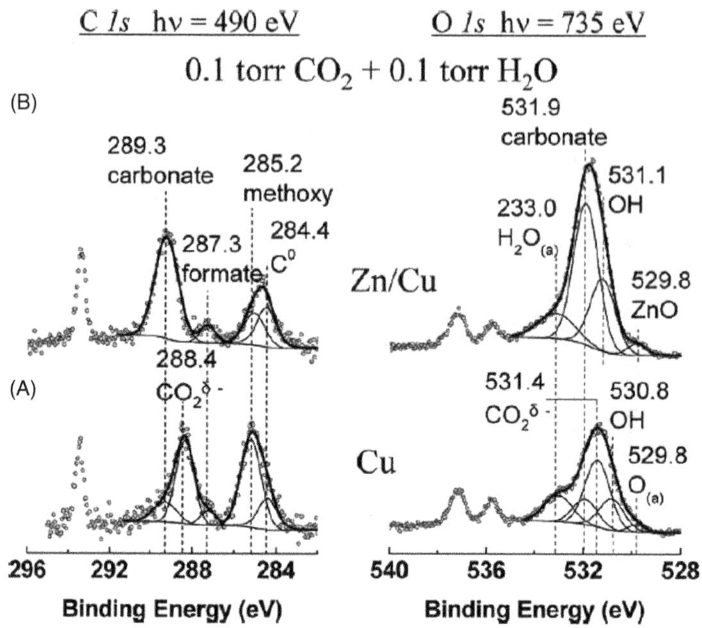

Figure I.7. Carbon and oxygen 1s XPS spectra of (A) pure Cu and (B) Zn/Cu (0.1 mL Zn) in the presence of 0.1 Torr of CO_2 and 0.1 Torr of H_2O at room temperature. Two carbonaceous species, formate and methoxy, are shown to form on both surfaces. Activated CO_2 and carbonate species present in pure CO_2 remain visible on each surface. In addition, molecularly adsorbed H_2O is also observed in both spectra. The presence of Zn makes carbonate the majority species on the surface. Reprinted with permission from Reference 20. Copyright (2009) Elsevier.

techniques which take advantage of synchrotrons or neutron facilities. It is in these facilities where major advances have been made in the development of techniques which allow us the structural characterization of crystalline or amorphous materials [22–26], dealing with diluted samples [26, 27], the fast acquisition of data [27–29], and catalyst imaging [24, 26, 30]. There has also been substantial progress in the integration of techniques for the simultaneous study of the catalyst properties and reaction mechanism [25, 30–34].

This book is divided in 16 chapters which cover recent advances in an array of techniques which can be used to study *in-situ* catalytic processes taking place on high-surface area powders or on well-defined model catalysts. Using these techniques, one can perform fundamental studies of the physical and chemical properties of a catalyst obtaining information about composition, structural features, surface morphology, electronic states, and chemical reactivity. Figure I.8 shows a typical approach followed nowadays in the study of catalytic processes. First, one must identify and characterize the active phase

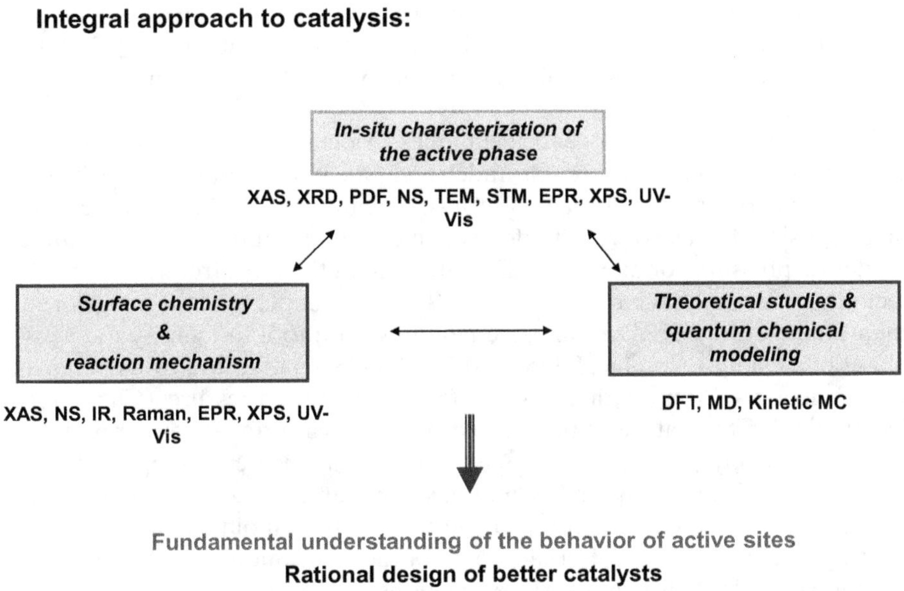

Figure I.8. Different areas of research during the study of catalytic processes. A fundamental understanding is obtained by combining *in-situ* characterization of the active phase in the catalyst with experimental and theoretical studies of the surface chemistry associated with the catalytic process. NS, neutron scattering; TEM, transmission electron microscopy; EPR, electron paramagnetic resonance; UV-Vis, ultraviolet–visible spectroscopy; IR, infrared; DFT, density functional theory; MD; molecular dynamics; MC, Monte Carlo.

of the catalyst. This is usually followed by studies of surface chemistry aimed at obtaining the details of the reaction mechanism. Theoretical studies can provide a conceptual frame for understanding the properties of the catalyst or the steps in the reaction mechanism [35, 36]. The integral approach shown in Figure I.8 eventually can yield a fundamental understanding of the behavior of active sites which can be used for the rational design of better catalysts. This approach usually takes advantage of the *in-situ* techniques described in this book.

CATALYST COMPOSITION AND ACTIVE PHASE IDENTIFICATION

The fist step in any catalytic study is to identify the composition and active phase of the catalyst under reaction conditions. XAS is perhaps the technique most frequently used for this task. The first three chapters in the book show different versions of XAS, making emphasis in time and spatial resolution.

XAS is one of the most powerful techniques available to follow the composition of catalysts as a function of time in kinetic or transient studies [27–29]. It also can be used to study spatial gradients of composition in chemical reactors [24, 26, 30].

XPS is among the most used techniques to measure the composition and chemical state of catalysts *ex situ* under high-vacuum conditions [6]. Chapter 12 describes recent developments in instrumentation [17, 20, 21] which now make possible the use of XPS to determine the composition of catalysts under moderate pressure conditions. XRD is also one of the most frequently applied techniques in catalyst characterization [37]. If the sample has long-range order, then time-resolved XRD is an extremely powerful tool to identify the active phase of a catalyst (see Fig. I.3 [5] and Fig. I.9 [38]) and to follow its behavior as a function of time with changing temperature or pressure (Chapter 4). Depending on the nature of the sample, neutron scattering (Chapter 6), infrared spectroscopy (Chapter 8 and Chapter 9), Raman spectroscopy (Chapter 10, Chapter 15, and Chapter 16), and electron paramagnetic resonance spectroscopy (Chapter 11) also can provide information about the nature of the active phase in a catalyst, although some of these techniques are better suited for studying adsorption and surface reactions.

STRUCTURAL FEATURES AND SPATIAL ARRAY OF THE ACTIVE PHASE IN THE CATALYST

The actual structure of the active phase of a catalyst can range all the way from crystalline to amorphous [1–3]. Furthermore, a part of the catalyst can be crystalline and another amorphous. This is a frequent situation seen in metal/oxide catalysts where the oxide support is mainly crystalline and small metal particles are amorphous (Fig. I.1B). In a common terminology used in the area of crystallography [37], the structure of a catalyst can have local, medium-, or long-range order. This book describes four techniques that can be used to obtain detailed information about the structure of catalysts: extended X-ray absorption fine structure (EXAFS; Chapter 1, Chapter 2, and Chapter 3), X-ray powder diffraction (Chapter 4), pair distribution function (PDF) analysis (Chapter 5), and neutron diffraction (Chapter 6). These techniques respond to different ranges of crystallinity or order in the sample.

X-ray powder diffraction and neutron diffraction require long-range crystalline order. XRD is one of the oldest and most frequently applied techniques for the structural characterization of catalysts [6, 37]. It is frequently used to identify crystalline phases inside catalysts, by means of diffraction lines and the corresponding lattice structural parameters, and to obtain an indication of particle size [6, 37]. Furthermore, time-resolved XRD (Chapter 4) provides a convenient way to study lattice variations in the catalyst structure as a function of temperature and pressure of the reactants (Fig. I.4). A large penetration depth and a high sensitivity to lighter elements, especially mobile species such

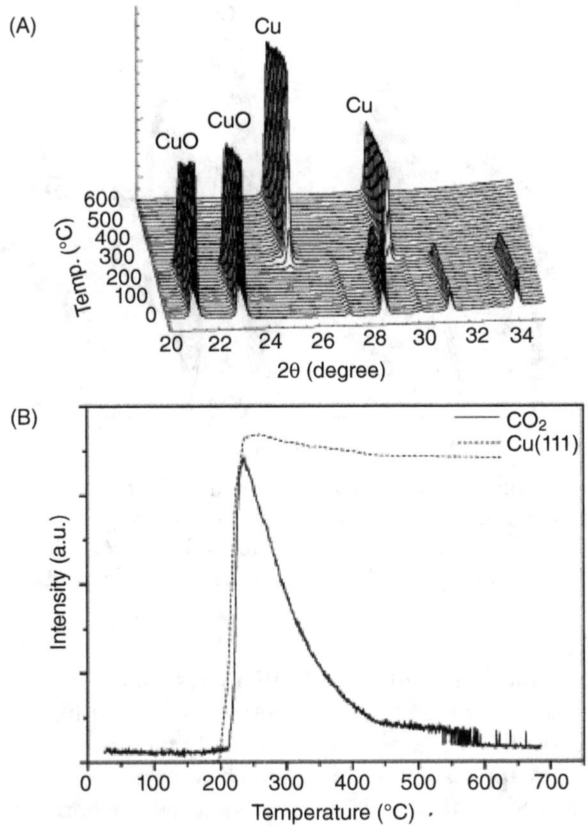

Figure I.9. Temperature-programmed reduction of CuO. (Temperature increase rate = 350°C/h or 5.8°C/min; 5% CO/95% He; gas flow rate ~15–20 mL/min). (A) Time-resolved X-ray diffraction (TR-XRD) patterns. (B) Intensities for the Cu(111) diffraction line in XRD and the CO2 signal (mass-to-charge ratio, *m*/*z*) 44 in quadrupole mass spectrometry (QMS) at the exit of the reactor. Reprinted with permission from Reference 38. Copyright (2004) American Chemical Society.

as hydrogen and oxygen, make neutron diffraction an excellent tool to probe heterogeneous catalysts in operational conditions (Chapter 6). Traditionally the use of neutron diffraction in the field of catalysis has been quite limited. However, recent advances in data acquisition and processing together with the availability of higher flux neutron instruments at various facilities now make the study of the structure and dynamics of heterogeneous catalysts using neutrons a realistic goal (Chapter 6).

Traditionally, EXAFS has been the technique of choice for obtaining structural information for samples with a low degree of order (Chapter 1, Chapter

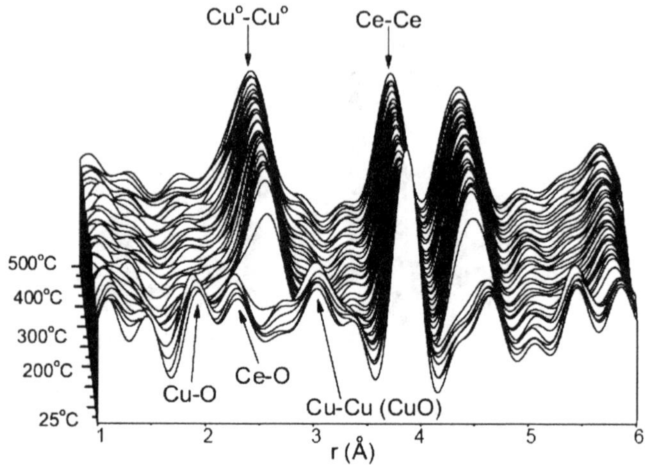

Figure I.10. Time-resolved PDF analysis of an inverse CeO_2/CuO catalyst under water–gas shift reaction conditions [39]. The CuO initially present in the catalyst is reduced to metallic Cu, whereas the ceria transforms into CeO_{2-x}. Reprinted with permission from Reference 39. Copyright (2010) American Chemical Society.

2, and Chapter 3), but in recent years PDF analysis has emerged as a complimentary technique (Chapter 5). The structural insights within the PDF extend from the local coordination environment, usually measured by EXAFS, up to several nanometers approaching the length scale of particles relevant to most catalytic processes. Since the early 2000s, advances in experimental approach, instrumentation, and detector technologies have increased access to PDF data, providing improved data quality in substantially reduced measurement times (Chapter 5). Nowadays, EXAFS and PDF structural studies can be performed to follow the behavior of catalysts in a wide range of pressures and temperatures (see Fig. I.10) [39]. To measure at the same time the structural parameters of crystalline and amorphous regions of a catalyst sample, one can combine XRD with PDF (Chapter 5) or XRD and EXAFS (Chapter 13). The power of combining XRD and EXAFS is illustrated by a recent study of the reduction of a water–gas shift $CuFe_2O_4$ catalyst in CO [32]. For $CuFe_2O_4$ under reducing conditions, the combined use of the two techniques allowed the acquisition of accurate data for the kinetics of nucleation and growth of metallic Cu (Fig. I.11), the active phase in the catalytic process.

THE SURFACE MORPHOLOGY OF THE CATALYST ACTIVE PHASE

On the basis of studies using well-defined single-crystal surfaces, it is possible to classify catalytic processes into structure-sensitive and structure-insensitive

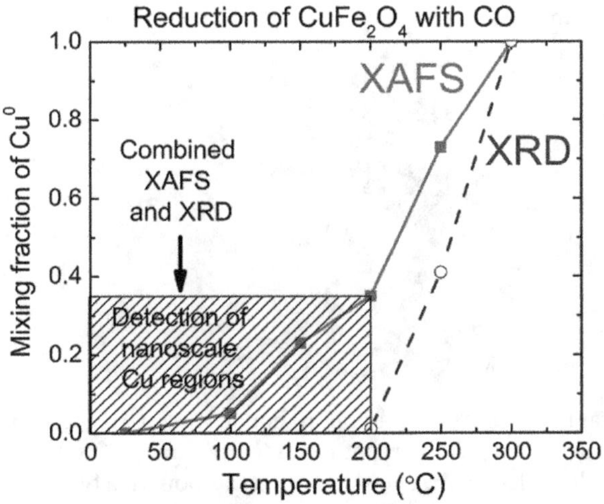

Figure I.11. Comparison of the Cu^0 fractions determined by X-ray diffraction (XRD) and X-ray absorption fine structure (XAFS) for the reduction of a water–gas shift $CuFe_2O_4$ catalysts in CO. Reprinted with permission from Reference 32. Copyright (2011) American Chemical Society.

reactions [9, 10]. If the reaction needs a specific array of sites to turn over, the number of sites may be varied by exposing different crystallographic orientations [6, 9, 10]. In catalysis, a main goal is to induce changes in surface morphology to enhance catalytic activity [10, 18, 36]. This is a very difficult task due to a general lack of knowledge of the behavior of catalytic surfaces under reaction conditions.

Environmental transmission electron microscopy (E-TEM) and HP-STM are two important techniques for studying the surface morphology of powder and model catalysts under reacting gases. Transmission electron microscopy is one the techniques most often used for the *ex-situ* characterization of catalysts [6, 40]. Using this technique the determination of particle size distribution and dispersion is a routine matter. The detection of supported particles is possible if there is sufficient contrast between particles and support (Fig. I.1A). The major limitation of conventional transmission electron microscopy is that it is not an *in-situ* technique [6, 40]. Recent technical developments have made it possible to study the behavior of catalysts under low pressures of a gas [25, 41–43], as shown in Figure I.12. Using aberration-corrected E-TEM, it has been shown that the morphology of an oxide substrate or supported metal particles can depend dramatically on the gaseous environment [25, 42, 43]. Since the 1980s, scanning tunneling microscopy (STM) has been a dominant tool for imaging the surface of single-crystal surfaces (Chapter 7). Studies with STM are usually performed under UHV conditions [10] but, in principle, this

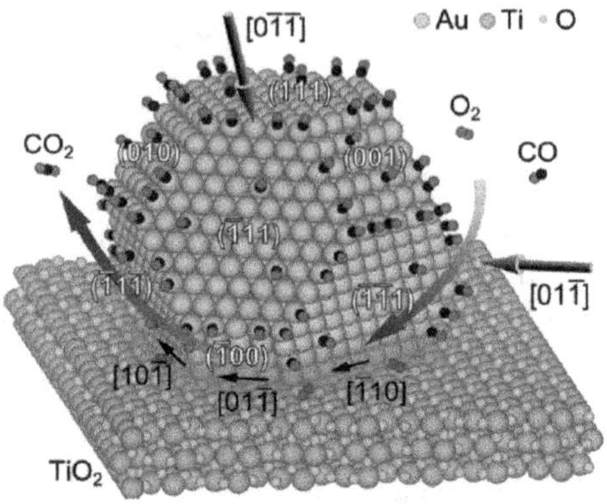

Figure I.12. Instantaneous view of a gold nanoparticle on TiO_2 under reaction conditions (in 1 vol% CO in air [0.75 Torr] at room temperature). The structural model was derived using results of E-TEM. A gold nanoparticle, bounded by {111} and {100} facets, has a polygonal interface with the TiO_2 support. The edges of the polygonal, or perimeter interface, are parallel to the <110> directions. Reprinted with permission from Reference 43. Copyright (2012) John Wiley & Sons.

technique can be applied in a pressure range from UHV to atmospheric or high pressures since the tunneling process between the sample and tip only occurs in a very close range of 3–5 Å. In recent years, remarkable progress has been made in the instrumentation for HP-STM [17, 44]. HP-STM has the unique capability of examining the structure of adsorbates and surface sites of a catalyst *atom by atom* to address several crucial issues of heterogeneous catalysis such as the restructuring of the catalyst surface, the ensemble on surfaces of alloy catalysts, binding site, and even configuration of adsorbates under reaction conditions (Chapter 7).

ELECTRONIC PROPERTIES AND CHEMICAL STATE OF THE ACTIVE PHASE IN THE CATALYST

The reactivity of an atom in a molecule, nanoparticle, or solid is strongly linked to its electronic properties [36, 45]. The identification of the active phase of a catalyst usually involves a determination of the chemical state and electronic properties of the elements present in the system. Over the years several techniques have been developed to probe the core and valence states of solids [6, 46, 47]. XAS (Chapter 1, Chapter 2, and Chapter 3), XPS (Chapter 12), ultraviolet–visible–near infrared spectroscopy (Chapter 16), and electron paramagnetic resonance spectroscopy (Chapter 11) are used widely for probing the electronic properties of atoms in solid catalysts. When acquiring data of XAS and XPS, the high intensity that a synchrotron source provides

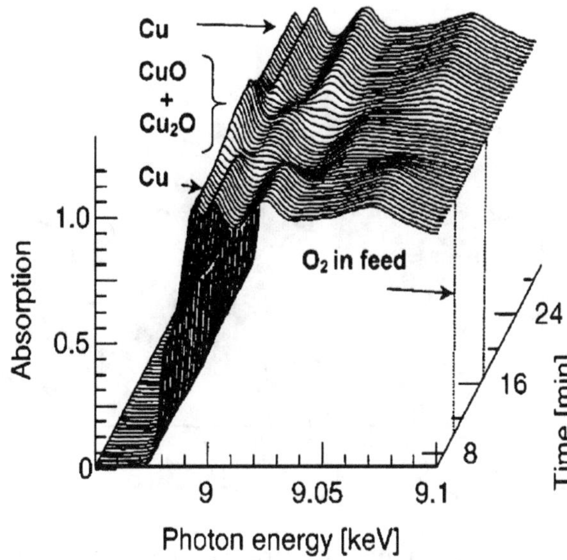

Figure I.13. Evolution of the Cu K-edge XANES of CuO/ZnO at 523 K during oxygen addition to a methanol steam reforming feed (*c*(MeOH) ~6 vol%, *c*(H_2O) ~1.5 vol% in 25 mL/min He). Different Cu phases are indicated. Reprinted with permission from Reference 48. Copyright (2001) Elsevier.

together with highly efficient detection devices leads to a high-energy resolution that improves our ability to separate different chemical states of an element in a catalyst (see Fig. I.13) [48]. The high intensity of the synchrotron radiation also makes it possible to track fast changes in the electronic properties during transient experiments (Chapter 1, Chapter 3, and Chapter 16) and sample mapping showing spatial gradients of concentration as a function of chemical state (Chapter 2 and Chapter 16). It is known that variations in the geometrical structure of a material usually produce changes in its electronic properties that modify the way in which the material responds to the presence of adsorbates [36]. The integration of instrumentation for XRD and XAS in the same experimental setup (Chapter 13) allows the direct study of possible correlations among the structural, electronic, and catalytic properties of a material.

PROPERTIES OF ADSORBATES, SURFACE CHEMISTRY, AND REACTION MECHANISM

In the scheme of Figure I.8, the determination of the active phase of the catalyst is followed by experimental and theoretical studies on the surface

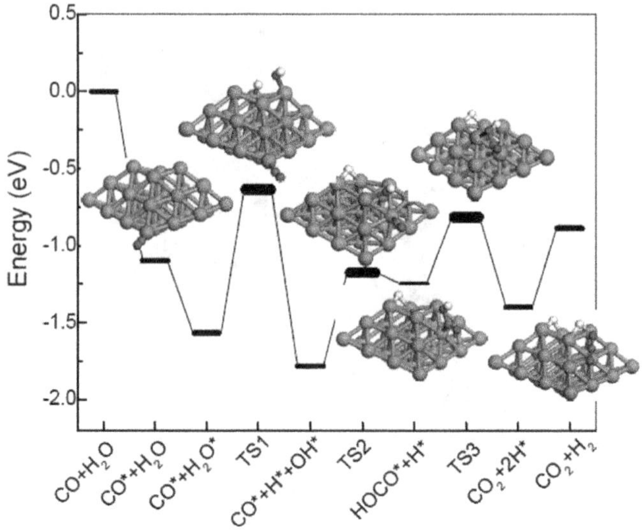

Figure I.14. Calculated energy profile and structures for the water–gas shift reaction ($CO + H_2O \rightarrow H_2 + CO_2$) on a Cu_{29} nanoparticle [50]. The label "TS" is used to denote a transition state. Reprinted with permission from Reference 50. Copyright (2007) American Institute of Physics.

chemistry and reaction mechanism associated with the catalytic process. The beauty of knowing a reaction mechanism is that it gives us a rational way to improve catalytic activity and selectivity [49]. Figure I.14 shows the calculated energy profile for a possible mechanism of the water–gas shift reaction ($CO + H_2O \rightarrow H_2 + CO_2$) on a Cu_{29} nanoparticle [50]. The Cu_{29} particle was present in $Cu/CeO_2(111)$ catalysts [51] and the calculations were performed using density functional theory [50]. The interaction of the reactants with Cu_{29} produces significant perturbations in the geometrical structure of the nanocatalyst. Chemical bonds are broken (O–H) and new ones are formed (O–C; H–H). The theoretical study predicts reaction intermediates (OH, H, HOCO) that must be validated by experimental measurements. Ideally, one should be able to monitor simultaneously the changes that occur in the structure of the metal particle and the chemical species that are being transformed on the catalyst surface. Proof of a mechanism for a catalyzed reaction lies in directly observing the chemical species involved in the elementary steps and the kinetics for their transformations [49].

Techniques such as XAS (Chapter 1, Chapter 2, Chapter 3, and Chapter 16), neutron scattering (Chapter 6), infrared spectroscopy (Chapter 8, Chapter 9, and Chapter 16), Raman spectroscopy (Chapter 10 and Chapter 16), electron paramagnetic resonance spectroscopy (Chapter 11), and ambient-pressure

XPS (Chapter 12) can be used to study the adsorption and transformation of chemical species on the surface of a catalyst under reaction conditions. Infrared spectroscopy is perhaps the most commonly used technique in catalysis to identify adsorbed species and to study the way in which these species are chemisorbed in the surface of the catalyst [6]. It is a frequently used technique in mechanistic studies [7, 15, 52–55], having one the longest histories among the methods for *in-situ* characterization [6]. When combined with probe molecules (CO, NO, etc.), infrared spectroscopy can yield valuable information about the adsorption sites in the catalyst [7, 52, 56–58]. Raman spectroscopy has also become popular for *in-situ* studies because it also can give information about the active sites and adsorbed species [52, 59]. Since the selection rules for vibrational transitions in infrared and Raman spectroscopies are different [60], the information obtained from a Raman spectrum often complements that obtained from an infrared study and thus provides valuable structural information (Chapter 10, Chapter 15, and Chapter 16). Inelastic neutron spectroscopy has been used in catalysis research principally due to its high sensitivity to the vibrations of hydrogen atoms (Chapter 6). Hydrogen due to its large incoherent scattering cross section is far more visible by interacting with neutrons than other elements [61], whereas Raman and infrared spectroscopies are more sensitive to the motion of heavier atoms.

Over the years, XPS has proven to be a very useful technique for studying the transformation of molecules on surfaces [10, 46]. The kinetic energy of the photoemitted electron is not only element specific but also contains chemical information because the energy levels of core electrons depend on the chemical state of the atom [46]. With the development of ambient-pressure XPS, it is now possible to study intermediates in common catalytic (Fig. I.7) and electrocatalytic processes (Chapter 12). In many cases, intermediates that are difficult to identify with infrared spectroscopy can be easily identified using XPS and vice versa. Electron paramagnetic resonance spectroscopy is not frequently used in mechanistic studies because it requires paramagnetic species containing unpaired electrons to be present in the system under study [60]. However, when this condition is satisfied, the technique can provide rich information on the local environment of paramagnetic species and how this local environment changes as a reaction progresses (Chapter 11).

The techniques mentioned in the previous two paragraphs, in addition to probe adsorbed species on the surface of a catalyst, can also yield simultaneous information about some properties of the catalytic material during reaction. The last three chapters in the book (Chapter 14, Chapter 15, and Chapter 16) aim at the integration of techniques that provide information on the structural properties of the catalysts, XRD, and X-ray absorption fine structure, with techniques that allow the study of surface reactions: infrared and Raman spectroscopy. A typical example of this type of studies is presented in Figure I.15 [62]. Such integration, a very challenging task, is a significant step forward toward the grand goal of being able to obtain a full picture of all the components in a catalytic process [24, 31, 34, 54, 62].

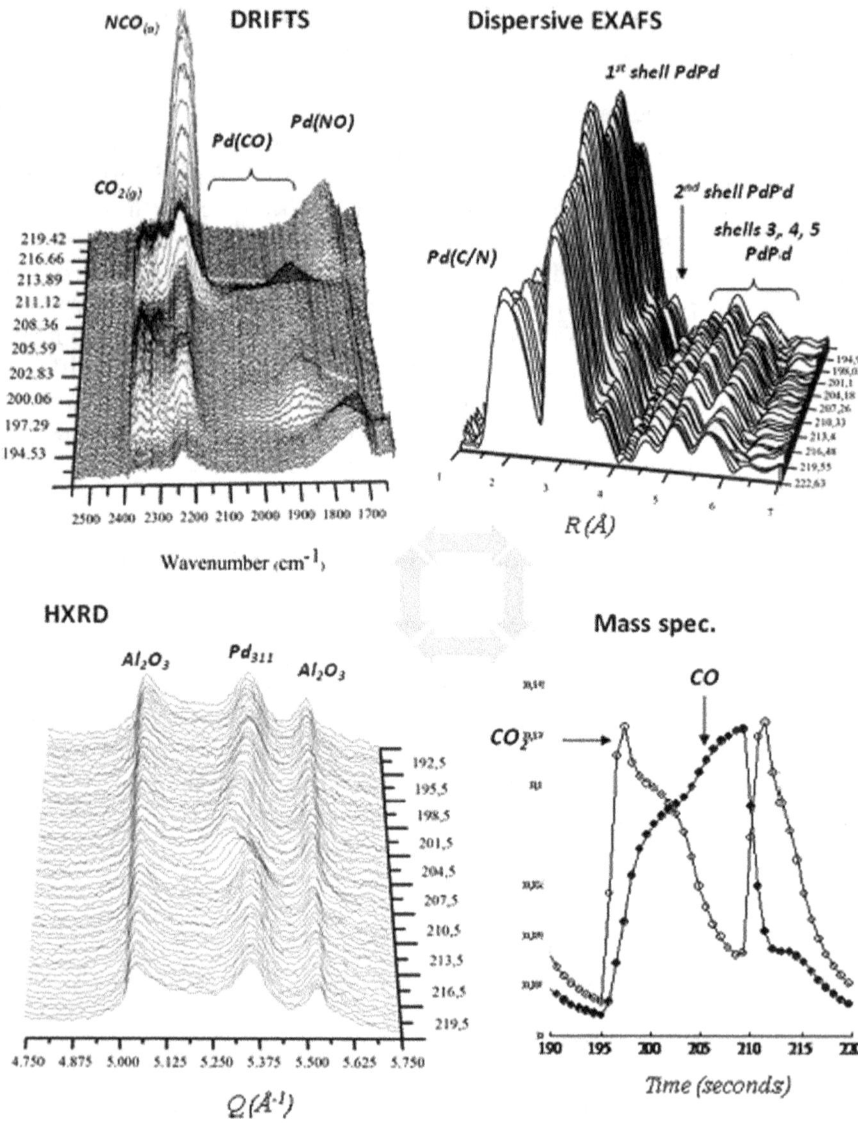

Figure I.15. Multitechnique approach to illuminate CO dissociation and transient carbon storage by Pd/Al_2O_3 during CO/NO cycling. Each of the X-ray techniques may be used in simultaneous time-resolved conjunction with DRIFTS and mass spectrometry (MS). The data from each technique pertain to the structure reactivity behavior of a 2 wt% Pd/Al_2O_3 catalyst during a single switch from 5% CO/He to 5% NO/He at 673 K. Reprinted with permission from Reference 62. Copyright (2010) American Chemical Society.

ACKNOWLEDGMENT

The authors are grateful for the financial support of the U.S. Department of Energy, Office of Basic Energy Sciences.

REFERENCES

[1] Heinemann, H. (1997) Development of industrial catalysis, in Ertl G., Knözinger H., Weitkamp J. (eds.), *Handbook of Heterogeneous Catalysis*, Wiley-VCH, Weinheim, Germany.

[2] Bartholomew, C. H., Farrauto, R. J. (2005) *Fundamentals of Industrial Catalytic Processes*, 2nd ed., Wiley-Aiche, New York.

[3] Hagen, J. (2006) *Industrial Catalysis: A Practical Approach*, Wiley-VCH, Weinheim, Germany.

[4] Boudart, M. (1997) Principles of heterogeneous catalysis, in Ertl G., Knözinger H., Weitkamp J. (eds.), *Handbook of Heterogeneous Catalysis*, Wiley-VCH, Weinheim, Germany.

[5] Clausen, B. S., et al. (1991) In situ cell for combined XRD and on-line catalysis tests: studies of Cu-based water gas shift and methanol catalysts, *J. Catal.*, 132, 524–535.

[6] Chorkendorff, I., Niemantsverdriet, J. W. (2007) *Concepts of Modern Catalysis and Kinetics*, 2nd ed., Wiley-VCH, Weinheim, Germany.

[7] Xu, W., et al. (2012) *In-situ* studies of CeO_2-supported Pt, Ru and Pt-Ru alloy catalysts for the water-gas shift reaction: active phases and reaction intermediates, *J. Catal.*, 291, 117–126.

[8] Kundu, et al. (2012) Exploring the structural and electronic properties of Pt/ceria-modified TiO2 and its photo-catalytic activity for water splitting under visible light, *J. Phys. Chem. C*, 116, 14062–14070.

[9] Rodriguez, J. A., Goodman, D. W. (1991) High-pressure catalytic reactions over single-crystal metal surfaces, *Surf. Sci. Rep.*, 14, 1–108.

[10] Somorjai, G. A., Li, Y. (2010) *Introduction to Surface Chemistry and Catalysis*, 2nd ed., John Wiley & Sons, New York.

[11] Rodriguez, J. A., Hrbek, J. (2010) Inverse oxide/metal catalysts: a versatile approach for activity tests and mechanistic studies, *Surf. Sci.*, 604, 241–244.

[12] Hornés, A., et al. (2010) Inverse CeO_2/CuO catalyst as an alternative to classical direct configurations for preferential oxidation of CO in hydrogen-rich stream, *J. Am. Chem. Soc.*, 132, 34–35.

[13] Yang, F., et al. (2011) CO oxidation on inverse CeO_x/Cu(111) catalysts: high catalytic activity and ceria-promoted dissociation of O_2, *J. Am. Chem. Soc.*, 133, 3444–3451.

[14] Schlögl, R. (1997) Bulk catalysts and supports, in Ertl G., Knözinger H., Weitkamp J. (eds.), *Handbook of Heterogeneous Catalysis*, Wiley-VCH, Weinheim, Germany.

[15] Wang, X., et al. (2006) *In-situ* studies of the active sites for the water gas shift reaction over Cu-CeO_2 catalysts: complex interaction between metallic copper and oxygen vacancies of ceria, *J. Phys. Chem. B*, 110, 428–434.

[16] Campbell, C. T. (1989) Studies of model catalysts with well-defined surfaces combining ultrahigh vacuum surface characterization with medium- and high-pressure kinetics, *Adv. Catal.*, 36, 1–54.

[17] Tao, F., Salmeron, M. (2011) *In-situ* studies of chemistry and structure of materials in reactive environments, *Science*, 331, 171–174.

[18] Somorjai, G. A. (1991) The flexible surface—correlation between reactivity and restructuring ability, *Langmuir*, 7, 3176–3182.

[19] Senanayake, S. D., et al. (2012) Nanopattering in CeOx/Cu(111): a new type of surface reconstruction and enhancement of catalytic activity, *J. Phys. Chem. Lett.*, 3, 839–843.

[20] Ogletree, D. F., et al. (2009) Photoelectron spectroscopy under ambient and temperature conditions, *Nucl. Instrum. Methods Phys. Res. A*, 601, 151–160.

[21] Salmeron, M., Schlögl, R. (2008) Ambient pressure photoelectron spectroscopy: a new tool for surface science and nanotechnology, *Surf. Sci. Rep.*, 63, 169–199.

[22] Chupas, P. J., et al. (2003) Rapid-acquisition pair distribution function (RA-PDF) analysis, *J. Appl. Crystallogr.*, 36, 1342–1347.

[23] Chupas, P. J., et al. (2007) Applications of an amorphous silicon-based area detector for high-resolution, high-sensitivity and fast time-resolved pair distribution function measurements, *J. Appl. Crystallogr.*, 40, 463–470.

[24] Schroer, C. G., et al. (2003) Mapping the chemical states of an element inside a sample using tomographic X-ray absorption spectroscopy, *Appl. Phys. Lett.*, 82, 3360–3362.

[25] Ciston, J., et al. (2011) Morphological and structural changes during the reduction and reoxidation of CuO/CeO_2 and $Ce_{1-x}Cu_xO_2$ nanocatalysts: *in-situ* studies with enviromnental TEM, XRD and XAS, *J. Phys. Chem. C*, 115, 13851–13859.

[26] Schroer, C. G., et al. (2010) Hard X-ray microscopy with elemental, chemical, and structural contrast, *Acta Phys. Pol. A*, 117, 357–368.

[27] Newton, M. A., et al. (2002) Bringing time resolution to EXAFS: recent developments and application to chemical systems, *Chem. Soc. Rev.*, 31, 83–95.

[28] Ferri, D., et al. (2010) First steps in combining concentration modulation techniques with synchronous dispersive EXAFS/DRIFTS/mass spectrometry for in situ time resolved study of heterogeneous catalysts, *Phys. Chem. Chem. Phys.*, 12, 5634–5646.

[29] Ressler, T. (2009) Solid state kinetics and catalytic behaviour of selective oxidation catalysts from time-resolved EXAFS investigations, *Catal. Today*, 145, 258–266.

[30] Kox, M. H. F., et al. (2009) Label-free chemical imaging of catalytic solids by coherent anti-Stokes Raman scattering and synchrotron-based infrared microscopy, *Angew. Chem. Intl. Ed.*, 48, 8990–8993.

[31] Beale, A. M., et al. (2005) Adding a third dimension to operand spectroscopy: a combined UV-Vis, Raman, and XAFS setup to study heterogeneous catalysts under working conditions, *Chem. Commun. (Camb.)*, 3015–3016.

[32] Frenkel, A., et al. (2011) Combining X-ray absorption and X-ray diffraction techniques for in situ studies of chemical transformations in heterogeneous catalysis: advantages and limitations, *J. Phys. Chem. C*, 115, 17884–17890.

[33] Boccaleri, E., et al. (2007) In situ simultaneous Raman/high-resolution X-ray powder diffraction study of transformations occurring in materials at non-ambient conditions, *J. Appl. Crystallogr.*, 40, 684–693.

[34] Newton, M. A., van Beek, W. (2010) Combining synchrotron-based X-ray techniques with vibrational spectroscopies for the in situ study of heterogeneous catalysts: a view from a bridge, *Chem. Soc. Rev.*, 39, 4845–4863.

[35] van Santen, R. A., Neurock, M. (2006) *Molecular Heterogeneous Catalysis: A Conceptual and Computational Approach*, Wiley-VCH, Weinheim, Germany.

[36] Nørskov, J. K., et al. (2008) The nature of the active site in heterogeneous metal catalysis, *Chem. Soc. Rev.*, 37, 2163–2171.

[37] Bergeret, G. (1997) Structure and morphology. X-ray powder diffraction, in Ertl G., Knözinger H., Weitkamp J. (eds.), *Handbook of Heterogeneous Catalysis*, Wiley-VCH, Weinheim, Germany.

[38] Wang, X., et al. (2004) Time-resolved studies for the mechanism of reduction of copper oxides with carbon monoxide: complex behavior of lattice oxygen and the formation of suboxides, *J. Phys. Chem. B*, 108, 13667–13673.

[39] Barrio, L., et al. (2010) Unraveling the active site in copper-ceria systems for the water-gas shift reaction: *in-situ* characterization of an inverse powder CeO_{2-x}/CuO-Cu catalyst, *J. Phys. Chem. C*, 114, 3580–3587.

[40] Datye, A. K. (1997) Electron microscopy and diffraction, in Ertl G., Knözinger H., Weitkamp J. (eds.), *Handbook of Heterogeneous Catalysis*, Wiley-VCH, Weinheim, Germany.

[41] Hansen, P. L., et al. (2002) Atom-resolved imaging of dynamic shape changes in supported copper nanocrystals, *Science*, 295, 2053–2055.

[42] Yoshida, H., et al. (2012) Visualizing gas molecules interacting with supported nanoparticulate catalysts at reaction conditions, *Science*, 335, 317–319.

[43] Kuwauchi, Y., et al. (2012) Intrinsic catalytic structure of gold nanoparticles supported on TiO_2, *Angew. Chem. Intl. Ed.*, 51, 7729–7733.

[44] Rasmussen, P. B., et al. (1998) The reactor STM: a scanning tunneling microscope for investigation of catalytic surfaces at semi-industrial reaction conditions, *Rev. Sci. Instrum.*, 69, 3879–3884.

[45] Sholl, D., Steckel, J. A. (2009) *Density Functional Theory: A Practical Introduction*, Wiley-Interscience, New York.

[46] Woodruff, D. P., Delchar, T. A. (1994) *Modern Techniques of Surface Science*, 2nd ed., Cambridge University Press, Cambridge, UK.

[47] Che, M., Bozon-Verduraz, F. (1997) UV-Vis-NIR and EPR spectroscopies, in Ertl G., Knözinger H., Weitkamp J. (eds.), *Handbook of Heterogeneous Catalysis*, Wiley-VCH, Weinheim, Germany.

[48] Günter, M. M., et al. (2001) Redox behavior of copper oxide/zinc oxide catalysts in the steam reforming of methanol studied by in situ X-ray diffraction and absorption spectroscopy, *J. Catal.*, 203, 133–149.

[49] Campbell, C. T. (1994) Micro- and macro-kinetics: their relationship in heterogeneous catalysis, *Top. Catal.*, 1, 353–366.

[50] Liu, P., Rodriguez, J. A. (2007) Water-gas shift reaction on Metal nanoparticles and surfaces, *J. Chem. Phys.*, 126, 164705.

[51] Rodriguez, J. A., et al. (2007) Water-gas shift reaction on Cu and Au nanoparticles supported on $CeO_2(111)$ and $ZnO(000\bar{1})$: intrinsic activity and importance of support interactions, *Angew. Chem. Intl. Ed.*, 46, 1329–1332.

[52] Stavitski, E., Weckhuysen, B. M. (2010) Infrared and Raman imaging of heterogeneous catalysts, *Chem. Soc. Rev.*, 39, 4615–4625.

[53] Chuang, S. C., Guzman, F. (2009) Mechanistic investigation of heterogeneous catalysis by transient infrared methods, *Top. Catal.*, 52, 1448–1458.

[54] Frenkel, A., et al. (2012) Synchrotron techniques for in situ catalytic studies: capabilities, challenges and opportunities, *ACS Catal.*, 2, 2269–2280.

[55] Burch, R., et al. (2011) A critical analysis of the experimental evidence for and against a formate mechanism for high activity water-gas shift catalysts, *Appl. Catal. A: Gen.*, 409–410, 3–12.

[56] Rodriguez, J. A., et al. (1992) Infrared vibrational studies of CO adsorption on Cu/Pt(111) and CuPt(111) surfaces, *J. Chem. Phys.*, 96, 7814–7825.

[57] Xu, M., et al. (2010) Monitoring electronic structure changes of TiO_2(110) via sign reversal of adsorbate vibrational bands, *Phys. Chem. Chem. Phys.*, 12, 3649–3652.

[58] Baron, M., et al. (2009) Resolving the atomic structure of vanadia monolayer catalysts: monomers, trimers, and oligomers on ceria, *Angew. Chem. Intl. Ed.*, 48, 8006–8009.

[59] Bañares, M. A., Mestl, G. (2009) Structural characterization of operating catalysts by Raman spectroscopy, *Adv. Catal.*, 52, 43–128.

[60] Drago, R. S. (1992) *Physical Methods for Chemist*, 2nd ed., Saunders, New York.

[61] Mitchell, P. C. H., et al. (2005) *Vibrational Spectroscopy with Neutrons, with Applications in Chemistry, Biology, Materials Science and Catalysis*, World Scientific Press, New York.

[62] Newton, M. A., et al. (2010) Combining time-resolved hard X-ray diffraction and diffuse reflectance infrared spectroscopy to illuminate CO dissociation and transient carbon storage by supported Pd nanoparticles during CO/NO cycling, *J. Am. Chem. Soc.*, 132, 4540–4541.

1

QEXAFS IN CATALYSIS RESEARCH: PRINCIPLES, DATA ANALYSIS, AND APPLICATIONS

Anatoly I. Frenkel, Syed Khalid, Jonathan C. Hanson, and Maarten Nachtegaal

1.1 INTRODUCTION

Mechanisms of catalytic reactions are often very complex and elusive, due, in part, to the multiple length and timescales that characterize chemical transformations [1]. Characteristic length scales relevant for catalysis range from several millimeters (typical sample sizes) to micrometers (the size of the support) to nanometers (the typical size of catalytic nanoparticles) to picometers (the root mean square bond length disorder). The timescales range from minutes (e.g., reduction–oxidation [redox] reactions) to milliseconds (typical turnover rates) to pico- and femtoseconds (photoexcitation processes). Furthermore, the only possibility to have a glimpse at the activity of a catalyst is to follow the chemical reaction in real time, as opposed to prenatal and postmortem investigations [2]. In that sense, catalytic investigations in chemical and energy sciences are akin to biological catalysis studies that have similar challenges due to the large range of typical length and timescales, and similar solutions, for example, the use of *in-situ* spectroscopic and scattering methods. Mechanistic investigations of enzymatic catalysis have a unique advantage

In-situ *Characterization of Heterogeneous Catalysts*, First Edition.
Edited by José A. Rodriguez, Jonathan C. Hanson, and Peter J. Chupas.

over nanoparticle catalysis due to the well-defined positions, and a small number, of active sites (metal ions) in the enzymes as opposed to a much larger number of active surface sites in the nanoparticles. Therefore, the modeling of many enzymatic processes, including the mapping of their energy landscapes, characterizing transition states, and studying reaction kinetics, can be successfully handled by first-principle calculations [3].

Among the indirect methods capable of resolving structural environment and electronic properties of active sites in catalytic materials, synchrotron-based X-ray absorption spectroscopy (XAS) has become one of the methods of choice, due to its excellent spatial, temporal, and energy resolutions [4]. The extended X-ray absorption fine structure (EXAFS) region refers to the oscillations observed in the X-ray absorption coefficient measured within 1000–1500 eV of the X-ray absorption edge energy, that is, the excitation energy of the core-level electron. In the EXAFS region, information about the local structural environment of the X-ray absorbing atom is extracted from the fine structure oscillations of the absorption coefficient. This fine structure signal is adequately described by the photoelectron scatterings from the neighboring atoms. The EXAFS measurement is, therefore, capable of probing atomic structure within the distance range of approximately 6–8 Å from the resonant (X-ray absorbing) atom [5]. Being a local structural probe, XAS does not require a long-range order. In addition, it is element specific (every element has its own core-level excitation energies) and can be used to study elements at low concentrations (such as promoters or noble metals) and low metal loading. The X-ray absorption near-edge structure (XANES) portion of the XAS signal is assigned to the energies between the first symmetry-allowed unoccupied state and the continuum states, that is, approximately, from 30 eV below to 40 eV past the absorption edge (e.g., Fig. 1.2a). XANES contains information about the electronic structure, density of unoccupied states, and bonding geometry around the absorbing atom.

XAS experiments have extremely short (10^{-16}–10^{-15} s) characteristic time (which is the duration of the interference between the outgoing and the incoming photoelectron waves, limited by the core hole lifetime and other losses) and thus can be used to probe catalytic processes in the course of the reaction. Due to the relatively large penetration depth of hard X-rays, this technique can also be applied *in situ*, under reaction conditions (e.g., controlled gas atmosphere or pressure, or sample temperature) [6–8].

Crystallographic techniques are superior in quantitative and direct determination of the sample structure in comparison with XAS. However, the applications of these techniques are limited to ordered phases. These techniques would therefore be useless when reaction intermediates are dilute, strongly disordered, or dispersed [9]. However, these are the most common conditions that occur in chemical or enzymatic catalysis. XAS applications are particularly powerful in these rapidly emerging areas of science, since they are capable to determine the transient states, and their evolution, in the process of catalytic reaction.

Timescales in homo- and heterogeneous catalysis range from milliseconds (nucleation and growth, sintering, particle morphology changes) to seconds (reaction turnover times) to minutes (reduction and oxidation reactions) to hours to days (catalyst degradation and aging). Time-resolved X-ray absorption spectroscopy (TR-XAS) is particularly useful in these studies since it contains relevant information about real-time catalyst structure in the course of reaction. The *in-situ* TR-XAS technique is among the most commonly used structural methods to date for probing intermediate states during real-time transformations in a large variety of systems of interest in structural biology [9, 10] and materials science [11–15]. In structural biology studies, TR-XAS is often carried out by the freeze–quench method [16], which enables access to similar reaction rates (from milliseconds to seconds) but better data quality compared to the capabilities of the alternative, energy-scanning, or dispersive XAS modes. In the case of inorganic catalysts used in the industry and in model studies, the main effort in the TR-XAS methodology has been the push for fast data collection methods due to the need for *in-situ* or *operando* investigations [1].

To study the structural changes in the reactions that take place in the subsecond regime, TR-XAS was developed in energy-dispersive EXAFS (EDE) and energy-scanning, or quick EXAFS (QEXAFS) modes. In EDE, described in greater detail in Chapter 3 of this book, the sample is illuminated by the polychromatic beam and thus one can take repeated snapshots of the entire EXAFS spectrum at a timescale limited by the detector readout time and the photon flux, a few milliseconds [17–22]. EDE can be efficiently used in transmission mode only, and is thus limited to samples that have a relatively large concentration of absorbing atoms (in the percentage range) and are uniform in thickness. Detector normalization problems, temporal changes in beam flux, and spatial beam stability can decrease data quality. QEXAFS was developed by Frahm and coworkers [23] and applied to solving structures of materials over the past couple of decades by many groups [1, 23–31]. The newest version of eccentric cam-driven monochromators allow for much larger spectral ranges than previously used piezo-driven ones [32] and can collect EXAFS data with a time resolution as fast as a few tens of milliseconds per spectrum [33]. Ultrafast spectroscopic methods (in the femto- and picosecond ranges) by means of pump–probe schemes have also been recently developed [34–36], but are not yet applied to *in-situ* studies of catalytic systems and thus fall outside the scope of this chapter.

In Section 1.2, we will give an overview of the existing implementations of the QEXAFS technology. Section 1.3 will highlight the most commonly used methods in processing and analyzing TR-XAS data. In Section 1.4, we will survey the different applications of QEXAFS to the problems of heterogeneous catalysis that will demonstrate the advantage of *in-situ* and *operando* investigations by combined techniques versus static (e.g., only prenatal and postmortem) measurements. Finally, Section 1.5 will present the summary and future directions.

1.2 IMPLEMENTATION

In its simplest form, conventional QEXAFS or, alternatively, "on-the-fly XAS," can be performed on all XAS beamlines when moving the double-crystal monochromator or a channel-cut monochromator (semi-)continuously through the energy range (Bragg angle) of interest, while the encoder readout of the Bragg angle and the detectors are sampled simultaneously. This mode of QEXAFS has been introduced by Frahm [23] at Hasylab in Germany and is now implemented at many beamlines, for example, at the DUBBLE beamline of the European Synchrotron Radiation Facility (ESRF) [37] in France, at BL18 [38] at Diamond in the United Kingdom, or at BL01 at the SPring-8 [39] in Japan, to name a few. With conventional QEXAFS, one typically obtains a time resolution in the range of a few seconds for the XANES region and up to a few minutes for the EXAFS region. In this chapter we will discuss only the implementation and technical details of dedicated QEXAFS systems that reach a subsecond resolution for a full EXAFS scan.

A beamline for subsecond QEXAFS is characterized by an intense and continuous photon source, a channel-cut crystal monochromator that moves in an oscillatory motion, and a data acquisition system that simultaneously samples several detectors and encoders.

The ideal X-ray source for QEXAFS is a tapered or helical undulator, a wiggler, or a supercooled bending magnet. What is important is that the source delivers a continuous spectrum, a flux at the sample of a minimum of 10^{11} photons/second and, ideally, deposits not too much heat on the monochromator crystal. After the source, the beam is collimated with a collimating mirror in order to obtain the best possible energy resolution. The best place to install a QEXAFS monochromator is downstream of a collimating mirror and upstream of a focusing mirror unit to minimize the movement of the beam caused by moving the channel-cut crystal. Alternatively, the channel-cut crystal of the QEXAFS monochromator could be made with a very small gap and be placed right before the sample.

The heart of a QEXAFS setup is the channel-cut monochromator [40]. In all dedicated QEXAFS monochromators that will be described later, the crystal moves in an oscillatory motion around a preselected Bragg angle, driven by an actuator such as an eccentric cam or galvano scanner. The angular range is chosen to cover the spectral range of the element of interest. The first developments of QEXAFS monochromators come from the Frahm group at the University of Wuppertal, Germany. Their latest "Frahm-type" monochromator [33] consists of a channel-cut crystal, mounted on a tilt table or cradle that is connected to an eccentric cam mechanism to oscillate the crystal with an angular amplitude that can be tuned dynamically between 0 and $\pm 1.5°$ and with a speed up to 40 Hz.

The main Bragg angle is selected on a goniometer, the tilt table is rocked around this angle, and the angular offsets of the crystal table are determined with an angular encoder [41]. The energy calibration is determined from a

QEXAFS spectrum covering two metal absorption edges. The apparent angular distance between both edges can be used to determine the step width of the angular encoder. Finally, from the signal of the angular encoder and a reference foil, the absolute energy can be back-calculated for each spectrum [41]. The channel-cut crystal is cooled indirectly using water or liquid nitrogen.

A first commercial version of this monochromator has been installed at the SuperXAS beamline of the Swiss Light Source (SLS) [42]. A QEXAFS monochromator based on the "Frahm design" has been constructed at the SAMBA beamline at the Soleil synchrotron facility in France. Both at SuperXAS and SAMBA, the QEXAFS monochromators are installed at (supercooled) bending magnet beamlines, where a final monochromatric flux of ~10^{11}–10^{12} photons/second is achieved. At both beamlines, the QEXAFS monochromators are placed parallel to a conventional double-crystal monochromator, so that switching between QEXAFS and conventional XAS or QEXAFS in the second-to-minute range takes less then 5 min.

At two bending magnet beamlines, X18A and X18B, of the National Synchrotron Light Source in the United States, a dual-purpose (QEXAFS or conventional X-ray absorption fine structure [XAFS]) monochromator has been installed that achieves an oscillatory movement of the channel-cut crystal through an external cam system [28, 30, 43]. Collimation and, consequently, energy resolution are achieved by slitting down the beam upstream of the monochromator. At X18B, beam movement at the sample is minimized by placing the monochromator near the sample while beam movement on the sample at X18A is reduced through a focusing mirror placed downstream of the QEXAFS monochromator. The spindle of the monochromator is driven by a 30-cm-long stainless steel arm. The spindle is connected to an optical encoder, which gives the absolute angular position and thus the energy. The system, driven by a direct current (dc) motor, ensures that different angular ranges of the channel-cut crystal, from ±0.22 to ±2.25°, are covered. With this setup, a maximum speed of 5 Hz can be achieved. The Heidenhain optical encoder, the moving arm, and the motorized micrometer mechanism are all located outside the monochromator chamber vacuum, and for this reason it is easy to change the QEXAFS system to a conventional XAFS system in a few minutes without venting the monochromator chamber. A monochromator flux of ~10^{10} photons/second is achieved at the sample position.

The last published mechanical version of a dedicated QEXAFS monochromator is implemented at SPring-8 [44] in Japan. Here, a helical undulator delivers on axis fundamental radiation, where off-axis higher harmonics are slitted off and a high flux, but low power quasimonochromatic beam is achieved. Accordingly, a compact channel-cut monochromator, without a cooling unit, could be constructed. This is a big advantage since a cooling unit slows down the maximum speed that could be achieved and might also be a possible source for unwanted vibrations. In this monochromator, the main Bragg angle is selected with a goniometer and the channel-cut crystal is moved around the main Bragg angle using a galvano scanner stage that can move up to a speed of 100 Hz. The channel-cut crystal has a gap of 3 mm and is placed before the experiment to minimize beam movement during a scan. The total number of monochromatic photons was estimated to be ~10^{13} photons/second.

For transmission-type measurements, the intensity of the beam before and after the sample, and after a reference foil, is detected with ionization chambers. More than the mechanics of the QEXAFS monochromator, these ionization chambers are currently the limiting factor in terms of speed. It has been generally observed at all QEXAFS facilities that the edge features are smeared out when one moves at a speed of 10 Hz or more. This can be attributed to the ionization chamber acting in some ways as a capacitor (bringing away charges in a gas takes time). Developments are currently being undertaken to improve detector technology for QEXAFS. The big advantage of QEXAFS over energy-dispersive XAS is that it can be combined with fluorescence mode detection using, for example, a Lytle detector [43], a photomultiplier [44], or a PIPS detector [43], all in combination with filters to reduce the elastic scattering peak background signal, and, in the future, with energy-dispersive detectors. Currents from the different detectors are amplified and converted to a voltage with a large dynamic range ±10 V by current amplifiers.

Different data acquisition systems are available, which are either standalone systems, using multifunctional data acquisition boards [43, 45], or are integrated into the beamline hard- and software environment. The major requirements for a general QEXAFS data acquisition system include: (1) the capability to acquire, store, and visualize huge amounts of data [45]; and (2) the ability to read in analogue (or, alternatively, the analogue signals need to be digitized and transferred) and digital signals (detectors) simultaneously at frequencies up to 100 kHz.

When photon statistics allow (this is currently only theoretically possible at SPring-8) full EXAFS, spectra can be collected up to a speed of 40 Hz. Each revolution includes one upward and one downward movement of the channel-cut crystal, resulting in 80 spectra per second. When a full EXAFS scan is about 1 keV long and one wants an energy resolution of 1 eV, then the data acquisition system needs to sample at a minimum speed of 80 kHz. When one does not know exactly when a reaction triggers or when one follows an oscillating reaction, one could be collecting data for a few hours. Furthermore, one might want to not only read in the different detectors signals and angular encoder signal but also a time stamp, signals of the experiment (position of a valve, voltage), or signals from complementary techniques such as infrared spectroscopy, mass spectrometry, or ultraviolet–visible spectroscopy (UV-VIS) simultaneously. Accordingly, the data acquisition system should be able to collect up to 10 different channels at 100 kHz and store all data over a long collection time. Finally, in order to check on the experiment, a real-time representation of the data being collected (e.g., every 5 s) is largely recommended.

1.3 DATA ANALYSIS METHODS

The main objective of the TR-XAS data analysis is threefold: (1) detection of reaction intermediates, (2) structural analysis of the intermediates and active species, and (3) extraction of the time-dependent mixing coefficients.

Data analysis methods for TR-XAS have been developed in parallel with instrumentation [11, 46–48]. Traditional methods of EXAFS data analysis based on the fitting of EXAFS data with theory based on the model structure are not efficient for TR-XAS data analysis because of the large number of fitting parameters involved [46]. For example, Figure 1.1 shows a schematic of structural transformations that are theoretically possible for the reduction of CuO: it can be a one-step, two-step, or three-step reaction, involving zero, one, or two intermediates, respectively. The crystal structures of CuO, Cu_4O_3, and Cu_2O have significant differences, and even the structure of a single-phase material may be complicated. For example, in Cu_4O_3, half of the Cu atoms have four O neighbors, while the other half of the Cu atoms has only two O neighbors. Since EXAFS is an ensemble-average technique, all local environments of Cu contribute to the signal, and fitting analysis of the Cu–O coordination number will be inconclusive, for phase speciation purposes. This limitation does not apply to the XANES region, which is much more sensitive to the unique geometry of different species than the first shell EXAFS signal. Thus, XANES spectra are often used to quantitatively deconvolute mixtures of different phases [48].

The prevailing approach in TR-XAS data analysis is to rely on a variety of algebraic methods. The techniques that will be described later (linear combination analysis, principal component analysis, and residual phase analysis) all assume that the phases that are present in the heterogeneous sample are not changing with time, only their mixing fractions do. The simplest scenario for

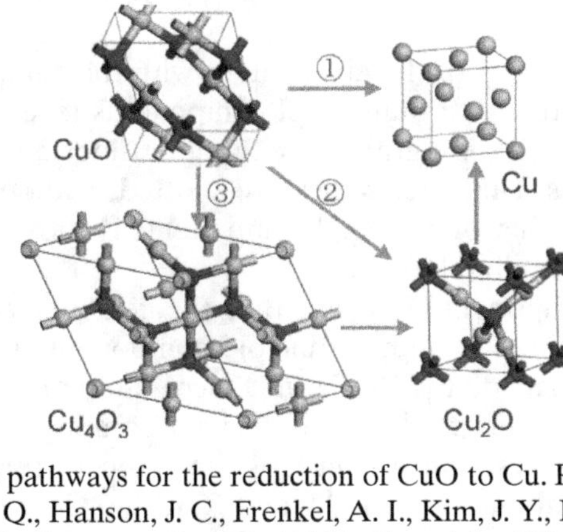

Figure 1.1. Possible pathways for the reduction of CuO to Cu. Reprinted with permission from Wang, X. Q., Hanson, J. C., Frenkel, A. I., Kim, J. Y., Rodriguez, J. A. (2004) *Journal of Physical Chemistry B*, 108, 13667. Copyright 2004 American Chemical Society.

the XAS (EXAFS or XANES) data analysis is when the sample contains only a two-component mixture at all times. The hint that this may be the case is the presence of isosbestic points in TR-XAS spectra. (Isosbestic points are those where all spectra taken at different stages of the reaction intersect each other [49].) The presence of one or more isosbestic points is a sign of a direct transformation of reactants to products while the absence of isosbestic points indicates the formation of one or more intermediate phases. These different situations are demonstrated in the succeeding sections. Although this information is by itself valuable, the absence of the isosbestic points merely guarantees that one or more reaction intermediates are present but offers no quantitative information about the number of intermediates and their structure. Furthermore, even the presence of the isosbestic points does not guarantee that there are exactly two components mixed within the sample at all times. In all cases, a quantitative analysis is needed.

Principal component analysis (PCA) is a robust quantitative method of linear algebra which allows the determination of the number of linearly independent components in the series of experimental spectra without making any model-dependent assumptions of their chemical nature or structure. The PCA scheme represents each experimental spectrum as a vector $\boldsymbol{x}_i$ ($i = 1, \ldots M$) in the N-dimensional space, where N is the number of data points in each spectrum and M is the number of spectra. The data matrix $\boldsymbol{D}$, of the dimension $M \times N$, is constructed from all the data sets. By finding the M eigenvectors and eigenvalues of $\boldsymbol{D}$, and by arranging the eigenvectors in the descending order of eigenvalues, one can construct an ordered orthogonal basis. Each original spectrum can be represented as a linear combination of M basic vectors or *abstract components*. By selecting the eigenvectors having the *largest* eigenvalues and neglecting those with the *smallest* ones, one can represent all the data sets by using a linear combination of just a few (M_c) principal components (eigenvectors). Because $M_c < M < N$, (in most practical cases, $M_c << N$), the PCA provides a convenient way to reduce the dimension of the representation.

By examining the decay of the eigenvalues with the component number, it is possible to obtain the least number of components (species in the sample) using the "scree test." This is a graphic method for determining the number of principal components. The eigenvalues are plotted in the sequence of their decrease, and the number of principal components is chosen where the curve levels off to a linear decline [46].

Two principal components indicate that the reaction occurs without an intermediate, while three or more principle components indicate that there are one or more intermediates. While first successful applications of PCA to phase speciation were reported for gas chromatography, mass spectrometry, and nuclear magnetic resonance data [50], more recent reports demonstrate the power of the combination of PCA and XAS [46, 51–57]. PCA is a more powerful XAFS data analysis method than a simple linear combination of standard compounds for several reasons. First, the number N of principal components, that is, independent species, is found by the PCA model independently, while a linear combination fit *assumes* a certain number of species. Next, the mixing fractions of all species contributing to the experimental data are obtained by a *linear* combination of their spectra (see later), which affords greater confidence in the analysis result than the EXAFS data fitting method. The latter is based on a nonlinear least squares algorithm and thus contains multiple minima in the parameters space, not a single minimum that is found by PCA.

Identification of the intermediate phases and determination of their time-resolved mixing fractions is carried out by fitting the abstract principal components obtained by PCA to various standard compounds, also known as the *target transform* procedure in PCA. It was successfully applied to a number of TR-XAS data in catalytic systems and resulted in the determination of the following phases: Ni + NiO for the H_2 reduction of NiO [58]; CuO + Cu_2O + Cu for the CO [11] and H_2 [59] reduction of CuO, the same three standards for the H_2 reduction of CuO/ZnO [60]; and MoO_3 + $Mo_{18}O_{52}$ + MoO_2 for MoO_3 reduction in propene and MoO_2 oxidation in oxygen [61]. We will show how such analysis can be conducted using two different examples, for the one-step and the two-step reactions.

1.3.1 One-Step Reaction

As a reaction progresses, changes in the XANES and/or EXAFS spectra reflect the changes in the absorbing atom's electronic and atomic structure during the reaction. One-step transformations, without an intermediate, would result in a two-component system:

$$f(E,t) = x(t) f_R(E) + (1 - x(t)) f_P(E), \qquad (1.1)$$

where $f(E,t)$ is the time-dependent XANES or EXAFS spectrum, $f_R(E)$ and $f_P(E)$ are the reactant and product spectra, and $x(t)$ is the time-dependent mixing fraction. PCA analysis is needed to prove the absence of intermediates, as previously described. Since the identities of the reactants and the products are usually known from the steady-state measurements, or just from the first and the last spectra in the TR-XAS measurement, $x(t)$ can then be obtained by a simple linear combination fit. An illustration of this approach can be found in Figure 1.2.

1.3.2 Two-Step Reactions

For two-step reactions with one intermediate (the following methods can be easily extended to multistep reactions), $f(E,t)$ can be presented as:

$$f(E,t) = x(t) f_R(E) + y(t) f_I(E) + z(t) f_P(E), \text{ where } x(t) + y(t) + z(t) = 1. \qquad (1.2)$$

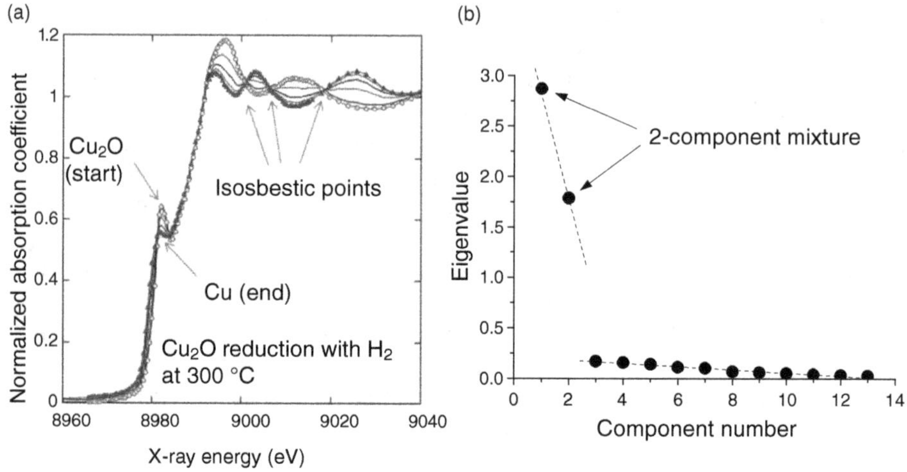

Figure 1.2. (a) Plots of time-resolved Cu K-edge XANES data for Cu_2O reduction at 300°C with H_2/He mixture. The open circles correspond to the initial oxide, whereas the filled symbols denote the final product (metallic copper). Reprinted with permission from Kim, J. Y., Rodriguez, J. A., Hanson, J. C., Frenkel, A. I., Lee, P. L. (2003) *Journal of the American Chemical Society*, 125, 10684. Copyright 2003 American Chemical Society. (b) PCA-obtained eigenvalues and the "scree test" for the Cu K-edge XANES data showing that only two principal components are needed to represent all the spectra.

If the intermediate phases are crystallographically ordered, as, for example, in CuO decomposition, their XAS data can be mimicked by relevant experimental standards, usually measured in the same experimental conditions, for example, temperature and/or nanoscale dimensions, to best match the physical state of the unknown intermediates. The standards are tested, one by one, by linear combination of the principal components obtained by PCA until a good fit is obtained and the best standard, or standards, are chosen. Then, the target transform is performed, of the eigenvectors (abstract components) in the basis obtained by the PCA, onto the new basis (that corresponds to the XAS data of the standards), and the mixing fractions x, y, and z are obtained for all steps of the reaction (Fig. 1.3 and Fig. 1.4). To minimize the ambiguity of the standard selection, the first and the last data set in the series are often taken as the standards for the target transform, which is a good strategy when the reaction (e.g., reduction or oxidation) is completed during the QEXAFS measurement.

When the nature of the intermediate is not known a priori, or cannot be found in any of the standard compounds due to its metastable, low-dimensional, and/or disordered nature, other methods are needed, in addition to the PCA. The need in the new analysis methods, for standardless reaction phases, is especially timely due to the increasing use of QEXAFS, dispersive XAFS, as

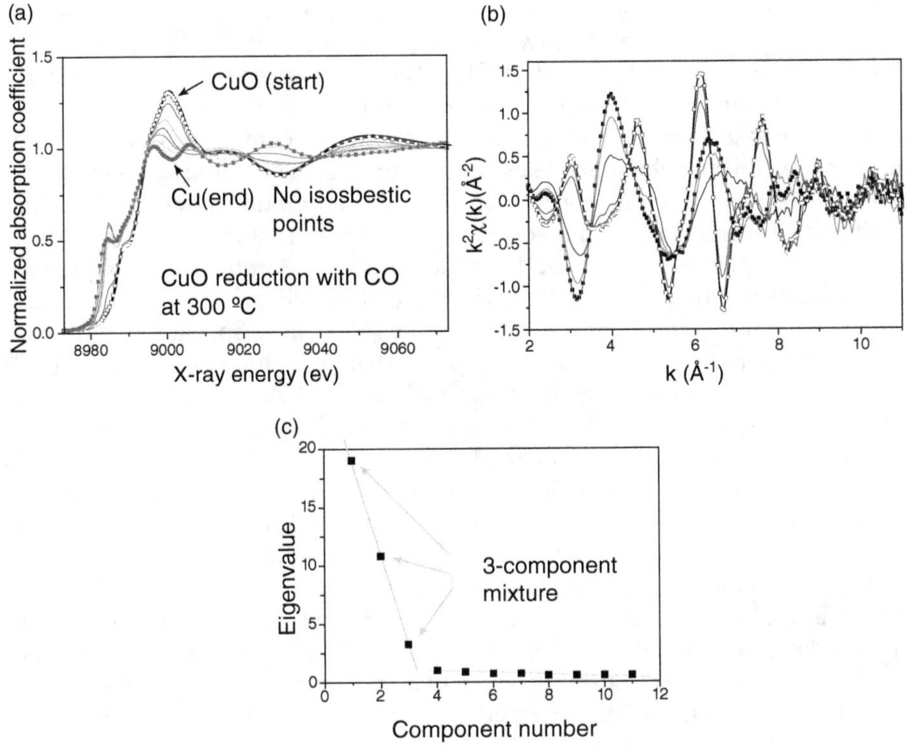

Figure 1.3. Time-resolved Cu K-edge (a) XANES and (b) EXAFS spectra during the isothermal reduction of CuO at 300°C with CO/He mixture. (c) PCA scree test for the Cu K-edge EXAFS data. Reprinted with permission from Wang, X. Q., Hanson, J. C., Frenkel, A. I., Kim, J. Y., Rodriguez, J. A. (2004) *Journal of Physical Chemistry B*, 108, 13667. Copyright 2004 American Chemical Society.

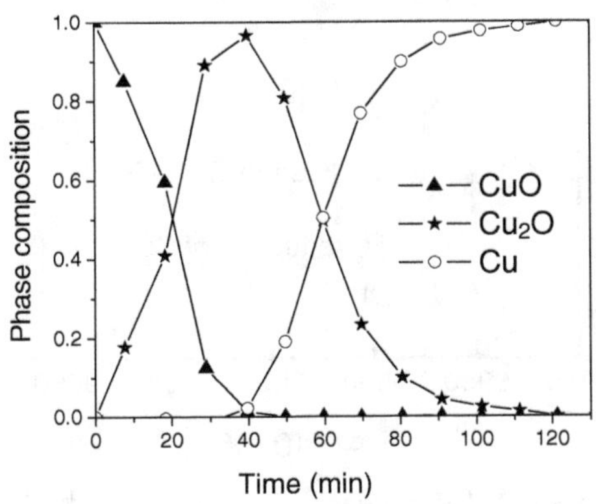

Figure 1.4. Weight fractions of the CuO, Cu_2O, and Cu standards during the time-resolved Cu K-edge XANES spectra of the isothermal reduction of CuO at 300°C with CO/He mixture as obtained from PCA. Reprinted with permission from Wang, X. Q., Hanson, J. C., Frenkel, A. I., Kim, J. Y., Rodriguez, J. A. (2004) *Journal of Physical Chemistry B*, 108, 13667. Copyright 2004 American Chemical Society.

well as methods of ultrafast X-ray spectroscopy, for example, pump–probe methods [62]. One such possibility is a technique called residual phase analysis (RPA) developed by Frenkel et al. [46]. The RPA approach, described here using the example of a two-step reaction, utilizes two of the known species (usually the start and the end phases of the reacting compounds that contain the X-ray absorbing element). These two components are subtracted from the raw data at each time point with incrementally, and independently, changing mixing fractions x and z. The residual, properly normalized, is a candidate intermediate phase that can be fit by a suitable theoretical local structure model using a fitting method [16]. The best fit would correspond to unique values of x and z, and hence, y, which solves the phase speciation problem. Due to the complexity of any intermediate structure that cannot be derived from the known standards, the RPA approach of TR-XAS data analysis is most suitable to enzymatic catalysis, where the structure in question is that an active site in the protein, which is a relatively simple task.

In the case of multiple reaction sites present in heterogeneous catalysts, for example, partially reduced oxides or metal nanoparticles, PCA can still be used in its capacity of obtaining the number of intermediate phases. In QEXAFS analysis of time-resolved reduction with carbon monoxide of Cu–ceria catalysts, Wang et al. [47] used PCA to obtain evidence of the formation of an intermediate phase. In QEXAFS data of Cu K-edge XANES during the hydrogen reduction of Cu–ceria at 200°C, the absence of isosbestic points indicates the presence of an intermediate phase during the reaction (Fig. 1.5).

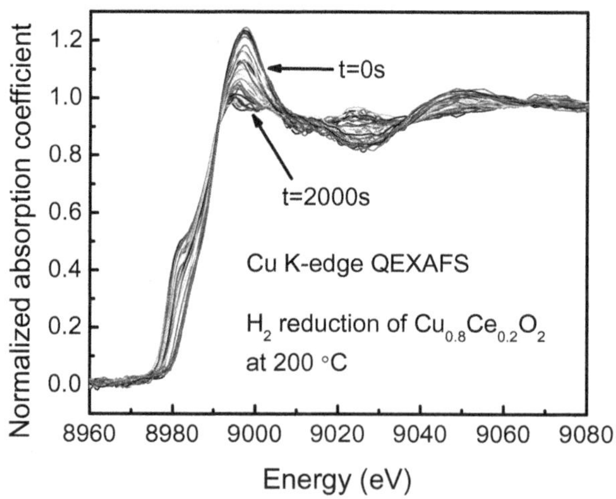

Figure 1.5. QEXAFS data of Cu K-edge XANES during the hydrogen reduction of Cu–ceria at 200°C. The absence of isosbestic points indicates the presence of an intermediate phase during the reaction. Reprinted with permission from Wang, Q., Hanson, J. C., Frenkel, A. I. (2008) *Journal of Chemical Physics*, 129, 234502. Copyright 2008, American Institute of Physics.

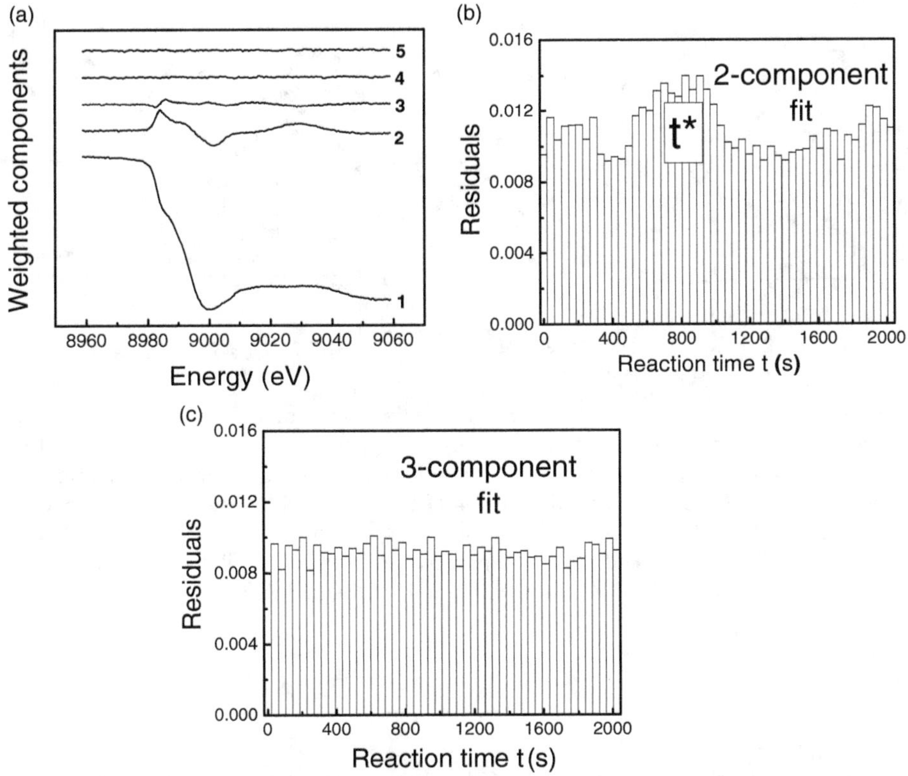

Figure 1.6. (a) Principal components (from 1 through 5) of the Cu–ceria QEXAFS data (shown in Fig. 1.5) obtained by PCA. Residuals of the linear combination fit of the data using (b) two and (c) three principal components. The peak in the residual at $t = t^*$ in (b) and its disappearance in (c) both indicate that there is one intermediate phase during the reaction. Reprinted with permission from Wang, Q., Hanson, J. C., Frenkel, A. I. (2008) *Journal of Chemical Physics*, 129, 234502. Copyright 2008, American Institute of Physics.

Figure 1.6a shows principal components (from 1 through 5) obtained by PCA processing of these data. It is evident from the linear combination fit of the data using two and three principal components that the latter model provides smaller residuals (Fig. 1.6c), while the former features a systematic increase in the middle of the reaction (Fig. 1.6b). A combination of these observations indicates that there is one intermediate phase during the reaction.

1.4 EXAMPLES OF QEXAFS IN CATALYSIS RESEARCH

In this section, we highlight a few examples showing the usefulness of QEXAFS, complemented by other techniques, in studying catalytic reactions. The first

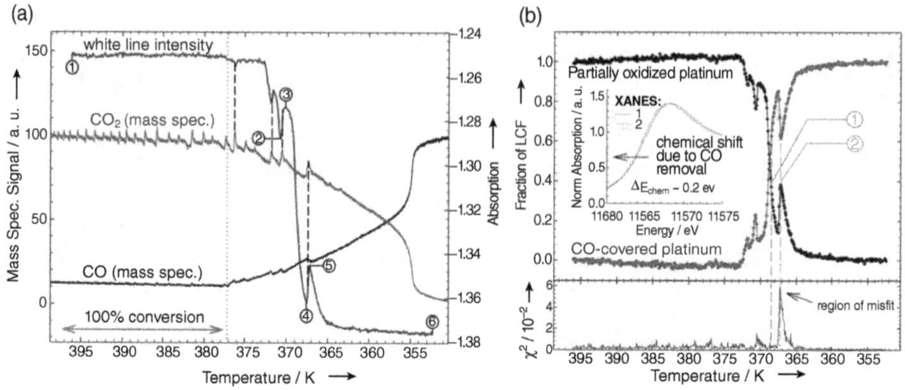

Figure 1.7. (a) Mass spectrometer traces of CO (bottom) and CO_2 (middle) compared to the white-line intensity (top) of the Pt L_3-edge spectra measured 1 mm below the top of a plug flow reactor during the oscillations of CO oxidation as a function of temperature. (b) Linear combination fitting (LCF) of the XANES data using spectra collected at two positions within the reactor corresponding to partially oxidized Pt and fully reduced Pt. Reproduced with kind permission from John Wiley & Sons: Singh, J., Nachtegaal, M., Alayon, E. M. C., Stötzel, J., van Bokhoven, J. A. (2010) *ChemCatChem*, 2, 653.

example comes from Singh et al. [63], where a spatially resolved QEXAFS ($100 \times 100\ \mu m^2$-sized beam, 2 spectra per second, SuperXAS beamline of the SLS) at the Pt L_3-edge was used to follow the state of the catalyst during the rapidly oscillating CO oxidation on a technical catalyst (2 wt% Pt on γ-Al_2O_3) at different heights in a plug flow reactor. CO conversion was achieved by heating a mixture of CO and O_2 (19:1) to 398 K. Full conversion was obtained at this temperature as can be seen in Figure 1.7a, where the mass spectrometer signals are plotted as a function of temperature. Upon decreasing temperature, small oscillations in CO_2 conversion were observed. Then, the cooling of the reactor was started. The reaction extinguished at approximately 377 K, which is characterized by an increase of the CO and a decrease of the CO_2 signal. Also shown in Figure 1.7a is the height of the Pt white line of each quick XAS spectrum, collected 1 mm below the inlet of the plug flow reactor. Since full QEXAFS spectra were collected at each point, EXAFS analysis allowed determining the local structure at each point in Figure 1.7a. Full EXAFS analysis indicated that high intensity of the Pt white line corresponds to partially oxidized platinum, whereas low intensity indicates fully reduced platinum particles. When CO conversion decreased, oxidized Pt was converted to reduced Pt, suggesting that partially oxidized Pt is the more active phase. In addition to the reduction of the catalyst during extinction, oscillations in the white-line intensity can also be observed.

Figure 1.7b shows that most of the spectra collected during the few oscillations can be well described by a combination of spectra of surface-oxidized Pt and CO-covered reduced Pt. Within an oscillation, the initial loss of activity was accompanied by an increased amount of CO-covered surface at the expense of the surface-oxidized platinum. During the sharp rise and excess formation of CO_2, the linear combination fit showed misfits, which suggests that surface-oxidized and CO-covered surfaces are not the only components during this part of the oscillation. Comparing two spectra with identical white-line intensity (Fig. 1.7b, inset) showed a small shift in energy, which is characteristic of the difference between a bare surface and a surface with adsorbed CO. These linear combination fits provide evidence that after freeing the surface of CO, there exist sites for dissociative O_2 adsorption. In summary, these space-resolved QEXAFS studies allowed to show the dynamic structure of a technical catalyst in a plug flow reactor and allowed to identify the mechanism of oscillations in CO oxidation in a packed bed reactor.

Reimann et al. [64] performed QEXAFS studies in the subsecond timescale to understand the reaction mechanism of Heck-type C–C coupling reactions in the presence of supported Pd-based catalysts (5 wt% Pd/γ-Al_2O_3). The reactants are phenyl bromide and styrene and the primary product is trans-stilbene. This is a homogenous reaction which takes place in the solution, but in this system the soluble palladium catalyst is leached from the solid heterogeneous catalysts. Therefore, the experiment must measure both the palladium XAFS in the solution and the palladium clusters on the alumina support to fully understand the reaction. In this work, the batch process becomes steady state in less than 20 min, so subminute XAFS spectra are required. At the SLS SuperXAS beamline, an EXAFS spectrum for Pd clusters was obtained in 0.5 s. The Pd species in solution were much more dilute (~0.3 mol%) and an average of 400 half-second spectra was required to get a spectrum which provided meaningful EXAFS analysis.

Using a specially designed *in-situ* XAS cell, both the solid catalyst and the liquid reaction mixture during the reaction of phenyl bromide (PhBr) with styrene were monitored. Soluble Pd species were only, and rapidly, detected in the liquid reaction phase once the reaction temperature of 150°C was reached. At the same time, the conversion of PhBr started, and during the following "active phase" of the catalyst, hardly any changes in the corresponding XANES (Fig. 1.8a) and EXAFS (Fig. 1.8b) spectra were observed. The solution species could be identified as colloidal Pd^0 clusters with a size of ~2 nm estimated from EXAFS analysis. When the reaction rate started to decrease, pronounced changes in the EXAFS spectra were observed, which were attributed to an increased formation of bromopalladates ($[PdBr_4]^{2-}$, $[Pd_2Br_6]^{2-}$). In addition to the liquid-phase species, significant changes were observed for the solid catalyst that was also probed *in situ* during the reaction. The originally oxidized Pd catalyst on alumina was efficiently reduced upon heating. Additionally, growth of the supported Pd particles was observed by both EXAFS (Fig. 1.8b) and STEM.

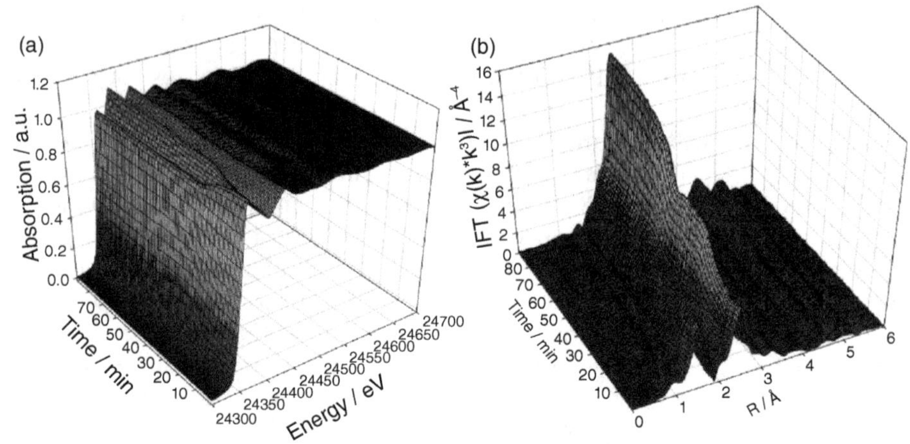

Figure 1.8. (a) Normalized XANES spectra of the 5 wt % Pd/Al_2O_3 catalyst as a function of time. (b) Fourier-transformed EXAFS spectra of the 5 wt % Pd/Al_2O_3 catalyst as a function of time. Reprinted with permission from Reimann, S., Stötzel, J., Frahm, R., Kleist, W., Grunwaldt, J.-D., Baiker, A. (2011) *Journal of the American Chemical Society*, 133, 3921–3930. Copyright 2011 American Chemical Society. (See color insert.)

The results described earlier confirm the role of the soluble molecular Pd species as the catalytically active species and clarify their conjunction with the *in-situ* formed Pd colloids. Furthermore, the investigation demonstrated the potential of QEXAFS not only for monitoring rapid changes during catalysis but also for gaining deeper insight into the mechanism of such complex industrially important systems under relevant reaction conditions. QEXAFS allowed monitoring rapid changes in the seconds-to-minutes timescale and permitted the minimization of effects caused by the heterogeneity of the systems. The study indicates the high potential of QEXAFS to elucidate complex reaction dynamics in industrially important reaction mixtures, not only interesting for C–C couplings with other aryl halides but also for related reaction systems.

The strength of QEXAFS is also demonstrated for the *in-situ* analysis of the thermal decomposition of Co-oxalate-hydrate in different gas atmospheres [65]. A number of solid–solid transformations such as decomposition reactions or the redox of heterogeneous catalysts occur relatively fast, that is, within less than a minute. Time-resolved investigations are thus important in order to gain insight into the mechanisms of such solid–solid transformations [65]. These experiments were performed at the Advanced Photon Source (beamline 1-ID) at Argonne National Laboratory using a temporary installation of the "Frahm-type" QEXAFS setup. The *in-situ* cell for the experiments consists of a quartz glass capillary tube loaded with Co-oxalate-dihydrate. Different premixed gases from a gas manifold could pass through the capillary reactor cell. The raw data show the shift between the Co metal reference and Co-oxalate-

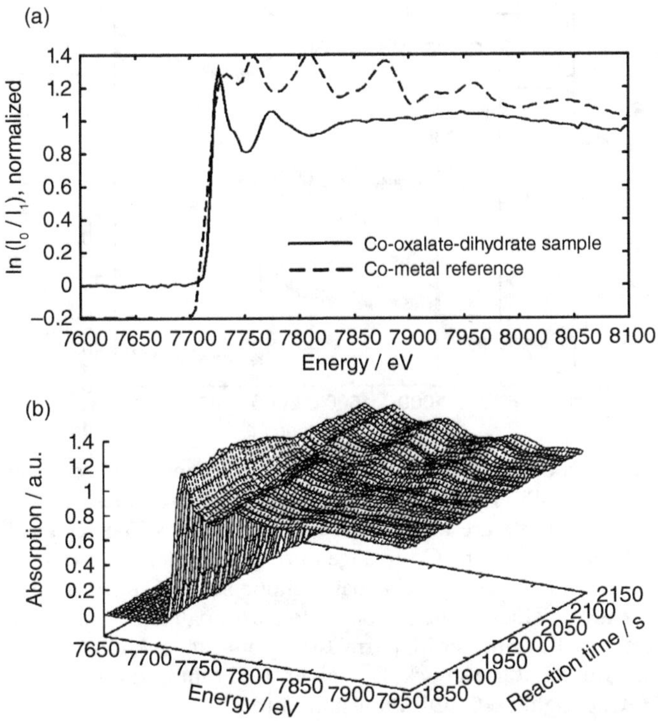

Figure 1.9. (a) Co K-edge EXAFS spectrum of $CoC_2O_4 \cdot 2H_2O$ measured *in situ* within 250 ms prior to the initiation of a decomposition reaction together with the Co metal reference measured simultaneously. (b) *In-situ* Co K-edge absorption spectra measured during the thermal decomposition of $CoC_2O_4 \cdot 2H_2O$ in Ar + 4% H_2. Each spectrum was measured within 250 ms. The displayed spectra belong to the temperature region between approximately 340 and 400°C. Reprinted with permission from Lutzenkirchen-Hecht, D., Grunwaldt, J. D., Richwin, M., Griesebock, B., Baiker, A., Frahm, R. (2005) *Physica Scripta*, T115, 831–833. Copyright 2005 IOP Publishing Ltd.

dihydrate. This shift and the XANES shape changes (Fig. 1.9) are both indicative of the changing oxidation state of Co. PCA resulted in only two distinct phases during the decomposition which could be identified as the Co(II) oxalate and Co-metal (Fig. 1.10).

1.5 SUMMARY AND OUTLOOK

In this chapter we reviewed the latest developments in QEXAFS technique and its applications to catalysis. This method of rapid XAS data collection, first proposed by Frahm in 1988, not only grew into a premier time-resolved measurement technique in catalysis science, it also stimulated the development of

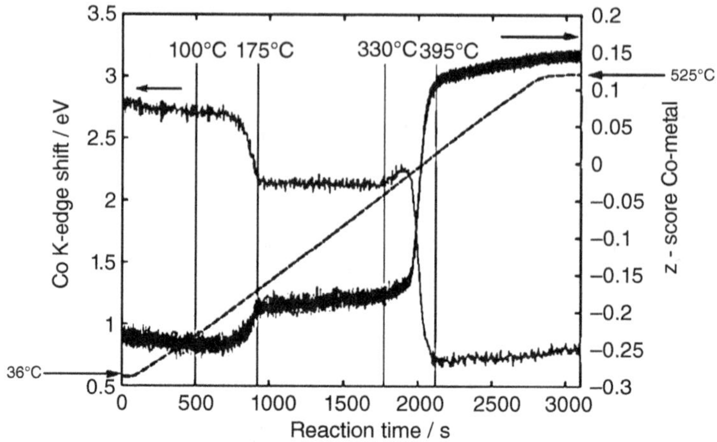

Figure 1.10. XANES data analysis for the *in-situ* decomposition reaction of Co-oxalate-dihydrate in Ar + 4% H_2. The dashed line shows the temperate as a function of time; some essential points are highlighted by vertical lines. The regular line belongs to the observed edge shift of the Co K-edge during the experiment (left ordinate) and the bold line corresponds to the principal component of Co-metal (right ordinate). Negative and positive values of the z-score belong to small and high Co-metal concentrations, respectively. Reprinted with permission from Lutzenkirchen-Hecht, D., Grunwaldt, J. D., Richwin, M., Griesebock, B., Baiker, A., Frahm, R. (2005) *Physica Scripta*, T115, 831–833. Copyright 2005 IOP Publishing Ltd.

new data analysis methods, new reactors for *in-situ* and *operando* investigations [66–69], and new synchrotron beamlines all over the world. Furthermore, it rapidly became a multidisciplinary tool, easily adaptable to solve problems not just in catalysis but also in environmental and materials sciences.

A recent increase in interest in resonant inelastic X-ray scattering (RIXS) and X-ray emission spectroscopy (XES), which need emission detectors with high energy resolution, is caused by the ability of these techniques to probe the electronic structure, including valence band structures, of nanocatalysts [70]. Developments are currently underway to combine RIXS with QEXAFS to obtain information on the dynamic changes in the electronic structure of the catalysts *in situ*. Analytical power of *in-situ* XAS, X-ray diffraction (XRD), and other techniques used in studying chemical transformations involved in heterogeneous catalysis is greatly improved when these methods are cointerpreted, whether they are applied together, in a single experiment [1, 7, 71, 72], or separately, in the same process under similar reaction conditions [73, 74].

Using TR-XAS, one is just scratching the tip of the iceberg when dealing with the possibility of really determining the structure of the active site, which normally makes up only a very small fraction of the atoms present in the catalyst. A recent development, so-called modulation excitation spectroscopy, offers the possibility to select out the XAS signal of the active species [75, 76]. Here, the sample is excited with a periodically alternating external simulation, such as gas atmosphere, pressure, temperature, voltage, and so on. The measured system response, all TR-XAS spectra, are then filtered with the excitation frequency, so that only the signal of the species that responded to the excitation is left over and can be analyzed.

Where will QEXAFS technique be in 5–10 years from now? Will it reach a submillisecond timescale? Currently, the internal diffusion limit in the most commonly used catalytic reactors is the reason that 10-ms time resolution is sufficient for most time-resolved studies in liquid and gas phases [77]. However, with the ongoing improvement of instrumentation for catalytic measurements, including those relying on micro- and nanofluidic reactors, the mass transport limitation can be overcome, at least for now. Vacuum motors can be used with Golovchenko-type (constant exit height) monochromators [78] to eliminate the beam movement for low energy elements where we deal with large Bragg angles. Among other bottlenecks in speeding up QEXAFS experiments, we mention the lack of fast, capacitor-free X-ray detectors and amplifier-free current measurement methods. Together with the photon statistics that is not adequate for measuring real catalytic processes with dilute metal content at high reaction rates, these are the main challenges toward catalytic studies with a submillisecond time resolution.

ACKNOWLEDGMENTS

A.I.F. acknowledges the support of this work by the U.S. Department of Energy (DOE) Grant No. DE-FG02-03ER15476. The use of the National Synchrotron Light Source (NSLS) beamlines was supported by U.S. DOE Contract No. DE-AC02-98CH10886. Beamlines X18A and X18B at the NSLS are supported in part by the Synchrotron Catalysis Consortium, U.S. DOE Grant No DE-FG02-05ER15688.

REFERENCES

[1] Clausen, B. S., Grabaek, L., Steffensen, G., Hansen, P. L., Topsoe, H. (1993) A combined QEXAFS XRD method for online, in-situ studies of catalysts—examples of dynamic measurements of Cu-based methanol catalysts, *Catalysis Letters*, 20, 23–36.

[2] Haw, J. F. (ed.) (2000) *In-situ Spectroscopy in Heterogeneous Catalysis*, John Wiley & Sons, Chichester, UK.

[3] Schwartz, S. D., Schramm, V. L. (2009) Enzymatic transition states and dynamic motion in barrier crossing, *Nature Chemical Biology*, 5, 552–559.

[4] Sinfelt, J. H., Via, G. H., Lytle, F. W. (1984) Application of EXAFS in catalysis. Structure of bimetallic cluster catalysts, *Catalysis Reviews*, 26, 81–140.

[5] Sayers, D. E., Stern, E. A., Lytle, F. W. (1971) New technique for investigating noncrystalline structures—Fourier analysis of extended X-ray absorption fine structure, *Physical Review Letters*, 27, 1204–1207.

[6] Grunwaldt, J. D., Caravati, M., Hannemann, S., Baiker, A. (2004) X-ray absorption spectroscopy under reaction conditions: suitability of different reaction cells for combined catalyst characterization and time-resolved studies, *Physical Chemistry Chemical Physics*, 6, 3037–3047.

[7] Grunwaldt, J.-D., Clausen, B. S. (2002) Combining XRD and EXAFS with on-line catalytic studies for in situ characterization of catalysts, *Topics in Catalysis*, 18, 37–43.

[8] Newton, M. A., Dent, A. J., Evans, J. (2002) Bringing time resolution to EXAFS: recent developments and application to chemical systems, *Chemical Society Reviews*, 31, 83–95.

[9] Penner-Hahn, J. E. (2005) Characterization of "spectroscopically quiet" metals in biology, *Coordination Chemistry Reviews*, 249, 161–177.

[10] Dau, H., Haumann, M. (2003) X-ray absorption spectroscopy to watch catalysis by metalloenzymes: status and perspectives discussed for the water-splitting manganese complex of photosynthesis, *Journal of Synchrotron Radiation*, 10, 76–85.

[11] Wang, X. Q., Hanson, J. C., Frenkel, A. I., Kim, J. Y., Rodriguez, J. A. (2004) Time-resolved studies for the mechanism of reduction of copper oxides with carbon monoxide: complex behavior of lattice oxygen and the formation of suboxides, *Journal of Physical Chemistry. B, Materials, Surfaces, Interfaces & Biophysical*, 108, 13667–13673.

[12] Tibiletti, D., Amieiro-Fonseca, A., Burch, R., Chen, Y., Fisher, J. M., Goguet, A., Hardacre, C., Hu, P., Thompsett, A. (2005) DFT and in situ EXAFS investigation of gold/ceria-zirconia low-temperature water gas shift catalysts: identification of the nature of the active form of gold, *Journal of Physical Chemistry. B, Materials, Surfaces, Interfaces & Biophysical*, 109, 22553–22559.

[13] Wang, X., Rodriguez, J. A., Hanson, J. C., Perez, M., Evans, J. (2005) In situ time-resolved characterization of Au-CeO2 and AuOx-CeO2 catalysts during the water-gas shift reaction: presence of Au and O vacancies in the active phase, *Journal of Chemical Physics*, 123, 221101–221106.

[14] Rodriguez, J. A., Wang, X., Liu, P., Wen, W., Hanson, J. C., Hrbek, J., Perez, M., Evans, J. (2007) Gold nanoparticles on ceria: importance of O vacancies in the activation of gold, *Topics in Catalysis*, 44, 73–81.

[15] Wen, W., Jing, L., White, M. G., Marinkovic, N., Hanson, J. C., Rodriguez, J. A. (2007) In situ time-resolved characterization of novel Cu-MoO2 catalysts during the water-gas shift reaction, *Catalysis Letters*, 113, 1–6.

[16] Kleifeld, O., Frenkel, A., Martin, J. M. L., Sagi, I. (2003) Active site electronic structure and dynamics during metalloenzyme catalysis, *Nature Structural Biology*, 10, 98–103.

[17] Tolentino, H., Baudelet, F., Dartyge, E., Fontaine, A., Lena, A., Tourillo, G. (1990) Aberration-free and harmonic-free optics for time-resolved X-ray absorption spectroscopy using synchrotron radiation, *Nuclear Instruments and Methods in Physics Research. Section A, Accelerators, Spectrometers, Detectors and Associated Equipment*, 289, 307–316.

[18] Phizackerley, R. P., Rek, Z. U., Stephenson, G. B., Conradson, S. D., Hodgson, K. O., Matsushita, T., Oyanagi, H. (1983) An energy-dispersive spectrometer for the rapid measurement of X-ray absorption-spectra using synchrotron radiation, *Journal of Applied Crystallography*, 16, 220–232.

[19] Allen, P. G., Conradson, S. D., Pennerhahn, J. E. (1993) A 4-point crystal bender for dispersive-X-ray absorption-spectroscopy, *Journal of Applied Crystallography*, 26, 172–179.

[20] Aquilanti, G., Pascarelli, S. (2005) EXAFS study of the local structure of InAs up to 80 GPa, *Journal of Physics. Condensed Matter: An Institute of Physics Journal*, 17, 1811–1824.

[21] Dent, A., Evans, J., Newton, M., Corker, J., Russell, A., Abdul Rahman, M. B., Fiddy, S., Mathew, R., Farrow, R., Salvini, G., Atkinson, P. (1999) High-quality energy-dispersive XAFS on the 1 s timescale applied to electrochemical and catalyst systems, *Journal of Synchrotron Radiation*, 6, 381–383.

[22] Fiddy, S. G., Newton, M. A., Dent, A. J., Salvini, G., Corker, J. M., Turin, S., Campbell, T., Evans, J. (1999) In situ energy dispersive EXAFS (EDE) of low loaded Pt(acac)(2)/H-I SiO2 catalyst precursors on a timescale of seconds and below, *Chemical Communications*, 851–852.

[23] Frahm, R. (1988) Quick scanning EXAFS—1st experiments, *Nuclear Instruments and Methods in Physics Research. Section A, Accelerators, Spectrometers, Detectors and Associated Equipment*, 270, 578–581.

[24] Grunwaldt, J. D., Lutzenkirchen-Hecht, D., Richwin, M., Grundmann, S., Clausen, B. S., Frahm, R. (2001) Piezo X-ray absorption spectroscopy for the investigation of solid-state transformations in the millisecond range, *Journal of Physical Chemistry. B, Materials, Surfaces, Interfaces & Biophysical*, 105, 5161–5168.

[25] Richwin, M., Zaeper, R., Lutzenkirchen-Hecht, D., Frahm, R. (2001) Piezo-QEXAFS: advances in time-resolved X-ray absorption spectroscopy, *Journal of Synchrotron Radiation*, 8, 354–356.

[26] Frahm, R. (1989) New method for time-dependent X-ray absorption studies, *Review of Scientific Instruments*, 60, 2515–2518.

[27] Frahm, R., Richwin, M., Lutzenkirchen-Hecht, D. (2005) Recent advances and new applications of time-resolved X-ray absorption spectroscopy, *Physica Scripta*, T115, 974–976.

[28] Caliebe, W. A., So, I., Lenhard, A., Siddons, D. P. (2006) Cam-driven monochromator for QEXAFS, *Radiation Physics and Chemistry*, 75, 1962–1965.

[29] So, I., Siddons, D. P., Caliebe, W. A., Khalid, S. (2007) Hard real-time quick EXAFS data acquisition with all open source software on a commodity personal computer, *Nuclear Instruments and Methods in Physics Research. Section A, Accelerators, Spectrometers, Detectors and Associated Equipment*, 582, 190–192.

[30] Khalid, S., Caliebe, W., Siddons, P., So, I., Clay, B., Lenhard, T., Hanson, J., Wang, Q., Frenkel, A. I., Marinkovic, N., Hould, N., Ginder-Vogel, M., Landrot, G. L., Sparks, D. L., Ganjoo, A. (2010) Quick extended x-ray absorption fine structure instrument with millisecond time scale, optimized for in situ applications, *Review of Scientific Instruments*, 81, 015105–015112.

[31] Lutzenkirchen-Hecht, D., Grundmann, S., Frahm, R. (2001) Piezo-QEXAFS with fluorescence detection: fast time-resolved investigations of dilute specimens, *Journal of Synchrotron Radiation*, 8, 6–9.

[32] Briois, V., Lutzenkirchen-Hecht, D., Villain, F., Fonda, E., Belin, S., Griesebock, B., Frahm, R. (2005) Time-resolved study of the oxidation of ethanol bycerium(IV) using combined Quick-XANES, UV-Vis, and Raman spectroscopies, *Journal of Physical Chemistry. A, Molecules, Spectroscopy, Kinetics, Environment & General Theory*, 109, 320–329.

[33] Stotzel, J., Lutzenkirchen-Hecht, D., Frahm, R. (2010) A new flexible monochromator setup for quick scanning x-ray absorption spectroscopy, *Review of Scientific Instruments*, 81, 073109.

[34] Chen, L. X. (2004) Taking snapshots of photoexcited molecules in disordered media by using pulsed synchrotron X-rays, *Angewandte Chemie International Edition*, 43, 2886–2905.

[35] Saes, M., Bressler, C., van Mourik, F., Gawelda, W., Kaiser, M., Chergui, M., Bressler, C., Grolimund, D., Abela, R., Glover, T. E., Heimann, P. A., Schoenlein, R. W., Johnson, S. L., Lindenberg, A. M., Falcone, R. W. (2004) A setup for ultrafast time-resolved x-ray absorption spectroscopy, *Review of Scientific Instruments*, 75, 24–30.

[36] Cavalleri, A., Rini, M., Chong, H. H. W., Fourmaux, S., Glover, T. E., Heimann, P. A., Kieffer, J. C., Schoenlein, R. W. (2005) Band-selective measurements of electron dynamics in VO2 using femtosecond near-edge x-ray absorption, *Physical Review Letters*, 95, 067405–067409.

[37] Nikitenko, S., Beale, A. M., van der Eerden, A. M. J., Jacques, S. D. M., Leynaud, O., O'Brien, M. G., Detollenaere, D., Kaptein, R., Weckhuysen, B. M., Bras, W. (2008) Implementation of a combined SAXS/WAXS/QEXAFS set-up for time-resolved in situ experiments, *Journal of Synchrotron Radiation*, 15, 632–640.

[38] Dent, A. J., Cibin, G., Ramos, S., Smith, A. D., Scott, S. M., Varandas, L., Pearson, M. R., Krumpa, N. A., Jones, C. P., Robbins, P. E. (2009) B18: a core XAS spectroscopy beamline for diamond, *Journal of Physics: Conference Series*, 190, 012039.

[39] Tada, M., Murata, S., Asakoka, T., Hiroshima, K., Okumura, K., Tanida, H., Uruga, T., Nakanishi, H., Matsumoto, S.-I., Inada, Y., Nomura, M., Iwasawa, Y. (2007) In situ time-resolved dynamic surface events on the Pt/C cathode in a fuel cell under operando conditions, *Angewandte Chemie International Edition*, 46, 4310–4315.

[40] Grunwaldt, J.-D., Beier, M., Kimmerle, B., Baiker, A., Nachtegaal, M., Griesebock, B., Lutzenkirchen-Hecht, D., Stotzel, J., Frahm, R. (2009) Structural changes of noble metal catalysts during ignition and extinction of the partial oxidation of methane studied by advanced QEXAFS techniques, *Physical Chemistry Chemical Physics*, 11, 8779–8789.

[41] Stotzel, J., Lutzenkirchen-Hecht, D., Fonda, E., De Oliveira, N., Briois, V., Frahm, R. (2008) Novel angular encoder for a quick-extended x-ray absorption fine structure monochromator, *Review of Scientific Instruments*, 79, 083107.

[42] Frahm, R., Nachtegaal, M., Stotzel, J., Harfouche, M., Bokhoven, J. A., Grunwaldt, J.-D. (2010) The dedicated QEXAFS facility at the SLS: performance and scientific opportunities, *AIP Conference Proceedings*, 1234, 251–255.

[43] Khalid, S., Ehrlich, S. N., Lenhard, A., Clay, B. (2010) Hard X-rays QEXAFS instrumentation with scan range 20->4000 eV, *Nuclear Instruments and Methods in Physics Research. Section A, Accelerators, Spectrometers, Detectors and Associated Equipment*, 649, 64–66.

[44] Uruga, T., Tanida, H., Inoue, K., Yamazaki, H., Irie, T. (2007) Quick XAFS system using quasimonochromatic undulator radiation at SPring-8, *AIP Conference Proceedings*, 882, 914–916.

[45] Stotzel, J., Lutzenkirchen-Hecht, D., Frahm, R. (2011) A new stand-alone QEXAFS data acquisition system for in situ studies, *Journal of Synchrotron Radiation*, 18, 165–175.

[46] Frenkel, A. I., Kleifeld, O., Wasserman, S. R., Sagi, I. (2002) Phase speciation by extended x-ray absorption fine structure spectroscopy, *Journal of Chemical Physics*, 116, 9449–9456.

[47] Wang, Q., Hanson, J. C., Frenkel, A. I. (2008) Solving the structure of reaction intermediates by time-resolved synchrotron x-ray absorption spectroscopy, *Journal of Chemical Physics*, 129, 234502.

[48] Smolentsev, G., Guilera, G., Tromp, M., Pascarelli, S., Soldatov, A. V. (2009) Local structure of reaction intermediates probed by time-resolved x-ray absorption near edge structure spectroscopy, *Journal of Chemical Physics*, 130, 174508–174516.

[49] McNaught, A. D., Wilkinson, A. (1997) *IUPAC Compendium of Chemical Terminology, the Gold Book*, 2nd ed., Blackwell Science, Oxford, UK.

[50] Mallinoswki, E. R. (1991) *Factor Analysis in Chemistry*, 2nd ed., John Wiley & Sons, Inc., New York.

[51] Beauchemin, S., Hesterberg, D., Beauchemin, M. (2002) Principal component analysis approach for modeling sulfur K-XANES spectra of humic acids, *Soil Science Society of America Journal*, 66, 83–91.

[52] Ressler, T., Wong, J., Roos, J., Smith, I. L. (2000) Quantitative speciation of Mn-bearing particulates emitted from autos burning (methylcyclopentadienyl) manganese tricarbonyl-added gasolines using XANES spectroscopy, *Environmental Science & Technology*, 34, 950–958.

[53] Scheinost, A. C., Kretzschmar, R., Pfister, S., Roberts, D. R. (2002) Combining selective sequential extractions, X-ray absorption spectroscopy, and principal component analysis for quantitative zinc speciation in soil, *Environmental Science and Technology*, 36, 5021–5028.

[54] Terzano, R., Spagnuolo, M., Vekemans, B., De Nolf, W., Janssens, K., Falkenberg, G., Flore, S., Ruggiero, P. (2007) Assessing the origin and fate of Cr, Ni, Cu, Zn, Ph, and V in industrial polluted soil by combined microspectroscopic techniques and bulk extraction methods, *Environmental Science & Technology*, 41, 6762–6769.

[55] Wasserman, S. R., Allen, P. G., Shuh, D. K., Bucher, J. J., Edelstein, N. M. (1999) EXAFS and principal component analysis: a new shell game, *Journal of Synchrotron Radiation*, 6, 284–286.

[56] Jentoft, R. E., Hahn, A., Jentoft, F. C., Ressler, T. (2001) Manganese, iron and sulfur K edge XAFS of promoted sulfated zirconia catalysts, *Journal of Synchrotron Radiation*, 8, 563–565.

[57] Manceau, A., Marcus, M. A., Tamura, N. (2002) Quantitative speciation of heavy metals in soils and sediments by synchrotron X-ray techniques, *Reviews in Mineralogy and Geochemistry*, 49, 341–428.

[58] Rodriguez, J. A., Hanson, J. C., Frenkel, A. I., Kim, J. Y., Perez, M. (2002) Experimental and theoretical studies on the reaction of H-2 with NiO: role of O vacancies

and mechanism for oxide reduction, *Journal of the American Chemical Society*, 124, 346–354.

[59] Kim, J. Y., Hanson, J. C., Frenkel, A. I., Lee, P. L., Rodriguez, J. A. (2004) Reaction of CuO with hydrogen studied by using synchrotron-based x-ray diffraction, *Journal of Physics. Condensed Matter: An Institute of Physics Journal*, 16, S3479–S3484.

[60] Ressler, T., Wienold, J., Jentoft, R. E., Neisius, T., Gunter, M. M. (2002) Kinetics of solid-state reactions in heterogeneous catalysis from time-resolved X-ray absorption spectroscopy, *Topics in Catalysis*, 18, 45–52.

[61] Ressler, T. (2003) Application of time-resolved in-situ X-ray absorption spectroscopy in solid-state chemistry, *Analytical and Bioanalytical Chemistry*, 376, 584–593.

[62] Bressler, C., Chergui, M. (2004) Ultrafast X-ray absorption spectroscopy, *Chemical Reviews*, 104, 1781–1812.

[63] Singh, J., Nachtegaal, M., Alayon, E. M. C., Stötzel, J., van Bokhoven, J. A. (2010) Dynamic structure changes of a heterogeneous catalyst within a reactor: oscillations in CO oxidation over a supported platinum catalyst, *ChemCatChem*, 2, 653–657.

[64] Reimann, S., Sto¨Tzel, J., Frahm, R., Kleist, W., Grunwaldt, J.-D., Baiker, A. (2011) Identification of the active species generated from supported Pd catalysts in heck reactions: an in situ quick scanning EXAFS investigation, *Journal of the American Chemical Society*, 133, 3921–3930.

[65] Lutzenkirchen-Hecht, D., Grunwaldt, J. D., Richwin, M., Griesebock, B., Baiker, A., Frahm, R. (2005) Monitoring of fast transformations in solid state chemistry and heterogeneous catalysis by QEXAFS in the second scale, *Physica Scripta*, T115, 831–833.

[66] Chupas, P. J., Chapman, K. W., Kurtz, C., Hanson, J. C., Lee, P. L., Grey, C. P. (2008) A versatile sample-environment cell for non-ambient X-ray scattering experiments, *Journal of Applied Crystallography*, 41, 822–824.

[67] Grunwaldt, J. D., Ramin, M., Rohr, M., Michailovski, A., Patzke, G. R., Baiker, A. (2005) High pressure in situ x-ray absorption spectroscopy cell for studying simultaneously the liquid phase and the solid/liquid interface, *Review of Scientific Instruments*, 76, 054104–054110.

[68] Bare, S. R., Kelly, S. D., Ravel, B., Greenlay, N., King, L., Mickelson, G. E. (2010) Characterizing industrial catalysts using in situ XAFS under identical conditions, *Physical Chemistry Chemical Physics*, 12, 7702–7711.

[69] Bare, S. R., Yang, N., Kelly, S. D., Mickelson, G. E., Modica, F. S. (2007) Design and operation of a high pressure reaction cell for in situ X-ray absorption spectroscopy, *Catalysis Today*, 126, 18–26.

[70] Glatzel, P., Bergmann, U. (2005) High resolution 1s core hole X-ray spectroscopy in 3d transition metal complexes—electronic and structural information, *Coordination Chemistry Reviews*, 249, 65–95.

[71] Tinnemans, S. J., Mesu, J. G., Kervinen, K., Visser, T., Nijhuis, T. A., Beale, A. M., Keller, D. E., van der Eerden, A. M. J., Weckhuysen, B. M. (2006) Combining operando techniques in one spectroscopic-reaction cell: new opportunities for elucidating the active site and related reaction mechanism in catalysis, *Catalysis Today*, 113, 3–15.

[72] Newton, M. A., Belver-Coldeira, C., Martinez-Arias, A., Fernandez-Garcia, M. (2007) Dynamic in situ observation of rapid size and shape change of supported Pd nanoparticles during CO/NO cycling, *Nature Materials*, 6, 528–532.

[73] Ressler, T., Jentoft, R. E., Wienold, J., Günter, M. M., Timpe, O. (2000) In situ XAS and XRD studies on the formation of Mo suboxides during reduction of MoO3†, *Journal of Physical Chemistry. B, Materials, Surfaces, Interfaces & Biophysical*, 104, 6360–6370.

[74] Piovano, A., Agostini, G., Frenkel, A. I., Bertier, T., Prestipino, C., Ceretti, M., Paulus, W., Lamberti, C. (2011) Time resolved in situ XAFS study of the electrochemical oxygen intercalation in SrFeO2.5 brownmillerite structure: comparison with the homologous SrCoO2.5 system, *Journal of Physical Chemistry C*, 115, 1311–1322.

[75] Ferri, D., Kumar, M. S., Wirz, R., Eyssler, A., Korsak, O., Hug, P., Weidenkaff, A., Newton, M. A. (2010) First steps in combining modulation excitation spectroscopy with synchronous dispersive EXAFS/DRIFTS/mass spectrometry for in situ time resolved study of heterogeneous catalysts, *Physical Chemistry Chemical Physics*, 12, 5634–5646.

[76] Eyssler, A., Kleymenov, E., Kupferschmid, A., Nachtegaal, M., Kumar, M. S., Hug, P., Weidenkaff, A., Ferri, D. (2010) Improvement of catalytic activity of LaFe0.95Pd0.05O3 for methane oxidation under transient conditions, *The Journal of Physical Chemistry C*, 115, 1231–1239.

[77] Grunwaldt, J.-D., Baiker, A. (2007) Time-resolved and operando XAS studies on heterogeneous catalysts—from the gas phase towards reactions in supercritical fluids, *ChemInform*, 38, 577–581.

[78] Golovchenko, J. A., Levesque, R. A., Cowan, P. L. (1981) X-ray monochromator system for use with synchrotron radiation sources, *Review of Scientific Instruments*, 52, 509–516.

2

SPATIALLY RESOLVED X-RAY ABSORPTION SPECTROSCOPY

CHRISTIAN G. SCHROER AND JAN-DIERK GRUNWALDT

2.1 INTRODUCTION

The design of heterogeneous catalysts is manifold. On the one side, it requires the design of a catalytic center or an ensemble of catalytic centers on an atomic scale. On the other side, its macroscopic structure inside reactors or special devices is important. The structure on an atomic scale directly influences the physical and electronic properties and thereby the catalytic performance of the exposed surfaces or active centers. However, designing a macroscopic solid with certain pore structure, mechanical strength, and high efficiency is equally important. Figure 2.1 illustrates the large variability of catalyst structures on the nanoscale. A very good example where nano- and macroscale both play an important role is an exhaust gas catalyst: in three-way catalysts, the oxidation of CO or hydrocarbons, but also the reduction of NO_x, occur on noble metals like Pt, Pd, and Rh. Decisive is here the adsorption of the entitled gas molecules on the metallic surfaces. A high dispersion of the metallic particles and their stability is important to efficiently use the noble metals. Equally important is the design of the catalyst body, that is, allowing a high gas flow through the catalyst bed and supporting the metal particles on ceramics with a highly porous structure. Usually a thin layer (washcoat) is supported on a ceramic honeycomb material, allowing a high mechanical stability, a high gas flow, and the exploitation of most of the noble metal particles (underlying layers may not fully contribute due to mass transport limitations). This interplay

In-situ *Characterization of Heterogeneous Catalysts*, First Edition.
Edited by José A. Rodriguez, Jonathan C. Hanson, and Peter J. Chupas.

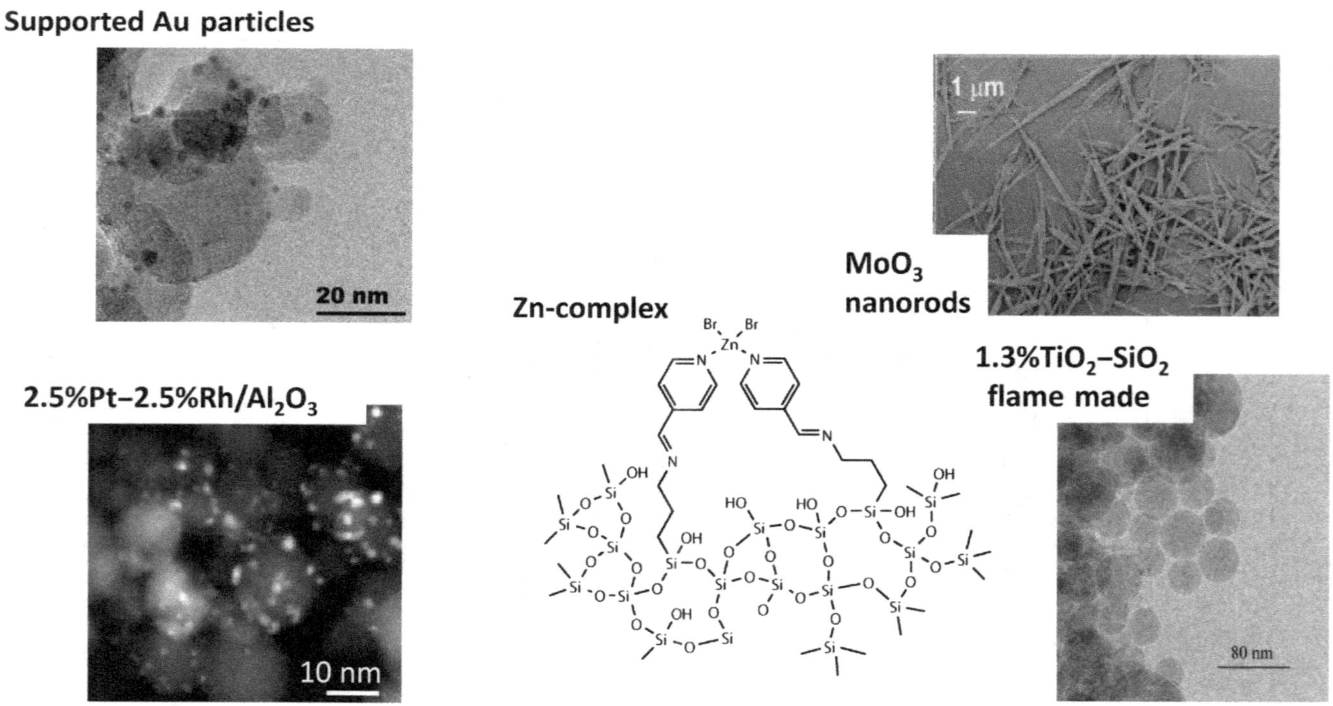

Figure 2.1. Different examples of the structure of heterogeneous catalysts ranging from monometallic (gold particles supported on alumina for CO-oxidation) and bimetallic (2.5% Pt–2.5% Rh/Al_2O_3 for partial oxidation of methane) supported nanoparticles to mixed oxides (flame-made TiO_2–SiO_2 for epoxidation reactions), MoO_3-nanorods (for selective oxidation), and immobilized complexes (Zn-complexes for CO_2-activation). Reprinted with permission from Grunwaldt, J.-D., Schroer, C. G. (2010) Hard and soft x-ray microscopy and tomography in catalysis: bridging the different time and length scales, *Chem. Soc. Rev.*, 39, 4741. Reproduced by permission of The Royal Society of Chemistry.

between micro- and macroscale is a general challenge in catalysis, since the catalyst body always has to be put into a reactor that comes in a large variety of designs [1–3].

Another equally important aspect is that many of the catalysts are pretreated before use and undergo strong structural changes when approaching reaction conditions [4–9]. Hence, for establishing structure–activity relationships the structure has to be determined at *in-situ* (process) conditions, that is, in catalytic reactors that are closely related to the respective catalytic process [10–12]. Many of the techniques described in this book focus on the identification of the active centers and the adsorption sites under reaction conditions. But as outlined earlier, the macroscale of catalysts also plays an important role, and, more importantly, there may also be structural changes on macroscopic length scales within a catalytic reactor. This requires spatially resolved techniques that allow probing atomic-scale information in a two-dimensional (2-D) or even three-dimensional (3-D) manner under static or dynamic reaction conditions on a nano-, micro-, and even millimeter scale [2, 13–15].

For this purpose, X-ray-based techniques are excellent tools [10, 11, 16]. Structural information can be gathered by X-ray absorption spectroscopy (XAS), X-ray diffraction, and small-angle X-ray scattering. Due to the often highly amorphous character of the catalytic ensembles or even single centers, X-ray absorption spectroscopy in terms of X-ray absorption near-edge structure (XANES) and extended X-ray absorption fine structure (EXAFS), in particular, have proved to be powerful ways to gain atomic scale information about catalysts and their structural changes [11, 17–20]. By combining XAS with X-ray microscopy, the spatial changes within the catalyst bodies or in the catalytic converters can be studied, on both the mesoscopic and macroscopic scales. As X-rays have a large penetration depth and can probe the inside of samples or special sample environments, X-ray microscopy is ideal for *in-situ* and *in-operando* studies that are very challenging for other microscopy techniques.[1] In combination with tomography, the 3-D inner structure of an object can be reconstructed.

X-ray microscopy can be combined with various X-ray analytical techniques, such as X-ray fluorescence, diffraction, and absorption spectroscopy (XAS), yielding elemental, structural, and chemical contrast, respectively [23]. As outlined earlier, in the context of catalysis, XAS is the most important contrast, but one of the key strengths of X-ray microscopy is combining it with the other techniques to gain additional information on the spatial distribution of the elements and crystalline components. In XAS, the atomic species under study defines the energy of the X-rays that may range from very soft (e.g., 284 eV at the carbon K-edge) to rather hard X-rays (e.g., 24,350 eV at the

[1]Tremendous progress in the field of electron microscopy has been made [14, 21, 22], but very small and special sample environments that cannot fully reproduce the conditions inside a larger reactor are still required.

palladium K-edge or even at higher energies). Probing the local environment of different elements involved in a given chemical reaction yields valuable complementary information.

In this chapter, we give an overview of the application of X-ray microscopy to the study of catalysts. In Section 2.2, we introduce XAS in general, including the aspects needed for imaging and *in-situ* studies of catalytic reactors. Full-field microscopy and tomography is described in Section 2.3. Section 2.4 reviews scanning microscopy and its applications to catalysis.

2.2 X-RAY ABSORPTION SPECTROSCOPY

Absorption spectroscopy is a powerful tool to analyze the chemical state and local environment of a given element without the need for long-range order in the specimen. In the simplest case, a homogeneous specimen of well-defined thickness z is illuminated by monochromatic X-rays, whose energy is tunable around the absorption edge of the element of interest. The experimental setup is shown in Figure 2.2a. Two ionization chambers I_0 and I_1 measure the intensity of the radiation before and behind the sample, respectively. From the ratio of I_1 and I_0 measured at a given energy E, the attenuation coefficient $\mu(E)$ can be extracted from Lambert–Beer's law:

$$\mu(E) = \frac{1}{z}\ln\frac{I_0}{I_1}.$$

An absorption spectrum is acquired by scanning the energy E in an interval around the absorption edge. From these data, information about the chemistry of the element of interest can be extracted [24]:

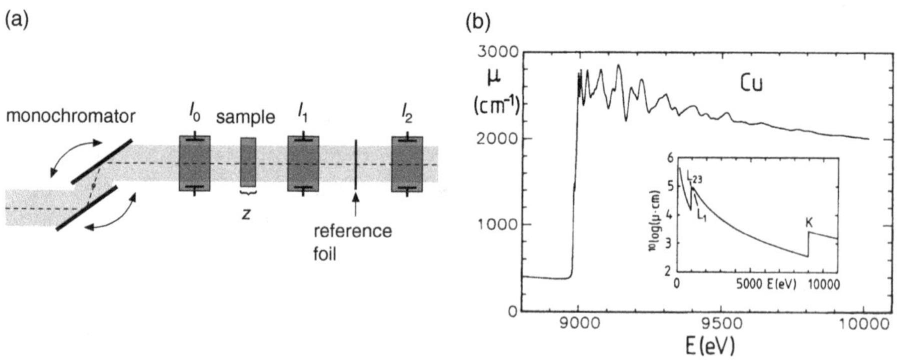

Figure 2.2. (a) Experimental setup for absorption spectroscopy at a synchrotron radiation source. (b) Absorption spectrum of metallic Cu. The inset shows the absorption spectrum of atomic copper (spectrum courtesy of B. Lengeler, RWTH Aachen).

- The exact position of the absorption edge, for example, defined by the first inflection point, yields the oxidation state of the element. The more positively charged the ion of the atomic species of interest is, the more the absorption edge shifts toward higher energies.
- The XANES extending from below the edge to about 50 eV above the edge probes the local projected free density of states for the given element. Each local configuration around the atomic species of interest has a characteristic near-edge structure that can serve as a fingerprint to identify it in a heterogeneous sample; that is, it is strongly dependent on the local symmetry and the oxidation state (see Section 2.3.2 or Section 2.4.3).
- The extended EXAFS that extends to about 1 keV above the edge gives information about the local chemical neighborhood of the element, yielding the interatomic distances with the neighbors and coordination numbers.

As an example, the EXAFS spectrum, that is, μ as a function of E, of metallic copper is shown in Figure 2.2b. For comparison, the inset in Figure 2.2b shows the spectrum of copper atoms in the gas phase. In the gas phase, the fine structure at and above the absorption edge is not present, as the free atoms have no neighbors and the free density of states around the atom is featureless. As opposed to that, the absorption spectrum of metallic copper shows a lot of structure, both in the near-edge and in the extended fine-structure region. Detailed analysis of this structure yields the local projected free density of states and the distance and number of neighboring atoms in several coordination shells [24].

Another important aspect is that X-ray absorption spectroscopy allows the monitoring of the structure under conditions that are close to the reaction conditions. To underline the importance of the *in-situ* aspect, Figure 2.3 shows the structure of a 2.5% Pt–2.5% Rh/Al_2O_3 catalyst while it is catalyzing the catalytic partial oxidation (CPO) of methane to hydrogen and carbon monoxide. If the catalyst is monitored at the Rh K-edge before and after heating in the reaction mixture, the Rh-component (and also the Pt-component; see Reference [25]) is mostly in an oxidized state as uncovered by the fingerprint in the XANES region (strong white line, i.e., the first peak immediately at the absorption edge). However, when the catalyst is heated to the reaction temperature, it is reduced at the moment when the reaction is ignited. Further *in-situ* XANES measurements show that it is reoxidized when the catalyst extinguishes while heating to room temperature. In this experiment, a large $8 \times 1\ mm^2$ beam was used, and when analyzing the oxidation state more properly by linear combination fits of oxidized and metallic Rh, it turned out that the catalyst is not fully reduced at the ignition temperature [25]. This is a result of Rh having different oxidation states along the catalyst bed, as discussed in Section 2.3.2.

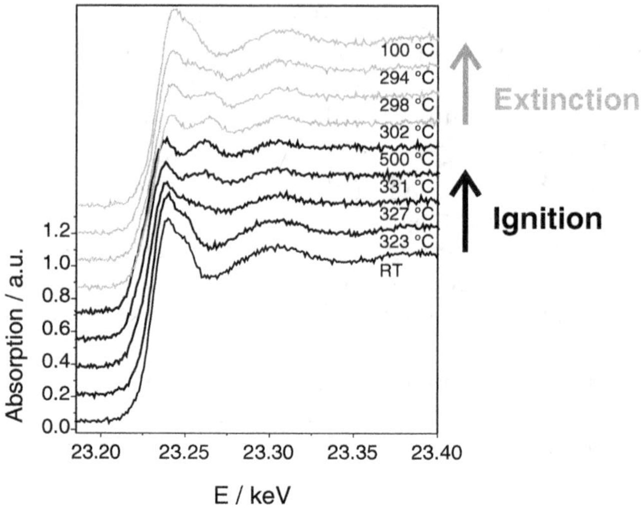

Figure 2.3. XANES spectra at the Rh K-edge of 2.5% Pt–2.5% Rh/Al_2O_3 during the heating (ignition of the CPO of methane) and cooling (extinction) in 6% CH_4/3% O_2/He (see also Reference [25]).

The example highlights two issues: the structure of catalysts in certain cases should be analyzed under reaction conditions. Furthermore, microscopy studies are required. It is also not sufficient to apply transmission X-ray microscopy but the combination with X-ray absorption spectroscopy is needed. For the purpose of imaging chemical species of a given element in X-ray microscopy with absorption spectroscopic contrast, the measured spectrum at a given point in the specimen is decomposed into known spectra of different chemical species. For this analysis, the spectra of different chemical species serve as a fingerprint to identify them at different locations inside a sample. In the following sections, several examples of this type of analysis are given.

2.3 FULL-FIELD MICROSCOPY AND TOMOGRAPHY

X-ray imaging in projection was the first application of X-rays and is still the most common application of X-rays today, in particular for medicine, nondestructive testing, and security. X-ray microscopy in full-field mode aims at imaging an object in projection, recording the transmission image at once by a position-sensitive detector, as opposed to scanning microscopy, where the sample is scanned through a confined X-ray beam.

X-ray microscopy requires a very brilliant X-ray source, especially for highest spatial resolutions. Therefore, this technique is best implemented at modern synchrotron radiation sources, such as the European Synchrotron Radiation Facility (ESRF) in Grenoble, France; the Advanced Photon Source (APS) near Chicago, Illinois; the SPring-8 in Japan; or the synchrotron radiation source PETRA III at Deutsches Elektronen-Synchrotron (DESY) in Hamburg, Germany. These large-scale facilities are built to generate brilliant X-ray beams for a number of experiments. X-ray beams from synchrotron radiation sources are up to 12 orders of magnitude more brilliant than those from an X-ray tube. In addition, they are tunable in energy—a property that is crucial to elemental contrast based on absorption spectroscopy.

2.3.1 Full-Field Imaging with High Spatial Resolution

In the simplest form of X-ray microscopy, the sample is illuminated by X-rays from a distant source and imaged in projection onto a 2-D position-sensitive detector. Figure 2.4a depicts the setup that is usually used at a synchrotron radiation source. The sample is illuminated by nearly parallel X-rays (submillimeter source at 50–100 m distance), and its projection image is recorded on a position-sensitive detector placed a distance L behind the object. The spatial

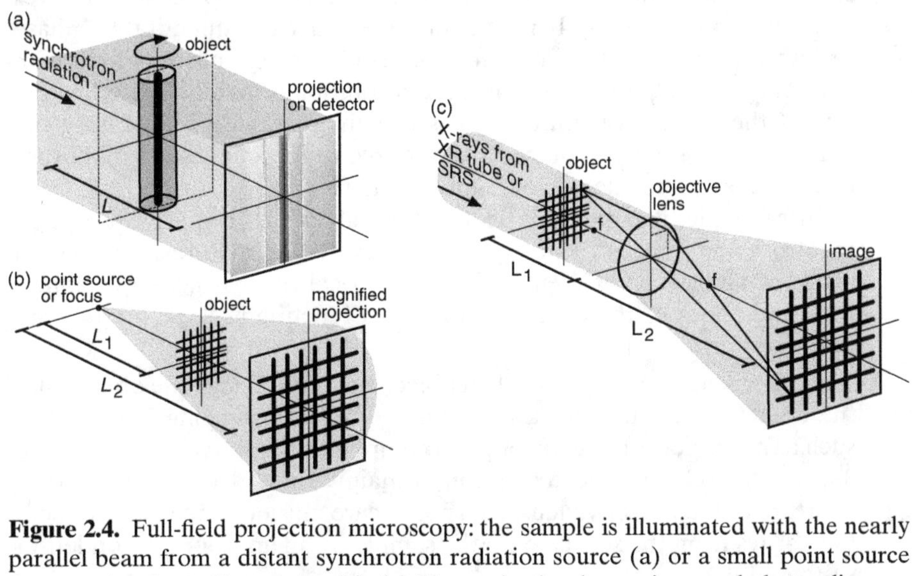

Figure 2.4. Full-field projection microscopy: the sample is illuminated with the nearly parallel beam from a distant synchrotron radiation source (a) or a small point source from a microfocus X-ray tube (b). (a) The projection image is recorded at a distance L behind the object with a position-sensitive area detector. (b) With the object placed at a distance L_1 from the point source, the image is recorded by the detector placed a distance L_2 away from the source. The image is magnified by $m = L_2/L_1$. (c) Magnified imaging using an X-ray optic: the sample is illuminated by X-rays from behind. Using an X-ray optic as an objective lens, the magnified image (magnification L_2/L_1) is projected onto the detector in a magnifying geometry. XR, X-ray; SRS, storage ring source.

resolution is mainly limited by that of the detector and can range down to slightly less than 1 μm [26, 27].

The distance L between the sample and the detector (see Fig. 2.4a) can be varied to adjust the contrast: In contact, that is, $L \approx 0$ mm, the contrast from the sample is purely given by absorption. When the detector is moved away from the sample, the near-field phase contrast becomes more and more pronounced. A quantitative evaluation of the near-field phase contrast yields information about the refraction inside the sample in addition to the absorption [28].

To combine this method with absorption spectroscopy, the detector is usually placed as close to the sample as possible to reduce phase contrast effects.

In order to improve the spatial resolution further into the submicrometer range, X-ray imaging with magnification is needed. This can be realized in different ways, as shown in Figure 2.4b,c:

- One scheme is to generate a small secondary X-ray source (Fig. 2.4b) and project the sample onto the detector with magnification. The latter depends on the imaging geometry and is given by $m = L_2/L_1$, where L_1 is the source-to-object and L_2 is the source-to-detector distance. As m needs to be large to achieve high resolution, this method intrinsically involves phase-contrast imaging. It is possible to extract the quantitative phase and absorption information from a series of micrographs recorded with a different L_1 [29]. In this scheme, the resolution is mainly limited by the size of the secondary source that is usually generated by focusing the X-rays from a synchrotron radiation source to the smallest possible spot (see Section 2.4 for details on nanofocusing).

  In principle, it is possible to use this scheme to acquire absorption spectroscopic images of an object. However, this requires recording a series of images at different L_1 and numerical reconstruction of amplitude and phase of the object transmission function [29]. So far, this has not been carried out.
- The other scheme makes use of an objective lens to image the object onto the detector with magnification. This requires an imaging X-ray optic, such as a Fresnel zone plate [30] or a compound refractive X-ray lens [31, 32]. In this scheme, the resolution is mainly limited by the numerical aperture of the objective lens, but special care must also be taken to focus the X-rays from the source onto the sample by an appropriate condenser system [30].

The objective lens must be a true imaging optic. In the soft X-ray range, Fresnel zone plates are the only type of optics used for direct imaging. In this energy range, spatial resolutions in the range of 10 nm have been achieved [33, 34]. For harder X-rays, these optics lose their efficiency, but significant advances have been made in this range with Fresnel zone plates, too [30]. For hard X-rays, parabolic compound refractive lenses are also well suited for direct imaging [31, 32]. Using beryllium as a lens material, spatial resolutions down to about 100 nm have been achieved [35].

As both types of imaging optics are chromatic, changing the X-ray energy requires refocusing of the X-ray microscope. While this is in principle possible, this technique has not been used for absorption spectroscopic microscopy so far.

2.3.2 Full-Field Imaging of a Catalyst during Partial Oxidation of Methane

In order to study the local chemical state of a catalyst inside a reactor, full-field imaging (Fig. 2.4a) can be combined with XAS. To this end, a series of full-field transmission images is recorded as a function of energy E around the absorption edge of the element of interest. With the detector placed very near to the sample ($L \approx 0$; see Section 2.3.1), the contrast in each pixel can be interpreted according to Lambert–Beer's law:

$$I(E,x,y) = I_0(E,x,y)\cdot \exp\left\{-\int \mu(E,x,y,z)\,\mathrm{d}z\right\}. \tag{2.1}$$

Here, x and y are the coordinates of a given pixel, and $I_0(E,x,y)$ is the incident intensity. The incident beam at the pixel (x,y) is attenuated along the beam (along the z-direction) as described by the integral over the linear absorption coefficient $\mu(E,x,y,z)$ inside the object. For an inhomogeneous sample, it can vary in three dimensions. $I_0(E,x,y)$ is measured for each energy E by recording an image of the flat incident beam without a sample.

The experimental setup for an *in-operando* experiment is shown in Figure 2.5a [36]. This experiment was carried out at the synchrotron radiation source DORIS III (beamline X1) at DESY in Hamburg. The example shows the CPO of methane over a Rh/Al_2O_3 catalyst inside a capillary microreactor with a diameter of about 1 mm. The catalytic reactor was designed for *in-operando* X-ray absorption studies. A capillary containing the catalyst is heated by a hot air stream from below the catalyst bed and shielded by Kapton foil (DuPont, Circleville, OH) [36]. The experiment corresponds to the one in Section 2.2 after the ignition of the CPO of methane to CO and H_2. In the experiment with an integral beam (8×1 mm^2), it was observed that during the ignition the Rh-component was not completely but only partly reduced. Similar to that experiment, also here, the flow of reactants was adjusted by mass flow controllers and the product gas composition was monitored by mass spectrometry. Temperature and gas flow was adjusted here in a way to establish the partial oxidation of methane in a steady state at an oven temperature that was slightly above the ignition temperature (note that inside the catalytic reactor the temperature is much higher).

X-ray absorption images of the capillary microreactor were recorded with a high-resolution X-ray camera placed as closely as possible behind the furnace.

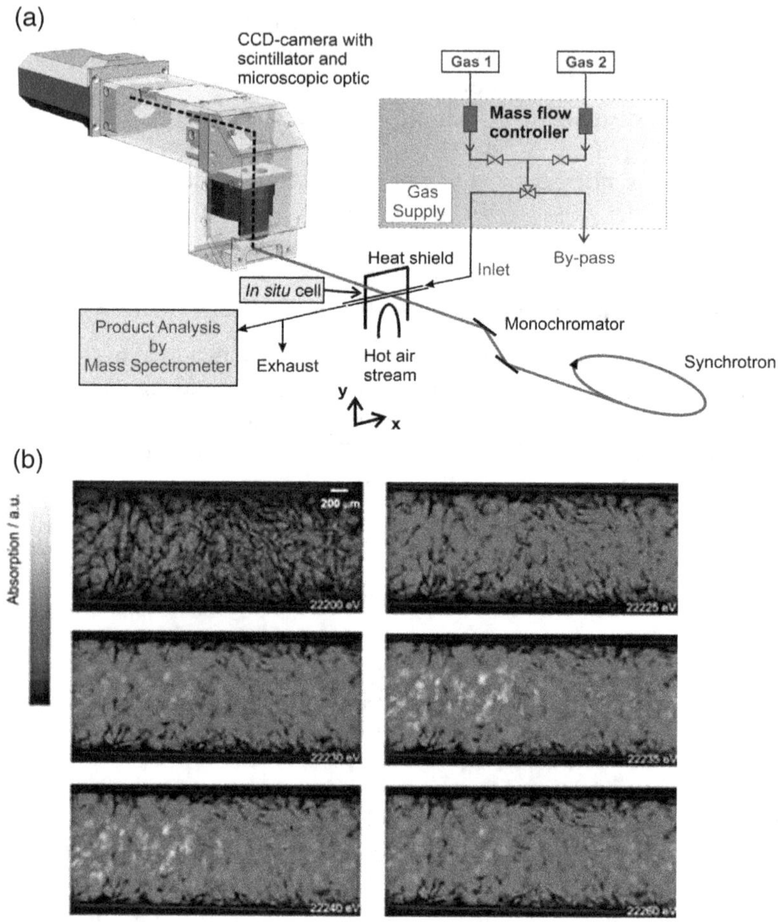

Figure 2.5. (a) Experimental setup for *in-operando* imaging of the catalyst bed during partial oxidation of methane. (b) X-ray absorption in the catalyst bed as a function of X-ray energy (see Reference [36] for details). Reprinted with permission from Grunwaldt, J.-D., Hannemann, S., Schroer, C. G., Baiker, A. (2006) 2D-mapping of the catalyst structure inside a catalytic microreactor at work: partial oxidation of methane over Rh/Al_2O_3, *J. Phys. Chem. B*, 110, 8674–8680. Copyright 2006 by American Chemical Society. (See color insert.)

The energy was varied around the Rh K-edge from 23,190 to 23,375 eV in steps of 1 eV. Figure 2.5b shows the integral absorption

$$\int \mu(E,x,y,z)\,dz = \ln \frac{I_0(E,x,y)}{I(E,x,y)}$$

as a function of position for a few selected X-ray energies. For each pixel, a full XANES spectrum of the Rh was obtained that can be compared with reference spectra for different oxidation states of Rh.

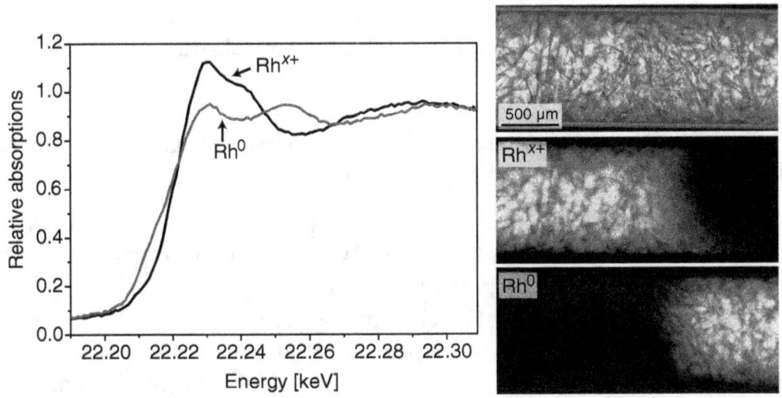

Figure 2.6. Left: Absorption spectra at the Rh K-edge extracted from the stack of absorption images at the far left (Rh^{x+}) and the far right (Rh^0). These spectra together with a flat spectrum were used to fit the spectra in each pixel of the image, resulting in the distributions shown on the right. Right: absorption due to elements other than Rh (featureless spectrum) (top), distribution of oxidized (middle), and reduced (metallic) Rh (bottom) (see Reference [36]). Reprinted with permission from Grunwaldt, J.-D., Hannemann, S., Schroer, C. G., Baiker, A. (2006) 2D-mapping of the catalyst structure inside a catalytic microreactor at work: partial oxidation of methane over Rh/Al_2O_3, *J. Phys. Chem. B*, 110, 8674–8680. Copyright 2006 by American Chemical Society. (See color insert.)

The XANES spectra at the entrance (left) and exit (right) of the reactor are characteristic of oxidized (Rh^{x+}) and metallic rhodium (Rh^0), respectively. Note that the Rh particles are not fully reoxidized because the core remains metallic. The corresponding spectra are shown in Figure 2.6 on the left. By fitting these spectra to those in each pixel, the distribution of these two rhodium species inside the capillary can be determined. Figure 2.6 (right) shows the distribution of these species with a strong gradient in between them. The oxidation state changes within less than 200 μm.

This leads to the conclusion that partial oxidation of methane occurs here as a result of a total combustion and reforming mechanism, which is also supported by other studies [37, 38]. In fact, a similar oxidation state distribution was found for Pt- and Pt/Rh-based catalysts [39, 40], whereas Pd/Al_2O_3 catalysts were not stable under these conditions, resulting in oscillations between the ignition and extinction of the CPO reaction [40]. The gradient in structure is accompanied by strong temperature gradients with a strong hot spot at the beginning of the catalyst bed due to total combustion, as parallel imaging by infrared (IR) thermography has shown [40].

This experimental technique requires the catalytic reaction to be stationary over the full acquisition time. In this case, approximately 3 h of data acquisition were needed on the bending magnet beamline X1 at DORIS III. Transferring this experiment to an undulator beamline at a third-generation synchrotron radiation source reduces the acquisition time by several orders of magnitude.

2.3.3 Time-Resolved Imaging of a Catalyst during the Ignition of Partial Oxidation of Methane

Time-resolved imaging requires the imaging process to be fast compared to the changes of the system under study. Typically, recording full absorption spectra by the method described in Section 2.3.2 is too slow. Instead, it is sometimes possible to monitor the changes of the state of a catalyst at a fixed energy, for example, when the absorption at that energy depends very sensitively on the oxidation state or coordination geometry of the catalyst. At modern synchrotron radiation sources, imaging with monochromatic X-rays requires exposure times from milliseconds to seconds, depending on the spatial resolution needed.

This scheme was used to image the ignition of the partial oxidation of methane over a Rh–Pt/Al_2O_3 catalyst [41]. The experiment was carried out at the insertion device beamline ID26 at ESRF. The experimental setup was very similar to that described in Section 2.3.2, the main difference being a fast readout low noise detector (FReLoN) used for image acquisition [42]. The catalytic reaction was carried out inside a heated glass capillary (Fig. 2.7, left). The X-ray energy was set to the white line of the Pt absorption spectrum (E = 11,596 eV), where the absorption of platinum is highly sensitive to the oxidation state (Fig. 2.7, left). While ramping up the temperature to ignite the partial oxidation reaction, the X-ray transmission images of the reactor capillary are recorded at a rate of 4 Hz. Figure 2.7 (right) shows a time series of the difference in the attenuation induced by changes of the oxidation state of Pt at the exit of the catalyst bed. Before the reaction ignites, Pt is in the oxidized state (Fig. 2.7, right). When the reaction ignites, the Pt is successively reduced along the reactor, starting from the end of the catalyst bed (from right to left; see Fig. 2.7a–d). The reduction of the catalyst particles starts at some individual particles first and then evolves from the inside of the particle to its periphery. The results are in line with complementary time-resolved studies of the ignition by quick EXAFS (QEXAFS), which show that the catalyst is reduced from the end of the catalyst bed within less than a second [43].

2.3.4 Tomography

One of the great strengths of hard X-rays is their large penetration depth in matter. A single full-field image is a projection of the object along the beam. These projections are not easily interpretable if too many features overlap along the beam. To overcome this issue, tomographic imaging can be used.

In the case of parallel illumination (see Fig. 2.4a) the sample is scanned in rotation around an axis perpendicular to the beam in an angular range of 180°.

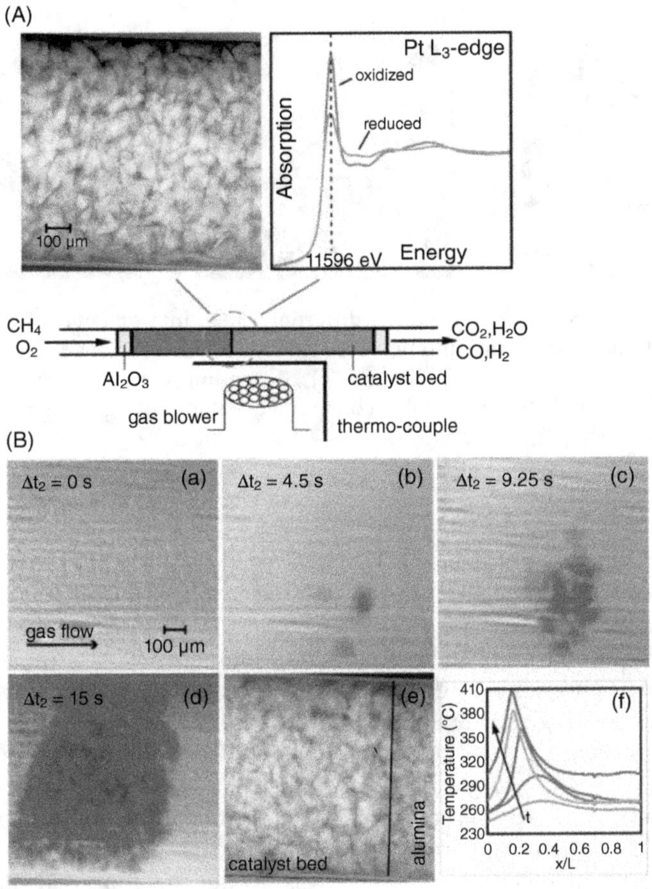

Figure 2.7. (A) Schematic sketch of the reactor capillary for partial oxidation of methane over a $Rh/Pt/Al_2O_3$ catalyst inside a heated reactor capillary together with a transmission image through the capillary and absorption spectra of reduced and oxidized Pt at the Pt L_3-edge. (B) (a–d) Time series of the oxidation state of Pt (yellow oxidized, orange-red reduced) during the ignition of the partial oxidation reaction. (e) Transmission image at the end of the catalyst bed. The black line delineates the end of the catalyst bed. (f) Evolution of the temperature along the capillary during ignition [41]. Reprinted with permission from Kimmerle, B., Grunwaldt, J.-D., Baiker, A., Glatzel, P., Boye, P., Stephan, S., Schroer, C. G. (2009) Visualizing a catalyst at work during the ignition of the catalytic partial oxidation of methane, *J. Phys. Chem.*, C, 113, 3037–3040. Copyright 2009 American Chemical Society. (See color insert.)

At each position of the scan, a projection image is recorded. From these data, the 3-D inner structure of the object is numerically reconstructed [44].

An example tomographic reconstruction of $CuCl_2$-impregnated alumina pellets with a cylindrical shape (3.2 mm diameter, 3.6 mm length) measured at beamline BW2 of DORIS III is shown in Figure 2.8. The photon energy was

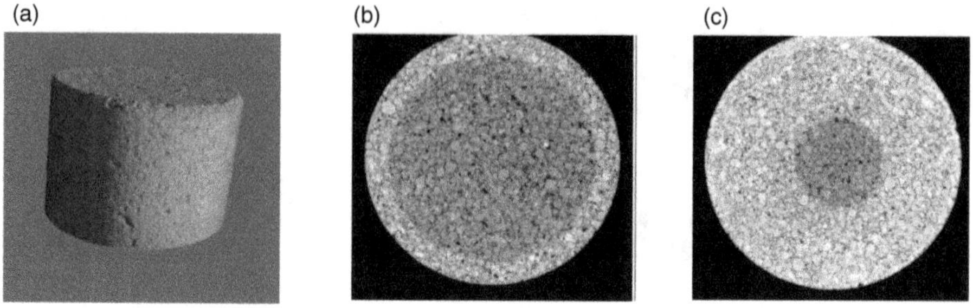

Figure 2.8. 3-D reconstruction of two different $CuCl_2$-impregnated alumina pellets after 1 and 10 min impregnation time, respectively: (a) surface view as extracted from the tomography, and (b) after 1 min and (c) 10 min impregnation time, respectively; field of view: $3.54 \times 2.36\ mm^2$; total reconstructed volume of $3.54 \times 3.54 \times 2.36\ mm^3$ (see also Reference [15]).

set to 18 keV and a pixel size of 2.3 μm was used (resulting in a spatial resolution of about 5 μm). The sample was rotated using a special stage described in Reference 45.

Spectroscopic contrast can be obtained in tomography by acquiring several transmission tomograms at different X-ray energies around an absorption edge. With two energies, one above and one below the absorption edge of an element of interest, this method is known as differential attenuation contrast tomography [46, 47] and yields the distribution of the element of interest. Rau et al. [48] recorded tomograms at more than two energies, distinguishing between several chemical compounds of Mo in 3-D reconstructions. Recording full absorption spectra in this way was, so far, too time consuming. With optimized modern data acquisition, this may become feasible soon.

2.4 SCANNING MICROSCOPY AND TOMOGRAPHY

In scanning microscopy the sample is scanned through a small focused X-ray beam. The main advantage of this microscopy scheme is the variety of contrast mechanisms available, giving local information about the element distribution, the local chemical state of an atomic species, or the local mesoscopic or atomic structure. Therefore, it can also be applied to image the structure and the local chemical state of a catalyst. As this method requires a high brilliance of the X-ray source, it is mainly applied at modern synchrotron radiation sources. Scanning microscopy is widely used in the hard and soft X-ray range. In the soft X-ray range, scanning transmission microscopy was very successfully combined with absorption spectroscopy, for example, to identify different organic compounds by absorption spectroscopy at the carbon edge [49]. In the context of catalyst imaging, soft X-ray scanning microscopy was successfully applied to investigate a Fischer–Tropsch catalyst [13]. In the hard X-ray range, a 2-D map similar to that described in Figure 2.6 was obtained in scanning mode at the μXAS beamline at the Swiss Light Source [15].

2.4.1 Experimental Setup and Contrast Mechanisms

Figure 2.9 shows a schematic sketch of a scanning microscope. The X-rays from a synchrotron radiation source are focused by an X-ray optic onto the sample in a strongly reducing geometry. Today, a variety of hard X-ray optics is available, reaching focal spot sizes down to well below 100 nm [33, 50–54]. These optics are based on reflection, refraction, and diffraction. For absorption spectroscopy, focusing mirrors based on total external reflection are best suited, as they are achromatic, that is, the location of their focus is independent of the X-ray energy. Thus, the energy can be scanned over large ranges with the sample staying in focus. For the other types of optics, small energy changes are tolerable, as long as the focal distance changes by less than a depth of focus (see Section 2.4.3). Larger changes of energy have to be accompanied by refocusing of the sample. Today, most of the optics are technology limited. As the fabrication techniques are improved further, they will eventually approach the physical limits of focusing that lie in the range of about 10 nm for reflective and refractive optics [55–57]. The highest theoretical spatial resolution is expected for diffractive optics, in particular so-called multilayer Laue lenses or thick tilted zone plates [58–60].

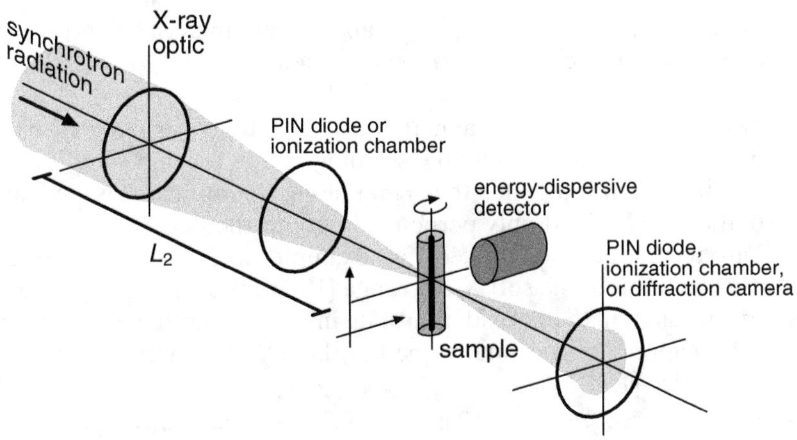

Figure 2.9. Scheme of a hard X-ray scanning microscope. The X-rays from a synchrotron radiation source are focused onto the sample. The X-ray fluorescence from the sample is recorded by an energy-dispersive detector from the side, the absorption is measured by a PIN diode or ionization chamber behind the sample, and the diffraction can be acquired by a 2-D diffraction camera. The incident intensity is recorded by a diode or ionization chamber.

While the sample is scanned through the focus, a variety of contrasts can be obtained from the sample. The most common are:

- *X-ray fluorescence:* Atoms along the focused beam that absorbed an X-ray photon can emit X-ray fluorescence. It is characteristic of the atomic species, allowing one to identify the elemental composition along the focused beam. The fluorescence radiation is recorded with an energy-dispersive detector (Fig. 2.9).
- *X-ray absorption:* Atoms along the focused beam absorb the X-rays, attenuating the transmitted beam according to the Lambert–Beer's law in Equation 2.1. The transmission is measured by two detectors before and after the sample, for example, two ionization chambers (Fig. 2.9). If the energy of the focused beam is varied, local absorption spectra of the sample can be recorded. The absorption spectroscopic signal can also be measured indirectly in fluorescence. This is of great advantage for dilute samples, where the contrast in the attenuation signal is small [61].
- *X-ray diffraction:* The X-rays can be scattered by the atomic structure along the focused beam and recorded by a diffraction camera behind the sample (Fig. 2.9). Depending on the distance of the detector from the sample, wide-angle atomic structure information or small-angle mesoscopic structure information can be obtained from the sample in this way.

2.4.2 Scanning Modes in Microscopy and Tomography

Depending on the scanning mode, 2-D maps, tomographic slices, or even 3-D tomographic maps of the sample can be obtained:

- *2-D maps:* The sample is scanned in two directions perpendicular to the beam (Fig. 2.9). This is a suitable scanning mode for thin samples, as the pencil-like beam integrates the signal along the optical axis, and spatial resolution is obtained only perpendicularly to the beam.

  This technique was applied, for example, to investigate a Fischer–Tropsch catalyst in the soft X-ray range [13]. In the hard X-ray range, the strong gradient in the Pt oxidation state in CPO of methane was observed by 2-D mapping of the catalyst bed at the μXAS beamline at the Swiss Light Source [15].
- *2-D scanning tomography:* The sample is scanned through the beam along one direction perpendicular to the beam, for example, the horizontal direction (Fig. 2.9). After completing this scan, the sample is rotated by an integer fraction of 180° around the axis orthogonal to both the beam and the translational scanning direction (i.e., the vertical axis in the given example, Fig. 2.9) and the translational scan is repeated. The procedure ends when the sample has completed a rotation over 180 or 360°.

- *Full 3-D tomographic map:* A series of 2-D maps is recorded at equidistant angular positions of the sample over 180 or 360°.

Generally, hard X-ray scanning microscopy and tomography is slow, as the data for each pixel are acquired sequentially. Therefore, the size of scanning micrographs is mainly limited by the overall acquisition time, and the technique is less well suited for time-resolved studies.

2.4.3 Example: XANES Tomography of a Cu/ZnO Catalyst inside a Reactor Capillary

As one example of a hard X-ray scanning microscopy technique, XANES tomography is described here and applied to finding the local oxidation state of a Cu/ZnO catalyst inside a quartz glass microreactor with 500 μm diameter.

In XANES tomography, the X-ray absorption near-edge structure is used as contrast. At each position of the 2-D tomographic scan, a full near-edge structure spectrum is recorded. Figure 2.10 illustrates the schematic setup. The X-rays from a synchrotron radiation source are monochromatized using a fast-scanning monochromator, as shown in Figure 2.10a. With a conventional monochromator, scanning the energy as an additional degree of freedom

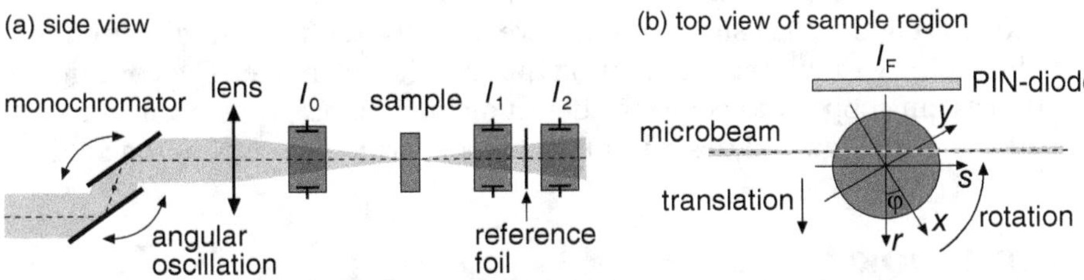

Figure 2.10. Setup for scanning XANES tomography. (a) The X-rays from a synchrotron radiation source are monochromatized by a fast-scanning monochromator and focused onto the sample. The incident intensity is monitored by an ionization chamber I_0. The transmitted intensity is monitored by the ionization chamber I_1, and the transmission through a reference foil of the given atomic species is recorded in parallel with I_2. (b) Top view of the sample region for XANES tomography. The sample is scanned in translation and rotation, recording at each position a full absorption spectrum in transmission and fluorescence. The fluorescence intensity I_F is recorded by the PIN diode. Reprinted with permission from Schroer, C. G., Kuhlmann, M., Günzler, T. F., Lengeler, B., Richwin, M., Griesebock, B., Lützenkirchen-Hecht, D., Frahm, R., Ziegler, E., Mashayekhi, A., Haeffner, D., Grunwaldt, J.-D., Baiker, A. (2003) Mapping the chemical states of an element inside a sample using tomographic x-ray absorption spectroscopy, *Appl. Phys. Lett.*, 82, 3360–3362. Copyright 2003 by American Institute of Physics.

would be much too time-consuming. Therefore, this technique requires a fast-scanning monochromator with a synchronized data acquisition system [62]. In this monochromator, a channel-cut monochromator crystal is rocked back and forth, covering the XANES range of an element of interest. After the monochromator, the X-rays are focused to the sample. The intensities before and after the sample are measured by ionization chambers. A third ionization chamber is used to measure the transmission through a reference foil (Fig. 2.10a) and a positive–intrinsic–negative (PIN) diode records the fluorescence radiation from the sample (Fig. 2.10b). The data acquisition is synchronized with the motion of the monochromator. In this way, several full XANES spectra can be acquired each second. In the example given, the oscillation frequency of the monochromator was 10 Hz, acquiring data at a rate of 100 kHz.

The sample was scanned in translation and rotation, acquiring 101 projections with 90 translational steps of 10 μm each. At each position of the scan, 10 full XANES spectra were recorded in 1 s and averaged. For each energy in the XANES spectrum, a complete tomographic reconstruction was made. Figure 2.11a,b shows the tomographic reconstructions of a cross section through the microreactor at an energy below and above the absorption edge of Cu, respectively. At each location in the reconstruction, a full XANES spectrum is available. Figure 2.11c shows the spectra at six selected locations in Figure 2.11b together with reference spectra for metallic, mono-, and bivalent copper. The latter can be fitted to the spectra at each location of the reconstruction to yield the relative distribution of the different copper species. Figure 2.11d,e shows the distributions of metallic and monovalent copper, respectively, Figure 2.11f shows the difference between the two distributions, and Figure 2.11g shows the attenuation due to other elements. The concentration of bivalent copper was below the detection limit in this given example and is not shown. For a detailed description of this experiment, see Reference 62.

2.5 OUTLOOK

The examples show that X-ray microscopy, in particular with absorption spectroscopic contrast, yields valuable information on the structure and function of catalysts in a large range of length scales. While absorption spectroscopy probes the local atomic structure around a catalytic center, imaging with this contrast yields spatial variations on longer length scales, so far with spatial resolutions typically in the range of micrometers and, in exceptional cases, down to a few tens of nanometers in the soft X-ray range. Pushing the spatial resolution into the range of few nanometers will provide new insight into the porous structure of catalysts, changes in catalytic reactors, and structural information on a scale that cannot be reached by electron microscopies (too large penetration depth) or other spectroscopic techniques (IR, Raman, etc.) that allow monitoring down to a microscale.

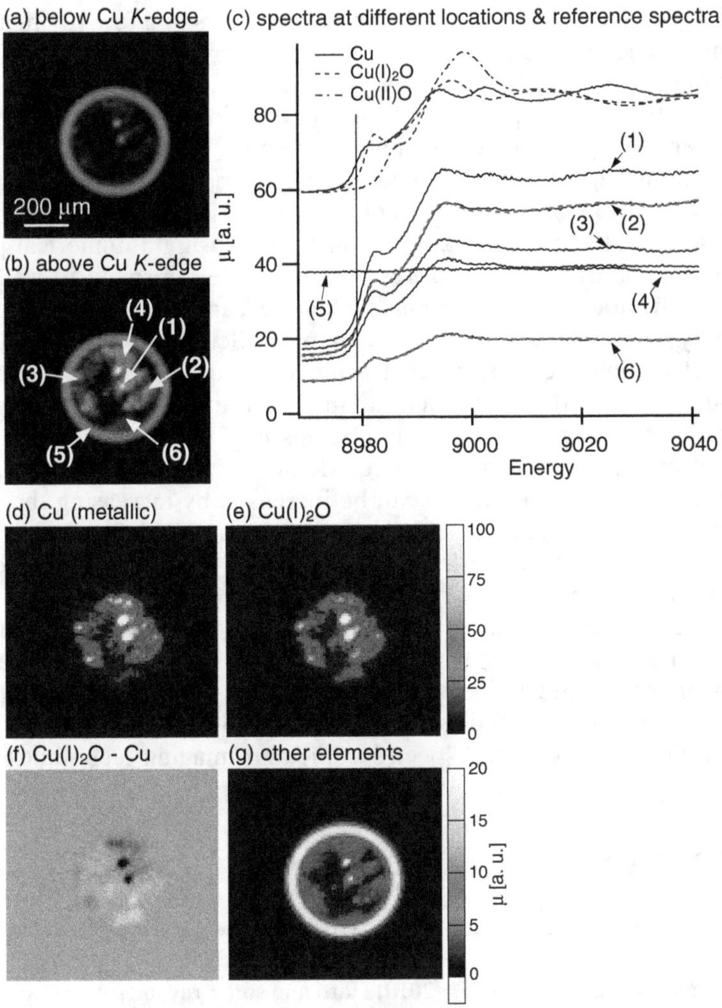

Figure 2.11. XANES tomogram of a Cu/ZnO catalyst mixed with inert material inside a quartz microreactor taken after several reduction/reoxidation cycles between 200 and 400°C [62]. Panels (a) and (b) show the tomographic reconstruction of the sample at an energy below and above the absorption edge of copper, respectively. (c) Absorption spectra reconstructed on the tomographic slice through the sample at different locations depicted in (b) and reference spectra of copper in different oxidation states. The reference spectra are fitted to the measured spectra at each location on the reconstructed slice. In this way, the distribution of copper in different oxidation states can be obtained, as shown in panels (d) and (e), together with the distribution of other elements (g). Panel (f) shows the difference between monovalent and metallic copper. The concentration of bivalent copper lies below the detection limit. Reprinted with permission from Schroer, C. G., Kuhlmann, M., Günzler, T. F., Lengeler, B., Richwin, M., Griesebock, B., Lützenkirchen-Hecht, D., Frahm, R., Ziegler, E., Mashayekhi, A., Haeffner, D., Grunwaldt, J.-D., Baiker, A. (2003) Mapping the chemical states of an element inside a sample using tomographic x-ray absorption spectroscopy, *Appl. Phys. Lett.*, 82, 3360–3362. Copyright 2003 by American Institute of Physics. (See color insert.)

For catalysis, the length scales of a few nanometers to a few 10 nm are very important, too. Therefore, the current aim is to improve the spatial resolution in X-ray microscopy to this range. A promising route to pushing the spatial resolution in the mesoscopic range is the combination of scanning microscopy with coherent X-ray diffraction imaging, also known as ptychography [63–65]. In this technique, the object is scanned through the nanofocused beam, recording at each position of the scan a coherent small-angle diffraction pattern. From these data, the object is then reconstructed using a numerical phasing algorithm. In this way, the spatial resolution can be improved to well below the size of the focus. This scanning technique can also be combined with tomography (in a full 3-D tomographic map; see Section 2.4.2); the 3-D inner structure of an object can be reconstructed with high spatial resolution [66]. By recording a series of ptychograms at an absorption edge, chemical imaging at a very high spatial resolution should be possible. For single coherent X-ray diffraction images, this has already been demonstrated [67]. In principle, the spatial resolution in these images can be increased by increasing the fluence on the sample, for example, by efficient nanofocusing. This has been demonstrated in a coherent X-ray diffraction experiment with a nanofocused beam [68], where a spatial resolution of 5 nm was reached.

Besides improving the imaging techniques and analysis tools themselves, the development of appropriate chemical reactors is of particular importance, especially in environmental electron microscopy [69] or, for example, for *in-operando* tomographic studies. Extreme temperature stability, small design, and a combination with other spectroscopic and imaging techniques will be required.

REFERENCES

[1] Bell, A. T. (2003) The impact of nanoscience on heterogeneous catalysis, *Science*, 299, 1688–1691.

[2] Grunwaldt, J.-D., Schroer, C. G. (2010) Hard and soft x-ray microscopy and tomography in catalysis: bridging the different time and length scales, *Chem. Soc. Rev.*, 39, 4741.

[3] Christensen, K. A. (2007) Design of industrial catalysts, in Ng, K. M., Gani, R., Dam-Johansen, K. (eds.), *Chemical Product Design*, Elsevier, Amsterdam, The Netherlands, p. 311.

[4] Weckhuysen, B. M. (2003) Determining the active site in a catalytic process: operando spectroscopy is more than a buzzword, *Phys. Chem. Chem. Phys.*, 5, 4351–4360.

[5] Topsøe, H. (2003) Developments in operando studies and in situ characterization of heterogeneous catalysts, *J. Catal.*, 216, 155–164.

[6] Grunwaldt, J.-D., Clausen, B. S. (2002) Combining XRD and EXAFS with on-line catalytic studies for in situ characterization of catalysts, *Topics Catal.*, 18, 37–43.

[7] Bañares, M. A. (2005) Operando methodology: combination of in situ spectroscopy and simultaneous activity measurements under catalytic reaction conditions, *Catal. Today*, 100, 71–77.

[8] Helveg, S., Lopez-Cartes, C., Sehested, J., Hansen, P. L., Clausen, B. S., Rostrup-Nielsen, J. R., Abild-Pedersen, F., Norskov, J. K. (2004) Atomic-scale imaging of carbon nanofibre growth, *Nature*, 427, 426–429.

[9] Thomas, J. M. (1999) Design, synthesis and in situ characterisation of new solid catalysts, *Angew. Chem. Int. Ed.*, 38, 3588.

[10] Bare, S. R., Ressler, T. (2009) Characterization of catalysts in reactive atmospheres by x-ray absorption spectroscopy, *Adv. Catal.*, 52, 339–465.

[11] Grunwaldt, J.-D., Caravati, M., Hannemann, S., Baiker, A. (2004) X-ray absorption spectroscopy under reaction conditions: suitability of different reaction cells for combined catalyst characterization and time-resolved studies, *Phys. Chem. Chem. Phys.*, 6, 3037–3047.

[12] Meunier, F. C. (2010) The design and testing of kinetically-appropriate operando spectroscopic cells for investigating heterogeneous catalytic reactions, *Chem. Soc. Rev.*, 39, 4602–4614.

[13] de Smit, E., Swart, I., Creemer, J. F., Hoveling, G. H., Gilles, M. K., Tyliszczak, T., Kooyman, P. J., Zandbergen, H. W., Morin, C., Weckhuysen, B. M., de Groot, F. M. F. (2008) Nanoscale chemical imaging of a working catalyst by scanning transmission X-ray microscopy, *Nature*, 456, 222.

[14] Thomas, J. M., Hernandez-Garrido, J.-C. (2009) Probing solid catalysts under operating conditions: electrons or x-rays? *Angew. Chem. Int. Ed.*, 48, 3904–3907.

[15] Grunwaldt, J.-D., Kimmerle, B., Baiker, A., Boye, P., Schroer, C. G., Glatzel, P., Borca, C. N., Beckmann, F. (2009) Catalysts at work: from integral to spatially resolved x-ray absorption spectroscopy, *Catal. Today*, 145, 267–278.

[16] Grunwaldt, J.-D., Frenkel, A. (2009) Hard x-rays and catalysis, *Synchrotron Radiat. News*, Special Issues 1 and 2.

[17] Iwasawa, Y. (1996) *X-ray Absorption Fine Structure for Catalysts and Surfaces*, World Scientific, Singapore.

[18] Clausen, B. S., Topsoe, H., Frahm, R. (1998) Application of combined x-ray diffraction and absorption techniques for in situ catalyst characterization, *Adv. Catal.*, 42, 315–344.

[19] Sankar, G., Thomas, J. M. (1999) In situ combined x-ray absorption spectroscopic and x-ray diffractometric studies of solid catalysts, *Topics Catal.*, 8, 1–21.

[20] Bazin, D., Lynch, J., Rarnos-Fernandez, M. (2003) X-ray absorption spectroscopy and anomalous wide angle x-ray scattering: two basic tools in the analysis of heterogeneous catalysts, *Oil Gas Sci. Technol.*, 58, 667–683.

[21] Boyes, E. D., Gai, P. L. (1997) Environmental high resolution electron microscopy and applications to chemical science, *Ultramicroscopy*, 67, 219–232.

[22] Creemer, J. F., Helveg, S., Hoveling, G. H., Ullmann, S., Molenbroek, A. M., Sarro, P. M., Zandbergen, H. W. (2008) Atomic-scale electron microscopy at ambient pressure, *Ultramicroscopy*, 108, 993–998.

[23] Quitmann, C., David, C., Nolting, F., Pfeiffer, F., Stampanoni, M. (2009) *Proceedings of the 9th International Conference on X-ray Microscopy, Journal of Physics: Conference Series*, Vol. 186, IOP, Bristol, UK.

[24] Koningsberger, D. C., Prins, R. (1988) *X-ray Absorption*, John Wiley & Sons, New York.

[25] Grunwaldt, J.-D., Baiker, A. (2005) Axial variation of the oxidation state of Pt-Rh/Al_2O_3 during partial methane oxidation in a fixed-bed reactor: an in situ x-ray absorption spectroscopy study, *Catal. Lett.*, 99, 5–12.

[26] Koch, A., Raven, C., Spanne, P., Snigirev, A. (1998) X-ray imaging with submicrometer resolution employing transparent luminescent screens, *J. Opt. Soc. Am. A Opt. Image Sci. Vis.*, 15, 1940–1951.

[27] Stampanoni, M., Borchert, G., Wyss, P., Abela, R., Patterson, B., Hunt, S., Vermeulen, D., Rüegsegger, P. (2002) High resolution X-ray detector for synchrotron-based microtomography, *Nucl. Instrum. Methods A*, 491, 291–301.

[28] Cloetens, P., Ludwig, W., Baruchel, J., Van Dyck, D., Van Landuyt, J., Guigay, J. P., Schlenker, M. (1999) Holotomography: quantitative phase tomography with micrometer resolution using hard synchrotron radiation x rays, *Appl. Phys. Lett.*, 75, 2912–2914.

[29] Mokso, R., Cloetens, P., Maire, E., Ludwig, W., Buffière, J.-Y. (2007) Nanoscale zoom tomography with hard x rays using Kirkpatrick-Baez optics, *Appl. Phys. Lett.*, 90, 144104.

[30] Chu, Y. S., Yi, J. M., De Carlo, F., Shen, Q., Lee, W.-K., Wu, H. J., Wang, C. L., Wang, J. Y., Liu, C. J., Wang, C. H., Wu, S. R., Chien, C. C., Hwu, Y., Tkachuk, A., Yun, W., Feser, M., Liang, K. S., Yang, C. S., Je, J. H., Margaritondo, G. (2008) Hard-x-ray microscopy with Fresnel zone plates reaches 40 nm Rayleigh resolution, *Appl. Phys. Lett.*, 92, 103119.

[31] Lengeler, B., Schroer, C. G., Richwin, M., Tümmler, J., Drakopoulos, M., Snigirev, A., Snigireva, I. (1999) A microscope for hard x-rays based on parabolic compound refractive lenses, *Appl. Phys. Lett.*, 74, 3924–3926.

[32] Lengeler, B., Schroer, C., Tümmler, J., Benner, B., Richwin, M., Snigirev, A., Snigireva, I., Drakopoulos, M. (1999) Imaging by parabolic refractive lenses in the hard x-ray range, *J. Synchrotron Radiat.*, 6, 1153–1167.

[33] Chao, W., Harteneck, B. D., Liddle, J. A., Anderson, E. H., Attwood, D. T. (2005) Soft x-ray microscopy at a spatial resolution better than 15 nm, *Nature*, 435, 1210–1213.

[34] Vila-Comamala, J., Jefimovs, K., Raabe, J., Pilvi, T., Fink, R. H., Senoner, M., Maassdorf, A., Ritala, M., David, C. (2009) Advanced thin film technology for ultrahigh resolution X-ray microscopy, *Ultramicroscopy*, 109, 1360–1364.

[35] Schroer, C. G., Kuhlmann, M., Lengeler, B., Günzler, T. F., Kurapova, O., Benner, B., Rau, C., Simionovici, A. S., Snigirev, A., Snigireva, I. (2002) Beryllium parabolic refractive x-ray lenses, *Proceedings of SPIE 4783*, Design and Microfabrication of Novel X-Ray Optics, Bellingham, WA, pp. 10–18.

[36] Grunwaldt, J.-D., Hannemann, S., Schroer, C. G., Baiker, A. (2006) 2D-mapping of the catalyst structure inside a catalytic microreactor at work: partial oxidation of methane over Rh/Al_2O_3, *J. Phys. Chem. B*, 110, 8674–8680.

[37] Schwiedernoch, R., Tischer, S., Correa, C., Deutschmann, O. (2003) Experimental and numerical study on the transient behavior of partial oxidation of methane in a catalytic monolith, *Chem. Eng. Sci.*, 58, 633–642.

[38] Dalle Nogare, D., Degenstein, N. J., Horn, R., Canu, P., Schmidt, L. D. (2008) Modeling spatially resolved profiles of methane partial oxidation on a Rh foam catalyst with detailed chemistry, *J. Catal.*, 258, 131–142.

[39] Hannemann, S., Grunwaldt, J.-D., van Vegten, N., Baiker, A., Boye, P., Schroer, C. G. (2007) Distinct spatial changes of the catalyst structure inside a fixed-bed microreactor during the partial oxidation of methane over Rh/Al_2O_3, *Catal. Today*, 126, 54–63.

[40] Kimmerle, B., Baiker, A., Grunwaldt, J.-D. (2010) Oscillatory behaviour of catalytic properties, structure and temperature during the catalytic partial oxidation of methane on Pd/Al_2O_3, *Phys. Chem. Chem. Phys.*, 12, 2288–2291.

[41] Kimmerle, B., Grunwaldt, J.-D., Baiker, A., Glatzel, P., Boye, P., Stephan, S., Schroer, C. G. (2009) Visualizing a catalyst at work during the ignition of the catalytic partial oxidation of methane, *J. Phys. Chem.*, C(113), 3037–3040.

[42] Labiche, J.-C., Mathon, O., Pascarelli, S., Newton, M. A., Guilera Ferre, G., Curfs, C., Vaughan, G. B. M., Homs, A., Fernandez Carreiras, D. (2007) Invited article: the fast readout low noise camera as a versatile x-ray detector for time resolved dispersive extended x-ray absorption fine structure and diffraction studies of dynamic problems in materials science, chemistry, and catalysis, *Rev. Sci. Instrum.*, 78, 091301.

[43] Grunwaldt, J.-D., Beier, M., Kimmerle, B., Baiker, A., Nachtegaal, M., Griesebock, B., Lützenkirchen-Hecht, D., Stötzel, J., Frahm, R. (2009) Structural changes of noble metal catalysts during ignition and extinction of the partial oxidation of methane studied by advanced QEXAFS techniques, *Phys. Chem. Chem. Phys.*, 11, 8779–8789.

[44] Kak, A. C., Slaney, M. (1988) *Principles of Computerized Tomographic Imaging*, IEEE Press, New York.

[45] Beckmann, F., Donath, T., Fischer, J., Dose, T., Lippmann, T., Lottermoser, L., Martins, R. V., Schreyer, A. (2006) New developments for synchrotron-radiation-based microtomography at DESY, *Proceedings of SPIE 6318*, Developments in X-ray Tomography V, Bellingham, WA, pp. 631810–1–11.

[46] Bonse, U., Johnson, H., Nichols, M., Nußhardt, R., Krasnicki, S., Kinney, J. (1986) High resolution tomography with chemical specificity, *Nucl. Instrum. Methods A*, 246, 644–648.

[47] Kinney, J., Johnson, Q., Bonse, U., Nichols, M., Saroyan, R., Nußhardt, R., Pahl, R. (1988) Energy modulated x-ray microtomography, *Rev. Sci. Instrum.*, 59, 196–197.

[48] Rau, C., Somogyi, A., Bytchkov, A., Simionovici, A. (2002) XANES microimaging and tomography, *Proceedings of SPIE 4503*, Developments in X-ray Tomography III, Bellingham, WA, pp. 249–255.

[49] Ade, H., Zhang, X., Cameron, S., Costello, C., Kirz, J., Williams, S. (1992) Chemical contrast in x-ray microscopy and spatially resolved XANES spectroscopy of organic specimens, *Science*, 258, 972–975.

[50] Kang, H. C., Maser, J., Stephenson, G. B., Liu, C., Conley, R., Macrander, A. T., Vogt, S. (2006) Nanometer linear focusing of hard x rays by a multilayer Laue lens, *Phys. Rev. Lett.*, 96, 127401.

[51] Mimura, H., Yumoto, H., Matsuyama, S., Sano, Y., Yamamura, K., Mori, Y., Yabashi, M., Nishino, Y., Tamasaku, K., Ishikawa, T., Yamauchi, K. (2007) Efficient focusing of hard x rays to 25 nm by a total reflection mirror, *Appl. Phys. Lett.*, 90, 051903.

[52] Kang, H. C., Yan, H., Winarski, R. P., Holt, M. V., Maser, J., Liu, C., Conley, R., Vogt, S., Macrander, A. T., Stephenson, G. B. (2008) Focusing of hard x-rays to 16 nanometers with a multilayer Laue lens, *Appl. Phys. Lett.*, 92, 221114.

[53] Mimura, H., Handa, S., Kimura, T., Yumoto, H., Yamakawa, D., Yokoyama, H., Matsuyama, S., Inagaki, K., Yamamura, K., Sano, Y., Tamasaku, K., Nishino, Y., Yabashi, M., Ishikawa, T., Yamauchi, K. (2010) Breaking the 10 nm barrier in hard-X-ray focusing, *Nat. Phys.*, 6, 122–125.

[54] Schroer, C. G., Kurapova, O., Patommel, J., Boye, P., Feldkamp, J., Lengeler, B., Burghammer, M., Riekel, C., Vincze, L., van der Hart, A., Küchler, M. (2005) Hard x-ray nanoprobe based on refractive x-ray lenses, *Appl. Phys. Lett.*, 87, 124103.

[55] Bergemann, C., Keymeulen, H., van der Veen, J. F. (2003) Focusing X-ray beams to nanometer dimensions, *Phys. Rev. Lett.*, 91, 204801.

[56] Suzuki, Y. (2004) Resolution limit of refractive lens and Fresnel lens in x-ray region, *Jpn. J. Appl. Phys.*, 43, 7311–7314.

[57] Schroer, C. G., Lengeler, B. (2005) Focusing hard x rays to nanometer dimensions by adiabatically focusing lenses, *Phys. Rev. Lett.*, 94, 054802.

[58] Pfeiffer, F., David, C., van der Veen, J. F., Bergemann, C. (2006) Nanometer focusing properties of Fesnel zone plates described by dynamical diffraction theory, *Phys. Rev. B*, 73, 245331.

[59] Schroer, C. G. (2006) Focusing hard x rays to nanometer dimensions using Fresnel zone plates, *Phys. Rev. B*, 74, 033405.

[60] Yan, H., Maser, J., Macrander, A., Shen, Q., Vogt, S., Stephenson, G. B., Kang, H. C. (2007) Takagi-Taupin description of x-ray dynamical diffraction from diffractive optics with large numerical aperture, *Phys. Rev. B*, 76, 115438.

[61] Schroer, C. G., Boye, P., Feldkamp, J. M., Patommel, J., Schropp, A., Samberg, D., Stephan, S., Burghammer, M., Schöder, S., Riekel, C., Lengeler, B., Falkenberg, G., Wellenreuther, G., Kuhlmann, M., Frahm, R., Lützenkirchen-Hecht, D., Schroeder, W. H. (2010) Hard X-ray microscopy with elemental, chemical, and structural contrast, *Acta Phys. Pol. A*, 117, 357–368.

[62] Schroer, C. G., Kuhlmann, M., Günzler, T. F., Lengeler, B., Richwin, M., Griesebock, B., Lützenkirchen-Hecht, D., Frahm, R., Ziegler, E., Mashayekhi, A., Haeffner, D., Grunwaldt, J.-D., Baiker, A. (2003) Mapping the chemical states of an element inside a sample using tomographic x-ray absorption spectroscopy, *Appl. Phys. Lett.*, 82, 3360–3362.

[63] Rodenburg, J. M., Faulkner, H. M. L. (2004) A phase retrieval algorithm for shifting illumination, *Appl. Phys. Lett.*, 85, 4795–4797.

[64] Thibault, P., Dierolf, M., Menzel, A., Bunk, O., David, C., Pfeiffer, F. (2008) High-resolution scanning x-ray diffraction microscopy, *Science*, 321, 379–382.

[65] Schropp, A., Boye, P., Goldschmidt, A., Hönig, S., Hoppe, R., Patommel, J., Rakete, C., Samberg, D., Stephan, S., Schöder, S., Burghammer, M., Schroer, C. G. (2011) Non-destructive and quantitative imaging of a nano-structured microchip by ptychographic hard x-ray scanning microscopy, *J. Microsc.*, 241, 9–12.

[66] Dierolf, M., Menzel, A., Thibault, P., Schneider, P., Kewish, C. M., Wepf, R., Bunk, O., Pfeiffer, F. (2010) Ptychographic x-ray computed tomography at the nanoscale, *Nature*, 467, 436–440.

[67] Song, C., Bergstrom, R., Ramunno-Johnson, D., Jiang, H., Paterson, D., Jonge, M. D. D., McNulty, I., Lee, J., Wang, K. L., Miao, J. (2008) Nanoscale imaging of buried structures with elemental specificity using resonant x-ray diffraction microscopy, *Phys. Rev. Lett.*, 100, 025504.

[68] Schroer, C. G., Boye, P., Feldkamp, J., Patommel, J., Schropp, A., Schwab, A., Stephan, S., Burghammer, M., Schöder, S., Riekel, C. (2008) Coherent x-ray diffraction imaging with nanofocused illumination, *Phys. Rev. Lett.*, 101, 090801.

[69] Grunwaldt, J.-D., Wagner, J. B., Dunin-Borkowski, R. (2013) Imaging catalysts at work: a hierarchical approach from the macro- to the meso- and nano-scale, *Chem. Cat. Chem.*, 5, 62–80.

3

ENERGY-DISPERSIVE EXAFS: PRINCIPLES AND APPLICATION IN HETEROGENEOUS CATALYSIS

Mark A. Newton and Andrew J. Dent

3.1 INTRODUCTION

Chapter 1 and Chapter 2 of this book have described how scanning X-ray absorption spectroscopy (XAS) may be used to study working heterogeneous catalysts from both time and spatially resolving perspectives. In this chapter we introduce the energy-dispersive approach to XAS. The dispersive approach is a fundamentally different solution to the problem of how to obtain X-ray absorption data from a given sample; instead of scanning through the energy range required to build up an extended X-ray absorption fine structure (EXAFS) spectrum, the entire X-ray energy range required is directly and immediately applied. As such it comes with its own unique possibilities and caveats, which we describe during the course of this chapter.

The development of XAS, as the extremely powerful and commonly applied probe of local structure and chemical state that we know today, essentially began with the theoretical breakthroughs due to Sayers and coworkers in the early 1970s [1]. The progressive implementation and development of dedicated X-ray sources such as synchrotrons has resulted in XAS becoming a frontline structural probe in many disciplines. XAS does, however, have a much longer history, and to find the origins of energy-dispersive XAS we need to start in

In-situ *Characterization of Heterogeneous Catalysts*, First Edition.
Edited by José A. Rodriguez, Jonathan C. Hanson, and Peter J. Chupas.

the early 1930s [2–5]. At this time, certain pioneers of X-ray physics realized that using a curved crystal element in their X-ray optics might have distinct benefits for a number of things, among which were X-ray absorption fine structure and emission spectroscopies (XAFS and XES).

Yvette Cauchois [5] concluded the following in 1932: "Absorption spectra may be collected photographically in a very short time . . . this optical arrangement is well adapted for chemical analysis using both absorption and emission of X-rays. Finally, it is clear that a suitably placed slit transforms this arrangement into a powerful X-ray monochromator."[1]

Cauchois' 1932 conclusions contain all of the salient elements that have eventually become dispersive EXAFS as it is practiced and applied today in a range of disciplines, among which is the study of heterogeneous catalysis.

However, and alongside the theoretical developments due to Sayers and others [1], the practical implementation and exploitation of XAS at first- and second-generation synchrotron X-ray sources began, and has generally followed, a rather different path: that of scanning (mechanically) through the energy range required to obtain both near-edge (XANES or NEXAFS) and EXAFS. To this day, the use of mechanically actuated double-crystal monochromators remains the predominant mode of collection for X-ray absorption studies whether for static or time-resolved research (see Chapter 1 and Chapter 2).

Thirty years ago, however, and 49 years after Cauchois, Matsushita, and coworkers—working in Tokyo [6]—and Stanford [7] broke from that tradition and demonstrated an alternative methodology, that of energy-dispersive EXAFS collected using curved crystal optics. Their synchrotron setup [7] is shown in Figure 3.1 along with an exemplary absorption spectrum obtained simultaneously from Cu and Ni foils that had been taped together. The detector comprised a photographic film, and this spectrum took only 15 s to obtain; at the time, such a short timescale for a single-shot X-ray absorption measurement was unprecedented.

The use of a curved (or bent) "polychromator," which provides a focused X-ray beam containing all the energies required to make an XAS measurement, is the defining element of the dispersive approach. As we shall delineate, this approach to XAS is, on the one hand, extremely elegant, bringing with it many unique features and possibilities. Equally, it is in some senses capricious and demanding, especially in its application to materials such as catalysts that have an intrinsically reduced level of spatial homogeneity.

[1]The original extract reads: "*l'étude de spectres d'absorption peut être faite photographiquement avec de temps de pose très court . . . tout appareil construit sur cet principale conviendrait bien a l'analyse chimique par rayons X (émission et absorption). Enfin il est évident que l'emploi d'une fente fine convenablement placée sur la cylindre de focalisation transformerait l'appareil en un monochromator puissant.*" The translation is by M.A.N.

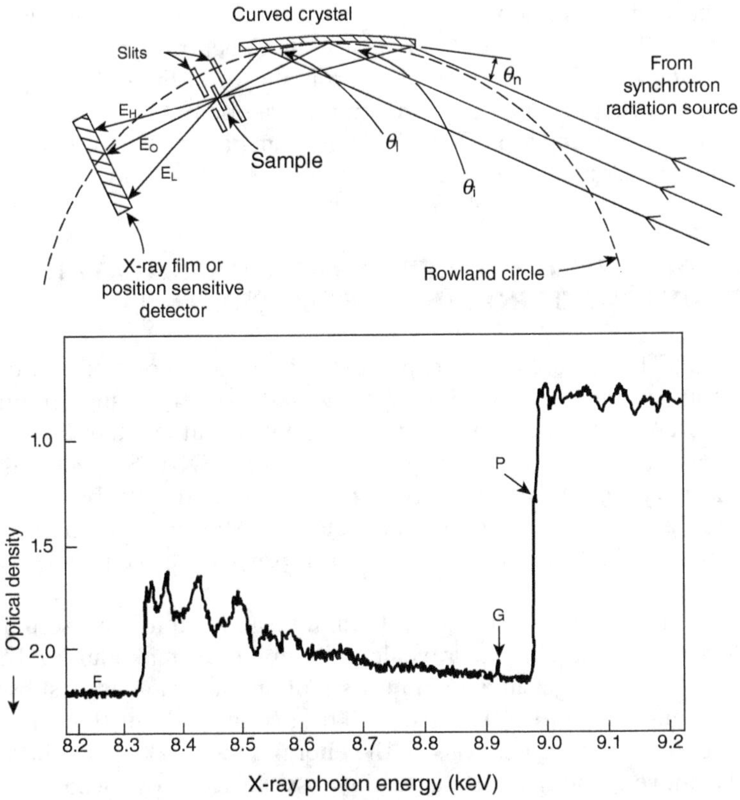

Figure 3.1. Top: The geometric arrangement of the first dispersive EXAFS setup demonstrated by Matsushita and Phizackerley at SSRL, Stanford, CA, USA [7]. Bottom: Dispersive EXAFS spectra obtained in 15 s from a 4 μm Ni foil and a 12.5 μm Cu foil placed together at the focal point of the polychromatic beam. The 2.2 GeV SSRL source was operated at 12 mA. Modern third-generation sources operate in the 100–500 mA range. Point F indicates a region of the spectrum where the photographic film was saturated within the overall exposure time; point G indicates the presence of a diffraction "glitch"; and point P points out the preedge feature for a Cu foil and which may only be clearly resolved if the energy resolution of the spectrometer is <2eV. Reprinted from Matsushita and Phizackerley, *Jpn. J. Appl. Phys.*, 20, 2223–2228, with permission from the Japan Society of Applied Physics.

A number of EXAFS resources based on the dispersive concept subsequently appeared in a variety of first- and then second-generation sources, such as Stanford [8], Daresbury [9], Lure [10], HasyLab [11], and National Synchrotron Light Source (NSLS) [12], among others. It was not, however, until the early 1990s that the unique properties of dispersive EXAFS were specifically turned toward problems in heterogeneous catalysis [13–16]. In particular, the work of Couves et al. [14] can be seen as defining in a number of ways. Their use of dispersive optics—to alternate between X-ray absorption and diffraction measurements during catalyst synthesis—is the first work that realizes in full the substance of the conclusions Cauchois had made almost 60 years previously. Their experimental setup is shown in Figure 3.2. As we will see later in this chapter, the use of moving slit can be put to a highly practical use in terms of obtaining rapid XANES data from particularly difficult catalyst samples in some demanding situations.

3.2 THE DISPERSIVE CONCEPT AND ITS RAMIFICATIONS FOR THE STUDY OF HETEROGENEOUS MATERIALS

Figure 3.1 and Figure 3.2 neatly capture some of the unique elements that a dispersive approach to XAS brings with it. As the energy range required to make the EXAFS measurement is directly applied, rather than having to be mechanically scanned, the measurement of entire EXAFS spectra is made potentially very fast: to a first approximation, limited only by the applied photon flux and the efficiency of the detector. Moreover, as there are no moving parts within the entire optical setup, the dispersive arrangement should be intrinsically stable.

Further, in its first demonstration [7], a unique and potentially useful feature of the dispersive approach was revealed: the ability to make an XAFS measurement from two edges in a genuinely simultaneous manner. A subsequent review of time-resolved EXAFS from 2001 [17] pointed out that there were almost 30 elemental combinations, in the energy range 7–30 keV, wherein this could be achieved. Though hypothetically possible, using two scanning monochromators operating side by side, the equivalent experiment using scanning optics has yet to be demonstrated 30 years onward.

What Figure 3.1 and Figure 3.2 do not readily convey, however, are a number of devils lurking in the detail of the dispersive method that are of direct relevance to the exploitation of this technique in the study of heterogeneous catalysts. For the most part these result in challenges to successful experimentation that the scanning approach to EXAFS does not have to contend with, or does so in a much reduced sense. They may be summarized as follows: the obtainable energetic bandwidth; an inability to perform inline normalization of obtained absorption data; the requirement for fast, spatially resolving, detectors; the intrinsic size and uniformity of the polychromatic focal spot; and the general limitation to a transmission geometry.

3.2.1 Detectors, Bandwidths, and Inline Normalization

From the initial use of film, photodiode arrays were already anticipated by Matsushita and coworkers [6, 7]. Such detectors were subsequently, and very successfully, employed at synchrotrons, remaining for a long time as the principal detectors for dispersive EXAFS (Chapter 1 and Chapter 2) [1–12, 18, 19]. More recently, however, fast charge-coupled device (CCD) cameras such

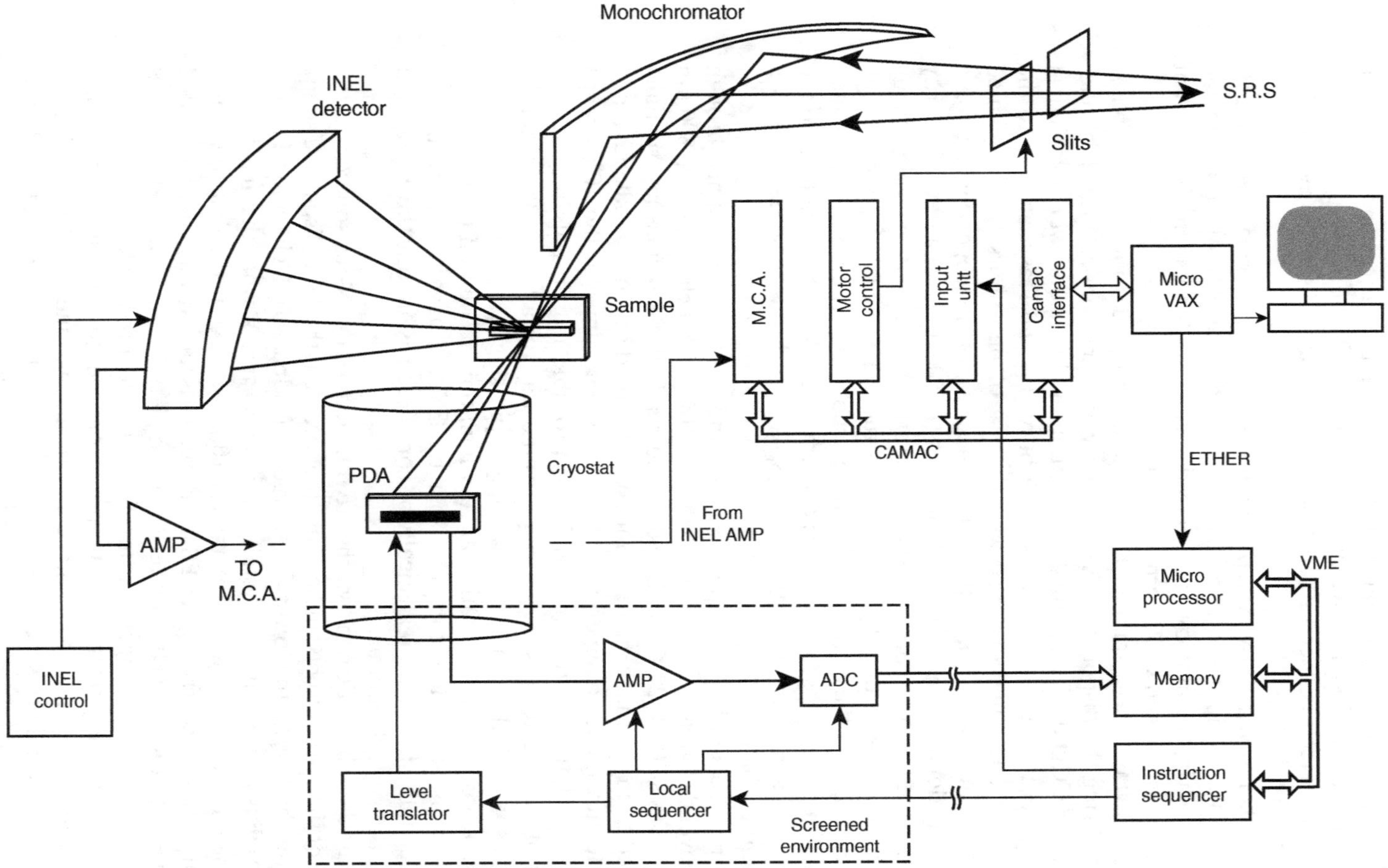

Figure 3.2. The experimental setup due to Couves et al. at station 7.4, SRS Daresbury, UK. The energy-dispersive X-ray optics provide a polychromatic X-ray beam focused at the sample position and used to collect dispersive EXAFS data in transmission. A movable slit is then used to "monochromate" these X-rays for the diffraction measurements using the INEL detector. By opening and closing the slits one alternates between the dispersive EXAFS and diffraction measurements. Figure from Dent et al., *Rev. Sci. Instrum.*, 63, 903–906. Reproduced with permission of the American Institute of Physics.

as the fast readout low noise detector (FReLoN) at the European Synchrotron Radiation Facility (ESRF) have progressively taken over this mantle [20, 21]. Using these detectors, spectral repetition rates in the millisecond regime have long been obtainable.

In the last few years, however, potentially paradigm-changing developments have come to pass in the form of detectors based on Si [22–24], or latterly Ge [25], "microstrip" technology. These new detectors are fundamentally different from diode arrays and CCD cameras in a number of ways.

These new detectors sample the X-rays directly and do not rely on phosphors to bring the dispersive EXAFS data within the (optical) range of photodiodes or CCD devices. Though some photodiode systems were evolved, at second-generation sources such as Daresbury [19] and HasyLab [11], to the point where one could dispense with the use of phosphors, they remained limited in a number of ways. Though the removal of the dependence upon a fluorescing element improved readout times (to 500 μs; [19]) and the linearity in response of the detectors, they remained serial devices with each pixel being read out in turn. Additionally, the readout structure was integral to the diodes. This meant that for the active—for X-ray detection—depth, each pixel presented was small and overall sensitivity was compromised. Moreover, damage to the integrated readout structure, from X-rays passing through the diodes, was a persistent problem.

The new generation of strip detectors avoids all these limitations through their novel design and electronics: active depths are much larger (ca. 300 μm for the Si–XSTRIP [22–24]); the advanced readout electronics are spatially removed from the pixel; and each microstrip is read out in a parallel manner. These features lead to superior linearity and dead times in the 10 μs range [22–25]; the latter, moreover, seeming improvable to sub-1 μs regime. The advent of these detectors therefore paves the way for a leap in the potential of dispersive EXAFS both for extremely rapid single-shot experiments and for pump–probe experimentation at very high time resolution (see later).

As such, issues of detectors for dispersive EXAFS have largely been overcome and they should not realistically be considered as limiting the application of the dispersive technique.

Obtainable energetic bandwidths are functions of the X-ray source size, the energy of the X-rays to be studied, and the focal length of the dispersive setup: the lower the energy, the lower the energetic bandpass for a given focal length (bend radius of the curved crystal). In practical terms, what this has by and large meant is that below approximately 10 keV (e.g., for the first-row transmission metals), dispersive EXAFS measurements have been generally limited to studies based on XANES rather than complete EXAFS, simply as a result of the limited energy ranges that may be obtained. (To our knowledge, there is but one exception to this observation, from the now defunct Laboratoire pour l'Utilisation du Rayonnement Electromagnetique (LURE) facility in France [26]. Even in this case the EXAFS reported has been Fourier filtered, suggestive in itself of not such high-quality raw data.)

This historical deficit is, however, also convoluted with the nature of the source and the capacity of the detector technology available at any given time. Some second-generation sources could clearly obtain bandwidths of up to approximately 1 keV in the 6–10 keV energy range required for the first-row transition metal K-edges but were no doubt limited in other ways, some of which will be outlined later in this chapter.

Ironically, these issues are not at all helped by having a very small source in the first place. For instance, though potentially much more powerful than any dispersive resource that has gone before it, ID24 at the ESRF, as it existed until 2009, has been somewhat disadvantaged in these respects when dealing with materials like catalysts. The combination of an undulator source at a very brilliant synchrotron along with other overall design aspects has, thus far, resulted in an inability to obtain the necessary bandwidth required for EXAFS in the sub-10 keV range. That said, even these issues are not insurmountable, providing that necessarily short (bend radii of the polychromator crystal) focal lengths can be achieved.

Similarly, the failure in dispersive EXAFS to realize a parallel normalization of the transmitted (by the sample) X-ray beam with respect to the incident (to the sample) turns out to be much less of a problem to successful measurement than it might at first appear. The stability of modern (third-generation) synchrotron sources is such that, compared with the timescales generally used in time-resolved catalysis experiments, temporal variations in the overall intensity of the applied X-rays do not limit the ability to make a successful measurement.

3.2.2 Issues Related to X-ray Scattering and Sample Uniformity

Far more pernicious and problematic than all of the above, however, are those problems that arise out of the collision between focused polychromatic X-ray beams and particulate samples such as catalysts. Such samples tend to be intrinsically nonuniform on just the wrong length scales. This is especially the case when one is trying to correctly parameterize the overall experiment, in terms of (non-X-ray) factors relevant to the catalysis at hand (e.g., sieve fractions, gas flows, and contact times) rather than simply trying to obtain an absorption spectrum from a solid experiencing a gaseous environment.

When this overall spatial nonuniformity is coupled to the "necessity" for quasi-two-dimensional (2-D) detection of polychromatic X-rays that have passed through the sample, some serious issues for a successful EXAFS experiment arise. These involve the very nature of the samples and their presentation, but can also severely test the character and stability of the synchrotron electron beam source as well as the optical setup of the entire dispersive instrument [27]. The essence of the problem is shown in Figure 3.3.

Figure 3.3 contrasts the overall spatial uniformity, in single X-ray energy absorption terms, from two very different samples. Figure 3.3A shows a laser heated Fe mineral sample within a diamond anvil cell (DAC) maintained

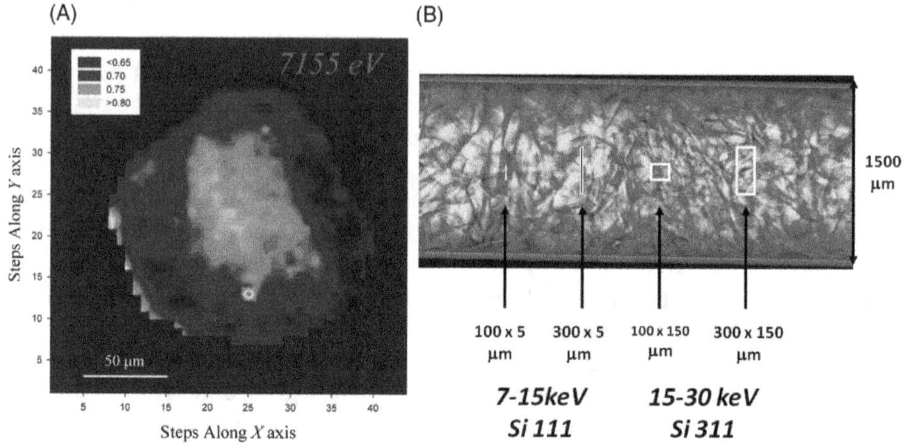

Figure 3.3. A comparison of sample uniformity for two types of sample commonly studied using dispersive EXAFS. (A) An Fe mineral sample contained within a DAC at 78 GPa mapped above the Fe K-edge using a 5×5 μm (FWHM) focal spot. The central (green) area maps a thermal transition to a new mineral phase as a result of laser heating. Reproduced with permission from the International Union of Crystallography (IUCR) [28]. (B) Full-field image of a sieved Rh/Al_2O_3 catalyst contained within a 1.5 mm internal diameter (i.d.) quartz capillary [29]. The white boxes refer to the focal spot sizes accessible on ID24 for the energy ranges and polychromator crystals indicated. This measurement is made using scanning EXAFS and at an energy just below the Rh K-edge. The different colors, therefore, directly refer to the differing levels of X-ray transmission observed simply as result of the spatial nonhomogeneity intrinsic to the catalyst bed. Reprinted with permission from Grunwaldt et al., *J. Phys. Chem. B*, 110, 8674–8680. Copyright (2006) American Chemical Society. (See color insert.)

under 78 GPa static pressure and mapped with approximately a 5×5 μm spatial resolution using dispersive EXAFS on ID24 at the ESRF [28]. Figure 3.3B shows what a typical, sieved powder catalyst bed (2.5 wt% Rh/Al_2O_3, contained within a 1.5 mm diameter quartz capillary) looks like in the same terms. This image was obtained using a full-field scanning EXAFS approach [29], with an energy just below the Rh K-edge. Superimposed upon Figure 3.3B are blocks indicative of typical dispersive focal spots achievable on ID24 (up to 2009) using symmetric Si[111] and [311] Bragg polychromators over the ID24 energy range (ca. 7–30 keV).

What can be easily seen from this comparison is that the sample in the DAC is extremely homogeneous on length scales significantly greater than the focal spot dimensions. In this case the two colors are indicative of two discrete Fe mineral phases, the central patch being the result of where the laser heating has effected a structural transition [28].

On the scale of the available beam sizes of the preupgrade ID24, it is readily evident that the reverse is true in the case of the catalyst sample wherein a high degree of spatial disorder is inherent. The relatively low density packing of the catalyst particles required to obtain gas flow characteristics that are central to a meaningful catalysis measurement means that this is unavoidable.

As the polychromatic X-ray beam has to traverse, and therefore sample, this inhomogeneity, a rather complex situation arises. Moreover, the smaller the X-ray beam is made, then the more and more sensitive one becomes to any potential displacements of either the X-ray beam or the disposition of the sample. As and when one starts to sample this intrinsic inhomogeneity, through movement of any kind (see Section 3.2.1), the subsequent normalization required to obtain absorption spectra becomes rapidly impossible and useable EXAFS data cannot be restored (no matter how many photons one throws at the problem).

A second, more subtle issue also arises out of the transmission of a polychromatic beam through a particulate sample and the resulting necessity for quasi-2-D detection in transmission. As the X-rays of different energies pass through the sample, they are scattered (in all directions) as well as being absorbed. Depending on the details of the geometry, those X-rays scattered through angles of close to 0° (within a small-angle scattering [SAXS] cone of ca. ±2° on ID24) will inevitably end up at the detector and can be counted. The degree to which this effect expresses itself will be sample-dependent and, a priori, not so easy to assess. However, we can qualitatively point to potential ways that this effect might manifest itself: first, that it might impose a "structure" upon the basic X-ray absorption spectrum over and above that that may arise from effects due to spatial nonuniformity; and second, it may result in a loss of energy resolution and a damping of the intensity of XANES and EXAFS features.

These issues were recognized by Hagelstein and coworkers at ID24 [30] as early as 1998. They demonstrated that by implementing a Bonze–Hart scattering filter [31] within the dispersive setup, such parasitic scattering from relatively simple samples (and from typical X-ray windows) could be largely ameliorated in and around the Pt L_3 K-edge (11.564 keV). Though elegant and successful, this approach to resolving these issues has not generally been adopted, most likely as it introduces a further optical complexity to what are often already quite complicated experimental setups.

As with the intrinsic convolution of focal spot size and intrinsic spatial homogeneity, these scattering-related effects impose "structure" rather than "noise," and no amount of counting or additional photons will stop them degrading the resulting absorption spectra.

We note that, in the typical example of the application of dispersive EXAFS to research into high-pressure physics/geosciences given in Figure 3.3 [28], neither spatial inhomogeneity nor parasitic SAXS are likely to be of any significance. As is evident from Figure 3.3, however, catalysts are very different entities: they are rarely bulk materials. Most often the elements of interest are present at low concentrations and dispersed on materials that are in themselves often very absorbing and/or rather good small-angle scatterers. As such, it is these two challenging issues that have to be accounted for to have any chance of a successful experiment.

Figure 3.4 systematically shows examples of how these effects manifest themselves; it also shows how they are currently dealt with on ID24 [32], in this case, for 1 wt% Pd catalysts dispersed upon a support that contains 10 wt% of a $CeZrO_4$ phase. This latter component exists in the form of nanoscale islands adsorbed upon the bulk γ-Al_2O_3 [33]. This sample, though far from being the worst one can think of, contains both a low loading of the element to be studied and a significant amount of a promoter phase that is both absorbing and has the potential to produce a significant parasitic scattering over and above that derived from the majority support material.

Figure 3.4 shows both the raw absorption data (obtained using 100 ms of X-ray exposure) and k^3-weighted and Fourier transform (FT) representations of the EXAFS that result from analysis, in three different cases: (a) where one makes no attempt to try and account for the uniformity and scattering of the sample itself (the reference measurement is made in air); (b) one tries to improve the averaging over the sample inhomogeneity by vertically defocusing the X-ray beam (from ca. 100 to 300 μm); and (c) if one then tries to further deal with the spatial inhomogeneity of the sample and the intrinsic scattering properties of the support material. This is achieved using a Pd-free sample of the support material (10 wt% $CeZrO_4/Al_2O_3$) as the reference [32].

Case (a) is clearly a waste of time: simply attempting to account for the overall instrument response function as the "air" reference does is completely inadequate. Attempting to better average over the intrinsic spatial nonuniformity of the sample by making the beam three times bigger in the vertical does improve the situation but fails to properly account for some severely deleterious glitch structure (e.g., ~24.7 keV). Though the apparent signal-to-noise ratio (S/N) of the resulting k-weighted data is considerably better than for case (a), the accessible k-range (crucial to any subsequent EXAFS analysis) is not improved and remains meager.

It is only when we combine a larger X-ray beam with a reference that attempts to mimic both the spatial and scattering properties of the support system that we get to what might be considered a successful end point, one wherein the apparent S/N and the accessible k-range are greatly improved (2.5–12.5 $Å^{-1}$) as reflected in the similarly improved FT. Within certain bounds, spectrum (c) is now dominated by statistical noise—as opposed to static "structure"—and we have reached the point where we can start to think about properly exploiting the dispersive arrangement for useful investigations on low loaded samples of this type [34, 35].

Typically, however, there are caveats. At the Pd K-edge on ID24 we are already in a relatively favorable position in terms of the net dimensions of the focal spot yielded to us by the optics of this beamline: the intrinsic horizontal

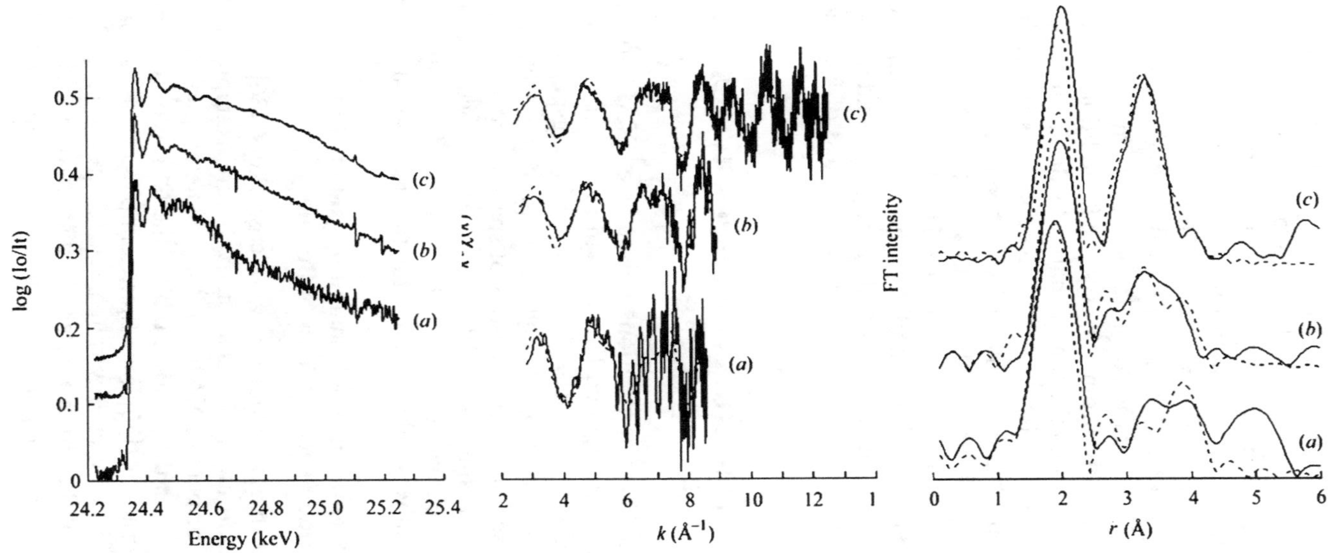

Figure 3.4. Raw, k^3-weighted and FT representations of Pd K-edge energy dispersive EXAFS derived from a 1 wt% Pd/10 wt% $CeZrO_4/Al_2O_3$ sieved powder sample (5-mm transmission path length) at room temperature. The three cases shown in each panel use differing vertical X-ray beam (VB) dimensions and differing approaches to normalizing the raw absorption data to yield the EXAFS data: case (a), VB = 100 μm, air reference; case (b), VB = 300 μm, air reference; and case (c), VB = 300 μm, sieved, 10 wt% $CeZrO_4/Al_2O_3$ as reference. Each spectrum was collected using 100 ms of X-ray exposure. Reproduced with permission from the International Union of Crystallography (IUCR) [32].

focus is, for instance, approximately 150 μm full width at half maximum (FWHM). Further at the Rh and Pd K-edge (23–25 keV) we are able to use a relatively large X-ray path length (5 mm), which, as long as any parasitic scattering is being accounted for, allows a reasonably effective averaging of spatial nonuniformity across the sample (50–100 μm sieve fraction with an effective bed density <1 g cm^3). At energies lower than 15 keV (on the preupgrade ID24) the horizontal focus is some 15–30 times smaller when using symmetric Si [111] or [311] polychromator crystals. In addition, the available transmission path lengths through the sample naturally diminish with decreasing X-ray energy. Achieving a satisfactory endpoint, such as shown in Figure 3.4, is made, therefore, much more difficult. As a result, and on top of the bandwidth limitations below approximately 10 keV that ID24 has historically suffered from, it would be fair to say that EXAFS data have not thus far been realizable on ID24 for these sorts of samples in the 7–15 keV energy range. Up to now studies of catalysts made within that energy have had to be content with time-resolved XANES which is, as a rule, is much less of a challenge to obtain.

3.3 EXAMPLES OF APPLICATIONS OF DISPERSIVE EXAFS TO PROBLEMS IN CATALYSIS

Following the initial surge in the application of dispersive EXAFS to problems in catalysis that arose from the United Kingdom in the very early 1990s [13–16], and the emergence of a number of dedicated beamlines, dispersive EXAFS has contributed significantly in many catalytic arenas. In this section, we will consider a number of case studies. For reasons of space this cannot be an all-encompassing review but it is hoped that this demonstrates various facets of how the dispersive approach may be used in the pursuit of a more profound understanding of how catalysts behave and catalytic processes are mediated.

3.3.1 Establishing the Significance of Vanadium Oxidation States during Selective Maleic Anhydride Synthesis from N-Butane

The first case study is due to Coulston and coworkers [36] working at Brookhaven National Laboratory. This is a relatively rare example of a highly successful application of dispersive XAFS to studying heterogeneous catalysis by elements of the first row in the periodic table—the NSLS source being well adapted for such energy ranges ([26]; for other examples of the use of dispersive XANES in the sub-10 keV energy range, see Reference [37]). It is also, somewhat ironically, one of the last examples of a dispersive EXAFS measurement made in the United States. Figure 3.5 summarizes the main results from this study.

At this dispersive beamline a total X-ray bandwidth of approximately 700–750 eV could be obtained even at the relatively low energy V K-edge (5.465 keV). This did yield the potential for obtaining some form of EXAFS.

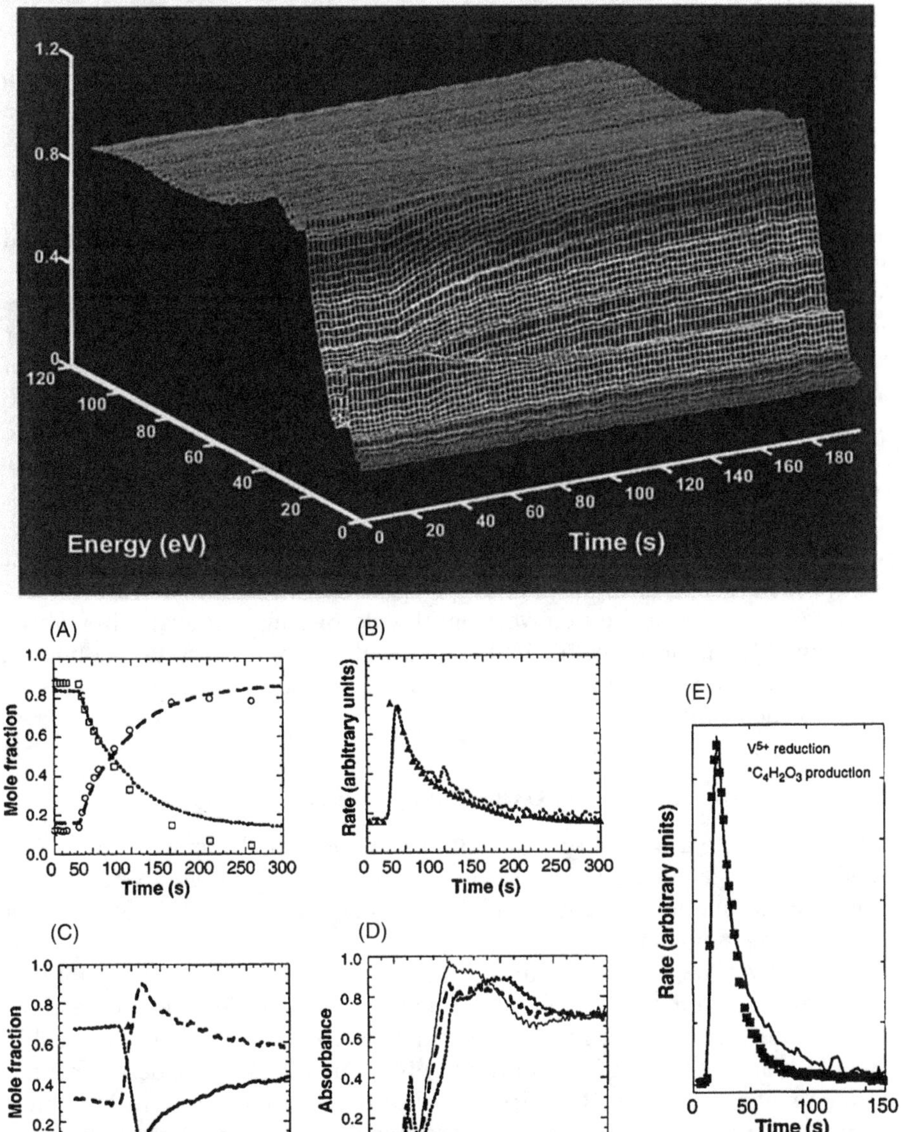

Figure 3.5. Top: V K-edge dispersive XANES obtained during reduction of a vanadium phosphate/SiO_2 catalyst at 593 K. At t = 40 s the sample is perturbed by a change in feedstock from 20% O_2/He to n-butane (C_4H_{10}). Bottom: (A) Temporal dependence of the [V^{5+}] (squares) and V^{4+} (circles) obtained through fitting of the XANES; (B) rate of change of [V^{5+}] compared to maleic anhydride formation (593 K); (C) concentration dependence of maleic anhydride production (solid line) compared with [V^{4+}] (dashed line) and [V^{3+}] (solid line); (D) reference spectra for V^{5+}, V^{4+}, and V^{3+} used for the PCA analysis of the XANES; and (E) reaction rate of V^{5+} centers in the $(VO)_2P_2O_7/SiO_2$ catalyst reduced in 10% C_4H_{10}/Ar compared to maleic anhydride production as determined by online mass spectrometry. Figure from Coulston et al., *Science*, 275, 191–193. Reprinted with permission from the American Association for the Advancement of Science.

Such EXAFS data did not, however, arise from this study, most likely as a result of issues of sample uniformity and X-ray scattering described in the previous section, but also simply the photon throughput/detector limitations of the era alluded to previously.

While this precluded a more complete structural analysis of the centers at work in the catalysis to hand, the selective oxidation of n-butane to yield maleic anhydride, a highly important result, can be obtained using the dispersive approach. By tracking the changes in near-edge structure, and simultaneously observing the course of the catalysis using an online mass spectrometer (MS), unambiguous determination of the role of different oxidation states of vanadium can be achieved. In vanadium, differing oxidation states (in this case from V^{5+} to V^{3+}) are readily separable and quantifiable using the V K-edge XANES coupled to a principal component analysis (PCA) [38]. The major conclusions are best illustrated in panel E of Figure 3.5, where the highly significant correlation between [V^{5+}] centers and the production of the selective oxidation product, provides very strong evidence for their direct involvement in maleic anhydride formation. [V^{4+}] centers, in contrast, are inversely correlated with the desired conversion: this despite aged industrial catalysts reporting only the presence of [V^{4+}] in terms of observable crystalline phases, a profound comment on the reliance on techniques such as diffraction to assess materials such as these.

3.3.2 Oscillations during CO Oxidation by O_2 over Pd Catalysts

At the same time as Coulston et al. [36] were applying dispersive EXAFS to vanadium, Ressler et al. [39] were investigating the possibilities for *in-situ* dispersive measurements at much higher energies (Pd K-edge, 24.35 keV), in their case to study spontaneous oscillations in CO oxidation over supported Pd catalysts. At the Pd K-edge, and using a Si 400 polychromator crystal in a Bragg configuration, the overall energetic bandwidth achievable increases to approximately 1.5 keV. A greater potential for rapidly obtaining analyzable EXAFS data therefore arises. Moreover, Ressler et al. [39] made their measurements at the Doppel Ring-Speicher (DORIS) storage ring in Hamburg operating at 4.5 GeV—a higher energy source than NSLS at Brookhaven and therefore better adapted (in terms of achievable photon flux) to measurements at such X-ray energies. Figure 3.6 shows some of the salient results from these investigations.

The (pressed disk) samples used contained 10 wt% Pd supported on carbon and therefore might be considered as not entirely typical of the sorts of catalysts one would really desire to be studying, and as we have seen in the previous section, 1 wt% Pd samples can be handled on much faster timescales nowadays [32]. Nonetheless, at the time (1998), one might consider these results as significant both in terms of the use of the dispersive technique, especially at energies approximately >15 keV, and for its study of oscillatory phenomena in catalytic systems. Such phenomena are of considerable general

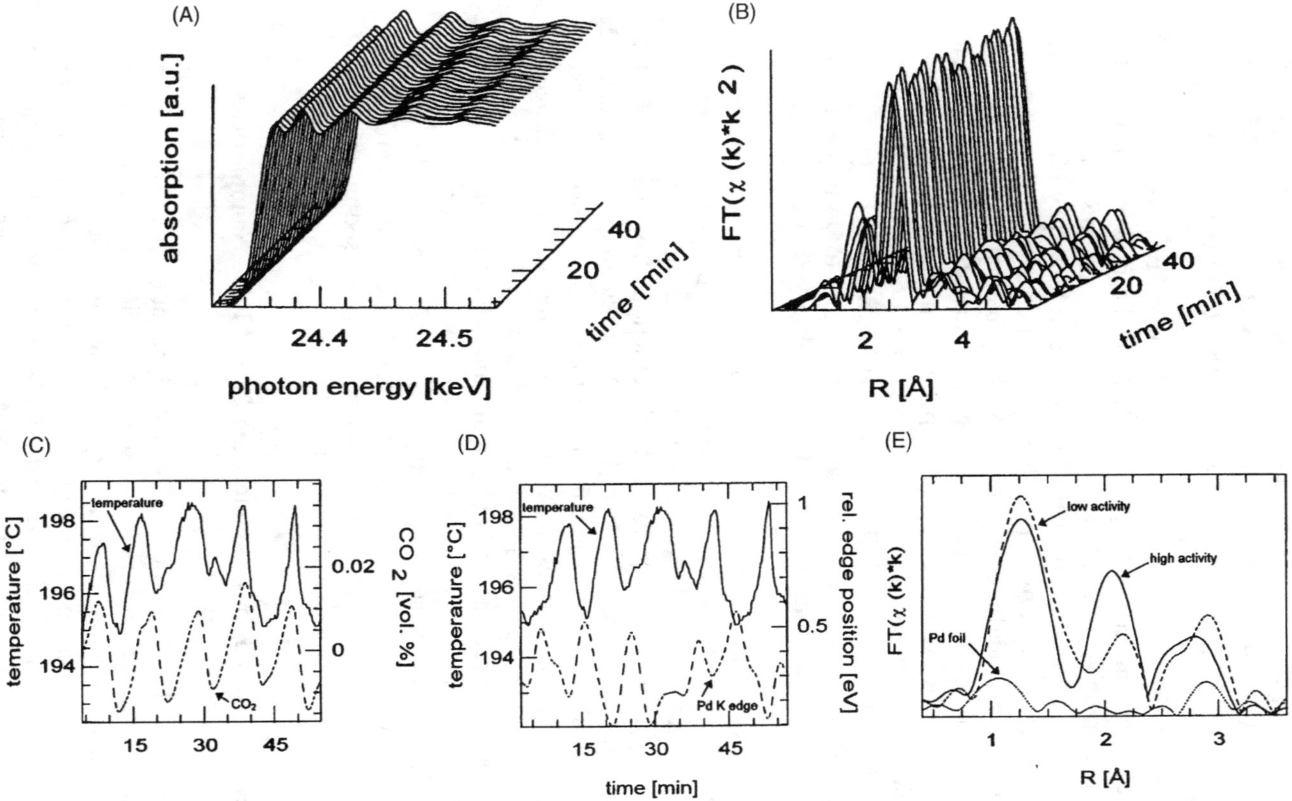

Figure 3.6. (A) Normalized Pd K-edge dispersive EXAFS from a 10 wt% Pd/C catalyst during oscillatory CO oxidation and with a time resolution of 15 s per spectrum; (B) same as (A) but in FT representation; (C) variations in CO_2 production and measured temperature of the pressed sample during oscillatory chemistry; (D) correlation between the measured position of the Pd K-edge and sample temperature; (E) FT spectra indicative of the state of the Pd in the catalyst at times of "high" and "low" activity for CO_2 production. Reprinted with permission from Ressler et al., *J. Phys. Chem. B*, 101, 6680–6687. Copyright (1997) American Chemical Society.

interest but are often characterized by a spontaneity/rapidity that makes them difficult to study: the ability of the dispersive approach to obtain EXAFS data on relatively short timescales, and in a single shot, makes it very well suited to these sorts of situations.

During the chemistry the catalyst oscillates between two apparent states of differing efficiency in converting CO to CO_2 in the presence of O_2. The application of dispersive EXAFS, in parallel with online MS, allows some insight into the nature of each of these Pd states and how they interconvert even though this is signaled by only relatively small changes in, for instance, the Pd K-edge position and the amplitudes of the principal EXAFS features. This highlights one of the potential benefits of the dispersive approach mentioned previously. As the dispersive optical arrangement is free from any mechanical movement, and should therefore be intrinsically stable, dispersive should be naturally adept at measuring subtle changes in, for instance, edge position. From this experiment the ability to do precisely this, and then correlate these changes directly to the observed chemistry, is well demonstrated.

Here the EXAFS (albeit shown only in FT representation) may also be assessed to complement the information coming from the XANES region. By and large this variation is to be found in small changes in the amplitude of the major scattering features due to Pd–Pd interactions from within the Pd nanoparticles and the associated low Z (Pd–C/O coordination) at the surface of these particles. This allowed the authors to formulate a possible reaction mechanism, relating the changes between states to the surface coverage of adsorbed oxygen and two forms (linear and bridged) of adsorbed CO species: increasing oxygen coverage and a concomitant augmentation of bridged carbonyl at the expense of linear CO adsorbates, coming together to stifle CO oxidation. While it could be seriously debated as to whether EXAFS (of any denomination) can, on its own, realistically yield such specificity regarding the precise nature and role of such surface species, the proposed mechanism is consistent with a range of other measurements.

3.3.3 Dispersive EXAFS for Characterizing Pt and PtGe Nanoparticles and Wires in Mesoporous Hosts

The previous examples both utilized dedicated bending magnet dispersive EXAFS resources at second-generation sources. While convenient in many ways for the dispersive setup (large source and low heatloads on the optics, for example), a bending magnet at any given synchrotron will be outperformed by insertion devices such as wigglers and undulators in terms of the photon flux that can be applied to any given problem. Given that one can achieve a situation (see earlier) where photon counting statistics dominate the S/N from the dispersive experiment, then this approach should clearly benefit from the use of insertion devices. The first attempt to achieve this was made using the wiggler-based station 9.3 at the SRS at Daresbury, UK [9].

To illustrate some aspects of the unique utility of dispersive EXAFS, and to fill in an energetic gap between those examples already given, we consider the work of Fiddy et al. [40–42], who used dispersive EXAFS to make, what was at that time, the first practical implementation of the ability to study two elements at the same time, as had been recognized much earlier by Matsushita and Phizackerley [7]. These studies did not explicitly consider catalytic chemistry but rather the genesis of PtGe catalysts for potential application in hydrocarbon reforming catalysis. Figure 3.7 summarizes a number of results that come from these studies.

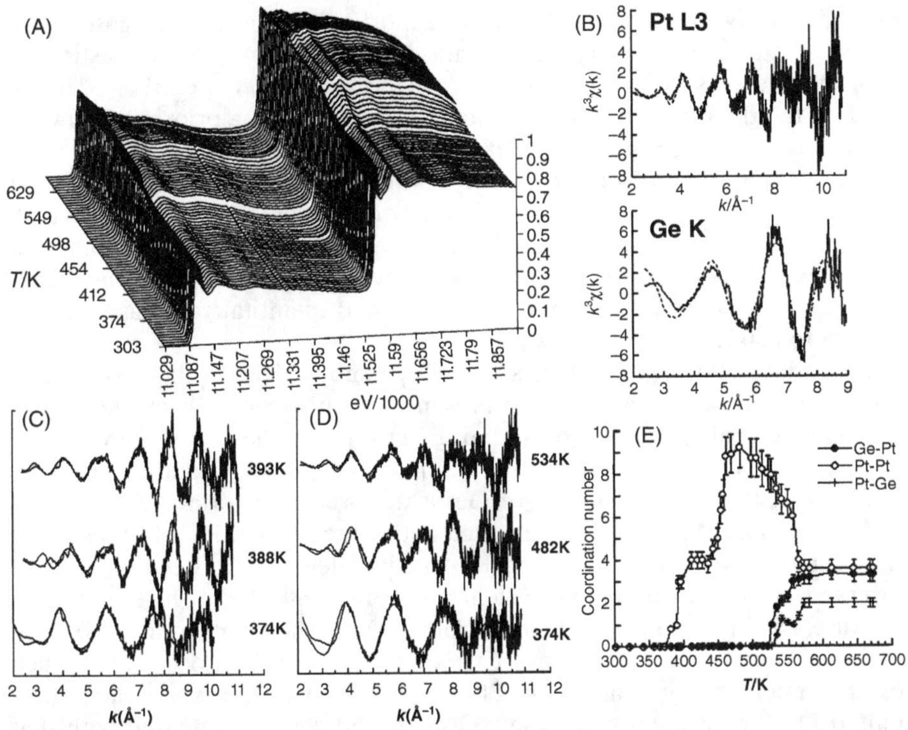

Figure 3.7. (A) Raw dispersive absorption spectra incorporating both Ge K- and Pt L_3-edges simultaneously from a 5 wt% Pt/1.9 wt% Ge/H_1SiO_2 catalyst during *in-situ* reduction under 10% H_2/N_2. Each spectrum is collected in 8 s. (B) Exemplary Pt L_3- and Ge K-edge k^3-weighted EXAFS extracted from spectra collected above 500 K. (C) Illustrative Pt L_3 EXAFS from a 5 wt% Pt/H_1SiO_2 sample during reduction at temperatures indicated. (D) Illustrative Pt L_3 EXAFS from a 5 wt% Pt/1.9 wt% Ge/H_1SiO_2 sample during reduction at temperatures indicated. (E) A summary of the local coordination derived from both analysis of Ge K- and Pt L_3-edge EXAFS as a function of temperature during *in-situ* reduction for the 5 wt% Pt/1.9 wt% Ge/H_1SiO_2 sample. Panels A and E from Fiddy et al. (2001), *Chem. Commun.*, 445–446. Reproduced by permission of The Royal Society of Chemistry. Panels B–D from Fiddy et al. (2002), *Phys. Chem. Chem. Phys.*, 4, 827–834. Reproduced by permission of the PCCP Owner Societies.

The first point to note is that the beamline character yields sufficient energetic bandwidth (ca. 1–1.5 keV) to permit both (Pt L_3- and Ge K-) edges to be comfortably studied in the 11–12 keV range for EXAFS. The proximity of the Pt L_3-edge to the Ge K-edge does mean that the EXAFS from the Ge is naturally restricted to approximately 9 K/Å but nonetheless the refinement of structural data from both edges was possible (Fig. 3.7) and yielded a considerable *in-situ* insight into the genesis and development of PtGe alloy nanoparticles. The second point to note is that the sample presentation in this experiment was somewhat different from the previous examples.

In both the work due to Coulston et al. [36] and Ressler et al. [39], the catalyst under study was pressed into a self-supporting disk over which gases were flowed. While improving the spatial uniformity of the samples in question, in doing this one loses control over the nature of the environment experienced inside the disk; the details of the gas flow through the disk are, a priori, unknown.

Fiddy et al. used a packed powder bed held within a quartz tube microreactor [40–42]. The gas feedstock is therefore obliged to pass through the catalyst bed and the basic requirements for plug-flow operation are met, whereas they are not in the previous examples. While not of particular relevance to any of the issues addressed in this and the two previous examples, this is a serious consideration when making more advanced and quantitative measurements relating to catalysis and kinetics [43].

These data show that the genesis of alloy nanoparticles in these systems is not a simple process, and that the presence of an organo-Ge species in the precursor formulation has a profound effect on how the organo-Pt complex reduces as a function of temperature. Comparing Figure 3.7C and Figure 3.7D, it is evident that in the absence of Ge in the precursor formulation (Figure 3.7C) that an EXAFS signature characteristic of the formation of Pt nanoparticles can be observed as low as 388 K. An equivalent EXAFS envelope is not observed in the Ge-containing system until practically 100 K higher in temperature. The more detailed collation from the data shown in Figure 3.7E does, however, tell us that some form of structural change does occur at a much lower temperature, just as in the Pt-only system, with the appearance of a limited Pt–Pt coordination number (CN) around 400 K: compared with that observed in the absence of Ge [40], it is considerably curtailed reaching an average CN of approximately 4, as opposed to approximately 9 at 400 K. It is only at much higher temperatures (>450 K) that the formation of similar Pt nanoparticles is observed in the Pt–Ge case. Between 450 and 500 K the system seems to comprise Pt nanoparticles that coexist with organo-Ge phases. It is only after this latter phase starts to decompose, possibly as a result of Pt-promoted hydrogen spillover, that any evidence for the formation of an Pt–Ge alloy phase starts to appear (from Ge–Pt (Ge K-edge) and Pt–Ge (Pt L_3-edge)) coordination. As this occurs the Pt–Pt CN plummets (from ca. 9 to 4), yet the Pt–Ge CN observed from the Pt L_3-edge only reaches a value of 2. At the same time, the viewpoint of the Ge K-edge indicates an average Ge-to-Pt CN of approximately 4, consistent with the Ge only seeing the surface of the Pt nanoparticle. This unusual situation can be resolved if, alongside the process of alloying, the Ge also causes a more global rearrangement of the Pt such that the average particle size is significantly reduced.

Simple interpretation of changes in apparent coordination numbers in EXAFS is always fraught with difficulties as a result of the intrinsic correlation of this parameter with thermal and static disorder, both of which could be playing roles. In this case, however, that such an alloy-induced change in the overall particle size distribution was indeed occurring could be verified postfactum using transmission electron microscopy (TEM) [42]. Unlike the EXAFS, however, TEM is also able to show that this particle shrinkage is at work irrespective of particle morphology: the size distribution of both the extrapore "spherical" particles and the "wires" formed inside the pores of the H_ISiO_2 are both affected in this intriguing manner.

As such, and as is often the case, the time-resolved EXAFS is an excellent "pathfinder," providing at that time unique insight into an unexpectedly complex nanoparticle genesis. The details of such processes, which in general are not reversible and therefore must be studied in a single shot, have some considerable ramification for how many catalysts might be best prepared. However, it must also be noted that the absolute verification of some of the events implied by the EXAFS require the application of other, sometimes *ex-situ*, techniques.

Perhaps just as important in the current context is, however, the simple observation that, though the packed powder bed must have a lower net spatial uniformity than the pressed disk, successful experimentation across practically the entire applied X-ray bandwidth applied was possible. That the disposition of the sample is changing in some way as a function of temperature can be deduced from the changes in background observable in Figure 3.7A; in this case, however, these changes did not prevent subsequent analysis. The major reasons for this lie (most likely) in the large focal spot (200 μm (h), 500 μm (v)), and lower coherence of the second-generation synchrotron wiggler source. We note that this experiment is simply not (preupgrade) currently possible at the undulator-based ID24 at the ESRF. While more powerful in many ways than 9.3 at Daresbury, the overall beamline design (up to 2009) means that the energetic bandwidth required to encompass EXAFS from the Ge K- and Pt L_3-edges could not be obtained. Even if it were obtainable, the exceptional small horizontal beam focus on ID24 at these energies made it very unlikely that EXAFS could be obtained for the reasons discussed in the previous section. Ameliorating these restrictive elements of the ID24 design is central to the upgrade of that line currently underway at the ESRF.

3.3.4 Dispersive EXAFS and the World of Auto Exhaust Catalysis for Pollution Abatement

Thus far we have shown how dispersive EXAFS may be applied to studying structural and oxidation state changes in catalytic systems on timescales of approximately 5–15 s. In many cases relevant to catalysis, this is more than sufficient. Equally, however, in many areas a much higher (<<1 s) single-shot capacity is required to fully access, kinetically quantify, and understand fundamental processes occurring in catalytic situations; this is especially true of support metal catalysts [44].

From this point on, therefore, we consider a number of highly time-resolved dispersive EXAFS studies of working catalysts centered on matters concerning the intrinsically dynamic and fluxional world of auto exhaust catalysis. Often, though not exclusively, these studies have been instigated and driven by industrial research groups (Toyota [45–48] and Daihatsu [49–51]). For reasons of space we cannot, however, address all these studies and the interested reader is directed elsewhere for Daihatsu's work regarding the behavior of the so-called intelligent La-peroskvite-based catalysts [49–51].

It would be fair to say that these unusually proactive and directly collaborative (with central facilities such as the ESRF, Spring 8, and the Photon factory) industrial efforts have provided a good deal of momentum for driving forward and improving the application dispersive EXAFS for the study of heterogeneous catalysts. Both in terms of the nature of the samples to be studied and the correct parameterization of the overall experiment to match the requirements of these industrial catalysis groups, improvements, innovations, and a general development of the dispersive beamlines has arisen, improvements which benefit researchers in all the areas wherein dispersive EXAFS is actively used [28, 52].

3.3.4.1 Studies of Rh Nanoparticles and their Interaction with CO, O_2, and in CO Oxidation Catalysis The behavior of Rh, Pt, and Pd nanoparticles in response to gases such as CO, O_2, and NO is of central interest within the paradigm of the primary usage of these metals within three-way auto exhaust catalysis. As three-way catalysts (TWCs) experience very rapidly changing feedstock conditions—on the hertz timescale—as a matter of their working life, *in-situ* and dynamic study of their behavior is highly pertinent. Here we will consider only studies concerning CO and O_2 with Rh catalysts. For further examples of other pertinent examples of the application of dispersive EXAFS, coupled to infrared spectroscopy, involving Rh, Pd, CO, NO, and O_2, the reader is, however, referred to Chapter 14 of this book.

For a very long time, it has been known that gases such as CO [53–55], NO [56], and O_2 [57] can severely change the character of Rh catalysts: O_2 simply through oxidation under a variety of circumstances; CO through oxidative disruption to form $Rh^{I}(CO)_2$ species; and NO potentially via both routes. What had been missing in any of these cases until the early 2000s, however, was any real understanding of exactly how fast these processes might be, what sort of kinetics characterized them, or how they integrated themselves into global catalytic processes.

This somewhat unsatisfactory state of affairs changed in 2003 with the application of dispersive EXAFS at the Photon factory in Japan in answer to some basic questions regarding what happens when small, support Rh nanoparticles interact with CO [58]. While some interpretive aspects of this work are contentious, and one could argue that the use of a pressed sample might be a weakness in terms of kinetic evaluations [43], this study nonetheless remains as the first experiment that really started to showcase what dispersive EXAFS could yield in terms of accessing and understanding fast, structural/kinetic change on a <<1 s timescale: this work very effectively dissects some (single-shot) reactive chemistry that is essentially complete in 3 s under the conditions used.

The application of dispersive EXAFS leads to the deconstruction of the oxidative dispersion process into three discrete stages (as shown schematically in Fig. 3.8) that occur with an overall apparent activation energy of approximately 17 kJ/mol. First, molecular CO adsorbs on the small Rh nanoparticles to a certain surface coverage ($Rh_{10}CO_3$ according to Reference [58]). Beyond this coverage the Rh particles are corroded into monodisperse $(Rh\text{-}CO)_{10}$ units. That is to say, the final $Rh^I(CO)_2$ species is not the primary product of this interaction. It is only subsequently within the obtained mechanistic model that more CO adds to these monomer units to yield the monodispersed $Rh^I(CO)_2$. In short, by applying dispersive EXAFS on the 100-ms timescale, an alternative mechanism, whereby the Rh nanoparticle is continuously and directly degraded directly into $Rh^I(CO)_2$ units, is ruled out.

A little while later, with the introduction of the FReLoN detector at the ESRF [20, 21], similar kinetics studies were made regarding how small Rh nanoparticles (average diameter ca. 11 Å) oxidized and reduced in the presence of O_2 and H_2, using a packed powder bed/plug flow approach rather than a pressed disk. Among other kinetic aspects, this uncovered a temperature-dependent induction time to the reduction of oxidized nanoparticulate Rh particles: though once initiated at the highest temperatures studied, the subsequent reduction was again extremely fast, being completed in <2 s.

The previously mentioned experiments addressed some of the basic aspects of the interaction between CO, O_2 and H_2, and supported Rh nanoparticles that underpin many aspects of the catalysis for which Rh finds most significant usage. However, how might some of these more basic processes come together during actual catalysis?

Figure 3.9 shows some of the results derived from a 2006 study from the ESRF [59]. This study draws together a plug flow microreactor configuration with downstream mass spectrometry to assess the catalytic function (in this case CO oxidation) with both the temporal and spatial (axial position in the catalyst bed) possibilities that come with dispersive EXAFS.

These data, along with subsequent fitting and interpretation of the EXAFS, reveal a number of aspects regarding how three phases of Rh (nanoparticulate Rh metal, an Rh^{III} oxide, and isolated $Rh^I(CO)_2Cl$) compete for existence within the catalyst under reaction conditions, and how they contribute to the end point of efficient CO elimination.

With a CO/O_2 ratio of 2.5 in the feed, we can see that at 473 K (Figure 3.9B) the Rh is found to exist in an essentially reduced nanoparticulate state; the

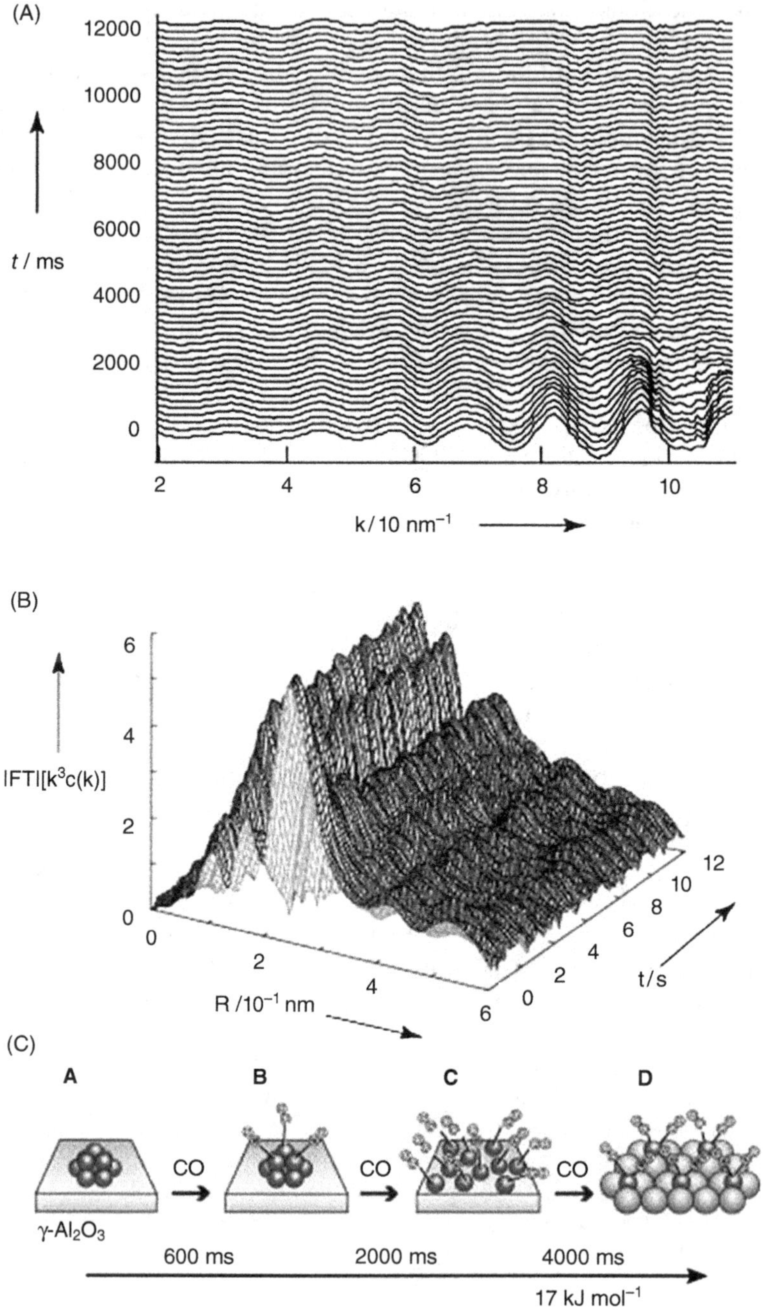

Figure 3.8. (a) *k*-Weighted EXAFS as a function of time obtained from 2 wt% Rh/Al_2O_3 catalyst during exposure to CO. (b) The resulting FT representations of this data. (c) The mechanism of the oxidative disruption of Rh nanoparticles by CO that is concluded from this study. Figure from Suzuki et al., *Angew. Chem. Int. Ed.*, 42, 4795–4799. Reproduced by permission of John Wiley & Sons.

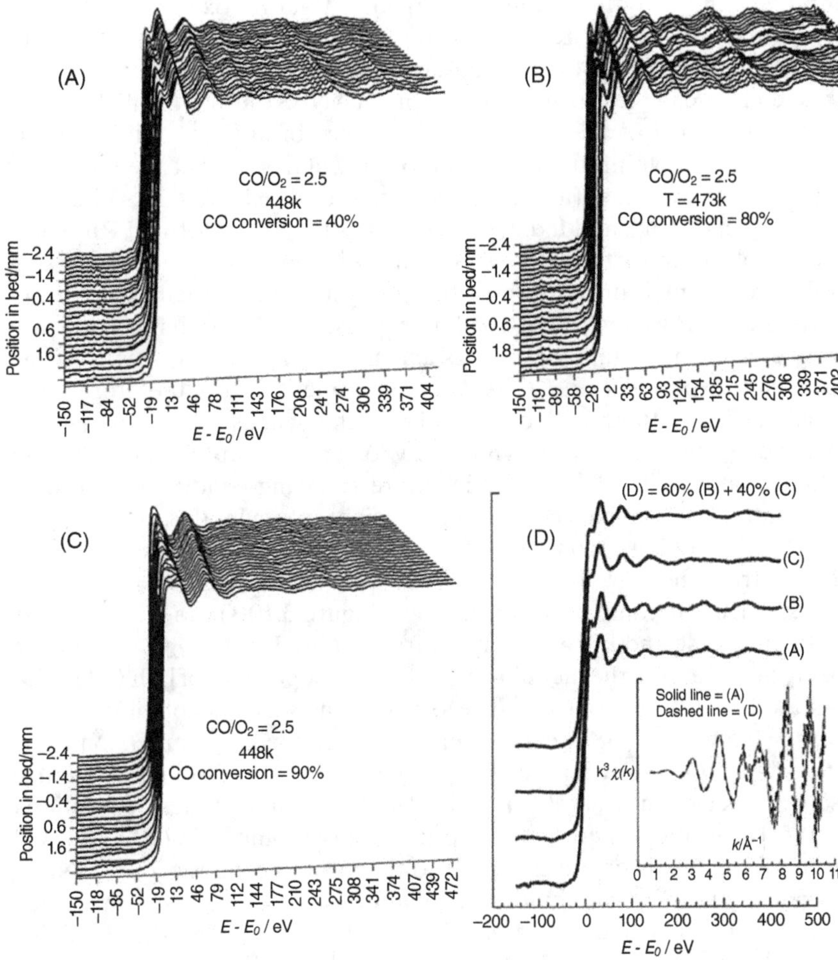

Figure 3.9. (A–C) Rh K-edge dispersive EXAFS data derived through rapidly sampling axially along a 5 wt% Rh/Al_2O_3 catalyst bed while the catalyst experience differing CO/O_2 feedstocks at the temperatures indicated. (D) Deconstruction of the Rh K-edge EXAFS obtained in case A into components due to reduced Rh nanoparticles and monodispersed $Rh^I(CO)_2Cl$ species. The inset shows a comparison, in terms of k^3-weighted EXAFS, of the obtained experimental data and a linear combination of 60% Rh metal nanoparticles and 40% $Rh^I(CO)_2Cl$. Figure from Newton et al., *Chem. Eur. J.*, 12, 1975–1985. Reproduced by permission of John Wiley & Sons.

entire length of the bed and the measured conversion of CO is 80%. If the temperature is lowered by 25 degrees (to 448 K), however, it is clear that the EXAFS signature is rather different and the overall conversion is diminishes to approximately 40% (Figure 3.9B). By and large, however, the overall spectral envelope, though changed from that observed at 473 K, is still essentially constant over the length of the bed. A much more oxidizing situation ($CO:O_2 = 0.11$) also at 448 K changes the appearance and performance (in terms of conversion, 80%) of the catalyst completely (Figure 3.9C). Moreover, we now see clear changes (principally in the XANES) in the axial character of the Rh. Over most of the bed, a predominance of an Rh^{III} oxidic state is indicated. However, within the first 2 mm of the catalyst bed at the feedstock inlet we see clear evidence (a diminished intensity in Rh K-edge white line) for a lower level of Rh^{III} and indications of a less than fully formed Rh oxide. The clear implication of the latter observation is that under these oxidizing conditions only a small proportion of the catalyst bed is actually engaged in CO conversion. However, the Rh phases present within that region are extremely efficient in eliminating CO, such that the remaining length of the bed essentially experiences a $CO_2 + O_2$ environment and is therefore completely oxidized. The thermal evolution of this chemistry [59] confirms that a more oxidized state of Rh has a significantly lower light-off temperature than the metallic states of Rh formed under more reducing conditions. The data shown in Figure 3.9D give us considerable insight into why this may be.

Panel D of Figure 3.9 deconstructs the experimental Rh EXAFS (spectrum A) obtained from the case shown in Figure 3.10A; that is, $CO/O_2 = 2.5$, 448 K. The total spectrum is compared to that from (Figure 3.10B) the fully reduced, nanoparticulate case and EXAFS derived from a pure $Rh^I(CO)_2Cl$ system, the latter being derived from the metallo-organic vapor deposition of $[Rh^I(CO)_2Cl]_2$ to dehydrated γ-Al_2O_3 [60, 61]. It can be seen that the spectrum obtained under the $CO/O_2 = 2.5$ feed at 448 K can be very well described (a comparison using k^3-weighted spectra is shown in the insert to Fig. 3.9D) by a weighted average of these two species in a 60:40 ratio. That is to say, under these conditions, 40% of the Rh in the system is tied up in the organometallic ($Rh^I(CO)_2Cl$) form of the Rh which is only a spectator to the desired catalysis [62]. As such, at least half of the Rh that could be available for CO conversion has been sidelined within an inactive state at 448 K and, at this point, the Rh remaining in the metallic state is simply not active enough to effect a complete CO elimination. As such, under real conditions and when dealing with nanoparticles on the approximately 11 Å average diameter scale, the effectiveness of the Rh catalyst is very much a function of the dynamic partitioning of the Rh into three competing phases; metallic, oxidic, and organometallic.

Moreover, on the basis of what was already known from standard scanning EXAFS measurements [54, 55, 58] regarding the limit to which CO on its own may effect a disintegration of supported Rh clusters, these data point to some form of synergism between CO and the oxidant (O_2) in the disruption process. Cross-referencing the measured (TEM) particle size distribution for this system with the dispersive EXAFS [59] shows that most of the Rh eventually found by the EXAFS to be in the form of the $Rh^I(CO)_2Cl$ species at 448 K ($CO:O_2 = 2.5$) must have come from particles that were much bigger (i.e., $N^{Rh} \gg$ ca. 20 atoms per particle) than had previously been observed to be disrupted by CO alone [54, 55, 58].

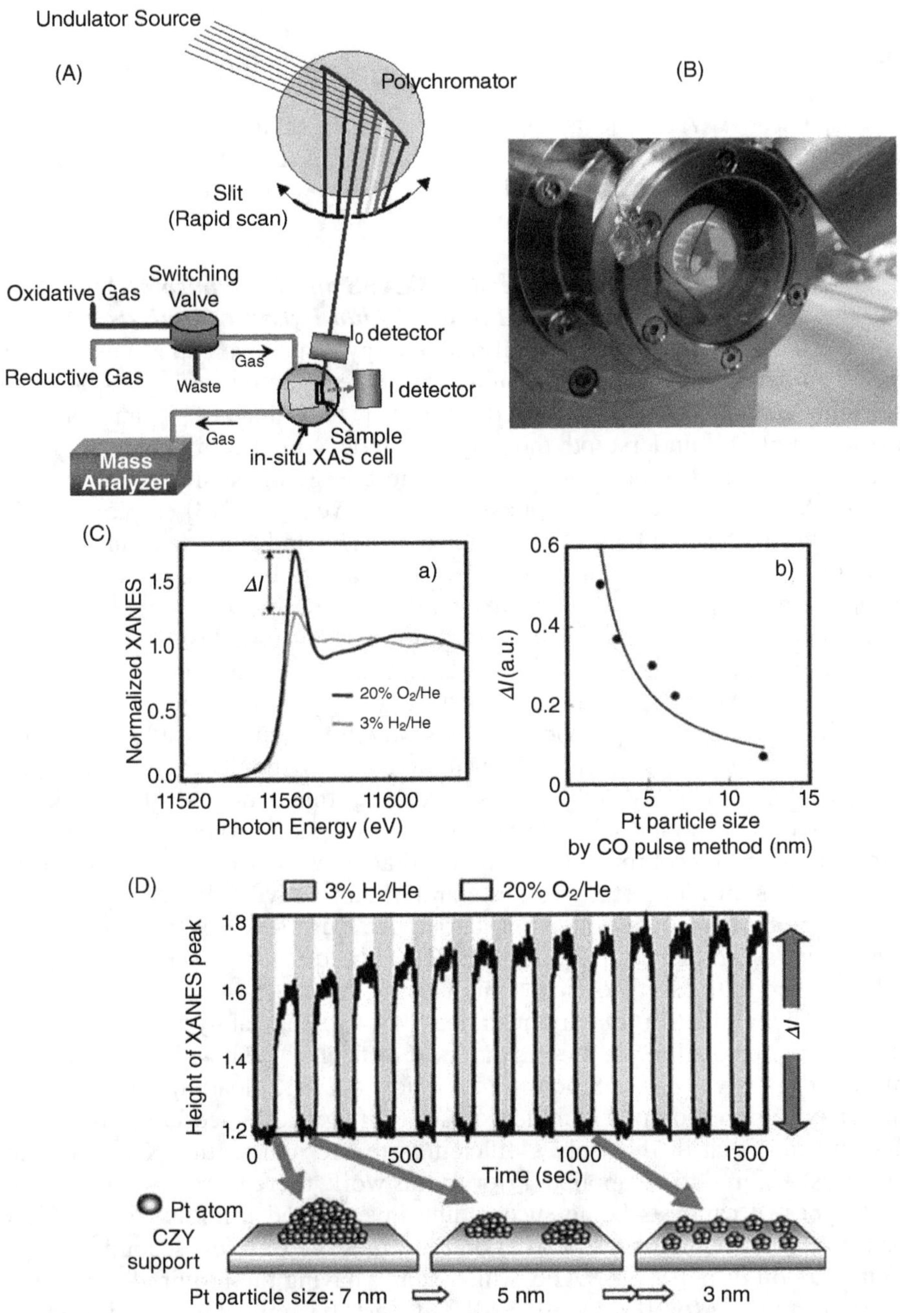

Figure 3.10. (A) Schematic of the overall experimental setup used to study Pt automotive exhaust catalysts *in situ* and using fluorescence yield dispersive XANES [45]. (B) A sample under study within the sample environment developed for this experiment [48]. (C) Left panel: the quantity (Pt L_3-edge white-line height) actually measured in the dispersive study and (right panel) how this is correlated to Pt particle size obtained by another (non-*in-situ*) technique. (D) A summary of the results obtained under redox cycling conditions and their interpretation in terms of a net redispersion of a sintered sample over time. Panel A from Suzuki et al., (2003), *Angew. Chem. Int. Ed.*, 42, 4795–4799. Reproduced by permission of John Wiley & Sons. Panel B from Nagai et al., (2007), *AIP Conference Proceedings*, 882, 594–596. Reproduced by permission of the American Institute of Physics.

Dispersive EXAFS allows this complex state of affairs to be observed and unraveled on the basis of the sensitivity of the EXAFS experiment to these differing phases of Rh, but also the speed within which it can be applied in what are often very dynamic situations.

3.3.4.2 Fluorescence Yield Dispersive EXAFS and Fundamental Aspects of the Behavior of Multicomponent Pt-Based Auto Exhaust Catalysts As has been noted previously, the conventional energy dispersive experiment is one based upon transmission: the use of fluorescence detection, as is possible with scanning EXAFS, is not permitted. However, as Cauchois [5], and latterly Couves et al. [13] understood, the polychromatic X-ray beam emanating from the curved crystal optics that comprise the energy dispersive setup can be "monochromated" using a simple slit, with Couves et al. [13] employing this approach to alternate between dispersive EXAFS and "monochromatic" diffraction measurements during catalyst generation. If, however, this slit is rapidly scanned across the polychromatic X-ray beam, one is able to use a dispersive X-ray beam to collect absorption data through fast stepwise accumulation of the energy bandwidth applied, and just as in a more conventional scanning EXAFS approach, fluorescence detection become possible. By reverting to a stepwise approach, the experiment becomes much less sensitive to many of the issues (such as overall stability of the X-ray beam/sample, spatial nonhomogeneity, and parasitic scattering) that bedevil the transmission experiment. This is especially the case in catalyst systems wherein the majority support material is comprised of elements that are very strong X-ray scatterers and the element under study is dilute and has a relative "soft" X-ray edge.

This method of data collection, christened Turbo-XAFS, was first introduced by Pascarelli et al. [63] at the ESRF. Subsequently it has been successfully applied to the study of how some very demanding Pt-based auto exhaust catalysts behave in certain circumstances [45–48]. The catalysts of interest in this sort of application are among the most difficult to deal with from a (scanning or dispersive) EXAFS point of view: low (<2wt%) loadings (in this case, of Pt) are supported upon complex substrates (e.g., CZY—cerium, zirconium, and yttrium) that in themselves efficiently absorb and scatter X-rays of the energies required to be applied. Prior to this work, there were no examples of such types of catalysts being successfully investigated using XAS (XANES or EXAFS, scanning or dispersive) on the timescale of a few seconds or less. Transmission dispersive EXAFS, while clearly having the required time resolution, cannot yet (on ID24 at the ESRF, at least) be applied in these instances, so deleterious is the interplay between the very coherent, very small X-ray beam and the parasitic scattering/uniformity of the samples (see earlier). As such a fluorescence yield approach was implemented that, while only yielding XANES on a timescale of approximately 1 s, opened up a window into the processes at hand that could not otherwise be achieved at that time.

Figure 3.10 shows the overall experiment setup used, the novel experimental cell developed for these studies [47], and summarizes the important insights that arise from the application of the fluorescence yield approach to dispersive EXAFS in this case.

The sintering of noble metal nanoparticles during catalytic operation has been a persistent industrial problem for a very long time: it decreases activity; can deleteriously alter selectivity; and curtails the useful lifetime of the catalyst. That some catalysts might be more resistant to sintering than others, or might be amenable to certain treatments that can reverse this process and "reactivate" the catalyst, has also been known for some time. Where this work goes for the first time is in quantitatively observing, and kinetically characterizing, sintering and redispersion phenomena as they happen, rather than (the much easier to achieve and conventional) pre- and postfactum observation that one or the other event has actually occurred. Within the highly dynamic operational paradigm of an automotive catalyst, this is extremely important. While this is an experiment based on a pressed sample of unknown gas permeability, the use of fluorescence means that this is far less of an impediment to the restoration of accurate kinetic process parameters. By contrast to transmission, the sampling depth of the experiment in this case is intrinsically limited by the escape depth of the Pt fluorescent X-rays (<10 keV). A calculation [64] of the attenuation length of such X-rays in Al_2O_3 and CZY matrices returns an approximate sampling depth of the order <100 μm into the pressed sample.

That said, to achieve all the desired ends there is (again) a necessity for *ex-situ* methods (such as are shown in Figure 3.10C) to provide the quantitative framework that links the achievable experimental observable (the Pt XANES and in particular the white-line intensity, Figure 3.10C) to the size of the Pt particles. As the XANES region of any absorption spectrum has (potentially) numerous factors that contribute to it, this basic groundwork is absolutely necessary. The axiom that results, and which allows the interpretation of the Pt L_3 XANES, is the following: that the oxidation of Pt nanoparticles proceeds only to a certain limit at their exterior at any given temperature. The white-line intensity quantifies the amount of oxidation, which in turn defines the overall dispersion of the Pt. In an isothermal situation, if the white-line height is changing, the dispersion and therefore the net particle size of the Pt, is also changing.

The results from a typical experiment are shown in Figure 3.10D. During cycling (between 3%H_2/He and 20%O_2/He), it is clear that over time more and more of the Pt appears to be oxidized during the same exposure to O_2. Within the previously given axiom, this must mean that the Pt nanoparticles are being progressively redispersed from their initial (sintered) starting state, as indicated in the cartoon in panel D of Figure 3.10. It should be noted that this does not happen in the case of Pt/Al_2O_3 samples of similar initial particle size: in the same circumstances only further sintering of the Pt is found in that case.

The veracity of the interpretive framework proposed by Toyota was subsequently confirmed using environmental TEM. This, latterly, permitted direct imaging of the disintegration large Pt nanoparticles at elevated temperature and in an oxidizing environment [45]. Comparing the two approaches, we first see that both methods can reveal the overall sintering and redispersion in real time. The dispersive EXAFS readily yields kinetic information, under conditions that can be far closer to the real situation in terms of applied gas partial pressures. Though the TEM could notionally provide some kinetic insight, it is much more difficult to extract and the microscopy is intrinsically limited in terms of the environment that can be applied as a result of it being bound to the detection of electrons. In contrast, the EXAFS is not deterministic in terms of the precise mechanism of the redispersion events: from the Pt XANES one cannot simply tell how the Pt particles are being eroded. The TEM is, however, quite clear in this respect, showing unambiguously that the Pt particles are progressively stripped of their outermost layers rather than splitting up as a result of oxygen adsorption.

3.3.4.3 The Dynamics of Oxygen Storage and Release in Pt/CeO_2-ZrO_2 Catalysts In many cases, not just auto exhaust applications, modern catalysts are multicomponent and multifunctional. With the exception of the very first example [36] we have considered in detail, all the subsequent studies we have reviewed have involved the study of noble metal nanoparticles and how they behave in a variety of circumstances. This is a rich vein [44], but when one considers what comprises a modern-day automotive catalyst, or indeed and many other types of catalyst, it is not the whole picture. While one example of the application of dispersive EXAFS, to study the behavior of a supported and oxidic catalyst, began this section, there exists a number of other examples, particularly in the case of selective oxidation by molybdenum catalysts [65–70]. Oxides provide a further test of the capacity of dispersive EXAFS simply as a result of the predominance of M–O (rather than M–M) scattering; the same can be said of a range of other catalytic materials where metal-low Z scattering predominates. This leads to intrinsically reduced amplitudes in the resulting EXAFS as compared to, for instance, the case for noble metal clusters. In some senses, though not all, they therefore provide a more exacting challenge for the dispersive approach.

Here, though, we consider a more recent exemplar, again due to the Toyota group [71], that concentrates in the behavior of the support material (CeO_2–ZrO_2) in a prototypical 1 wt% Pt/CeO_2–ZrO_2 catalyst. This support is not merely a dispersant for the noble metal but has active roles to play in facilitating and optimizing the catalysis to hand. As a result, quantifying the structural–kinetic reduction–oxidation (redox) properties of this mixed oxide, and how these result in effective oxygen storage and release within an automobile exhaust situation, are the central points of this study.

Figure 3.11 summarizes the salient results at the Photon factory in Japan and at both Ce L_3- (5.7 keV) and Zr K- (18 keV) edges. This study seeks to understand the rates at which Ce^{4+} and Ce^{3+} centers are created or consumed under model oxidizing and reducing conditions; it also seeks to understand the dynamics of Zr–O bond formation and scission that must also occur within

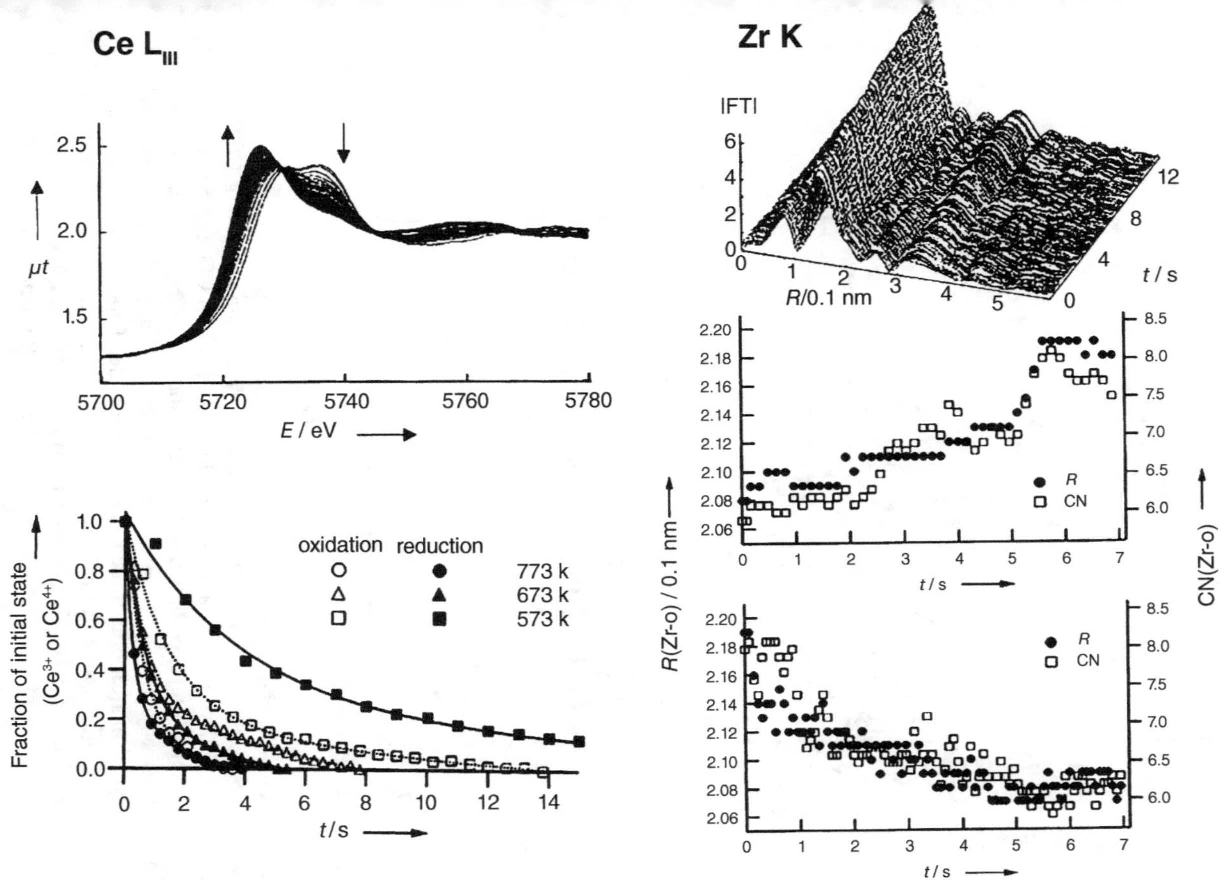

Figure 3.11. Dispersive XANES and EXAFS from Ce L_3- (absorbance, top left) and Zr K- (FT, top right) edges obtained during reduction (12.4 kPa H_2) and oxidation (12.4 kPa O_2) of a 1 wt% Pt/CeO_2–ZrO_2 catalysts. The graph summarizes various aspects of the structural–kinetics facets of the system that can be accessed from these measurements: [Ce^{3+}] versus time (bottom left); variation in Zr–O bond lengths and coordination numbers versus time during oxidation (middle right); and reduction (bottom right). Figure from Yamamoto et al. (2007), *Angew. Chem. Int. Ed.*, 46, 9253–9256. Reproduced by permission of John Wiley & Sons.

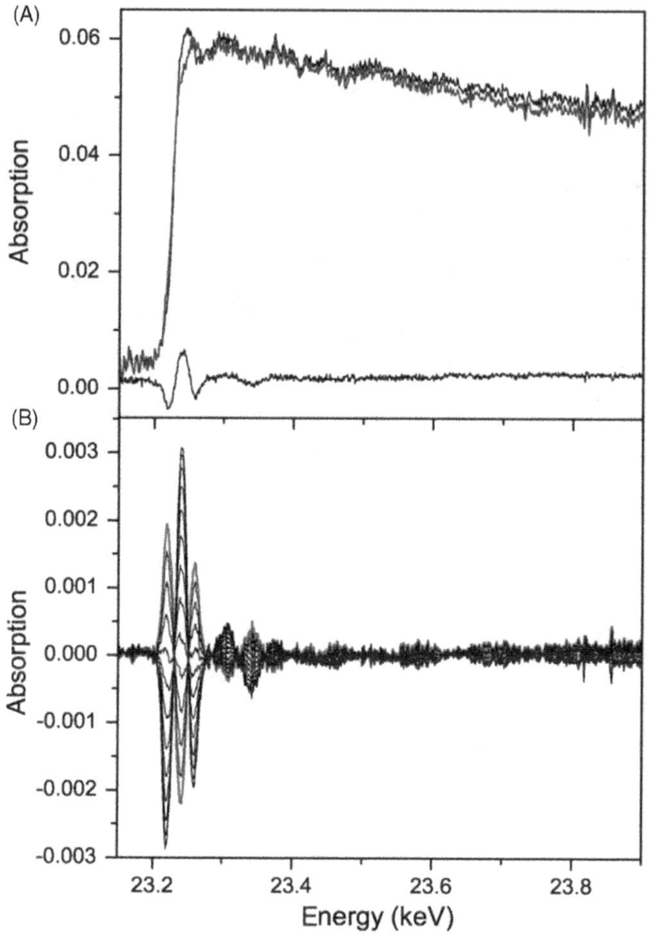

Figure 3.12. (A) Raw dispersive absorption spectra derived from a 0.3 wt% Rh/Al_2O_3 (edge jump ca. 0.05, 100 ms/spectrum, 4 Hz repetition rate) catalyst during CO/NO redox cycling at 573 K. Two spectra are shown taken from within CO and NO cycles, respectively. The simple difference between these two spectra is shown at the bottom of panel (A). (B) Spectra extracted from the same experiment after the application of a demodulation algorithm after only three gas switching cycles. Changes in the XANES and the EXAFS as a function of the reactive environment are now clearly resolved and amenable to further analysis. Figure from Ferri et al. (2010), *Phys. Chem. Chem. Phys.*, 12, 5634–5646. Reproduced by permission of the PCCP Owner Societies.

the mixed-oxide solid solution. The former objective is achieved at the Ce L_3-edge using XANES analysis with reference to Ce^{3+} and Ce^{4+} standards (left-hand side, Fig. 3.11).

To investigate changes in Zr–O coordination, however, the EXAFS has to be obtained. The FTs shown at the top right of Figure 3.12 are based upon an analysis of EXAFS data obtained in the $3 < k$ (Å^{-1}) ≤ 11.5 range, though this was supplemented by a linear combination analysis of the Zr K-edge XANES: though less informative than the Ce L_3-edge XANES, some distinct changes during oxygen storage and release processes could be accurately measured.

While a highly important system from the applied point of view, CeO_2–ZrO_2 is also intriguing in that one component of the oxide (Ce) changes its oxidation state (from 3+ to 4+) without changing its coordination, whereas the Zr does the reverse, changing coordination but not the oxidation state. What results from these studies is a quantitative assessment of the kinetics of the processes occurring during oxygen storage and release processes at elevated temperatures (from 573 to 773 K) from the perspective of both the Ce and the Zr components of the system. It is concluded that the "electronic and structural transformations at Ce and Zr sites . . . are not synchronised with each other in the solid solution" [71]. Instead there is a sequence of events wherein the Ce oxidation state changes first. This then stimulates a subsequent reorganization of the bonding local to the Zr sites.

Kinetically speaking, the activation energy for reoxidation of the Zr sites is five times smaller than that found from the perspective of the Ce, though the net rate constant is measured to be approximately 2.5 times less at the Zr site as opposed to the Ce site. Indeed, in more global terms the kinetic barriers derived from each edge, and for both oxygen storage and release processes, are also much lower than those known for, for instance, oxygen diffusion in pure $Ce_2Zr_2O_8$. Though this is not taken up by the authors, this clearly points to a significant role of the noble metal–oxide contact, due to the 1 wt% Pt present also in this system, in promoting the intrinsic oxygen storage and release properties of the $Ce_2Zr_2O_8$, a point that has been subsequently demonstrated and quantified using a combination of hard X-rays diffraction and dispersive EXAFS for related $Pd/CeZrO_4/Al_2O_3$ catalysts [72].

As such, the application of dispersive EXAFS in this case has provided a detailed structural–kinetic window into some very rapid processes occurring at high temperatures. A somewhat surprisingly complex interplay between the Ce and Zr components is revealed, one that no doubt lies at the very heart of the superior performance of this mixed oxide in auto exhaust gas applications than, for instance, CeO_2. As is so often the case, however, a definitive comprehension, by way of a specific mechanism, for the asynchronous behavior for the two elemental centers, is not forthcoming from these measurements. That said, it is difficult to see how any other *in-situ* technique could have provided the degree of insight that the dispersive EXAFS does.

3.4 FUTURE PERSPECTIVES AND POSSIBILITIES

We hope it can be seen that the dispersive technique has much to offer in the study of working catalysts; in certain quarters, what it may be able to offer is unique. Equally we have seen that this approach brings with it some challenges that need to be successfully overcome for this potential to be realized. Some of these challenges have been adequately met, while others are progressively (one would hope) being resolved.

From the perspective of the study of catalysts, we might point to a number of areas where dispersive can and will develop further. In some cases, these are well-known and exploited in other fields but have yet to be widely embraced by the catalysis community; in others, they will come from beamline-centered development and technical innovation. We might point to a temporal and spatially resolved study, but also the ability to deal with a more diverse range of difficult and dilute samples.

3.4.1 Difficult Samples, Dilute Samples, and Very High Time Resolution

As we have previously outlined, the conventional transmission-based dispersive EXAFS experiment is subject to potential issues arising from the intrinsic level of sample uniformity that can be achieved and the effects of parasitic scattering. Moreover, by being essentially limited to transmission, in terms of sample dilution, the traditional view is that dispersive experiment simply cannot go where scanning EXAFS may be able to, by virtue of fluorescence detection being possible in the latter case. That said, single-shot, subsecond, time-resolved, scanning fluorescence-yield EXAFS has yet to be demonstrated to our knowledge.

3.4.1.1 Modulations, Pumps, and Probes for Heterogeneous Catalysis A recent experiment [73], however, has opened the door to partial resolution of this issue in some senses, and at least in terms of how to acquire dispersive data in such difficult cases. This revolves around the use of a differential approach to data acquisition which is then augmented within the paradigm of reversible perturbations to a system and a so-called modulation de-excitation approach to data analysis [74]. This latter methodology has been known for a relatively long time within non-X-ray communities (principally infrared [74, 75]) but is only now starting to filter into the area of X-ray scattering and spectroscopy [73, 76–78]. Moreover, the reversibility in behavior required for this approach to be applied should be achievable in many catalytic situations—at least for any decent catalytic system, almost by definition.

Figure 3.12 shows the intriguing results of using this methodology to try and study the behavior of a 0.3 wt% Rh/Al_2O_3 sample during CO/NO cycling and at high time resolution (250 ms/spectrum–100 ms X-ray acquisition + 150 ms detector readout) [73]. As is clear from the raw absorption spectra shown in Figure 3.12A, and even with the protocols outlined in Figure 3.4 in place to try and combat issues due to scattering/nonuniformity, there is little that can be done with these spectra using a conventional approach to analysis: the precision with which we can match the reference to the sample, in terms of averaging over nonuniformity and matching the parasitic scattering from the support, is simply not good enough.

However, as is evident from the spectrum at the bottom of Figure 3.12A, by far the majority of the "structure" that obfuscates the underlying XANES and EXAFS is invariant and removable using a simple difference. The bottom panel shows what results from the further application of a demodulation algorithm for data averaged over only three modulations [73]. Both XANES and EXAFS details are now prized from the morass of scattering effects that encumbers them in the raw transmission data. This is sufficient to be able to show that in this 0.3wt% case, the Rh behaves in a manner that is, by and large, the same as found for a 2 wt% Rh/Al_2O_3 system [73], and we can start to probe the structural–reactive behavior of what would be an otherwise intractable system.

The fly in the ointment is that beyond an extremely sensitive fingerprinting of behavior and kinetic assessments, the quantitative structural detail contained within the differential EXAFS curves shown in Figure 3.12B is not simple to extract.

Differential EXAFS has been shown to be a very powerful way of addressing some areas of structural change [79–82]; it is fundamental to very high time-resolution pump–probe experiments, whose foundation also lies in a guaranteed reproducibility of events over many repetitions. However, and to this day, a quantitative interpretation of differential XAFS is only ever made on a system where one structural parameter (generally a bond distance) is varying as a result of the applied perturbation. To date, our understanding of how to comprehend differential EXAFS signals in cases where bond lengths, coordination numbers, atom types, or overall levels of disorder are all potentially changing is simply not yet sufficiently developed to fully exploit the potential of this approach to studying catalysts, wherein all these possibilities may coexist.

Nonetheless, this is an overall methodology that plays very well to one of the intrinsic strong points of the dispersive optical arrangement: its stability, especially in the energy scale, compared to a mechanically actuated system attempting to measure at these sorts of speeds. It is also an approach that, to a first approximation, essentially removes the sensitivity of the transmission experiment to parasitic scattering from the sample: nothing, not even the sample, need move at any time during such an experiment. As such, this sort of method could now be used in place of the recourse to fluorescence yield approach previously adopted in the case of the study of Pt catalysts by Toyota, for example [50–54, 56, 57]. See also Chapter 14.

As mentioned earlier, the use of a differential approach to XAS data is also central to pump–probe measurements for studying ultrafast processes (nanoseconds and below). Such fundamental experiments have been made using (step-)scanning EXAFS for some time [83–87] (though not, we might note, in the study of basic questions in heterogeneous catalysis).

It is only recently that dispersive EXAFS has been able to enter this particular fray (with nanosecond time-resolved reaction dynamics having been recently, for the first time, demonstrated for dispersive EXAFS [88]) as a result of the development of Si and Ge microstrip detectors [24–27], with their intrinsically low readout times (see earlier). The dispersive setup again offers much that is highly favorable for such measurements. Aside from stability, that one obtains the whole spectrum instantaneously, rather than having to step-scan through the energetic region of interest, should mean that the duty cycle of such experiments can be very significantly improved (see later).

It seems only a matter of time, therefore, before this sort of approach to studying more fundamental aspects of how functional materials such as catalysts behave, under photo- or thermal stimulation [89], is demonstrated using dispersive EXAFS.

3.4.1.2 Extreme Single-Shot Kinetics Pump–probe and demodulation experiments are of course generic and not the special preserve of any technique. Where dispersive EXAFS is, and for the foreseeable future seems likely to remain, unique is in the possibility it now offers for single-shot measurements on the approximately 10-μs timescales. Once again this potential comes directly out of the development of microstrip [24–27] detectors: though some CCD cameras such as the FReLoN [23] can grab frames in 100 μs, readout times, and therefore repetition rates, remain (currently) at approximately 1–2 ms. With these new "strip" detectors, therefore, comes the possibility to study, in a single shot, very fast processes. While for most potential systems it seems highly unlikely that sufficient statistics for an EXAFS analysis will arise from a single absorption measurement on a 10 μs timescale, if the measurements may be easily reproduced, then accessing of structure and kinetics in the same breath on the microsecond–millisecond timescale should be possible.

In terms of heterogeneous catalysis, reactor systems designed for temporal analysis of products (TAPs) [90] have been able to achieve comparable time resolutions for many years, and have been exploited on this basis for fundamental research into kinetics and mechanisms of heterogeneous process. Notionally, the only impediment that now exists to adding some form of structural input from dispersive EXAFS, to compliment the ubiquitous mass spectrometric assessment of gas phase product evolution, will be to design and implement a suitably X-ray-compatible reactor system.

3.4.2 Spatial Measurements and Tomography

As is shown in Chapter 2 of this book, it should be clear that, as with many techniques, absorption spectroscopy can be applied with high spatial resolution and used in a tomographic sense to study catalysts. Typically this is an area that has thus far been dominated by a scanning approach (be it full-field or point-by-point mapping); the only extant example of a (relatively simplistic use of) dispersive EXAFS in a rapid, spatially resolving sense, for a heterogeneous catalyst, was given as an example in Section 3.4 [62].

Dispersive offers a good deal of potential in these areas, even though the "full-field" option cannot be achieved with a dispersive X-ray beam. This potential is already evident from a number of studies in geophysics, where very small samples can now be very rapidly mapped with high spatial resolution and within difficult sample environments [30, 91, 92].

The ultimate focal spot dimensions that are achievable on a dispersive line are again very dependent on the nature of the line (e.g., source characteristics, bending magnet, or insertion device; wiggler or undulator). Postupgrade, one-half of ID24 should produce approximately a 2.5×2.5 μm focal spot (below 15 keV) and, as such, will be comparable with that achievable on conventional μ-XAS lines. As always, every dispersive measurement produces a complete XAS envelope, again making dispersive mapping/tomography potentially much faster than the scanning equivalent.

However, it does seem clear that the dispersive-specific convolution between the overall dimensions of the polychromatic beam and the inherent spatial character of catalyst samples, not to mention issues of sample scattering, will be highly problematic, certainly in the case of packed powder samples measured in transmission. Extrudates, and more structurally uniform catalysts bodies/carriers such as single-crystal zeolites, may, however, be more tractable, both in transmission and using the fluorescence yield variant (turbo-XAS) of the dispersive methodology.

3.4.3 Other Areas of Foreseeable or Ongoing Development

Even though our first case study in Section 3.4 revolved around a study of catalysis by vanadium, the first-row transition metals have been massively underrepresented within the general milieu of dispersive EXAFS as applied to catalysts. Recently, this overall trend has shown signs of changing mainly through the activity of the dispersive XAFS line at Laboratório Nacional de Luz Síncrotron (LNLS), Campinas, Brazil [39d–f, 93]. That said, while well suited in overall operational terms to this energy region, studies emanating from there have, thus far, been limited to XANES (though from our first example it is clear that this is often highly informative) and to relatively slow acquisition rates by modern standards.

Though ID24 and the ESRF spends a lot of its operational time in the sub-6–10 keV range, this is predominantly performed within the domains of high-pressure physics, geophysics, and magnetism (Fe K-edge). Some of the requirements of these disciplines, most notably the X-ray beam dimensions they prefer [30, 41, 92], are diametrically opposed to those required to study powders such as catalysts. As a result, the ID24 that existed up until the end of 2009 had an intrinsic horizontal focus of approximately 5 μm FWHM. As we have seen such a state of affairs that is in no way desirable for successful dispersive EXAFS on typical nonuniform, powder catalyst samples, ameliorating this restriction is one of the central objectives of the current ID24 upgrade that is due to be completed in 2012.

In more general terms, it should be noted that the degree of sensitivity to issues regarding sample scattering and uniformity will vary considerably: every synchrotron, and indeed every dispersive beamline, is different. We note that analyzable dispersive EXAFS in the 10–11 keV range (Pt L_3- and Ge K-edges, 1–10 s timescale; see earlier [38–40]) for low loaded heterogeneous catalysts was previously obtainable at the second-generation dispersive facility at Daresbury: a much larger, and less coherent source than the much more powerful (in photon flux terms) third-generation ESRF.

In many ways, ID24 at the ESRF probably represents the most difficult of (dispersive) situations that currently exists for research in these types of materials: the ESRF source is very small and coherent; the (three; four postupgrade) insertion devices are undulators, which further reinforce the small source/small spot/high coherence obtained; overall heatloads on the optics can be high (especially at low energies), with all the attendant potential for optical instability that this engenders; and the historical development of this beamline has been centered on achieving smaller and smaller possible focal spots [41, 92]. Moreover, whereas the newer third-generation synchrotrons are all built to sustain a "top-up" mode of operation, the ESRF has yet to adopt this possibility. It is difficult to see how this mode of operation will not improve overall levels of optical and X-ray stability, factors that are of the utmost importance to a dispersive experiment of any type, but especially those that involve materials such as catalysts.

As with the dispersive sources that existed at second-generation sources, other currently active dispersive resources that are based on bending magnet sources [94–97], and supplied by a "top-up" operational mode, should have a much easier time of things in these respects. In raw photon flux terms, they should, however, never compete with the potential power of the undulator-based ID24, or indeed wiggler (I20, Diamond Light Source; [97]) solution, and are more likely to be limited in other aspects of performance.

Indeed, a considerable part of the current upgrade of ID24 at the ESRF is based on the recognition that one focal spot size does not fit all the science that one might like to place on a dispersive beamline. One-half of the upgraded ID24 will be predicated on reversing the 10-year-old trend toward exceptionally small focal spots, using either upgraded versions of a Laue polychromator system [98] and/or asymmetric Bragg polychromators. The aim is to retroengineer the focal spot size (below 15 keV) such that it is far more amenable to studying samples such as heterogeneous catalysts, and to achieve this while retaining, controlling, and properly applying the full power of the undulator-based beamline configuration. Ironically this is, in many ways, more technically challenging at a source such as the ESRF than pushing for the smallest focal spot possible (which will be the role of the other half of what will be a new twin end-station resource).

Alongside the upgrade of ID24, a potentially very interesting development will be the completion and implementation of the wiggler-based dispersive branch of I20 at the Diamond Light source in the United Kingdom [97]: the larger source and lower coherence of the wiggler compared to the undulator should make this a very attractive proposition in terms of heterogeneous catalysts, as long as the extremely high heat loads that come with the wiggler can be successfully handled and the increased (relative to a bending magnet) photon throughput can be effectively applied.

3.5 CONCLUSIONS

Energy-dispersive EXAFS both complements and augments the overall palette of X-ray spectroscopic solutions for the study of heterogeneous catalysts that scanning EXAFS supplies. Side by side with the energy-scanning XAFS variant, these two conceptually very different methodologies yield a very powerful toolbox for studying many aspects of catalysis, especially when allied with other techniques and within single experiments.

The principal strengths of the dispersive approach, as regards the study of catalysts, lies in the intrinsic stability of the optical arrangement and the potential for very high single-shot time resolution (nowadays in the microsecond regime with "strip" detectors). Set against this is a susceptibility to deleterious interference from sample homogeneity issues and unwanted scattering due to the composition and texture of the samples. The degree to which these issues influence an experiment will be beamline dependent, as they are convolved with the nature of the X-ray source and numerous other beamline-specific parameters, as well as with the nature of the samples themselves.

The experience of the authors suggests that these effects rarely preclude the obtention of XANES data provided the sample presentation is adequate. EXAFS, however, being intrinsically more difficult to obtain, is another story, and to this day remains the most significant challenge in any experiment, be it using dispersive or quick scanning methods. However, we hope we have shown that when one gets things right a great deal can be achieved.

It is clear that dispersive EXAFS offers new possibilities for the study of very fast chemical processes. It offers an alternative to the traditional, scanning methods of making very fast pump–probe-type measurements. While at the present time these have not found their way into the pantheon of X-ray-based measurements used to understand heterogeneous systems, it seems only a matter of time before they do. Most uniquely dispersive EXAFS does now yield the possibility of studying systems in single shot in the microsecond regime through the advent of Si and Ge strip detectors.

As a final note, we would state that no matter what X-ray technique one is using, or indeed how brilliant or powerful the X-ray source is, what one may get from a catalysis experiment using X-rays is very much dependent upon the experiment that is placed in front of the X-ray beam and how it is parameterized. The continued development and fullest exploitation of what are, in their third generation, very powerful resources capable of bringing much to the understanding of heterogeneous catalysts and the processes they mediate, will therefore require ever increasing level of commitment and collaboration from both sides (synchrotron user and synchrotron scientist) to what is often a complex equation.

ACKNOWLEDGMENTS

The authors would like to thank all those who have, over the years, worked with them in applying dispersive EXAFS to problems in catalysis at SRS Daresbury and the ESRF. In particular, we would like to acknowledge the central and long-term vision of, and role played by, Prof. John Evans (University of Southampton) to the development and exploitation of dispersive EXAFS in this field.

In addition, M.A.N. would like to thank the following people: Sakura Pascarelli, Olivier Mathon, Florian Perrin, Trevor Mairs, Gemma Guilera (now at Alba, Barcelona), Anna Kroner (now at Diamond Light Source, Oxford), and all those user groups who have contributed, and continue to contribute, to the development of dispersive EXAFS as applied to heterogeneous catalysts on ID24. Toyota, whose 5-year collaboration with ID24 at the ESRF contributed much to the development of the beamline, is especially acknowledged in this respect. Such long-term commitments are central to the development of both research in this area and indeed the performance of the X-ray facility itself. Prof. Peter Lee (APS, Chicago) and Dr. Simon Bare (UOP) are also gratefully acknowledged for providing additional information regarding their dispersive EXAFS studies at the V K-edge [36]. Prof. Jan-Dierk Grunwaldt is thanked for the image of the catalyst bed used in Figure 3.3B.

A.J.D. would like to thank the detector groups at Daresbury (Gareth Derbyshire, Richard Farrow, Jon Headspith, Joe Salvini, Janet Groves, and William Helsby) who were crucial to early developments and continue to develop strip detectors for the new sources.

REFERENCES

[1] Sayers, D. E., Stern, E. A., Lytle, F. W. (1991) New technique for investigating non crystalline structures: Fourier analysis of the Extended X-ray absorption fine structure, *Phys. Rev. Lett.*, 27, 1204–1207.

[2] Johann, H. H. (1931) Die Erzeugung lichtstarker Röntgenspektren mit Hilfe von Konkavkristallen, *Zeit. Phys.*, 69, 183–206.

[3] Cauchois, Y. (1932) Une nouvelle method d'analyse les poudres crystalline par les rayons X, utilisant un monochromateur a cristal courbé, *C. R. Hebd. Seances Acad. Sci.*, 195, 228–231.

[4] Cauchois, Y. (1932) Spectrographie des rayons X par transmission d'un faiseau non canalisé à travers un cristal courbé, *C. R. Hebd. Seances Acad. Sci.*, 194, 1479–1482.

[5] Cauchois, Y. (1932) Spectrographie des rayons X par transmission d'un faiseau non canalisé à travers un cristal courbé, *J. Phys. Radium*, 3, 360–336; Cauchois, Y. (1933) Spectrographie des rayons X par transmission d'un faiseau non canalisé à travers un cristal courbé, *J. Phys. Radium*, 4, 61–72.

[6] Kaminaga, U., Matsushita, T., Kohra, K. (1981) A dispersive method of measuring extended X-ray absorption fine structure, *Jpn. J. Appl. Phys.*, 20, L355–L358.

[7] Matsushita, T., Phizackerley, R. P. (1981) A fast X-ray absorption spectrometer for use with synchrotron radiation, *Jpn. J. Appl. Phys.*, 20, 2223–2228.

[8] Phizackerley, R. P., Rek, Z. U., Stephenson, G. B., Conradson, S. D., Hodgson, K. O., Matsushita, T., Oyanagi, H. (1983) An energy-dispersive spectrometer for the rapid measurement of X-ray absorption spectra using synchrotron radiation, *J. Appl. Cryst.*, 16, 220–232.

[9] Derbyshire, G., Dobson, B., Greave, G. N., Harris, N., Mackle, P., Moore, P. R., Roberts, K. J., Allinson, N., Nicol, J., Doyle, S., Oldman, R. J. (1989) Integrated materials science facility on the SRS, *Rev. Sci. Intstrum.*, 60, 1897–1900.

[10] Flank, A. M., Fontaine, A., Jucha, A., Lemonnier, M., Raoux, D., Williams, C. (1983) EXAFS in dispersive mode, *Nucl. Instrum. Methods*, 208, 651–654.

[11] Hagelstein, M., Cunis, S., Frahm, R., Niemann, W., Rabe, P. (1989) The energy dispersive X-ray absorption spectrometer at Hasylab. *Physica B*, 158, 324–325.

[12] Lee, P. L., Beno, M. A., Jennings, G., Ramanathan, M., Knapp, G. S., Huang, K., Bai, J., Montano, P. A. (1994) An energy dispersive X-ray absorption spectroscopy beamline, X6A, at NSLS, *Rev. Sci. Instrum.*, 65, 1–6.

[13] Couves, J. W., Thomas, J. M., Catlow, C. R. A., Greaves, G. N., Baker, G., Dent, A. J. (1990) In situ studies of zeolitic catalysts by time resolved energy dispersive X-ray absorption spectroscopy, *J. Phys. Chem.*, 94, 6517–6519.

[14] Couves, J. W., Thomas, J. M., Waller, D., Jones, R. H., Dent, A. J., Derbyshire, G. E., Greaves, G. N. (1991) Tracing the conversion of Aurichalcite to a copper catalyst by combined X-ray absorption and diffraction, *Nature*, 354, 465–468.

[15] Keegan, M. B. T., Dent, A. J., Blake, A. B., Conyers, L., Moyes, R. B., Wells, P. B., Whan, D. A. (1991) An in situ energy dispersive EXAFS study of the calcinations and reduction of a PtZSM5 catalyst precursor, *Catal. Today*, 9, 183–188.

[16] Dent, A. J., Wells, M. P., Farrow, R. C., Ramsdale, C. A., Derbyshire, G. E., Greaves, G. N. (1992) Combined energy dispersive EXAFS and X-ray diffraction, *Rev. Sci. Instrum.*, 63, 903–906.

[17] Newton, M. A., Dent, A. J., Evans, J. (2002) Bringing time resolution to EXAFS: recent developments and application to chemical systems, *Chem. Soc. Rev.*, 31, 83–95.

[18] Allinson, N. M., Baker, G., Greaves, G. N., Nicoll, J. K. (1988) PDA system for Energy dispersive EXAFS, *Nucl. Instrum. Methods A*, 266, 592–597.

[19] Bogg, D., Dent, A. J., Derbyshire, G. E., Farrow, R. C., Ramsdale, C. A., Salvini, G. (1997) Characterisation of Reticon and Hamamatsu Photodiode array and the subsequent development of high performance VME-based detector systems optimised for Energy Dispersive EXAFS, *Nucl. Instrum. Methods A*, 392, 461–464.

[20] Newton, M. A., Fiddy, S. G., Guilera, G., Jyoti, B., Evans, J. (2005) Oxidation/reduction kinetics of supported Rh/Rh_2O_3 nanoparticles in plug flow conditions using dispersive EXAFS, *Chem. Commun.*, 118–119.

[21] Labiche, J.-C., Mathon, O., Pascarelli, S., Newton, M. A., Ferre, G. G., Curfs, C., Vaughan, G., Homs, A., Carrieras, D. F. (2007) The fast readout low noise camera as a versatile x-ray detector for time resolved dispersive extended x-ray absorption fine structure and diffraction studies of dynamic problems in materials science, chemistry, and catalysis, *Rev. Sci. Instrum.*, 78, 091301.

[22] Iles, G., Dent, A., Derbyshire, G., Farrow, R., Hall, G., Noyes, G., Raymond, M., Salvini, G., Seller, P., Smith, M., Thomas, S. (2000) A novel application of silicon microstrip technology for energy dispersive EXAFS studies, *J. Synchrotron Radiat.*, 7, 221–228.

[23] Headspith, J., Salvini, G., Thomas, S. L., Derbyshire, G. E., Dent, A., Rayment, T., Farrow, R., Anderson, C., Cliché, J., Doson, B. R. (2003) XSTRIP—a silicon micro strip-based X-ray detector for ultra-fast X-ray spectroscopy studies, *Nucl. Instrum. Methods A*, 512, 239–244.

[24] Salvini, G., Headspith, J., Thomas, S. L., Derbyshire, G. E., Dent, A., Rayment, T., Evans, J., Farrow, R., Diaz-Moreno, S., Ponchut, C. (2005) Detectors for energy-dispersive EXAFS (EDE) experiments, *Nucl. Instrum. Methods A*, 551, 27–34.

[25] Headspith, J., Groves, J., Luke, P. N., Kogirntzis, M., Salvini, G., Thomas, S. L., Farrow, R. C., Evans, J., Rayment, T., Lee, J. S., Goward, W. D., Amman, A., Mathon, O., Diaz-Moreno, S. (2007) First experimental data from XH, a fine pitch germanium microstrip detector for energy dispersive EXAFS (EDE), *IEEE Nucl. Sci. Symp. Conf. Record.*, 1–11, 241–2428.

[26] Groust, J. F., Pommier, C., Stievano, L., Villain, F., Giorgetti, C., Baudelet, F. & Massiani, P. (2005) Real time monitoring of the evolution of Ni2+ environment in faujasite upon rehydration by in situ dispersive-EXAFS, *Catal. Lett.*, 102, 257–260.

[27] Mairs, T. R., Mathon, O. (2010) Beam stability: benefits from concentrating on basics, 10th International Conference on Synchrotron Radiation Instrumentation, 2009, Australian Synchrotron, Melbourne, AUSTRALIA, *AIP Conference Proceedings*, 1234, 271–274.

[28] Aquilanti, G., Pascarelli, S., Mathon, O., Munõz, M., Naryginac, O., Dubrovinsky, L. (2009) Development of micro-XANES mapping in the diamond anvil cell, *J. Synchrotron Radiat.*, 16, 376–379.

[29] Grunwaldt, J. D., Hannemann, S., Schroer, C. G., Baiker, A. (2006) 2D-Mapping of the catalyst structure inside a catalytic microreactor at work: partial oxidation of methane over Rh/Al_2O_3, *J. Phys. Chem. B*, 110, 8674–8680.

[30] Hagelstein, M., Lienert, U., Ressler, T., San Miguel, A., Freund, A., Cunis, S., Schulze, C., Fontaine, A., Hodeau, J. L. (1998) A scattering filter for energy-dispersive optics, *J. Synchrotron Radiat.*, 5, 753–755.

[31] Bonse, U., Hart, M. (1965) An X-ray interferometer, *Appl. Phys. Lett.*, 7(9)238–240.

[32] Newton, M. A. (2007) Beamsize related phenomena and effective normalisation in energy dispersive EXAFS for the study of heterogeneous catalysts, powder materials, and the processes they mediate: observation and (some) solutions, *J. Synchrotron Radiat.*, 14, 372–381.

[33] Fernández-García, M., Martínez-Arias, A., Iglesias-Juez, A., Belver, C., Hungria, A. B., Conesa, J. C., Soria, J. (2000) Structural Characteristics and redox behaviour of CeO_2-ZrO_2/Al_2O_3 supports, *J. Catal.*, 194, 385–392.

[34] Newton, M. A. (2009) Synchronous Diffuse Reflectance infrared absorption spectroscopy (DRIFTS)/EXAFS to the structural reactive behaviour of dilute supported Rh/Al_2O_3 catalysts, *Top. Catal.*, 52, 1410–1424.

[35] Newton, M. A., Belver-Coldeira, C., Martínez-Arias, A., Fernández-García, M. (2007) Dynamic *in situ* observation of rapid size and shape change of supported Pd nanoparticles during CO/NO cycling, *Nat. Mater.*, 6, 528–532.

[36] Coulston, G. W., Bare, S. R., Kung, H., Birkeland, K., Bethke, G. K., Harlow, R., Herron, N., Lee, P. (1997) The kinetic significance of V5+ in n-Butane oxidation catalyzed by vanadium phosphates, *Science*, 275, 191–193.

[37] (a) Lamberti C., Prestipino C., Bonino F., Capello L., Bordiga S., Spoto G., Zecchina A., Moreno S. D., Cremaschi B, Garilli M, Marsella A, Carmello D, Vidotto S, (2002) Leofanti G, (2002), The chemistry of the oxychlorination catalyst: an in situ, time-resolved XANES study, *Angew. Chem. Int. Ed.*, 41, 2341–2344; (b) Lamberti C., Bordiga S., Bonino F., Prestipino C., Berlier G., Capello L., D'Acapito F., Llabrés i Xamena F. X. , Zecchina A., (2003) Determination of the oxidation and coordination state of copper on different Cu-based catalysts by XANES spectroscopy *in situ* or in *operando* conditions. *Phys. Chem. Chem. Phys.*, 5, 4502–4509; (c) Muddada N. B., Olsbye U., Caccialupi L., Cavani F., Leofanti G., Gianolio D., Bordiga S., Lamberti C., (2010), Influence of additives in defining the active phase of the ethylene oxychlorination catalyst, *Phys. Chem. Chem. Phys.*, 12, 5605–5618; (d) Alexopoulos K., Anilkumar M., Reyniers M. F., Poelman H., Cristol S., Balcaen V., Heynderickx P. M., Poelman D., Marin G. B., (2010). Time-resolved operando X-ray absorption study of $CuO\text{-}CeO_2/Al_2O_3$ catalyst during total oxidation of propane, *Appl. Catal. B*, 97, 381–388; (e) Profeti L. P. R., Ticianelli E. A., Assaf E. M., (2009), Production of hydrogen via steam reforming of biofuels on $Ni/CeO_2\text{–}Al_2O_3$ catalysts promoted by noble metals, *Int. J. Hydrogen Energy*, 34, 5049–5060; (f) Profeti L. P. R., Ticianelli E.A., Assaf E. M., (2009), Ethanol steam reforming for production of hydrogen on magnesium aluminate-supported cobalt catalysts promoted by noble metals, *Appl. Catal. A*, 360, 17–25; (g) Parizotto, N. V., Zanchet, D., Rocha, K. O., Marques, C. M. P., Bueno, J. M. C., (2009), The effects of Pt promotion on the oxi-reduction properties of alumina supported Ni catalysts for oxidative steam-reforming of methane: temperature resolved XAFS analysis, *Appl. Catal. A*, 366, 122–129; (h) Teixeira da Silva, V., Sousa, L. A., Amorim, R. M. Andrini, L., Figueroa, S. J. A., Requejo, F. G., Vicentini, F C., (2011), Lowering the synthesis temperature of Ni_2P/SiO_2 by Palladium addition, *J. Catal.*, 279, 88–102; (i) Figueroa, S. J. A., Stewart, S. J., Cano, L. A., Bengoa, J. F., Cagnoli, M. V., Marchetti, S. G. (2012) Fischer–Tropsch synthesis: optimization of the activation treatment with H_2 in a Fe/SBA-15 catalyst, in preparation.

[38] Fernandez-Garcia, M., Alvarez, C. M., Haller, G. L. (1995) XANES-TPR study of Cu-Pd bimetallic catalysts—application of factor analysis, *J. Phys. Chem.*, 99, 12565–12569.

[39] Ressler, T., Hagelstein, M., Hatje, U., Metz, W. (1997) In situ X-ray absorption spectroscopy studies on chemical oscillations in the CO/O_2 system on supported Pd catalysts, *J. Phys. Chem. B*, 101, 6680–6687.

[40] Fiddy, S. G., Newton, M. A., Dent, A. J., Salvini, G., Corker, J. M., Turin, S., Campbell, T., Evans, J. (1999) In situ energy dispersive EXAFS (EDE) of low loaded $Pt(acac)_2/H_I\ SiO_2$ catalyst precursors on a timescale of seconds and below, *Chem. Commun.*, 815–852.

[41] Fiddy, S. G., Newton, M. A., Dent, A. J., Salvini, G., Harvey, I., Turin, S., Campbell, T., Evans, J. (2001) *In situ*, time resolved, and simultaneous multi-edge determination of local order change during reduction of supported bimetallic (Pt–Ge) catalyst precursors using energy dispersive EXAFS, *Chem. Commun.*, 445–446.

[42] Fiddy, S. G., Newton, M. A., Dent, A. J., Salvini, G., Corker, J. M., Turin, S., Campbell, T., Evans, J. (2002) Particle development and characterisation in $Pt(acac)_2$ and $Pt(acac)_2/GeBu_4$ derived catalysts supported upon porous and mesoporous SiO_2 : effect of reductive environment, and support structure, *Phys. Chem. Chem. Phys.*, 4, 827–834.

[43] Grunwaldt, J. D., Caravati, M., Hannemann, S., Baiker, A. (2004) X-ray absorption spectroscopy under reaction conditions: suitability of different reaction cells for combined catalyst characterization and time-resolved studies, *Phys. Chem. Chem. Phys.*, 6, 3037–3047.

[44] Newton, M. A. (2008) Dynamic adsorbate/reaction induced structural change of supported metal nanoparticles: heterogeneous catalysis and beyond, *Chem. Soc. Rev.*, 37, 2644–2657.

[45] Nagai, Y., Dohmae, K., Ikeda, Y., Takagi, N., Tanabe, T., Hara, N., Guilera, G., Pascarelli, S., Newton, M. A., Kuno, O., Jiang, H., Matsumoto, H. S. (2008) In situ redispersion of Pt based, autoexhaust catalysts; towards and on line approach to increasing catalyst lifetimes? *Angew. Chem. Int. Ed.*, 48, 9303–9306.

[46] Nagai, Y., Dohmae, K., Teramura, K., Tanaka, T., Guilera, G., Kato, K., Nomura, M., Shijoh, H., Matsumoto, S. (2009) Dynamic in situ observation of automotive catalysts for emission control using X-ray absorption fine structure, *Catal. Today*, 145, 279–287.

[47] Guilera, G., Gorges, B., Pascarelli, S., Vitoux, H., Newton, M. A., Prestipino, C., Nagai, Y., Hara, N. (2009) Novel high-temperature reactors for in situ studies of three-way catalysts using turbo-XAS, *J. Synchrotron Radiat.*, 16, 628–634.

[48] Nagai, Y., Tajagi, N., Ikeda, Y., Dohmae, K., Tanabe, T., Guiera, G., Pascarelli, S., Newton, M., Shinjoh, H., Matsumoto, S. (2007) Real-time observation of platinum redispersion on ceria-based oxide by in-situ Turbo-XAS in fluorescence mode, *AIP Conf. Proc.*, 882, 594–596.

[49] Tanaka, H., Uenishi, M., Taniguchi, M., Tan, I., Narita, K., Kimura, M., Kaneko, K., Nishihata, Y., Mizuki, J. (2006) Intelligent catalyst having the self-regenerative function of Pd, Rh and Pt for automotive emissions control, *Catal. Today*, 117, 321–328.

[50] Uenishi, M., Tanaka, H., Taniguchi, M., Tan, I., Nishihata, Y., Mizuki, J., Kobayashi, T. (2008) Time evolution of palladium structure change with redox fluctuations in a $LaFePdO_3$ perovskite automotive catalyst by high-speed analysis with in situ DXAFS, *Catal. Commun.*, 9, 311–314.

[51] Matsumara, D., Nishihata, Y., Mizuki, J., Taniguchi, M., Uenishi, M., Tanaka, H. (2010) Dynamic structural change in Pd-perovskite automotive catalyst studied by time-resolved dispersive X-ray absorption fine structure, *J. Appl. Phys.*, 107, 124319.

[52] Pascarelli, S., Mathon, O. (2010) Advances in high brilliance energy dispersive X-ray absorption spectroscopy, *Phys. Chem. Chem. Phys.*, 12, 5535–5546.

[53] Yang, A. C., Garland, C. W. (1957) Infrared studies of carbon monoxide chemisorbed on Rhodium, *J. Phys. Chem.*, 61, 1504–1514.

[54] Van't Blik, H. F. J., van Zon, J. B. A. D., Huizinga, T., Vis, J. C., Koningsberger, D. C. (1985) Structure of Rhodium in an ultradispersed Rh/Al_2O_3 Catalyst, as Studied by EXAFS and Other Techniques, *J. Am. Chem. Soc.*, 109, 3139–3147.

[55] Van't Blik, H. F. J., van Zon, J. B. A. D., Huizinga, T., Vis, J. C., Koningsberger, D. C., Prins, R. (1983) An EXAFS study of highly dispersed Rh/Al_2O_3 catalyst- the influence of CO chemisorption on the topology of Rhodium, *J. Phys. Chem.*, 87, 2254–2267.

[56] Arai, H., Tominaga, H. (1976) Infrared study of nitric-oxide adsorbed on Rhodium-Alumina catalyst, *J. Catal.*, 43, 131–142.

[57] Martens, J. H. A., Prins, R., Zandbergen, H., Koningsberger, D. C. (1988) Structure of Rh/TiO_2 in the normal and SMSI state as determined by EXAFS and HRTEM, *J. Phys. Chem.*, 92, 1903–1916.

[58] Suzuki, A., Inada, Y., Yamaguchi, A., Chihara, T., Yuasa, M., Nomura, M., Iwasawa, Y. (2003) Time scale and elementary steps of CO-Induced disintegration of surface Rhodium Clusters, *Angew. Chem. Int. Ed.*, 42, 4795–4799.

[59] Newton, M. A., Dent, A. J., Diaz-Moreno, S., Fiddy, S. G., Jyoti, B., Evans, J. (2006) Rapid monitoring of the nature and interconversion of supported catalyst phases and of their Influence upon performance: CO oxidation to CO_2 by γ-Al_2O_3 Supported Rh Catalysts, *Chem. Eur. J.*, 12, 1975–1985.

[60] Newton, M. A., Burnaby, D. G., Dent, A. J., Diaz-Moreno, S., Evans, J., Fiddy, S. G., Neisius, T., Pascarelli, S., Turin, S. (2001) Simultaneous determination of structural and kinetic parameters in heterogeneous systems, *J. Phys. Chem. A*, 105, 5965–5970.

[61] Newton, M. A., Burnaby, D. G., Dent, A. J., Diaz-Moreno, S., Evans, J., Fiddy, S. G., Neisius, T., Turin, S. (2002) Dispersive EXAFS (EDE), mass spectrometric and diffuse reflectance infrared studies of the interaction of Al_2O_3 supported $RhI(CO)_2Cl$ species with NO and reformation under CO, *J. Phys. Chem. B*, 106, 4214–4222.

[62] Cavers, M., Davidson, J. M., Harkness, I. R., Rees, L. V. C., McDougall, G. S. (1999) Spectroscopic identification of the active site for CO oxidation on Rh/Al_2O_3 by concentration modulation in situ DRIFTS, *J. Catal.*, 188, 426–430.

[63] Pascarelli, S., Neisius, T., De Panfilis, S. (1999) Turbo-XAS: dispersive XAS using sequential acquisition, *J. Synchrotron Radiat.*, 6, 1044–1050.

[64] Center for X-ray optics, Lawrence Berkeley Laboratories, USA (http://www-cxro.lbl.gov, last accessed December 7, 2012).

[65] Ressler, T., Timpe, O., Neisius, T., Find, J., Mestl, G., Dieterle, M., Schlogl, R. (2000) Time-resolved XAS investigation of the reduction/oxidation of MoO_{3-x}, *J. Catal.*, 191, 75–85.

[66] Ressler, T., Wienhold, J., Jentoft, R. E., Timpe, O., Neisius, T. (2001) State kinetics of the oxidation of MoO_2 investigated by time-resolved X-ray absorption spectroscopy, *Solid State Commun.*, 19, 169–174.

[67] Ressler, T., Wienhold, J., Jentoft, R. E., Neisius, T. (2002) Bulk structural investigation of the reduction of MoO_3 with propene and the oxidation of MoO_2 with oxygen, *J. Catal.*, 210, 67–83.

[68] Ressler, T., Wienhold, J., Jentoft, R. E., Neisius, T., Gunther, M. M. (2002) Kinetics of Solid state reactions in heterogeneous catalysts from time resolved X-ray absorption spectroscopy, *Top. Catal.*, 18, 45–52.

[69] Ressler, T. (2009) Solid state kinetics and catalytic behaviour of selective oxidation catalysts from time-resolved EXAFS investigations, *Catal. Today*, 145, 258–266.

[70] Beale, A. M., van der Eeden, A. M. J., Kervinen, K., Newton, M. A., Weckhuysen, B. M. W. (2005) Adding a third dimension to operand spectroscopy: a combined UV-Vis, Raman, and XAFS setup to study heterogeneous catalysts under working conditions, *Chem. Commun.*, 3015–3016.

[71] Yamamoto, T., Suzuki, A., Nagai, Y., Tanabe, T., Dong, F., Inada, Y., Nomura, M., Tada, M., Iwasawa, Y. (2007) Origin and dynamics of oxygen storage/release in a Pt/ordered CeO_2-ZrO_2 catalyst studied by time resolved XAFS analysis, *Angew. Chem. Int. Ed.*, 46, 9253–9256.

[72] Newton, M. A., Di Michiel, M., Kubacka, A., Iglesias-Juez, A., Fernández-García, M. (2012) Seeing into oxygen storage and release and the synergies between noble metal nanoparticles and promoter oxides as they function, *Angew. Chem. Int. Ed.*, 51, 2363–2367.

[73] Ferri, D., Kumar, M. S., Eyssler, A., Wirz, R., Korsak, O., Hug, P., Weidenkaff, A., Newton, M. A. (2010) First steps in combining concentration modulation techniques with synchronous dispersive EXAFS/DRIFTS/mass spectrometry for in situ time resolved study of heterogeneous catalysts, *Phys. Chem. Chem. Phys.*, 12, 5634–5646.

[74] Baurecht, D., Fringeli, U. P. (2001) Quantitative modulated excitation Fourier transform infrared spectroscopy, *Rev. Sci. Instrum.*, 72, 3782–3792.

[75] Urakawa, A., Bürgi, T., Baiker, A. (2008) Sensitivity enhancement and dynamic behavior analysis by modulation excitation spectroscopy: Principle and application in heterogeneous catalysis, *Chem. Eng. Sci.*, 63, 4902–4909.

[76] Urakawa, A., Van Beek, W., Monrabal-Capilla, M., Galan-Mascaros, J. R., Palin, L., Milanesio, M. (2011) Combined modulation enhanced X-ray powder diffraction and Raman spectroscopic study of structural transitions in the spin crossover material [Fe($Htrz_2$(trz)(BF_4)], *J. Phys. Chem. C*, 115, 1323–1329.

[77] Chernyshov, D., Beek, W., Emerich, H., Milanesio, M., Urakawa, A., Viterbo, D., Palin, L., Caliandro, R. (2011) Kinematic diffraction on a structure with periodically varying scattering function, *Acta Crystallogr. A*, 67, 327–335.

[78] Ferri, D., Newton, M. A., Nachtegaal, M. (2011) Modulation excitation X-ray absorption spectroscopy to probe surface species on heterogeneous catalysts, *Top. Catal.*, 54, 1070–1078.

[79] Pettifer, R. F., Mathon, O., Pascarelli, S., Cooke, M. D., Gibbs, M. R. J. (2005) Measurement of femtometre-scale atomic displacements by X-ray absorption spectroscopy, *Nature*, 435, 78–81.

[80] Ruffoni, M. P., Pettifer, R. F., Pascarelli, S., Trapananti, A., Mathon, O. (2007) Probing atomic displacements with thermal differential EXAFS, *J. Synchrotron Radiat.*, 14, 421–425.

[81] Pascarelli, S., Ruffoni, M. P., Trapananti, A., Mathon, O., Aquilanti, G., Ostanin, S., Staunton, J. B., Pettifer, R. F. (2007) Effect of pressure on magnetoelastic coupling in 3d metal alloys studied with x-ray absorption spectroscopy, *Phys. Rev. Lett*, 99, 237204.

[82] Ruffoni, M. P., Pascarelli, S., Groessinger, R., Turtelli, R. S., Bormio-Nunes, C., Pettifer, R. F. (2008) Direct measurement of intrinsic atomic scale magnetostriction, *Phys. Rev. Lett*, 101, 147202.

[83] Bressler, C., Chergui, M. (2004) Utrafst X-ray absorption spectroscopy, *Chem. Rev.*, 104, 1781–1812.

[84] Chen, L. X. (2005) Probing transient molecular structures in photochemical processes using laser-initiated time-resolved X-ray absorption spectroscopy, *Ann. Rev. Phys. Chem.*, 56, 221–254.

[85] Mills, D. M., Lewis, A., Harootunian, A., Huang, J. B. (1984) Time resolved X-ray absorption spectroscopy of carbon-monoxide myoglobin recombination after laser photolysis, *Science*, 223, 811–813.

[86] Thiel, D. J., Livins, P., Stern, E. A., Lewis, A. (1993) Microsecond-resolved XAFS of the triplet excited-state of $Pt_2(P_2O_5H_2)_4$4-, *Nature*, 362, 40–43.

[87] Chen, L. X., Jäger, W. J. H., Jennings, G., Gosztola, D. J., Munkhilm, A., Hessler, J. P. (2001) Capturing a photoexcited molecular structure through time-domain X-ray absorption fine structure, *Science*, 292, 262–264.

[88] Evans, J., Dent, A. J., George, M., Tromp, M., Pascarelli, S., Mathon, O., George, M. J. To be published.

[89] Thibault-Starzyk, F., Seguin, E., Thomas, S., Daturi, M., Arnolds, H., King, D. A. (2009) Real-time infrared detection of cyanide flip on Silver-Alumina NO_x removal catalyst, *Science*, 324, 1048–1051.

[90] Gleaves, J. T., Ebner, J. R., Kuechler, T. C. (1988) Temporal analysis of products (TAP)—a unique catalyst evaluation system with sub millisecond time resolution, *Catal. Rev.-Sci. Eng.*, 30, 49–116.

[91] Pascarelli, S., Mathon, O., Munoz, M., Mairs, T., Susini, J. (2006) Energy-dispersive absorption spectroscopy for hard-X-ray micro-XAS applications, *J. Synchrotron Radiat.*, 13, 351–358.

[92] Pascarelli, S., Mathon, O. (2010) Advances in high brilliance energy dispersive X-ray absorption spectroscopy, *Phys. Chem. Chem. Phys.*, 12, 5535–5546.

[93] Tolentino, H. C. N., Cezar, J. C., Watanabe, N., Piamonteze, C., Souza-Neto, N. M., Tamura, E., Ramos, A., Neueschwander, R. (2005) The dispersive X-ray absorption spectroscopy beamline at LNLS, *Phys. Scr.*, T115, 977–979.

[94] Okajima, Y., Matsumara, D., Nishihata, Y., Konishi, H., Mizuki, J. (2007) Energy dispersive XAFS in the high energy region at BL14B1 in SPring-8, *AIP Conf. Proc.*, 879, 1234–1237.

[95] Bhattacharyya, D., Poswal, A. K., Jha, S. N., Sabharwal, S. C. (2009) First results from a dispersive EXAFS beamline developed at INDUS2-2 synchrotron source at RRCAT, Indore, India, *Nucl. Instrum. Methods A*, 609, 286–293.

[96] Baudelet, F., Kong, Q., Nataf, L., Cafun, J. D., Congeduti, A., Monza, A., Chagnot, S., Itie, J. (2011) ODE: a new beam line for high-pressure XAS and XMCD studies at SOLEIL, *High Press. Res.*, 31, 136–139.

[97] Diamond Light Source - Beamlines - I20 - LOLA: X-ray spectroscopy - Dispersive branch (http://www.diamond.ac.uk/Home/Beamlines/I20/dispersive.html, last accessed December 7, 2012).

[98] San-Miguel, A., Hagelstein, M., Borrel, J., Marot, G., Renier, M. (1998) An exchangeable Bragg/Laue polychromator for energy-dispersive EXAFS, *J. Synchrotron Radiat.*, 5, 1396–1397.

4

IN-SITU POWDER X-RAY DIFFRACTION IN HETEROGENEOUS CATALYSIS

JONATHAN HANSON AND POUL NORBY

4.1 INTRODUCTION

Powder X-ray diffraction (XRD) is a useful tool to characterize catalytic materials [1]. First of all, the technique discriminates between amorphous and crystalline samples. For crystalline materials, it allows for the determination of the component phases. Furthermore, it allows for the determination of particle size and other characteristics of the component crystallites. In most cases, it also allows for the determination of the atomic arrangement of the atoms in the crystallographic unit cell.

Heterogeneous catalysis takes place on the surface of the catalyst, which can be a bulk solid or, more frequently, nanoparticles of a catalytically active material dispersed on a solid support. When not dealing with nanoparticles, XRD primarily gives the bulk properties of the material. If one determines the dominant crystalline faces for the material, one can use this information together with crystal structure to derive the surface structure [1]. In the case of nanoparticles, the information provided by XRD serves as a good starting point for deriving the catalyst surface structure. Grazing incidence small-angle scattering together with other surface-sensitive techniques such as *in-situ* scanning tunneling microscopy (Chapter 7) and infrared spectroscopy (Chapter 9) can further elucidate the catalyst surface structure.

In-situ *Characterization of Heterogeneous Catalysts*, First Edition.
Edited by José A. Rodriguez, Jonathan C. Hanson, and Peter J. Chupas.

The next section will provide a brief introduction to XRD. This will be followed by a section on the techniques used for *in-situ* diffraction. Finally, there will be a section with specific applications of the *in-situ* technique to the characterization of active catalysts.

4.2 CRYSTAL STRUCTURE AND POWDER DIFFRACTION

The fundamental basis of powder XRD techniques is described in several books (e.g., Jenkins and Snyder [2], Dinnebier and Billinge [3]). For this introduction to powder diffraction, we will give some examples of features in powder patterns that arise from crystalline and structural features, which are important in catalysts. The powder diffraction pattern is a plot of the diffracted intensity versus the angle between the incident and diffracted beam (2θ). The powder pattern of a reduced copper–ceria inverse catalyst [4] shows how different phases appear in peaks at different 2θ angles (Fig. 4.1). This catalyst was designed to have nanoparticles of cerium oxide on larger particles of metallic

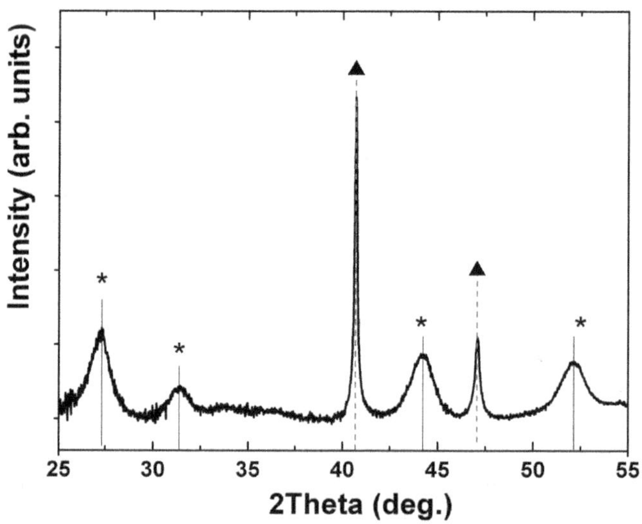

Figure 4.1. Powder pattern of an inverse copper/ceria catalyst obtained after reducing a CeO_2/CuO_x precursor. The nanoparticle ceria peaks are marked with an asterisk (*). The metallic copper peaks are sharper (marked with a triangle), indicating that they come from larger crystallites [38]. Reprinted with permission from Frenkel, A. I., Wang, Q., Marinkovic, N., Chen, J. G., Barrio, L., Si, R., Camara, A. L., Estrella, A. M., Rodriguez, J. A., Hanson, J. C. (2011) Combining X-ray absorption and X-ray diffraction techniques for in situ studies of chemical transformations in heterogeneous catalysis: advantages and limitations, *Journal of Physical Chemistry C*, 115, 17884–17890. Copyright (2006) American Chemical Society.

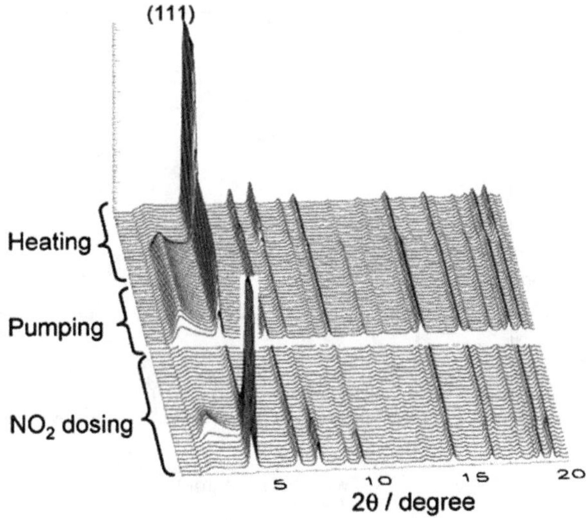

Figure 4.2. Time-resolved XRD patterns for NO_2 adsorption/desorption reactions over a dehydrated BaY sample. The sample was exposed to NO_2 gases at 25°C after dehydration. After the sample was saturated with NO_2, the gases around the sample were pumped away for 2 h, and then the sample was heated to 350°C in 1 h [5]. Reprinted with permission from Wang, X. Q., Hanson, J. C., Szanyi, J., Rodriguez, J. A. (2004) Interaction of H_2O and NO_2 with BaY faujasite: complex contraction/expansion behavior of the zeolite unit cell, *Journal of Physical Chemistry. B, Condensed Matter, Materials, Surfaces, Interfaces & Biophysical*, 108, 16613–16616. Copyright (2004) American Chemical Society.

copper. The peak width is wide for small particles and broad for large particles. The narrow metallic copper peaks and the broader cerium oxide peaks provide evidence that cerium oxide is present as nanoparticles and the metallic copper is present as larger crystallites.

Figure 4.2 shows changes in peak intensity for the powder pattern resulting from variations in the content of NO_2 and the barium positions in zeolite Y (BaY) during NO_2 absorption and desorption. The large intensity change in the 111 reflection has been shown to arise from adsorption of NO_2 molecules and barium cations leaving the supercage of the zeolite [5].

Changes to the intensity of the diffraction peaks can also occur when there are stacking faults in the catalyst. An example of this is found in a catalyst prepared by reduction of $NiMoO_4$ [6]. The catalysts were generated by two different routes and resulted in quite different powder patterns, which are shown in Figure 4.3. Catalyst I was prepared in two steps. First, the hydrate of $NiMoO_4$ was reduced in pure H_2 at 350°C to form Ni, NiO, and amorphous MoO_2. Then it was treated at 500°C in H_2O and CO. Catalyst II was prepared by first heating a hydrate of $NiMoO_4$ in He, followed by heating to 500°C CO and H_2O. Catalysts I and II were both mixtures of MoO_2 and Ni metal. In

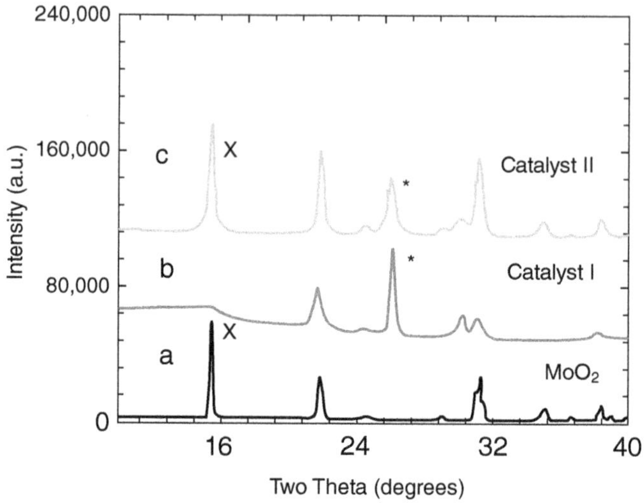

Figure 4.3. *Ex-situ* XRD patterns of (a) MoO_2, (b) catalyst I, and (c) catalyst II [6]. The Ni diffraction peak is marked with * and the low order MoO_2 peak is marked with x. For an explanation of the methodology followed to prepare the catalysts, see the main text.

catalyst I the Ni diffraction lines are sharper than in catalyst II, indicating that the Ni particles are smaller in catalyst II. Catalyst II has a diffraction pattern which is similar to the pattern for MoO_2 and Ni metal. However, the low-order diffraction peak from MoO_2 is extremely broad and very weak compared with the pure MoO_2 pattern. One model that would explain this feature is stacking faults in the MoO_2 structure, which would be expected to reduce the intensity of this reflection [6].

4.3 TOOLS FOR *IN-SITU* MEASUREMENTS

4.3.1 X-ray Sources

A large portion of recent *in-situ* X-ray measurements have taken place at synchrotron sources because of their high intensity. These sources are at least 100 times stronger than laboratory sources. The new third-generation sources provide even more intensity, and consequently intermediates with shorter lifetimes (<1 s) can be observed. Most measurements are carried out with monochromatic radiation, but significant results have also been obtained with white radiation [3].

4.3.2 Flow Control and Product Analysis

There are many commercial flow controllers, including some that allow ready flow control at pressures up to 10 bar. The product analysis is usually determined by delivering a small fraction of the gas to a mass spectrum analyzer of a gas chromatograph. The heating can be carried out with resistive heaters near the sample [7] or with blower heaters [8].

4.3.3 *In-situ* Cells

There have been many models of *in-situ* cells [9]. Cells used for energy-dispersive diffraction have the advantage that the opening needed to measure the diffraction is small. However, the diffraction peaks are not usually as resolved as with angle-dispersive diffraction [10, 11]. The cells used for dispersive diffraction are described in a recent review [9].

For angle-dispersive diffraction, a reaction cell has been developed consisting of a thin quartz glass capillary mounted in a Swagelok fitting using graphite or graphite/Vespel ferrules, containing the starting mixture/gel (Fig. 4.4). When a gas pressure is applied to the surface of the reaction mixture, it is possible to heat up a zone (using, e.g., a hot air blower) and maintain hydrothermal conditions. The maximum temperature for stable hydrothermal conditions is determined by the applied pressure, which must exceed the equilibrium vapor pressure for the reaction mixture. Standard capillaries for X-ray application (wall thickness 0.01 mm) may sustain internal pressures up to 50 bar. However, it must be noted that the chemistry of the reaction mixture may have a severe effect on the stability of the capillaries at higher temperatures. Especially basic

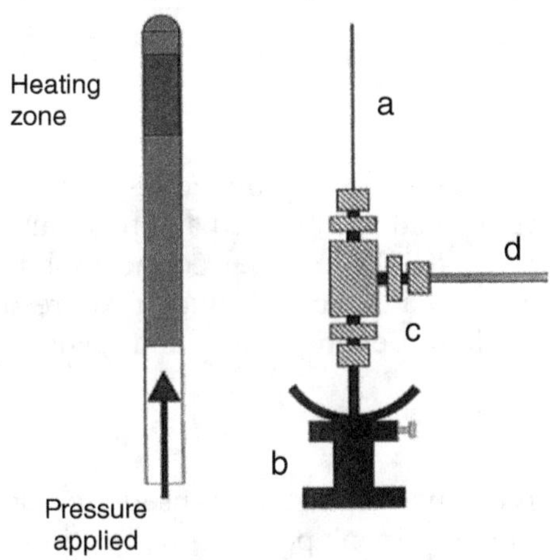

Figure 4.4. Sketch showing a capillary-based reaction cell for *in-situ* diffraction studies of hydrothermal synthesis. A quartz glass capillary (0.5–0.8 mm diameter) (a) is mounted on a goniometer head (b) using a Swagelok T-piece (c). A nitrogen pressure can be applied through the connected tubing (d). A detailed view of the capillary is shown to the left.

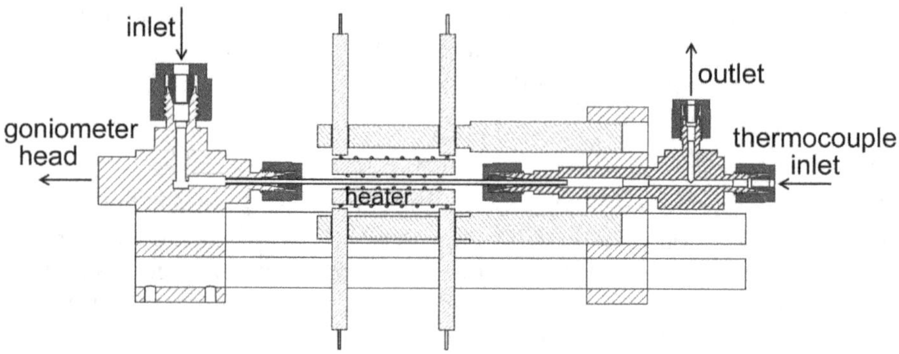

Figure 4.5. Cross section of the flow cell for *in-situ* measurements. The input and output gas flow and input for thermocouple are shown. In addition, the resistive heater and goniometer mount are shown [7].

conditions will cause dissolution of the quartz glass over time, resulting in failure. For higher pressures, it is possible to use steel tubes [12] or thicker-walled quartz or sapphire tubes. However, this requires a higher energy X-radiation in order to penetrate the sample container. An *in-situ* flow cell [7] is shown in Figure 4.5. This microreactor is related to a flow cell developed by Clausen et al. [8]. This cell has input and output for gases, a provision to place the thermocouple next to the sample, and a mount for a pair of resistive heaters.

4.3.4 Detectors

The advent of area detectors greatly facilitated *in-situ* measurements because the entire powder pattern could be collected and read out much more rapidly. Modern area detectors for XRD are reviewed in chapter 14 of Reference 3. The issues in choosing an area detector include pixel resolution, overall size of the detector, high sensitivity, energy range, and readout time.

4.3.5 Analysis

The qualitative analysis of a powder pattern is useful for identifying the phases of the material by matching the peak positions to positions of compounds that are in a database of known materials. Table17.3 in Reference 3 gives a list of search match software. Techniques for determining the lattice and the structure are more complex and are discussed in chapters 7 and 8 of Reference 3 and in Reference 13. If the rough structure is known, one can usually refine the structure to fit the powder pattern (see chapter 9 of Reference 3).

4.4 APPLICATIONS OF XRD TO THE *IN-SITU* CHARACTERIZATION OF CATALYSTS

In this section, a series of typical *in-situ* measurements on catalysts will be discussed. These applications of the XRD technique show its versatility covering different areas of heterogeneous catalysis and a wide range of catalytic materials.

4.4.1 *In-situ* Studies of Microporous Catalysts: Hydrothermal Synthesis, Adsorption, and Catalysis

An important aspect of the development and optimization of catalysts is control of the synthesis and preparation of materials. Morphology, size distribution, and surface properties are important parameters. For the crystalline microporous catalysts, which are used extensively in the petrochemical industry, knowledge and control over details in the crystal structure are crucial. Often, the properties of microporous catalysts are controlled by modifying synthesis conditions or by postsynthesis modifications. *In-situ* studies of hydrothermal synthesis of crystalline microporous materials may provide important information regarding structural and microstructural development during synthesis.

In-situ studies of microporous catalysts under operating conditions are a good example where detailed information from diffraction is necessary. The distribution of cations, location of active sites, as well as molecular interactions and dynamics inside the cages and channels of microporous materials are important parameters with a substantial effect on catalytic performance. Knowledge of structural parameters and their dependence on the chemical and physical environment is necessary, and *in-situ* studies must be performed at realistic operating conditions.

In the following section, we will give examples of *in-situ* experiments related to microporous catalysts, which includes studies of synthesis, cation migration, and interaction with adsorbed molecules, as well as studies of microporous catalysts during operation.

4.4.2 *In-situ* Studies of Hydrothermal/Solvothermal Synthesis of Zeolites and Microporous Catalysts

In-situ powder diffraction studies of hydrothermal synthesis of zeolites and other microporous materials provide information about nucleation and crystallization mechanisms and structural development during synthesis. This information may be used, for example, when searching for new microporous materials but also as a means to obtain improved control of crystal morphology, size, and size distribution. Often, the information gained from a single *in-situ* diffraction experiment is insufficient to clarify details in the formation process. Isothermal crystallization experiments at different temperatures and/

or at variable heating rates are often employed. In addition, information using other *in-situ* and *ex-situ* techniques (e.g., small-angle scattering, spectroscopy, scanning electron microscopy [SEM], transmission electron microscopy [TEM]) is necessary. For example, diffraction is a useful tool for studying crystallization kinetics, but processes preceding crystallization cannot be directly monitored by this technique. This includes reactions occurring in solution, gel formation, and nucleation. Combining experimental techniques in a single *in-situ* experiment—for example, XRD/small-angle X-ray scattering (SAXS), XRD/dynamic light scattering (DLS), and XRD/X-ray absorption spectroscopy (XAS)—is therefore often beneficial. In addition, hydrothermal synthesis involves very complex processes, and it is often not possible to recreate the exact same conditions required in order to compare information from two separate experiments.

The first reported *in-situ* studies of hydrothermal synthesis of microporous materials were performed using neutron powder diffraction by Polak et al. in 1990 [14]. The development of powerful synchrotron X-ray sources has advanced the possibility for studying chemical reactions and synthesis involving crystalline materials with very high time and spatial resolution using *in-situ* powder diffraction. A number of *in-situ* synchrotron XRD studies of hydrothermal synthesis of zeolites and microporous materials have been published since the first reported studies using energy-dispersive and angular dispersive methods (a few examples are given in References [10, 11, 15–19]). Using a microreaction cell consisting of a thin-walled quartz tube (see Fig. 4.4), it is possible to study hydrothermal synthesis *in situ* up to approximately 250°C using angular dispersive diffraction. At higher temperatures/pressures, it is necessary to use steel tubing (an example is given in Reference [12], studying hydrothermal synthesis at 350°C and 2–300 bar). Moreover, single-crystal sapphire tubes (Saint-Gobain Crystals, Milford, NH) may be used, as they can withstand high internal pressure and aggressive chemical environments.

As an example, the nonaqueous, solvothermal syntheses of microporous cobalt- and manganese-substituted aluminophosphates were studied in order to investigate the effect of the transition metal and added mineralizers on crystallization kinetics. Using isothermal heating at different temperatures, it is possible to extract some information about the crystallization mechanisms. Direct observation and identification of intermediate and transient phases are possible. The shape of the crystallization curves is influenced by the mechanism (Fig. 4.6), reflecting, for example, nucleation rate and dimensionality in crystallite growth. In addition, the temperature dependence of the crystallization rate may be used to extract apparent activation energies for the crystallization process.

It is important not to overinterpret the crystallization curves obtained by *in-situ* diffraction. These are complex systems, and there are many factors contributing to the crystallization kinetics; extracting specific crystallization mechanisms may not be meaningful. The apparent activation energies extracted must also be carefully evaluated, as they do not represent real activation energies

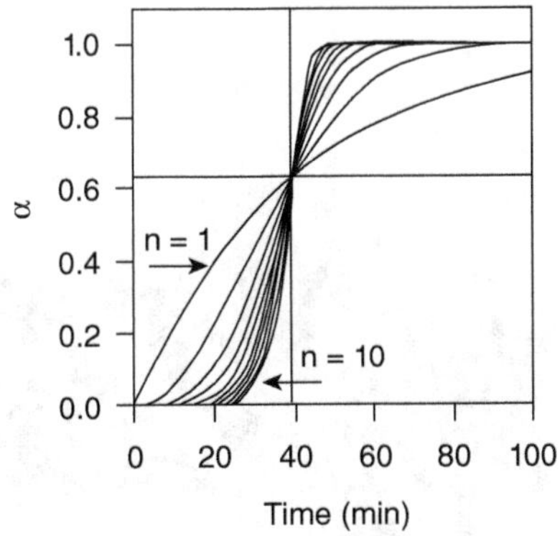

Figure 4.6. Calculated crystallization curves using Equation 4.1: $\alpha(t) = 1 - \exp(-(kt)^n)$. Curves were calculated using n-values from 1 to 10 with a constant value of k (0.025/min). The curves intersect at $\alpha = 1 - e^{-1} \sim 0.63$ at time $t = 1/k = 40$ min [19]. Reprinted with permission from Norby, P., Christensen, A. N., Hanson, J. C. (1999) Crystallization in nonaqueous media of Co- and Mn-substituted microporous aluminophosphates investigated by in situ synchrotron X-ray powder diffraction, *Inorganic Chemistry*, 38, 1216–1221. Copyright (1999) American Chemical Society.

as determined, for example, from reactions in homogeneous systems in solution. However, *in-situ* diffraction is a very strong tool in comparative analysis of similar systems or in investigation of effects of external perturbations or parameters. One challenge of using diffraction methods is that the earlier processes, for example, nucleation, are not probed. Therefore, it is also important to include other methods in the studies, preferably in combined *in-situ* experiments. One possibility is to extend the size domain probed by including information about small noncrystalline objects, for example, DLS and SAXS (e.g., Reference [20]).

One example of powder diffraction patterns collected during synthesis of Mn-substituted MAPO-5 at 125°C is shown in Figure 4.7. This figure shows a three-dimensional representation, where the powder diffraction patterns are stacked to show the evolution with time. In this example, the starting gel does not contain crystalline phases, and only very minor traces of intermediate phases are observed. By extracting the time dependence of the integrated intensity of some of the diffraction peaks, crystallization curves can be extracted at various temperatures and fitted to a kinetic expression, for example, an Avrami-type expression as shown in Equation 4.1 (where α is the crystal volume fraction):

$$\alpha(t) = 1 - \exp\left(-(kt)^n\right). \tag{4.1}$$

Figure 4.8 shows examples of crystallization curves for Mn-substituted MAPO-5, where a mineralizer (HF) is added to the synthesis mixture. The

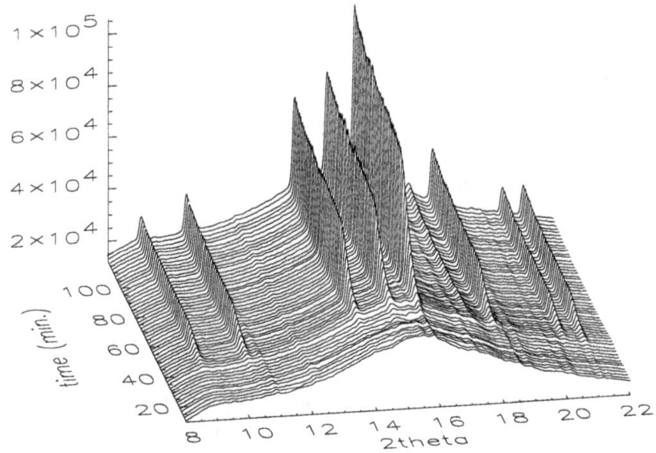

Figure 4.7. Stack of powder patterns obtained with the MnAPO-gel when kept at 125°C for 120 min. The reaction product is MnAPO-5, [AFI] [19]. Reprinted with permission from Norby, P., Christensen, A. N., Hanson, J. C. (1999) Crystallization in nonaqueous media of Co- and Mn-substituted microporous aluminophosphates investigated by in situ synchrotron X-ray powder diffraction, *Inorganic Chemistry*, 38, 1216–1221. Copyright (1999) American Chemical Society.

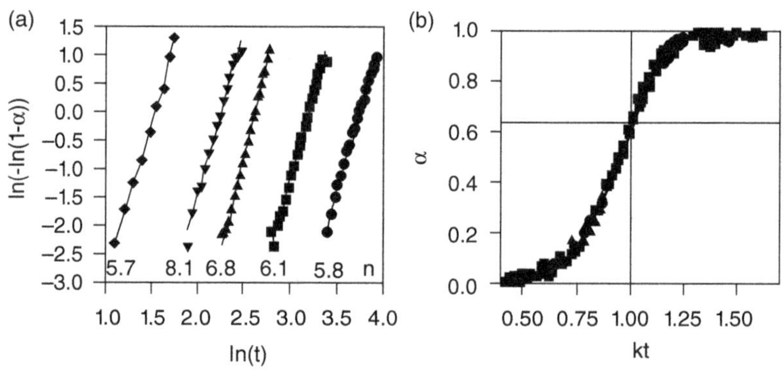

Figure 4.8. (a) Plot of $\ln(-\ln(1-\alpha))$ versus $\ln(t)$ for crystallization of MnAPO-5(HF) between 125 and 173°C. Solid lines are linear fits to each curve, and the value of n is the slope of the lines according to Equation 4.1. (b) The crystallization curves for MnAPO-5(HF) plotted on a normalized timescale, kt, where k is the individual rate constant obtained from fitting the curves to Equation 4.1. The point where $kt = 1$ and $\alpha(kt) = 1 - e^{-1}$ is indicated. Temperatures: ●, 125°C; ■, 135°C; ▲, 144°C; ▼, 154°C; and ◆, 173°C [19]. Reprinted with permission from Norby, P., Christensen, A. N., Hanson, J. C. (1999) Crystallization in nonaqueous media of Co- and Mn-substituted microporous aluminophosphates investigated by in situ synchrotron X-ray powder diffraction, *Inorganic Chemistry*, 38, 1216–1221. Copyright (1999) American Chemical Society.

crystallization curves are plotted as ln (−ln (1 − α)) versus ln(*t*), which should give straight lines according to Equation 4.1. The fits to the crystallization curves are also shown, and the extracted parameters are used to plot the crystallization curves together on a reduced timescale, *kt*. It is seen that the shape of the crystallization curves is constant throughout in the investigated temperature range, indicating that the crystallization mechanism does not change with temperature in this case.

Using Arrhenius-like plots, apparent activation energies were extracted for the manganese- and cobalt-substituted materials, as well as for the HF-mineralized manganese-substituted material. A decrease in the apparent activation energy for crystallization was observed when adding the mineralizer, with no indication of change in crystallization mechanism.

This example demonstrates the power of using *in-situ* diffraction to study crystallization. However, more experiments, including repetitions at identical conditions and a larger pool of substitution degrees, would be needed in order to make solid conclusions.

4.4.3 Cation Migration in Microporous Zeolites and Sorbate—Cation Interactions

Aluminosilicate zeolites are negatively charged framework materials that contain cations as counterions. These cations are usually localized on a set of sites that provide stabilization of their attachment. However, the relative occupancy of these sites depends on the temperature and the interaction with H_2O or other molecules and species. It is therefore not possible to know the structural details by extrapolating from room temperature structural information. This means that it is crucial to determine the position of the cations in microporous catalyst materials *in situ* at realistic operating conditions in order to improve the understanding of the mechanisms. *In-situ* synchrotron or neutron diffraction provides a good way to measure the previously mentioned changes under operating conditions.

This was demonstrated in an *in-situ* study of the structure of Cs(Na)–zeolite Y (faujasite-type framework) during heating under vacuum, where significant changes to the cation locations in the zeolite were observed [21]. Likewise, strong sorbate–cation interactions and cation migration have been observed upon adsorption of hydroflurocarbons in zeolite X and Y [22–24]. In these studies, synchrotron X-ray powder diffraction was combined with information from solid-state nuclear magnetic resonance (NMR) (magic angle spinning [MAS] NMR). Obtaining information from these two complementary techniques proved to be a very strong combination.

In the investigations of zeolite Cs(Na)–Y, partially Cs-exchanged materials were prepared by repeated cycles of aqueous Cs exchange/calcination. An increase in the Cs exchange capacity is obtained by the heating step, and one of the goals of the study was to understand the cation migration during dehydration and calcination. In addition, the study revealed how the cation

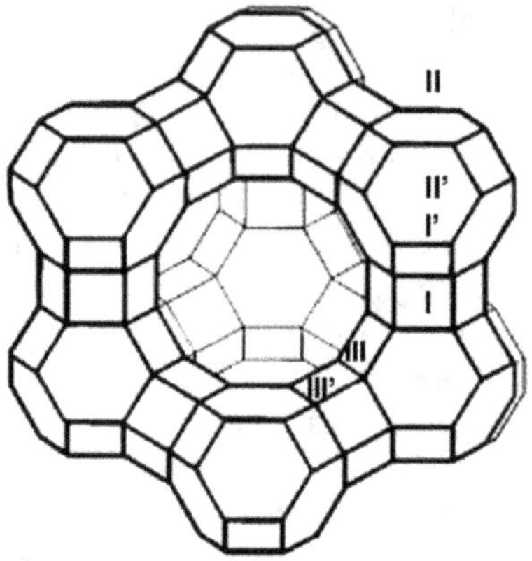

Figure 4.9. Representation of the faujasite framework structure showing only the tetrahedral atoms. The structure can be viewed as being built up from sodalite cages linked together to form double 6-rings, between interconnected supercages. The positions of the cation sites SI (double 6-ring), SI′, SII′ (in the sodalite cage), and SII, SIII, and SIII′ (in the supercage) are indicated. Adapted with permission from Grey, C. P., Poshni, F. I., Gualtieri, A. F., Norby, P., Hanson, J. C., Corbin, D. R. (1997) Combined MAS NMR and X-ray powder diffraction structural characterization of hydrofluorocarbon-134 adsorbed on zeolite NaY: observation of cation migration and strong sorbate-cation interactions, *Journal of the American Chemical Society*, 119, 1981–1989. Copyright (1997) American Chemical Society.

positions in the dehydrated material change with temperature. A combination of *ex*- and *in-situ* diffraction with ^{133}Cs MAS NMR spectroscopy was used in order to obtain very detailed structural information. Some of the findings are reported later. The commonly observed cation sites in the faujasite structure are shown in Figure 4.9. The distribution of the cations in the different sites depends on their interaction with the framework, adsorbed species, and other cations. In addition, the diffusion limitation imposed by the size of the rings connecting the various cages means that the thermal history of the sample may play a role, as migration through the six-ring openings is facilitated at high temperature.

A three-dimensional representation of the powder diffraction patterns as a function of time during heating under vacuum of a partially Cs-exchanged zeolite Y (Cs(Na)–Y) in vacuum are shown in Figure 4.10. The data were used in a sequential powder profile refinement to determine the positions and occupancies of Cs cations and the water molecules during dehydration. The amount of water molecules found from the refinement is shown in Figure 4.11. At 120°C, 140 water molecules were found and a steady decrease with temperature was observed. The water was mostly removed by 190°C.

The removal of the water molecules leads to a rearrangement of the occupancy of the cation sites. In Figure 4.12, the occupancy of the sodium and

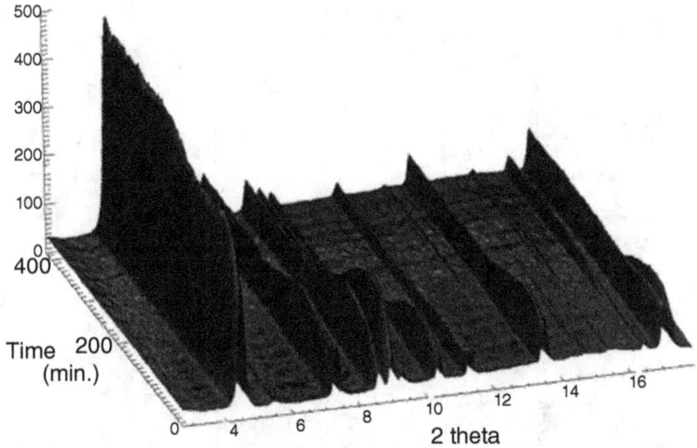

Figure 4.10. Three-dimensional plot of powder diffraction profiles obtained during the dehydration of zeolite Cs(Na)-Y from 90 to 300 min, followed by a hold at 500°C for another 240 min. Only the first 400 min are shown [21]. Reprinted with permission from Norby, P., Poshni, F. I., Gualtieri, A. F., Hanson, J. C., Grey, C. P. (1998) Cation migration in zeolites: an in situ powder diffraction and MAS NMR study of the structure of zeolite Cs(Na)-Y during dehydration, *Journal of Physical Chemistry. B, Condensed Matter, Materials, Surfaces, Interfaces & Biophysical*, 102, 839–856. Copyright (1998) American Chemical Society.

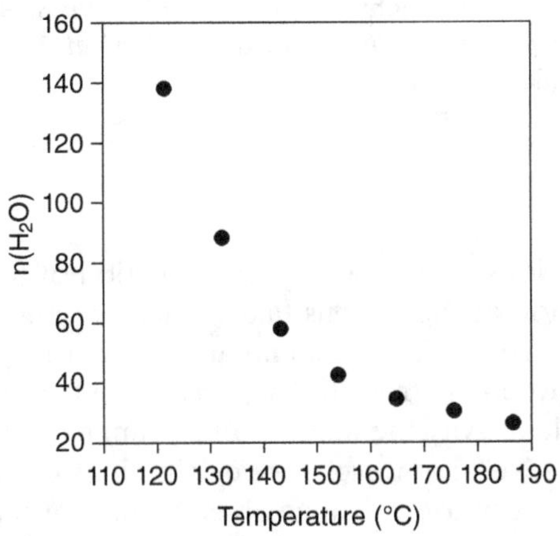

Figure 4.11. Number of water molecules per unit cell located by the Rietveld refinement during the dehydration of zeolite Cs-Y as a function of temperature [21]. Reprinted with permission from Norby, P., Poshni, F. I., Gualtieri, A. F., Hanson, J. C., Grey, C. P. (1998) Cation migration in zeolites: an in situ powder diffraction and MAS NMR study of the structure of zeolite Cs(Na)-Y during dehydration, *Journal of Physical Chemistry. B, Condensed Matter, Materials, Surfaces, Interfaces & Biophysical*, 102, 839–856. Copyright (1998) American Chemical Society.

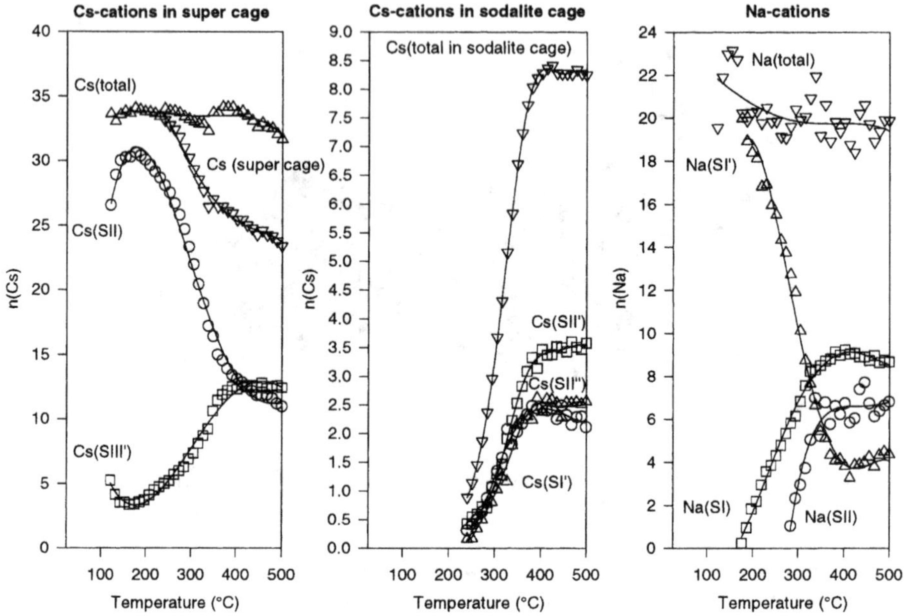

Figure 4.12. Changes in the cation populations of the SI, SI′, SII, SII′, SII″, and SIII′, sites with temperature during the dehydration of zeolite Cs(Na)-Y, as determined from the *in-situ* powder diffraction data by Rietveld refinement [21]. Reprinted with permission from Norby, P., Poshni, F. I., Gualtieri, A. F., Hanson, J. C., Grey, C. P. (1998) Cation migration in zeolites: an in situ powder diffraction and MAS NMR study of the structure of zeolite Cs(Na)-Y during dehydration, *Journal of Physical Chemistry. B, Condensed Matter, Materials, Surfaces, Interfaces & Biophysical*, 102, 839–856. Copyright (1998) American Chemical Society.

cesium cations in various sites is shown as a function of temperature. A considerable redistribution of the cations takes place even after the water molecules have been desorbed. The predominating migration observed was the cation exchange between Cs from the supercage and sodium in the sodalite cage. This result explains why the amount of cesium exchanged into zeolite Y can be increased by several ion exchange/calcinations cycles. During the first calcination, the amount of ion-exchangeable sodium cations in the supercage increases, which can then subsequently be ion-exchanged from aqueous solution. The cation distribution was followed as well during cooling of the dehydrated zeolite to room temperature under vacuum, shown in Figure 4.13. From this figure, it is seen that the distribution of cesium cations between sites, especially in the supercage, depends strongly on the temperature; even in the absence of adsorbed water molecules, no intercage cation migration was observed.

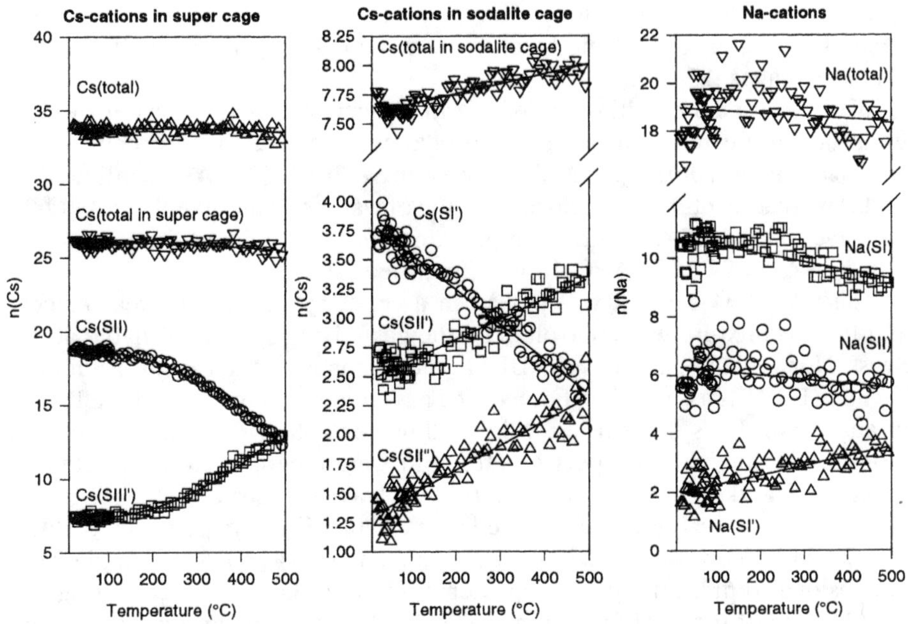

Figure 4.13. Changes in the cesium and sodium cation occupancies (per u.c.) for dehydrated zeolite Cs(Na)-Y, as determined by Rietveld refinement during cooling under vacuum from 500°C to room temperature [21]. Reprinted with permission from Norby, P., Poshni, F. I., Gualtieri, A. F., Hanson, J. C., Grey, C. P. (1998) Cation migration in zeolites: an in situ powder diffraction and MAS NMR study of the structure of zeolite Cs(Na)-Y during dehydration, *Journal of Physical Chemistry. B, Condensed Matter, Materials, Surfaces, Interfaces & Biophysical*, 102, 839–856. Copyright (1998) American Chemical Society.

4.4.4 *In-situ* Studies of Microporous Catalysts during Operation; the MTO Process

One of the examples where crystal structure information during a catalytic reaction is important is in catalytic conversions using microporous catalysts, for example, zeolites and zeo-type materials. In these materials, the catalytic reaction takes place in channels and voids in the crystalline structure, that is, the pores act as "nanoreactors." As seen in the preceding section, adsorbed molecules may interact strongly with the framework and/or extraframework cations. Moreover, the temperature dependence means that the crystal structure must be determined at operating conditions for the catalyst.

One example of such a catalytic system is the methanol-to-olefin (MTO) reaction, where a silicon-substituted microporous aluminophosphate (SAPO-34) has shown good catalytic activity and selectivity [25–27]. The mechanism for the reaction is not fully understood, but one of the most probable pathways is the "carbon pool" mechanism. Here, it is suggested that the active sites are created by building large methylated aromatic reactive intermediates in the chabazite cages [25–27].

We have used a combination of *in-situ* synchrotron X-ray powder diffraction, Raman spectroscopy, and mass spectrometry to investigate structural and chemical changes during the MTO process in SAPO-34 [28]. As a continuation of this work, the interaction and adsorption of water and methanol in SAPO-34 was investigated [25–27], and a comparison of *in-situ* studies of the MTO reaction on SAPO-34 and SAPO-18 was published [29].

The SAPO-34 catalyst was placed in a capillary-based microreactor cell, and the gas mixture was controlled by the gas-mixing system available at the Swiss–Norwegian beamlines (SNBL) at the European Synchrotron Radiation Facility (ESRF) in Grenoble, France. The gas stream was mixed with methanol in an evaporator system using a liquid flow controller. The exit stream was monitored using a mass spectrometer, and simultaneous XRD and Raman spectroscopy data were collected. In this way, the structural changes in SAPO-34 could be directly correlated to the catalytic activity and the buildup of coke and other molecules in the catalyst. Figure 4.14 shows two- and three-dimensional representations of powder diffraction data collected during the initial phase of the MTO reaction at 440°C. Dramatic changes in the diffraction pattern are observed when the reaction starts: very large intensity changes for some reflections and changes in the position of some of the reflections, indicating dramatic anisotropic changes in the unit cell parameters. Figure 4.15 shows the unit cell parameters in the first hour after the reaction starts. A very marked anisotropic expansion is seen, with the largest change along the *c*-axis in the hexagonal unit cell. The changes in the diffraction intensity of the reflections, some increasing and some decreasing, reflect a change in the electron

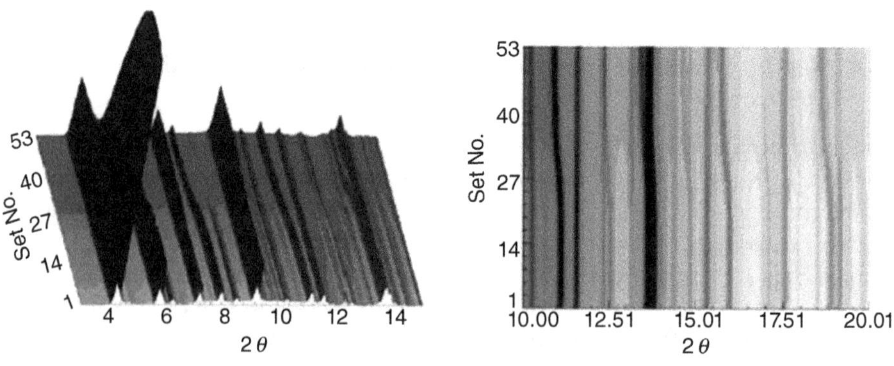

Figure 4.14. Left: Three-dimensional stack plot of SAPO-34 PXRD patterns (2θ range 3–15°) collected during the MTO process at 440°C; intensity changes are clearest in the initial (1 0 1) peak (2θ = 4.5°). Right: Film plot of the same data (2θ range 10–20°) showing peak position shifts which indicate anisotropic changes in the unit cell dimensions. Time increases with set number [28]. Copyright © 2009, Elsevier.

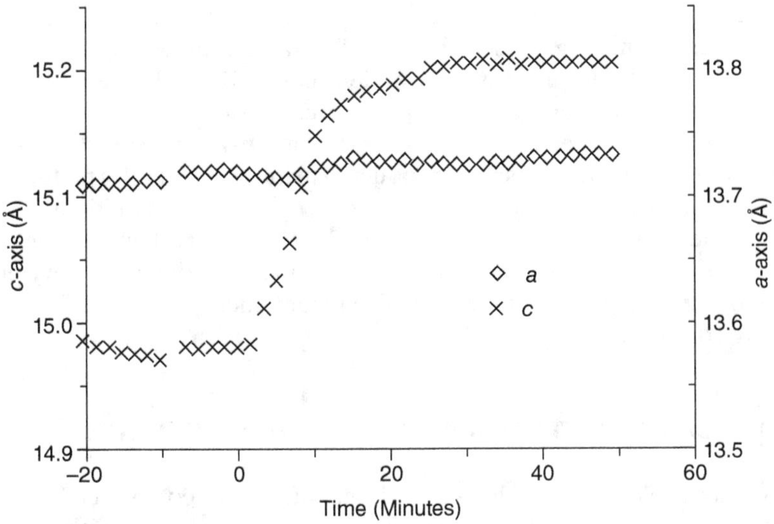

Figure 4.15. Variation of the *a*- and *c*-axes of SAPO-34 during the MTO process at 450°C. Time = zero corresponds to the first addition of methanol in this and all following graphs charting reaction progress against time. The estimated standard deviation error bars for the dimensions are smaller than the symbols used in the plots [28]. Copyright © 2009, Elsevier.

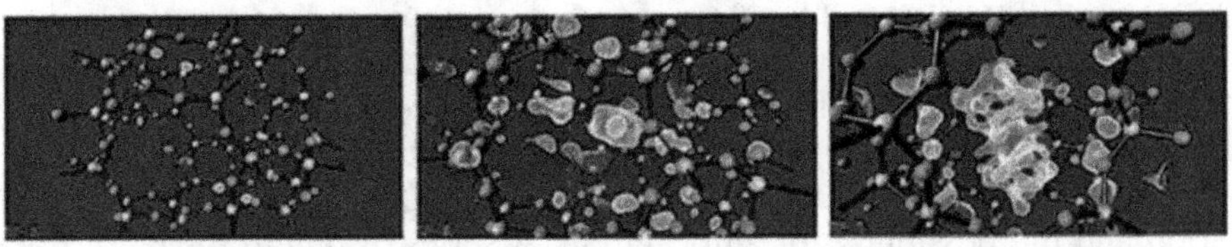

Figure 4.16. Images of the residual electron density in the SAPO-34 cage, determined from difference Fourier maps. From left to right: 440°C before reaction, and 4 and 20 min into reaction. Red spheres = oxygen, green = phosphorus, and blue = aluminum. The white contours represent residual electron density [28]. Copyright © 2009, Elsevier. (See color insert.)

density in the voids in the structure. The anisotropic change on the unit cell parameters is a result of morphological changes in the cages in the crystal structure, which also indicates that something is growing inside the voids. In order to investigate the changes in the crystal structure, Rietveld refinement and difference Fourier calculations were used to estimate the electron density in the previously empty chabazite cages. Figure 4.16 shows the residual electron density, as estimated from difference Fourier calculations, in the chabazite cages at various points in the MTO reaction at 440°C. It is evident that a localized residual electron density is appearing, reflecting the growth of some species in the cage. As to be expected at elevated temperatures, the species are disordered and not easily identifiable. Reoxidation experiments and pulsed catalytic reactions were performed in order to elucidate further the nature of the species formed. By correlating *in-situ* crystal structure and microstructure investigations with product distribution and information about the coke formation from Raman spectroscopy, the formation of highly reactive species in the chabazite cages seems plausible. The results clearly demonstrate that *in-situ* diffraction studies using synchrotron X-radiation are a very powerful tool for obtaining structural information from microporous catalysts under operating conditions.

4.5 SIMPLE REDUCTION AND ACTIVATION OF OXIDES

Metal oxides are used as catalysts for a large number of commercial processes for the conversion of hydrocarbons. The reduction of NiO with hydrogen provides a simple example of an *in-situ* measurement. These measurements contribute to an understanding of the mechanism of reduction [30]. The measurements were made at beamline X7B at the National Synchrotron Light Source (NSLS) where the wavelength was 0.92 Å. A sample of NiO was mounted in a sapphire tube in the *in-situ* cell described in Figure 4.5. A Mar345 image plate detector was used. The typical time for collecting an individual powder pattern was 0.5–2 min. The powder rings were integrated using the FIT2D code [31]. Rietveld refinements were performed with the General Structure Analysis System (GSAS) code [32].

There are two common kinetic models for the reduction of metal oxides: the "nucleation model" (N) and the "interface-controlled model" (IC) (see the references in Reference [30]). In the nucleation model there is an induction period and the possibility for autocatalysis. A series of time-resolved diffraction patterns during the reduction of NiO under isothermal conditions at 280°C is shown in Figure 4.17a. To determine the kinetics of reduction under isothermal conditions the normalized intensity was measured at several temperatures (Fig. 4.17b). The 270°C isotherm shows a nucleation time of 50 min while there was no nucleation time for the 300°C isotherm. This supports the nucleation model and is consistent with the surface science measurements [30]. Theoretical calculations suggest that hydrogen interacts poorly with a perfect NiO surface. However, it readily binds to defects formed by oxygen vacancies on the surface [30].

Recently, a similar study of NiO supported on $NiAl_2O_4$ was studied with a laboratory source and a point detector [33]. Narrow 2θ range patterns that were collected in about 4 min provided the information about the kinetics of reduction with 16% CH_4/He and reoxidation with air. The reduction in hydrogen at 800°C started immediately upon introduction of the hydrogen mixture. This is consistent with the synchrotron experiment where the nucleation time

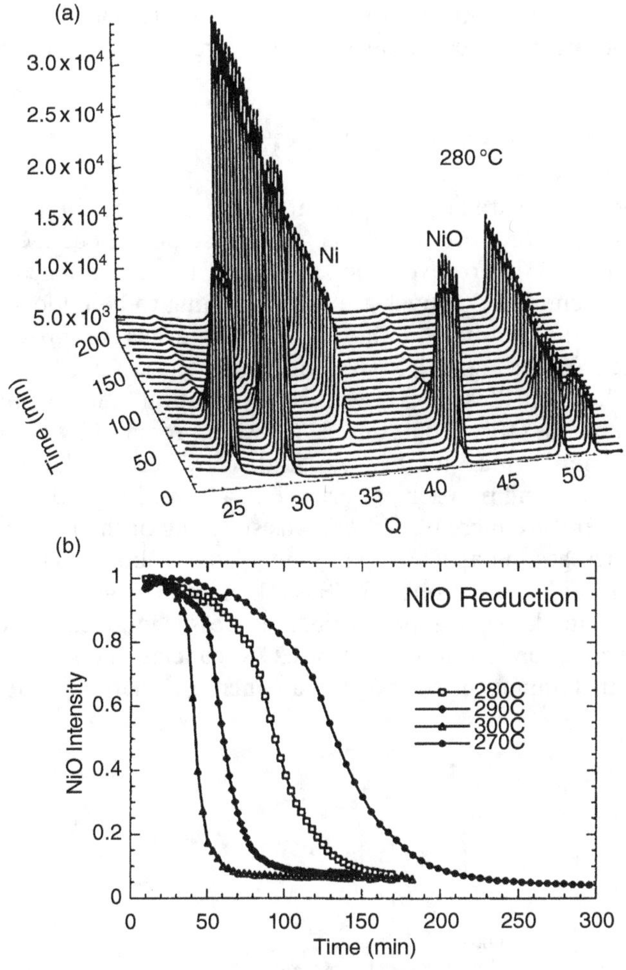

Figure 4.17. (a) A series of time-resolved diffraction patterns during the reduction of NiO under isothermal conditions at 280°C. The NiO phase converts directly to Ni. (b) A series normalized intensity of a diffraction intensity from a series of isothermal reductions of NiO at 270, 280, 290, and 300°C (15 cm/min 5% H_2/95% He) [30]. Reprinted with permission from Rodriguez, J. A., Hanson, J. C., Frenkel, A. I., Kim, J. Y., Perez, M. (2002) Experimental and theoretical studies on the reaction of H-2 with NiO: role of O vacancies and mechanism for oxide reduction, *Journal of the American Chemical Society*, 124, 346–354. Copyright (2002) American Chemical Society.

was negligible at temperatures greater than 300°C. This is a nice example of the use of laboratory source to measure *in-situ* reactions of catalysts.

4.6 SINTERING OF NANOCATALYSTS

Ceria and doped ceria are key components in the so-called three-way catalysts [34]. The properties of these materials as catalyst can be changed when they are nanoparticles [35]. However, the stability of these nanoparticles and the reason for their enhanced actives are just beginning to be understood. *In-situ* XRD studies of the thermal stability of these materials can provide information about the changes in particle size and strain. The study of Ca-doped ceria will be used here to illustrate how this information can be obtained [36]. Ca-doped nanoceria (10%, 20%, and 33%) were prepared by the microemulsion method. The thermal stability was studied by heating these materials to 800°C while measuring powder patterns (Fig. 4.18 and Fig. 4.19).

As the temperature increased, there was sintering of the nanoparticles and a significant increase in particle size at temperatures above 700°C (Fig. 4.19a). When the peak width variation with 2θ is taken into account, the lattice strain can also be obtained. In Ca-doped ceria, the stress is induced by doping of Ca or lattice imperfections from dislocations, O vacancies, and so on (Fig. 4.19b). An increase in temperature produces an increase in the particle size and

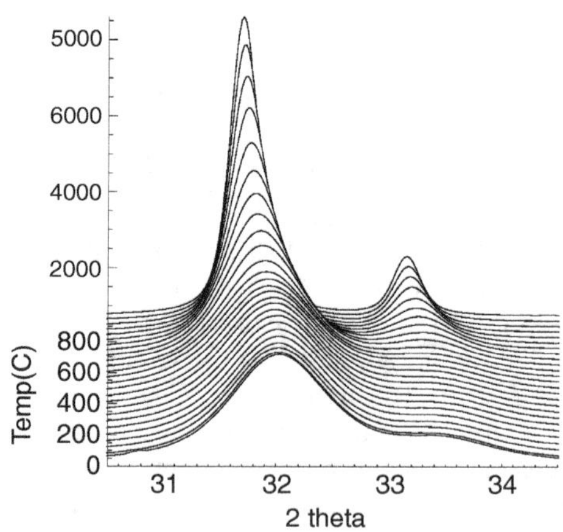

Figure 4.18. Time-resolved XRD pattern during heating of Ca-doped ceria [36]. The peaks start broad and get sharper as the temperature gets higher and the particles sinter to larger sizes. Reprinted with permission from Rodriguez, J. A., Wang, X. Q., Hanson, J. C., Liu, G., Iglesias-Juez, A., Fernandez-Garcia, M. (2003) The behavior of mixed-metal oxides: structural and electronic properties of $Ce_{1-x}Ca_xO_2$ and $Ce_{1-x}Ca_xO_{2-x}$, *Journal of Chemical Physics*, 119, 5659–5669. Copyright [2003], American Institute of Physics.

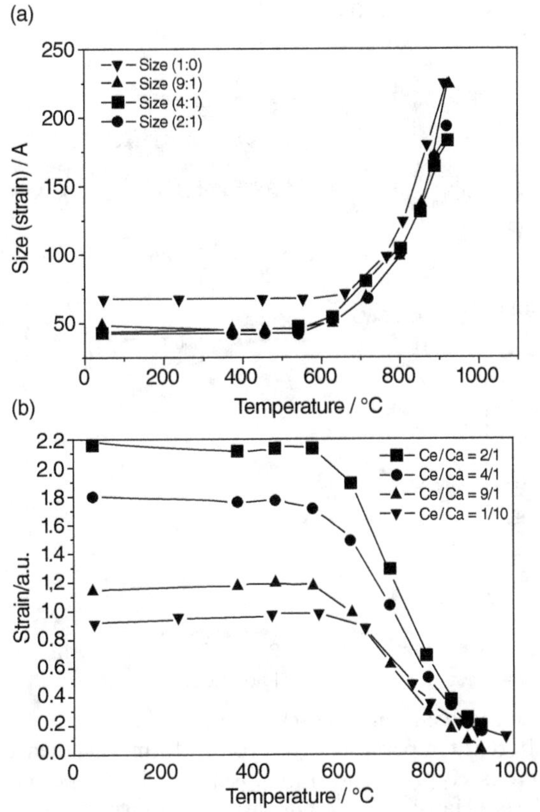

Figure 4.19. (a) Size of CeO_2 and $Ce_{1-x}Ca_xO_{2-x}$ particles as a function of temperature. (b) Strain in CeO_2 and $Ce_{1-x}Ca_xO_{2-x}$ particles as a function of temperature [36]. Reprinted with permission from Rodriguez, J. A., Wang, X. Q., Hanson, J. C., Liu, G., Iglesias-Juez, A., Fernandez-Garcia, M. (2003) The behavior of mixed-metal oxides: structural and electronic properties of $Ce_{1-x}Ca_xO_2$ and $Ce_{1-x}Ca_xO_{2-x}$, *Journal of Chemical Physics*, 119, 5659–5669. Copyright [2003], American Institute of Physics.

simultaneously imperfections are removed from the oxide lattice, decreasing the measured strain.

4.7 *IN-SITU* DETERMINATION OF ACTIVE PHASE IN METAL/ OXIDE CATALYST

The *in-situ* XRD study of Cu/ceria catalysts under water–gas shift (WGS) reaction conditions provides an example of the determination of an active phase. The WGS reaction ($CO + H_2O \rightarrow H_2 + CO_2$) is commonly used to produce pure H_2 and remove CO from hydrogen streams produced by the

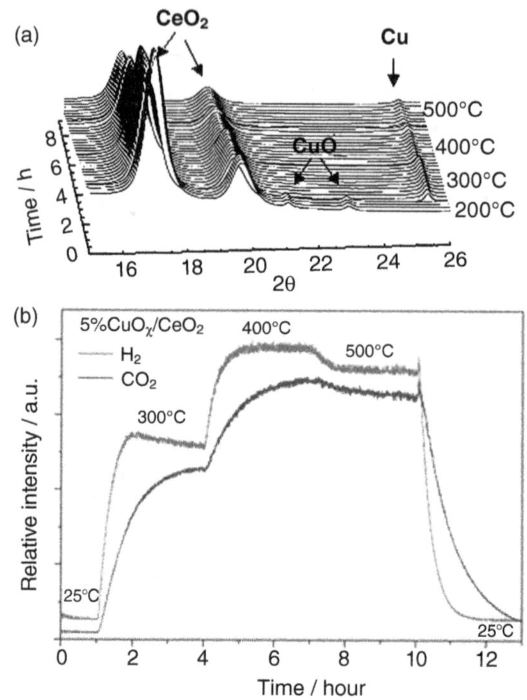

Figure 4.20. (a) Time-resolved XRD patterns for 5% CuO_x/CeO_2 catalysts during the WGS reaction at different temperatures. (b) Product formation measured during WGS reaction (upper curve is H_2 production) [37]. Reprinted with permission from Wang, X. Q., Rodriguez, J. A., Hanson, J. C., Gamarra, D., Martinez-Arias, A., Fernandez-Garcia, M. (2006) In situ studies of the active sites for the water gas shift reaction over Cu-CeO_2 catalysts: complex interaction between metallic copper and oxygen vacancies of ceria, *Journal of Physical Chemistry. B, Condensed Matter, Materials, Surfaces, Interfaces & Biophysical*, 110, 428–434. Copyright (2006) American Chemical Society.

steam reforming of hydrocarbons. In a recent study [37], nanoceria was impregnated with copper and a 5 wt% Cu/ceria catalyst was studied under *operando* conditions.

The sample was placed in the *in-situ* flow cell as shown in Figure 4.5 and diffraction patterns were collected while flowing 5% CO and 95% He that passed through a water bubbler. The WGS reaction was carried out isothermally at 200, 300, 400, and 500°C. The diffraction patterns are presented in Figure 4.20a. The relative concentrations of the products were measured with a residual gas analyzer (RGA) and are presented in Figure 4.20b. These data show that metallic copper is always present when the catalyst is active. A transformation from CuO to Cu takes place at temperatures below 250°C and the material becomes an active WGS catalyst. A Rietveld analysis of the ceria peaks [37] points to a lattice expansion of the oxide with a small amount of O vacancies and Ce^{3+} cations generated by reaction with the CO in the feed.

4.8 CONCLUSIONS

In this chapter we have shown examples where diffraction is used to determine the composition, crystalline properties, and three-dimensional structure of catalytic materials.

In addition, we have shown how *in-situ* diffraction can be used to determine the structure of catalysts under reaction conditions. Time-resolved XRD is an excellent tool to track chemical and structural changes as a function of pressure and temperature. We have shown that the kinetics of reaction during zeolite synthesis and under reduction conditions can be determined. The *in-situ* measurements of temperature-dependent changes of zeolite Cs-Y have shown that catalyst structures change significantly during heating. Furthermore, oxide catalysts undergo changes in composition and structure when present in reducing environments. We have also shown that active intermediates can be identified in the WGS reaction and the MTO reaction.

ACKNOWLEDGMENT

J.C. Hanson is grateful for the financial support of the U.S. Department of Energy, Office of Basic Energy Sciences under Contract No. DE-AC02-98CH10886.

REFERENCES

[1] Thomas, J. M., Thomas, W. J. (1997) *Principles and Practice of Heterogeneous Catalysts*, Wiley-VCH, Weinheim, Germany.

[2] Jenkins, R., Snyder, R. L. (1996) *Introduction to X-ray Powder Diffractometry*, John Wiley & Sons, New York.

[3] Dinnebier, R. E., Billinge, S. J. L. (eds.) (2008) *Powder Diffraction Theory and Practice*, Royal Society of Chemistry, Cambridge, UK.

[4] Barrio, L., Estrella, M., Zhou, G., Wen, W., Hanson, J. C., Hungria, A. B., Hornes, A., Fernandez-Garcia, M., Martinez-Arias, A., Rodriguez, J. A. (2010) Unraveling the active site in copper-ceria systems for the water-gas shift reaction: *in situ* characterization of an inverse powder CeO_{2-x}/CuO-Cu catalyst, *Journal of Physical Chemistry C*, 114, 3580–3587.

[5] Wang, X. Q., Hanson, J. C., Szanyi, J., Rodriguez, J. A. (2004) Interaction of H_2O and NO_2 with BaY faujasite: complex contraction/expansion behavior of the zeolite unit cell, *Journal of Physical Chemistry. B, Condensed Matter, Materials, Surfaces, Interfaces & Biophysical*, 108, 16613–16616.

[6] Wen, W., Calderon, J. E., Brito, J. L., Marinkovic, N., Hanson, J. C., Rodriguez, J. A. (2008) *In situ* time-resolved characterization of Ni-MoO_2 catalysts for the water-gas shift reaction, *Journal of Physical Chemistry C*, 112, 2121–2128.

[7] Chupas, P. J., Chapman, K. W., Kurtz, C., Hanson, J. C., Lee, P. L., Grey, C. P. (2008) A versatile sample-environment cell for non-ambient X-ray scattering experiments, *Journal of Applied Crystallography*, 41, 822–824.

[8] Clausen, B. S., Steffensen, G., Fabius, B., Villadsen, J., Feidenhansl, R., Topsoe, H. (1991) *In situ* cell for combined XRD and online catalysis tests—studies of Cu-based water gas shift and methanol catalysts, *Journal of Catalysis*, 132, 524–535.

[9] Norby, P. (2006) In-situ XRD as a tool to understanding zeolite crystallization, *Current Opinion in Colloid & Interface Science*, 11, 118–125.

[10] He, H. Y., Barnes, P., Munn, J., Turrillas, X., Klinowski, J. (1992) Autoclave synthesis and thermal transformations of the aluminophosphate molecular-sieve Vpi-5—an *in situ* X-ray-diffraction study, *Chemical Physics Letters*, 196, 267–273.

[11] Millange, F., Walton, R. I., Guillou, N., Loiseau, T., O'Hare, D., Ferey, G. (2002) Synthesis and structure of low-dimensional gallium fluorodiphosphates seen during the crystallization of the three-dimensional microporous gallium fluorophosphate ULM-3, *Chemistry of Materials*, 14, 4448–4459.

[12] Norby, P., Hanson, J. C., Fitch, A. N., Vaughan, G., Flaks, L., Gualtieri, A. (2000) Formation of alpha-eucryptite, LiAlSiO4: an in-situ synchrotron X-ray powder diffraction study of a high temperature hydrothermal synthesis, *Chemistry of Materials*, 12, 1473–1479.

[13] David, W. I. F., Shankland, K., McCusker, L., Baerlocher, C. H. (eds.) (2002) *Structure Determination from Powder Diffraction Data*, Oxford University Press Oxford.

[14] Polak, E., Munn, J., Barnes, P., Tarling, S. E., Ritter, C. (1990) Time-resolved neutron-diffraction analyses of hydrothermal syntheses using a novel autoclave cell, *Journal of Applied Crystallography*, 23, 258–262.

[15] Munn, J., Barnes, P., Hausermann, D., Axon, S. A., Klinowski, J. (1992) In situ studies of the hydrothermal synthesis of zeolites using synchrotron energy-dispersive X-ray-diffraction, *Phase Transitions*, 39, 129–134.

[16] Norby, P., Christensen, A. N., Hanson, J. C. (1994) In situ studies of zeolite syntheses using powder diffraction methods—crystallization of instant-zeolite-a powder and synthesis of Coapo-5, *Zeolites and Related Microporous Materials: State of the Art 1994*, 84, 179–186.

[17] Norby, P. (1997) Hydrothermal conversion of zeolites: an *in situ* synchrotron X-ray powder diffraction study, *Journal of the American Chemical Society*, 119, 5215–5221.

[18] Norby, P., Hanson, J. C. (1998) Hydrothermal synthesis of the microporous aluminophosphate CoAPO-5; *in situ* time-resolved synchrotron X-ray powder diffraction studies, *Catalysis Today*, 39, 301–309.

[19] Norby, P., Christensen, A. N., Hanson, J. C. (1999) Crystallization in nonaqueous media of Co- and Mn-substituted microporous aluminophosphates investigated by *in situ* synchrotron X-ray powder diffraction, *Inorganic Chemistry*, 38, 1216–1221.

[20] Carotenuto, L., Piccolo, C., Artioli, G., Norby, P., Grizzetti, R. (2000) Study of the initial stage of zeolite a synthesis by dynamic light scattering and synchrotron X-ray diffraction, *Photon Correlation and Scattering, Technical Digest*, 47, 86–88.

[21] Norby, P., Poshni, F. I., Gualtieri, A. F., Hanson, J. C., Grey, C. P. (1998) Cation migration in zeolites: an *in situ* powder diffraction and MAS NMR study of the structure of zeolite Cs(Na)-Y during dehydration, *Journal of Physical Chemistry. B, Condensed Matter, Materials, Surfaces, Interfaces & Biophysical*, 102, 839–856.

[22] Grey, C. P., Poshni, F. I., Gualtieri, A. F., Norby, P., Hanson, J. C., Corbin, D. R. (1997) Combined MAS NMR and X-ray powder diffraction structural characterization of hydrofluorocarbon-134 adsorbed on zeolite NaY: observation of cation migration and strong sorbate-cation interactions, *Journal of the American Chemical Society*, 119, 1981–1989.

[23] Ciraolo, M. F., Hanson, J. C., Norby, P., Grey, C. P. (2001) An in-situ X-ray powder diffraction study of the adsorption of hydrofluorocarbons in zeolites, *Journal of Physical Chemistry. B, Condensed Matter, Materials, Surfaces, Interfaces & Biophysical*, 105, 2604–2611.

[24] Ciraolo, M. F., Hanson, J. C., Toby, B. H., Grey, C. P. (2001) Combined X-ray and neutron powder refinement and NMR study of hydrochlorofluorocarbon HCFC-124a (CF2HCF2Cl) binding on NaX, *Journal of Physical Chemistry. B, Condensed Matter, Materials, Surfaces, Interfaces & Biophysical*, 105, 12330–12337.

[25] Dahl, I. M., Kolboe, S. (1993) On the reaction-mechanism for propene formation in the MTO reactionover SAPO-34, *Catalysis Letters*, 20, 329–336.

[26] Dahl, I. M., Kolboe, S. (1994) On the reaction-mechanism for hydrocarbon formation from methanol over SAPO-34 .1. Isotopic labeling studies of the co-reaction of ethylene AND methanol, *Journal of Catalysis*, 149, 458–464.

[27] Dahl, I. M., Kolboe, S. (1996) On the reaction mechanism for hydrocarbon formation from methanol over SAPO-34 .2. Isotopic labeling studies of the co-reaction of propene and methanol, *Journal of Catalysis*, 161, 304–309.

[28] Wragg, D. S., Johnsen, R. E., Balasundaram, M., Norby, P., Fjellvag, H., Gronvold, A., Fuglerud, T., Hafizovic, J., Vistad, O. B., Akporiaye, D. (2009) SAPO-34 methanol-to-olefin catalysts under working conditions: a combined *in situ* powder X-ray diffraction, mass spectrometry and Raman study, *Journal of Catalysis*, 268, 290–296.

[29] Wragg, D. S., Johnsen, R. E., Norby, P., Fjellvag, H. (2010) The adsorption of methanol and water on SAPO-34: in situ and ex situ X-ray diffraction studies, *Microporous and Mesoporous Materials*, 134, 210–215.

[30] Rodriguez, J. A., Hanson, J. C., Frenkel, A. I., Kim, J. Y., Perez, M. (2002) Experimental and theoretical studies on the reaction of H-2 with NiO: role of O vacancies and mechanism for oxide reduction, *Journal of the American Chemical Society*, 124, 346–354.

[31] Hammersley, A. P., Svensson, S. O., Thompson, A. (1994) Calibration and correction of spatial distortions in 2D detector systems, *Nuclear Instruments & Methods in Physics Research*, 346, 312–321.

[32] Larson, A. C., von Dreele, R. B. (1995) *GSAS General Structure Analysis Program*. Los Alamos National Laboratory: Los Alamos, NM.

[33] Readman, J. E., Olafsen, A., Smith, J. B., Blom, R. (2006) Chemical looping combustion using $NiO/NiAl_2O_4$: mechanisms and kinetics of reduction- oxidation

(Red-Ox) reactions from in situ powder X-ray diffraction and thermogravimetry experiments, *Energy and Fuels*, 20, 1382–1387.

[34] Trovarelli, A. (ed.) (2002) *Catalysis by Ceria and Related Materials*, World Scientific, London.

[35] Rodriguez, J. A., Fernandez-Garcia, M. (eds.) (2007) *Synthesis, Properties, and Applications of Oxide Nanomaterials*, John Wiley & Sons, Hoboken, NJ.

[36] Rodriguez, J. A., Wang, X. Q., Hanson, J. C., Liu, G., Iglesias-Juez, A., Fernandez-Garcia, M. (2003) The behavior of mixed-metal oxides: structural and electronic properties of $Ce_{1-x}Ca_xO_2$ and $Ce_{1-x}Ca_xO_{2-x}$, *Journal of Chemical Physics*, 119, 5659–5669.

[37] Wang, X. Q., Rodriguez, J. A., Hanson, J. C., Gamarra, D., Martinez-Arias, A., Fernandez-Garcia, M. (2006) *In situ* studies of the active sites for the water gas shift reaction over Cu-CeO_2 catalysts: complex interaction between metallic copper and oxygen vacancies of ceria, *Journal of Physical Chemistry. B, Condensed Matter, Materials, Surfaces, Interfaces & Biophysical*, 110, 428–434.

[38] Frenkel, A. I., Wang, Q., Marinkovic, N., Chen, J. G., Barrio, L., Si, R., Camara, A. L., Estrella, A. M., Rodriguez, J. A., Hanson, J. C. (2011) Combining X-ray absorption and X-ray diffraction techniques for *in situ* studies of chemical transformations in heterogeneous catalysis: advantages and limitations, *Journal of Physical Chemistry C*, 115, 17884–17890.

5

PAIR DISTRIBUTION FUNCTION ANALYSIS OF HIGH-ENERGY X-RAY SCATTERING DATA

KARENA W. CHAPMAN AND PETER J. CHUPAS

5.1 INTRODUCTION

Pair distribution function (PDF) analysis is an emerging structure characterization tool with the potential to provide valuable new insights into a broad range of problems in heterogeneous catalysis. The method yields atomic scale structural information with crystallographic resolution, without the need for crystalline order implicit in conventional Bragg diffraction analysis. Accordingly, it can be used to study the nanoscale and/or defect-rich materials systems often involved in heterogeneous catalysis.

While the methodology itself is not new [1–3], the application of PDF methods to address problems in catalysis, and complex materials in general, is less developed than many other techniques, due to previously limited access to high-quality data. Since the early 2000s, advances in experimental approach, instrumentation, and detector technologies have increased access to PDF data, providing improved data quality in substantially reduced measurement times [4, 5]. Accordingly, it has now become possible to conduct quantitative PDF measurements with appropriate sensitivity and on timescales relevant to the structural changes in heterogeneous catalysts.

The use of PDF methods to understand and optimize heterogeneous catalysts is rapidly expanding and the potential of this approach is only now being

In-situ *Characterization of Heterogeneous Catalysts*, First Edition.
Edited by José A. Rodriguez, Jonathan C. Hanson, and Peter J. Chupas.

fully explored [6–21]. Studies are being extended beyond determining the static structure of catalytic materials, and any changes that may occur following *ex-situ*/offline treatment, to *in-situ* studies under nonambient conditions relevant to catalytic synthesis, operations, or deactivation. This includes variable temperature, pressure, and reactive chemical environments. Demonstrated *in-situ* applications involve studying the formation and growth of catalytic-supported nanoparticles providing quantitative kinetic and mechanistic insights into the reaction with activation energies for reduction and surface diffusion [8, 16], probing the atomic structure of catalytic surface sites with crystallographic detail [22], understanding the dynamic changes that occur within the catalyst during operation to identify the active state [19], and probing the structural changes and conditions associated with catalyst deactivation through nanoparticle aggregation or coarsening, or poisoning the active sites [23].

5.1.1 Definition

The PDF describes atomic structure on the local and intermediate length scales. It is defined as the function $G(r) = 4\pi r[\rho(r) - \rho_o]$, where r is the interatomic distance, ρ_o is the average number density, and $\rho(r)$ is the local number density $\rho(r) = \rho_o[g(r)]$ [24, 25]. This incorporates the function $g(r)$, the direct probability of finding two atoms at a distance r. As such, the structural information is obtained as a histogram of the distance between all pairs of atoms within a material. This is weighted by the scattering power and the relative abundance of that atom pair. Historically, various nomenclatures and related definitions have been used to describe this type of atomic correlation function. This includes the term radial electron distribution (RED) the total correlation function, $T(r)$, and the radial distribution function, $R(r)$ [26, 27]. The form presented here, $G(r)$, is that most directly derived from the experimental data and most commonly used in modern X-ray PDF studies.

A complete description of the different atomic correlation functions can be found in Keen [27], and the mathematical foundation underlying PDF analysis has been described in detail by Warren in his book *X-Ray Diffraction* and, more recently, by Egami and Billinge in *Underneath the Bragg Peaks* [24, 25].

5.2 STRUCTURAL INSIGHTS FROM THE PDF DATA

The structural insights within the PDF extend from the local coordination environment up to several nanometers, thus spanning a range pertinent to the structure of active sites or reagent binding, approaching the length scale of the nanoparticles relevant to heterogeneous catalysis. In many ways, the structural information available from a PDF measurement is analogous to insights from extended X-ray absorption fine structure (EXAFS), diffraction, and small-angle X-ray scattering (SAXS) studies.

5.2.1 Model Independent Analysis

A wealth of information can be obtained directly from particular features in the PDF without the need for a structure model (see Fig. 5.1). By monitoring changes to particular features during *in-situ* studies, detailed insights can often be derived, including quantitative kinetics, changes in concentration, or changes in atomic structure. The large quantity and highly consistent nature of data possible using *in-situ* methods is key to enabling these types of studies and allows quantitative insights and correlated parameters to be identified *ab initio*, directly from the PDF data.

- The *position* of peaks in the PDF corresponds directly to the bond lengths and atomic distances within a material. Determining the position of a peak, for example, through fitting of a Gaussian function (see Fig. 5.2), provides an accurate measure of the atomic distances. In contrast, the peaks evident in an EXAFS analysis must be corrected for the phase shift, through development of a structure model, to determine the bond length. Furthermore, in the PDF, longer atomic distances can be readily determined. The atomic distances can be used to derive accurate coordination geometries, to develop structure models, or to identify changes in oxidation state.
- The *area* of a peak in the PDF is proportional to the abundance of the associated atom pair. By quantifying changes in peak area, for example, through fitting of a Gaussian function (see Fig. 5.2), changes in the abundance of the correlation can be monitored. For a well-defined cluster,

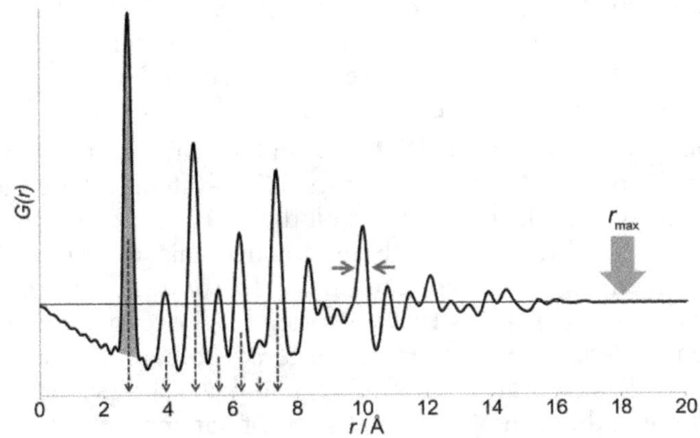

Figure 5.1. An example of a pair distribution function, $G(r)$. This PDF corresponds to nanoparticles with an fcc structure with dimension ~18 Å; no peaks are evident beyond this maximum r-value. The position, area, and width of peaks within the PDF are linked to atomic distances, coordination numbers/concentrations, and disorder within the material.

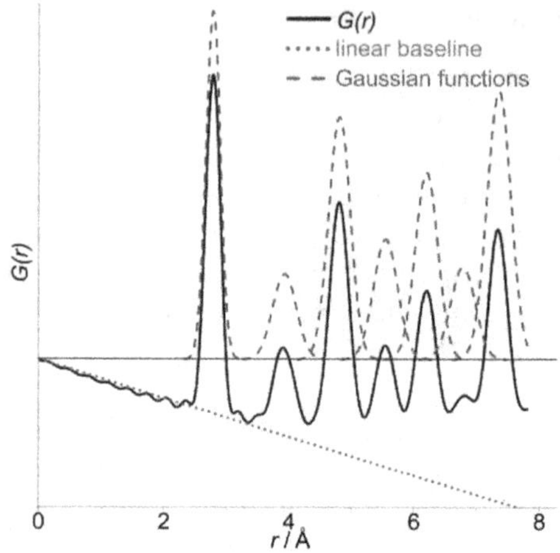

Figure 5.2. The position, area, and width of peaks within the PDF can be accurately determined by fitting of the Gaussian functions. A linear baseline fit to the data, with slope $-4\pi\rho_o$ and y-intercept = 0, provides an estimate of the density.

molecule, or polyatomic ion, the peak area is directly correlated to concentration. For nanoparticles, the peak area is additionally linked to particle size, with the average number of nearest neighbors decreasing for small particles with a higher surface area-to-volume ratio. This effect, however, is less pronounced for larger nanoparticles (beyond ~2 nm) than for small clusters (see Fig. 5.3). By evaluating the changes in peak area and, accordingly, species concentration though a reaction, models for the reaction mechanism can be developed and tested and rate constants accurately determined. Extension of this analysis to reactions performed at several temperatures allows activation energies for each reaction step to be accurately determined.

- Well-defined features in the PDF exist in the range over which the material is well ordered. Thus for bulk, crystalline systems, where the material is ordered to long length scales, well-defined features persist to high r (essentially infinite r, although in practice this is ultimately limited by instrument/experimental resolution). In contrast, for amorphous or nanoscale systems, where structural order exists over a finite range, peaks are clearly defined only over the ordered domain. Consequently, the *maximum distance r* at which well-defined peaks are evident in the PDF provides an indication of the dimensions of nanoparticles or the ordered domain size in amorphous systems. This can be modeled quantitatively using a spherical particle envelope [28, 29]. The insights into nanoparticle size, and how this can change during reactions, are valuable in understanding supported catalyst synthesis, catalyst deactivation through nanoparticle coarsening, and catalyst restructuring under reactive conditions. Maximal

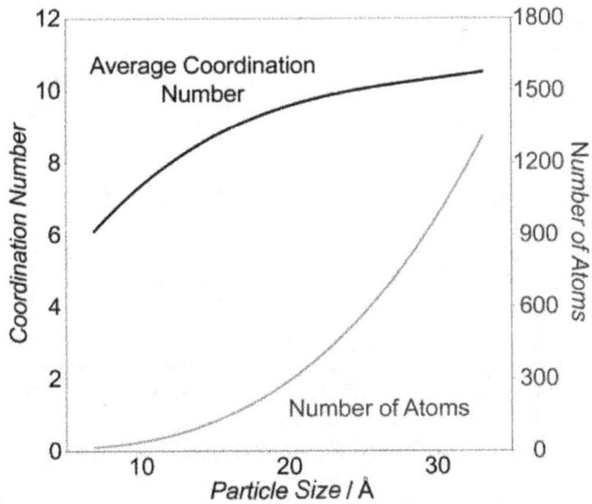

Figure 5.3. An example of how the average coordination number depends on particle size, in this case, for Pt nanoparticles with an fcc structure where the bulk coordination number is 12. Changes in the area of a peak within the PDF are not only correlated to the abundance of nanoparticles, but also to the particle size. This effect is less pronounced for larger particles.

particle dimensions up to ~10 nm can be accurately determined with no lower size limit. Thus, the PDF can be used to document the nucleation and growth of catalytic nanoparticles from the smallest initial multiatom clusters to larger nanoparticles in the range relevant to catalysis (~2–3 nm). The insights into nanoparticle size are similar to those available from an SAXS analysis, although for larger nanoparticles SAXS analysis provides greater accuracy.

- The *slope* of the PDF at low *r* is related to the sample density.

5.2.2 Modeling PDF

Separate from the insights that can be obtained directly from the PDF data, structural models can be tested or refined against the PDF data to determine the structure of the catalyst and how it may change under reactive conditions. Different approaches have been established to model PDF data. In modeling the data, it is possible to select the *r* range of the PDF used in the refinement, to focus on structural ordering over particular length scales. Further, based on models, it is possible to calculate the contributions to the PDF from different atom pairs—the partial PDFs—which can be valuable in identifying the origin of structural perturbations (see Fig. 5.4). Several programs are widely used to model PDF data, including PDFgui [30] and DISCUS [31].

- A real space analogue to Rietveld analysis of powder diffraction data is the most common approach to modeling PDF data. This applies periodic

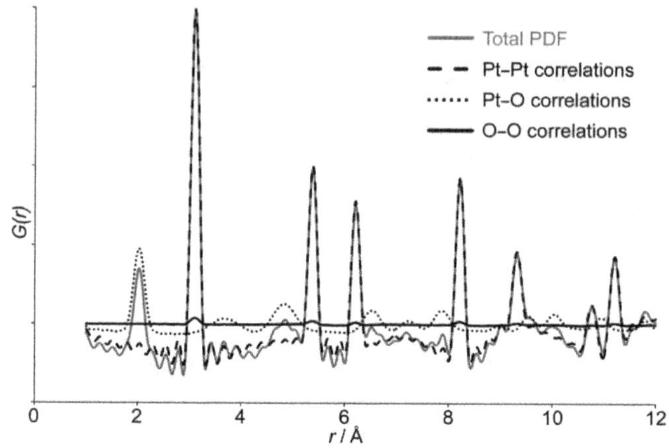

Figure 5.4. The partial pair contributions to the PDF, calculated from models fit to data obtained for a hexagonal PtO_2 phase, corresponding to correlations between different atom types. The impact of the scattering power of different species is evident, with the contributions from strongly scattering Pt–Pt correlations dominating the PDF.

boundary conditions to a single unit cell. Atomic positions are subject to space group symmetry constraints, atomic displacement parameters model static and dynamic disorder, and a spherical particle envelope function models the average dimension of nanoparticles or length scale of structural coherence in disordered systems. Models are optimized through least squares methods. This approach can be used to identify and quantify phases, refine atomic structure and stoichiometry, and estimate particle size, including for nanoscale systems where size broadening of Bragg peaks limits the crystallographic analysis. This can be applied to bulk phases and nanoscale materials; however, this approach has limited the ability to model discrete clusters or surface binding interactions. It is important to note that the absolute value of the R factors obtained from fits to PDF data are larger than conventional Rietveld fits of comparable quality, so to obtain the PDF, the scattering data are multiplied by a factor of Q, thus amplifying the noise.

- Extended models derived from multiple independent unit cells with periodic boundary conditions, not subject to symmetry constraints, can be optimized using reverse Monte Carlo algorithms [32]. This can be applied where the local defects or distortions cannot be described within a simple unit cell model. These models are often underconstrained by the PDF data and can benefit from additional constraints from complementary data (e.g., local structure constraints from EXAFS or nuclear magnetic resonance [NMR] spectroscopy).

- Structural models developed through computational simulations can be readily evaluated against experimentally obtained PDF data.
- Explicit atomistic models of discrete clusters or whole nanoparticles can be constructed. For nanoparticles, it is often important to construct an ensemble of nanoparticles, for example, with a distribution of defects [33, 34].

5.2.3 Differential Analysis

While not intrinsically an element-, chemical-, or surface-specific probe, by extending the PDF experiment to measure one or more additional related systems, such specificity can be recovered and insights greatly enhanced. Specifically using a differential analysis, the PDF data are treated to focus on particular species or features within the sample (Fig. 5.5). As the PDF is summed over all interatomic distances within a material, the contributions from multicomponent systems (e.g., the support and nanoparticles in a supported metal catalyst) are purely additive. Consequently, if the PDF from one component can be explicitly measured, this "reference PDF" can be subtracted from the total PDF for the multicomponent system, thus isolating the contribution from the second component, as well as any well-defined intercomponent interactions. For example, the PDF for a pristine support can be measured under the same conditions (of temperature, etc.) as the supported metal catalyst. This PDF can be directly subtracted from that measured for the catalyst system (under either static or reactive conditions), to recover the PDF for the

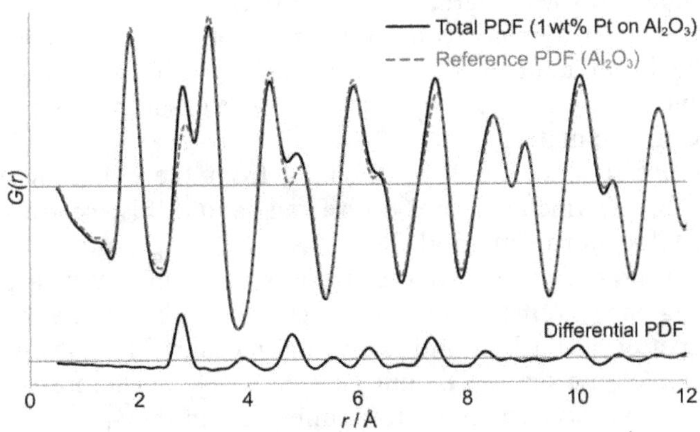

Figure 5.5. A differential approach can be used to recover chemically specific PDF information. In this case, for 1 wt% Pt nanoparticles on γ-Al_2O_3, a reference PDF measured for the γ-Al_2O_3 support alone can be subtracted from the total PDF to yield the differential PDF which isolate contributions from supported Pt particles (as well as any well-defined Pt–support interactions).

supported nanoparticles and any nanoparticle–support interactions. The differential PDF can then be analyzed as described previously.

The advantage of obtaining the differential PDF in this way, rather than calculating the difference in the scattering intensity $I(Q)$ before the Fourier transform, is that any artifacts in the PDF data are generally consistent between the total and reference PDF such that they cancel in the differential PDF. Notably, features at low r, where any artifacts in the PDF data are most pronounced, for example, from errors in normalization or the truncation of the Fourier transform, can be readily distinguished. This allows reliable analysis of dilute components within the system with demonstrated sensitivity to components at the 1 wt% level and below. This detection sensitivity is relevant to the concentration of the supported metal catalysis, typically 1 wt% or below, or the surface of high surface area catalysts.

In this way, and through careful experiment design and reference measurements, chemical specificity or surface sensitivity can be derived. While EXAFS offers intrinsic element specificity, explicitly only probing a single element within the sample, the differential approach to PDF is more broadly applicable, and not limited to systems with suitable absorption edges. For example, active catalytic sites in aluminosilicate zeolites are extremely difficult to probe using EXAFS.

5.3 MEASURING A PDF

The PDF is obtained experimentally from a scattering measurement. It is the sine Fourier transform of the reduced structure function $G(r) = 2/\pi \int Q[S(Q) - 1] \sin(Qr)dQ$, where Q is the momentum transfer (a parameter related to the scattering angle and wavelength, $Q = 2\pi \sin \theta/\lambda$).

PDF data can be derived from either neutron or X-ray scattering data. However, the lower neutron fluxes and much longer data collection times make neutrons less readily amenable to *in-situ* measurements on timescales relevant to catalytic materials.

The structure function $S(Q)$ is obtained from the coherent part of the scattering intensity, which contains both Bragg and diffuse scattering components. As a total scattering method, the diffuse scattering contains both thermal diffuse scattering and diffuse scattering arising from static displacements. The scattering measurement must be accurately corrected for incoherent (Compton) scattering, background scattering (measured for the empty sample holder or sample environment), multiple scattering, and fluorescence. Errors in these corrections, for example, from ambiguity in the sample composition (and, accordingly, the Compton scattering correction), produce slowly varying errors in $S(Q)$, which in the Fourier-transformed PDF are apparent as artifacts at low r, that is, at unphysical distances. In practice, the magnitude of these low r artifacts provides an indication of the accuracy of the data correction and PDF data.

The resolution of the PDF, being related to the scattering data via a Fourier transform, depends on the Q range measured and used in the Fourier transform. Thus, $S(Q)$ must be measured to a high value of Q, typically >20 $Å^{-1}$ for crystalline materials. The finite Q range that is adequate depends on the magnitude of the Debye–Waller factor. A higher Q range is needed for more ordered systems.

Historically, the Fourier transform calculation, the high Q range, and the excellent signal-to-noise ratio required have made PDF measurements an experimental challenge and limited the quality and quantity of X-ray PDF data available. Access to high-quality PDF data has been improved through a number of distinct technological and experimental innovations:

1. *Computing:* accelerates calculation of the Fourier transformation
2. *Synchrotron sources:* intense X-ray beams
3. *Experimental approach:* high-energy X-rays + area detectors
4. *Detector technologies:* improved sensitivity and fast readout.

5.3.1 Current Approach to PDF Measurements

A modern X-ray PDF experiment is essentially a specialized powder diffraction measurement in transmission geometry. Measurements can be performed for a broad range of materials, including crystalline and amorphous materials, heterogeneous or homogeneous materials, bulk or nanomaterials, or materials in solid, liquid, or other states. High-energy X-rays (>50 keV) are used to compress a broad Q range of X-ray scattering data into a finite 2θ range defined by the detection surface of a large two-dimensional detector. This detector is oriented perpendicular to the incident beam, and positioned a short distance from the sample (10–40 cm), to maximize the 2θ range (and Q range) intersected by the detector (Fig. 5.6). Using this approach, all data for an individual PDF are collected simultaneously, such that data rate and measurement times are limited only by the signal-to-noise ratio and the readout rate of the detector. Measurement times have been reduced by several orders of magnitude compared with traditional experimental approaches to X-ray PDF measurements—from hours to minutes or seconds. Historically, X-ray PDF measurements have used 2θ scans of a point counting detector, typically requiring ~12 h to accumulate sufficient counting statistics at each point. The solid detector angle afforded by the area detector provides increasing counting statistics at higher angles (where the X-ray form factor makes improved statistics most beneficial). The area detectors used for modern PDF measurements do not discriminate based on X-ray energy as was previously achieved using an analyzer-equipped scanning point counting detector. Consequently, the inelastic Compton scattering is measured and must be accurately calculated based on the known sample composition and subtracted from the measured intensities.

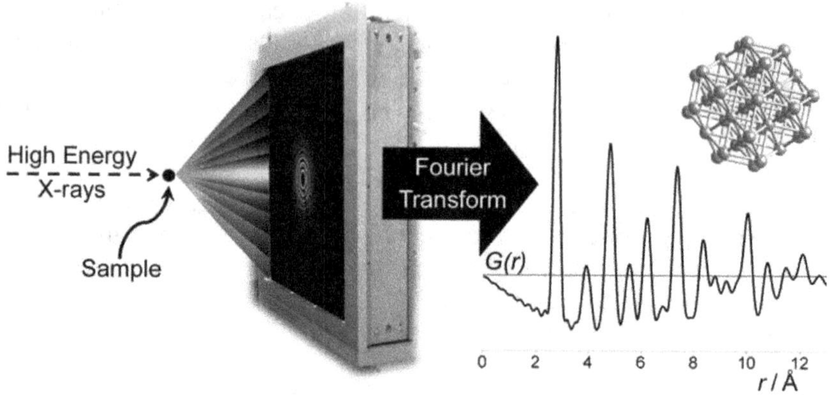

Figure 5.6. Modern PDF measurements combining high-energy X-rays (>50 keV) with large area detectors to efficiently measure scattering data over a wide Q-range.

Modern PDF measurements have benefited significantly from advances in detector technology, in particular, by adapting amorphous Si-based detectors originally developed for medical imaging applications [5]. Conventional detectors for X-ray scattering measurements are poorly sensitive to the high-energy X-rays needed for PDF measurements. Medical imaging applications use high-energy X-rays. To minimize patient doses, medical imaging detectors are optimized for high sensitivity to high-energy X-rays and for fast readout. Not only are these detectors more efficient than other X-ray area detectors at the high energies (80% cf. 2% efficient), but the fast detector readout allows a large number of images to be efficiently averaged, improving signal-to-noise ratio. The fast readout also allows fast time-resolved studies in *in-situ* catalytic processes and reactions.

5.3.2 Working with High-Energy X-rays

There are several consequences of using high-energy X-rays beyond the ability to access a high Q range:

- Absorption is minimized at high X-ray energies, and thus "beam damage" and sample heating in the beam, which scale with absorption, are negligible. In contrast, it has been shown for other methodologies, for example, EXAFS, that probing the sample can, in itself, modify the catalytic system. Corrections for sample geometry are unnecessary.
- High-energy X-rays are highly penetrating, and thus probe the whole sample. Accordingly, the method is not sensitive to inhomogeneities in the sample, which can be a concern in energy-dispersive X-ray absorption spectroscopy (XAS) measurements.

- High-*Z* elements have absorption/fluorescence edges at energies comparable to those used for the X-ray PDF measurements. For X-ray beam energies that are at higher energy edges for elements in the sample (within ~20 keV), fluorescence from the sample can have a large uniform contribution to the sample scattering. This inelastic scattering must be corrected before extracting the PDF or, more ideally, the X-ray beam energy chosen to avoid or minimize fluorescence effects. In practice, the X-ray beam energy should be below any sample fluorescence edges or as far above the sample fluorescence edge as is reasonable.
- Multiple scattering effects are minimized at high energy.

5.4 *IN-SITU* STUDIES

5.4.1 Sample Environment Needs

Catalysts are highly dynamic and respond to their environment. To understand their structure and activity under reaction conditions, sample environment cells are needed to control the sample temperature, pressure, and chemical atmosphere without compromising the X-ray PDF measurement. The highly penetrating nature of the high-energy X-rays allows versatility in the materials and design of the sample environment for *in-situ* studies. However, as the PDF makes use of the total scattering data, and not a selected portion of the data or sample (e.g., Bragg peaks in diffraction analysis, low-angle scattering in SAXS, or absorption in an individual element in EXAFS), more stringent criteria must be considered in sample environment design. A consequence of the more severe design constraints is that with little additional consideration, any sample environment optimized for PDF measurements can also be used for *in-situ* measurements involving other techniques including powder X-ray diffraction (PXRD), SAXS, and EXAFS.

Reproducible background: Unlike Bragg crystallographic analysis where the "background" scattering between the peaks can be treated empirically, PDF uses both Bragg and diffuse contributions to the measured scattering. The background scattering from the sample environment, air, and so on must be accurately measured for the empty cell and subtracted. Accordingly, the contribution of the sample environment to the background must be highly reproducible and, ideally, low.

Large scattering aperture: A sample environment must allow access to a broad 2θ range. It is possible to offset the aperture, or tilt the cell in the beam, to extend the 2θ range measured.

Physical constraints: The sample environment needs to be relatively compact along the beam direction such that the detector and beam stop can be placed close to the sample. This also allows background-minimizing shielding to be placed close to the sample.

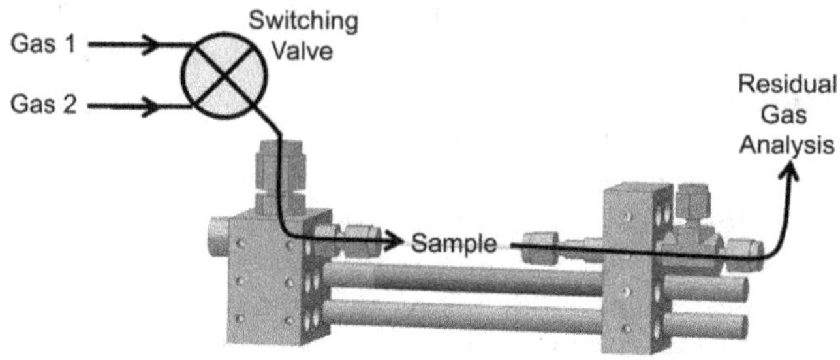

Figure 5.7. A capillary-based sample environment cell suitable for PDF measurements. The gas flow through the sample plug can be switched to initiate reactions or to emulate catalyst operating conditions. The exhaust gas can be diverted for subsequent analysis.

5.4.2 A Versatile Environment for Catalytic Studies: The Flow Cell–Furnace

For typical X-ray PDF measurements, powdered samples are contained within capillaries of ~1 mm diameter. Polyimide (Kapton®, DuPont, Wilmington, DE) or various glass capillaries can be used, depending on the temperature range of interest. Such capillary-loaded samples can be readily incorporated into sample environments for *in-situ* studies of catalytic systems, while retaining access to a broad 2θ range (see Fig. 5.7). Gases can be flowed through a short plug of the sample powder within the capillary. The temperature of the sample can be controlled using an open-flow cryostream or resistively heated furnace elements above or below the sample. Gas switching mechanisms can be incorporated in the inlet flow to initiate reactions or replicate on-stream catalytic conditions. The exhaust gas can be diverted for residual gas analysis to identify or quantify reaction products.

5.5 APPLICATIONS OF *IN-SITU* PDF METHODS TO PROBLEMS IN CATALYSIS

5.5.1 Reagent Binding and the Structure of Catalytically Active Surface Sites

Determining the interaction or binding of molecules at the active surface of a catalyst is important in understanding the catalytic mechanism. However, until recently, these surface interactions could be inferred only from vibrational spectroscopic measurements and others. Through the application of differential PDF methods, it is now possible to directly probe the atomic structure of molecules bound at the surface of nanoparticles or high surface area heterogeneous catalysts with crystallographic detail. In this type of experiment, data are collected for the pristine catalyst, as a "reference PDF," before exposure to the probe molecule and measurement of a PDF for the surface-bound system. The reference PDF contains only the correlations between all atoms in the bulk catalyst, so when subtracted from the PDF for the surface-bound system, the correlations within the probe molecule itself and between the probe molecule and surface-binding sites are selectively recovered. This approach represents a powerful method for probing the surface of nanomaterials that is ideally suited to studies of catalysts and retrieves structural information that is not possible to retrieve through other methods.

The surface differential approach has been used to understand the acid sites on high surface area γ-Al_2O_3 through binding of the basic probe molecule monomethyl-amine (CH_3NH_2, MMA) [22]. MMA was selected as an ideal probe molecule for this type of differential PDF analysis. The increased complexity of MMA compared with simpler molecules, such as NH_3, with a small number of well-defined intramolecular correlations, can be used to confirm and quantify the MMA loading. Larger probe molecules such as triphenylphosphine, with more numerous intramolecular correlations, would make it more difficult for the separate contributions from the inter- and intramolecular interactions to be more reliably assigned in the differential PDF.

In this experiment, the γ-Al_2O_3 sample was loaded into the flow cell/furnace and dehydrated at elevated temperature, before cooling to 80 K under inert (He) atmosphere. Data suitable for PDF analysis under inert atmosphere (the reference PDF) and following exposure to MMA were collected. The differential PDF, shown in Figure 5.8, was obtained by subtracting the reference PDF from that obtained for the MMA loaded sample. Thus, the differential PDF contains both the intramolecular correlations of the MMA molecule and the correlations arising from the interaction of the MMA molecule with the

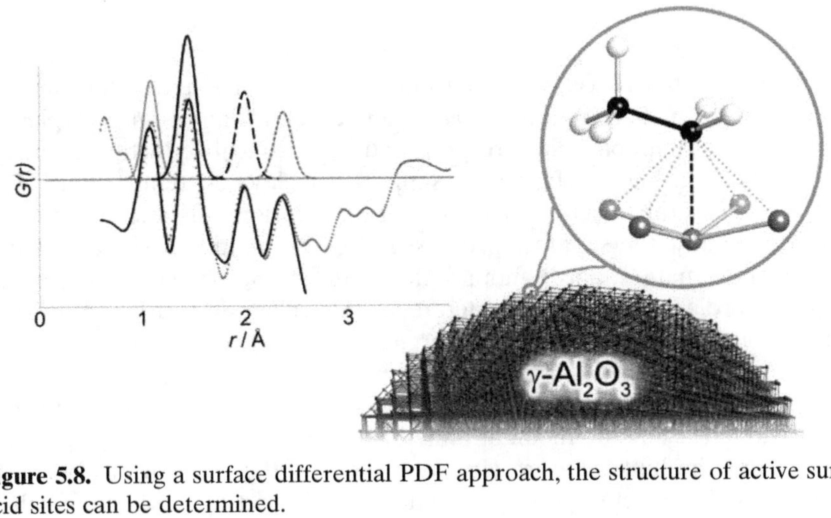

Figure 5.8. Using a surface differential PDF approach, the structure of active surface acid sites can be determined.

surface of the γ-Al_2O_3 particles. Several well-defined correlations are evident in the differential PDF below 3 Å, centered at ~0.9, 1.5, 2.0, and 2.5 Å. The MMA molecule itself has C/N–H and C–N bond lengths 0.95 and 1.45 Å, respectively, and second neighbor C–H and N–H distances of ~2.1 Å. The presence of additional distinct correlations in the differential PDF (e.g., at 2.5 Å), indicates that the MMA is bound strongly to a Lewis acid site, forming a well-defined bond to the catalyst surface. By contrast, the differential PDFs corresponding to physisorbed molecules are broad, owing to dynamic disorder in the weakly bound molecule, and are typically associated with longer distances [35].

The precise atomic distances determined from fitting Gaussian functions to the peaks in the differential PDF data were used to develop models of the active acid site. Based on a comparison of models with different AlO_x coordination, it was determined that the Lewis acid sites were five coordinate, forming a distorted octahedron upon binding MMA. This clear understanding of the structure of the acid site could not be obtained using molecular modeling, and, due to the low Z of the sample, an analogous experiment could not be performed using EXAFS.

This surface structure analysis was only made possible using the differential approach; direct refinement of models for the MMA-loaded sample would be nearly impossible due to the defect-rich structure of bulk γ-Al_2O_3, which is itself challenging to model, such that the contribution from the surface-bound MMA is smaller than the residual to the fit of the bulk phase. In this system, these acid sites account for 1 in 8 of the surface Al, or, approximately 1 in 50 of all Al, indicating sensitivity to MMA binding at a level of ~1.6 wt%.

5.5.2 Quantitative Reaction Kinetics and Mechanisms from Time-Resolved PDF Analysis

Supported metal nanoparticles find widespread application in catalysis. Controlling the size and structure of these catalytic nanoparticles is pivotal to controlling their reactivity; hence, an understanding of kinetics and mechanisms that govern their formation is of fundamental importance. Typically, supported metal nanoparticles are prepared by chemical reduction of metal salts dispersed on high surface area supports at elevated temperature. The nanoscale structural insights and fast time resolution possible using PDF methods can allow the separate processes of cation reduction, surface diffusion, cluster formation, and nanoparticle sintering and growth to be distinguished. Accordingly, detailed multistep reaction mechanisms can be delineated, and rate constants and activation energies estimated for different reaction steps. Using differential PDF methods, where the reference PDF is measured for the pure support, atom–atom correlations that arise from the supported material alone (i.e., metal–metal and metal–anion contacts) can be separated from those of the bulk support. (In these systems, the correlations between the support and the metal species are typically weak and not well defined; they do not contribute strongly to the PDF.) Accordingly, using time-resolved PDF measurements, it is possible to monitor changes in the abundance and structure of the precursor and eventual nanoparticles throughout the synthesis. Reactions can be initiated by switching from inert to reactive gas flows or by using a temperature impulse and collecting time-resolved data suitable for PDF analysis. The concentration of reagents or reaction products, determined by fitting Gaussian functions to peaks in the differential PDF, can be used to evaluate different models for the reaction mechanism and the associated rate constants. Using an Arrhenius plot for rate constants obtained at different temperatures, the activation energy for the reaction steps can be accurately determined.

Time-resolved differential PDF methods were used to investigate the formation of Pt nanoparticles supported on TiO_2 from a Pt^{4+} precursor salt ($PtCl_6^{2-}$). Initially, *ex-situ* PDF measurements were conducted for supported metal nanoparticles prepared offline with different loading levels to establish the sensitivity of the approach [9]. Even at 1 wt% level, the contribution from the Pt nanoparticles could be clearly determined. Time-resolved PDF data were collected at selected temperatures using a gas switching valve to initiate the reaction by switching from inert He gas to reductive H_2 (4% in He) gas flow. The temperatures for isothermal studies were selected based on earlier variable temperature (heating) measurements under reactive gas atmospheres, which helped identify the temperature range relevant to the reaction.

The differential PDF of the initial material features a peak at 2.5 Å corresponding to the Pt–Cl bond within the $PtCl_6^{2-}$ anion. This progressively disappears and a peak at 2.8 Å, corresponding to the Pt–Pt nearest-neighbor correlation, develops rapidly during the reaction (Fig. 5.9). The changes in intensities of these first shell Pt–Cl precursors and Pt–Pt product correlations, estimated from Gaussian fits to the peaks, reflect changes in species concentration and, in the case of the growing Pt nanoparticles, changes in coordination number. The intensity of the Pt–Cl correlation, from the discrete $PtCl_6^{2-}$ anion, is directly proportional to the concentration of $PtCl_6^{2-}$ ions and thus provides the most direct measure of the reaction progress.

Different kinetic models of the reduction mechanism were tested against the changes in the Pt–Cl peak intensity, that is, $PtCl_6^{2-}$ concentration. A zero-order kinetic model provided the best fit to the data. The rate constants from kinetic models fit to the data at different temperatures, indicating an activation energy for the $PtCl_6^{2-}$ reduction of 50 kJ/mol, in close agreement with established values. Zero-order kinetics, while rarely observed, can apply to reactions at surfaces with a fixed number of catalytic reaction sites (e.g., defects), and suggests that supported nanoparticle formation under reductive gas atmospheres is dominated by the support and existing surface defect sites. Importantly, using the PDF method, these insights can be derived for the gas–solid phase reactions directly relevant to industrial processes, rather than indirect solution-based studies for which different kinetic models apply [36].

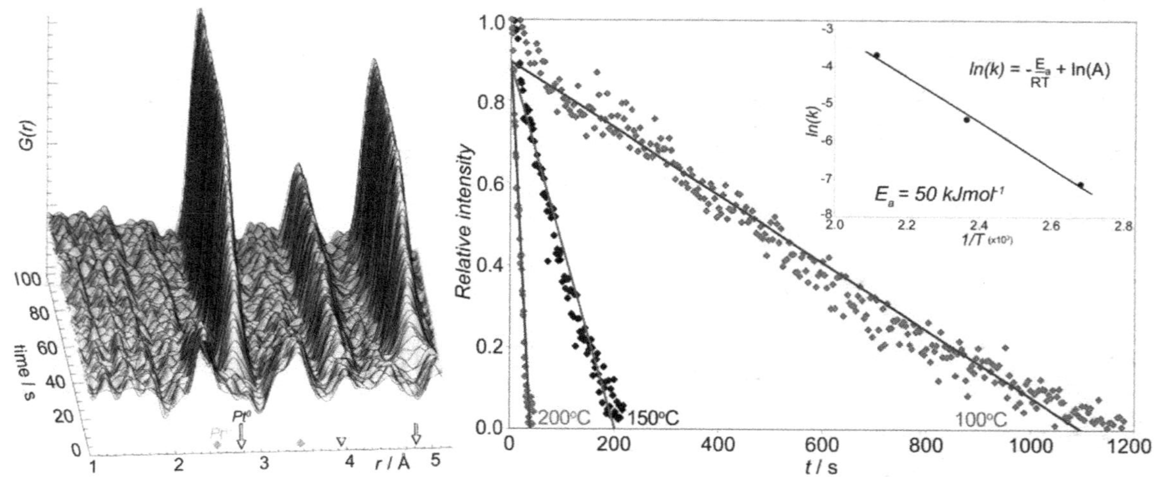

Figure 5.9. Left: The time-resolved differential PDFs during the reduction of $PtCl_6^{2-}$ to form supported Pt nanoparticles. Right: The changes in the intensity of the Pt–Cl correlation at 2.5 Å from the precursor anion reflect the reaction progress and indicate zero-order reaction kinetics. Studies at multiple temperatures indicate that the activation energy for the $PtCl_6^{2-}$ reduction is 50 kJ/mol.

5.5.3 Correlating Structure and Reactivity: Identifying Active Surface Species in Reactive Systems

Identifying the structure–activity relationship of realistic catalysts under realistic operating conditions is the first key step in identifying the active sites in a working catalyst. Supported metal systems that are applied industrially typically contain 1 wt% or less of the active metal catalyst, which form small nanoparticles of a few nanometer diameters. The highly reactive nature of these nanoparticles underlies their highly dynamic structure, which can vary in response to their changing chemical environment, including variations in feedstock composition and introduction of poisons. While attempts have been made to use extended metal surfaces, studied using surface science techniques, to simulate changes in the surface structure of real catalysts, it is not clear how representative the flat surface case is of realistic catalytic nanoparticles. Using differential PDF methods, with second or subsecond time resolution, the response of catalytic nanoparticles to a modulated chemical environment can be probed. By correlating the changes in feedstock and catalyst activity with changes in features in the PDF, the nature of active sites being formed and consumed within the catalytic process can be identified.

The *in-situ* time-resolved differential PDF measurements under modulated chemical environments have been used to characterize the active species in an archetypal catalytic system—supported Pt nanoparticles (1 wt% Pt on Al_2O_3) catalyzing CO oxidation ($CO + 1/2O_2 \rightarrow CO_2$). This catalyst is used widely and has been the subject of many studies based on other techniques including EXAFS and surface diffraction. At elevated temperature, the catalyst sample was subjected to feedstock that switched between gas mixtures containing CO and O_2 in either a stoichiometric ratio or with fivefold excess of O_2. This emulates conditions in a catalytic converter application. Feedstock switching occurred at 30-s intervals while collecting data suitable for PDF analysis every second. The time-resolved differential PDFs, where the reference PDF measured for the Al_2O_3 support was subtracted, contained features associated with the face-centered cubic (fcc) Pt nanoparticles as well as other non-fcc features, for example, at 3.3 Å. The variation in intensity and position of these features was evaluated by fitting Gaussian functions to the peaks in the PDF. The features at low r showed the most pronounced changes with feedstock, suggesting that the structural changes are localized at the nanoparticle surface, involving surface reconstruction or formation of surface species without significant change to the nanoparticle core. The modulation of features in the PDF with the alternating chemical feedstock was correlated with CO_2 production, the principal indicator of catalytic activity (Fig. 5.10). The steady-state CO_2 production was found to be strongly linked to the intensity of the non-fcc feature at approximately 3.3 Å; for a stoichiometric ratio of CO to O_2, where the 3.3 Å feature is smallest, CO_2 production was approximately two-thirds that under conditions of excess oxygen. The greater CO_2 productivity under oxidizing conditions occurs despite a lower CO concentration in the

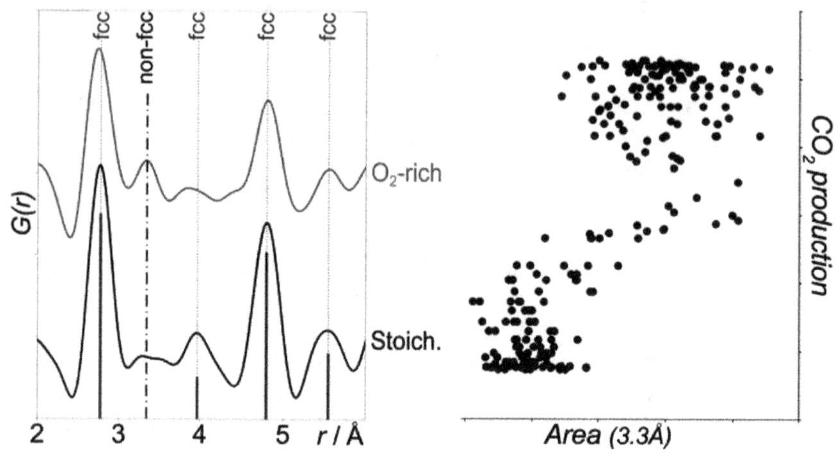

Figure 5.10. Left: Differential PDFs obtained under stoichiometric and oxygen-rich feedstocks show the most pronounced difference in the low *r* region. Right: The catalytic activity is correlated with the intensity of non-fcc features in the PDF associated with surface oxide formation.

feed. This suggests that under an excess of O_2, a surface oxide forms on the Pt nanoparticles, with this surface oxide being the catalytically active species.

To gain direct insight into the structure of this highly active surface oxide, a surface differential PDF was computed. In this case, the differential PDF obtained under stoichiometric feedstock, where the surface oxide is smallest, is treated as the reference PDF and subtracted from the differential PDF obtained under oxygen-rich conditions, where the surface oxide is largest. This surface (double) differential PDF essentially eliminates the invariant contribution from the nanoparticle core to the PDF, isolating only correlations between surface atoms and between the surface atoms and the underlying fcc lattice. Distinct structural features are evident in the surface differential at 3.3 Å and also at 2.0, 2.6, and 4.4 Å. This surface differential can be described by an oxidized Pt surface with close-packed Pt and O layers, not globally coherent with the bulk nanoparticle.

Using time-resolved differential PDF analysis, detailed insight can be gained into the dynamic nature of dilute nanoparticle catalysts under realistic operating conditions. By identifying correlations between peaks in the PDF, the feedstock switching, and the catalytic activity, features associated with the active catalytic sites can be identified *ab initio*. While the Pt nanoparticles studied here comprise 1 wt% of the sample, the surface of these nanoparticles, which represents an even smaller fraction of the total sample, could be probed using surface differential PDF methods, an indication of the extreme sensitivity of this approach.

5.6 OUTLOOK

Applications of PDF method to understand catalysis are still in a relative state of infancy. The significant advances in measurement strategy during the last decade have opened the possibility for new studies that allow catalysts to be probed under realistic reaction conditions with time resolutions that match the mass transfer rates in a gas/solid reaction. However, exciting opportunities remain to further optimize the methodology and expand insights and capabilities. These include:

- *Combined methodologies:* By combining PDF measurements with other complementary probes, more comprehensive insights can be gained. Characterization tools that are good candidates for combined measurements include SAXS (for particle size and size distribution), X-ray absorption spectroscopy (for chemical speciation), infrared (IR) spectroscopy (for molecular speciation at surfaces), and neutron scattering (for improved contrast of light elements). Some of these can be performed simultaneously such as PDF and SAXS, PDF and IR, and PDF and Raman; these combined capabilities are currently under development.
- *Streamline data reduction and analysis:* A consequence of the more rapid PDF measurements now possible is that data reduction and analysis requires many orders of magnitude longer times. This impediment to informed, real-time decision making during experiments could be addressed by software developments to match data reduction rates to current data acquisition rates.
- *Improved X-ray sources:* With upgrades planned for the high-energy synchrotrons that are best suited for X-ray PDF measurements, such as the Advanced Photon Source at Argonne National Laboratory, dramatic increases in beam intensity and reduction in beam size are anticipated. These will enable faster time-resolution measurements, enhance sensitivity and reduce lower detection limits, and open possibilities for nanoscale spatially resolved measurements.

The recent examples of PDF studies in catalysis presented here illustrate the valuable insights that can be readily accessed using the methodology. These current capabilities and potential future developments ensure that PDF methods will play a prominent role in the future of heterogeneous catalysis science.

ACKNOWLEDGMENT

Work done at Argonne and use of the Advanced Photon Source (APS), an Office of Science User Facility operated for the U.S. Department of Energy

(DOE) Office of Science by Argonne National Laboratory, was supported by the U.S. DOE under Contract No. DE-AC02-06CH11357.

REFERENCES

[1] Tarasov, L. P., Warren, B. E. (1936) X-ray diffraction study of liquid sodium, *J. Chem. Phys.*, 4, 236–238.

[2] Franklin, R. E. (1950) The interpretation of diffuse X-ray diagrams of carbon, *Acta Crystallogr.*, 3, 107–121.

[3] Shannon, R. D., Bergeret, G., Gallezot, P. (1985) Radial-distribution studies of non-framework aluminum species formed during dehydroxylation of zeolite HY, *Nature*, 315, 736–739.

[4] Chupas, P. J., Qiu, X., Hanson, J. C., Lee, P. L., Grey, C. P., Billinge, S. J. L. (2003) Rapid-acquisition pair distribution function (RA-PDF) analysis, *J. Appl. Crystallogr.*, 36, 1342–1347.

[5] Chupas, P. J., Chapman, K. W., Lee, P. L. (2007) Applications of an amorphous silicon-based area detector for high-resolution, high-sensitivity and fast time-resolved pair distribution function measurements, *J. Appl. Crystallogr.*, 40, 463–470.

[6] Chaudhuri, S., Chupas, P., Morgan, B. J., Madden, P. A., Grey, C. P. (2006) An atomistic MD simulation and pair-distribution-function study of disorder and reactivity of alpha-AlF3 nanoparticles, *Phys. Chem. Chem. Phys.*, 8, 5045–5055.

[7] Paglia, G., Bozin, E. S., Billinge, S. J. L. (2006) Fine-scale nanostructure in gamma-Al(2)O(3), *Chem. Mater.*, 18, 3242–3248.

[8] Chupas, P. J., Chapman, K. W., Jennings, G., Lee, P. L., Grey, C. P. (2007) Watching nanoparticles grow: the mechanism and kinetics for the formation of TiO2-supported platinum nanoparticles, *J. Am. Chem. Soc.*, 129, 13822–13824.

[9] Chupas, P. J., Chapman, K. W., Chen, H. L., Grey, C. P. (2009) Application of high-energy X-rays and Pair-Distribution-Function analysis to nano-scale structural studies in catalysis, *Catal. Today*, 145, 213–219.

[10] Dmowski, W., Yin, H. F., Dai, S., Overbury, S. H., Egami, T. (2010) Atomic Structure of Au Nanoparticles on a Silica Support by an X-ray PDF Study, *J. Phys. Chem. C*, 114, 6983–6988.

[11] Kurzman, J. A., Ouyang, X. Y., Bin Im, W., Li, J., Hu, J., Scott, S. L., Seshadri, R. (2010) La(4)LiAuO(8) and La(2)BaPdO(5): comparing two highly Stable d(8) square-planar oxides, *Inorg. Chem.*, 49, 4670–4680.

[12] Oxford, S. M., Lee, P. L., Chupas, P. J., Chapman, K. W., Kung, M. C., Kung, H. H. (2010) Study of supported PtCu and PdAu bimetallic nanoparticles using in-situ X-ray tools, *J. Phys. Chem. C*, 114, 17085–17091.

[13] Petkov, V., Shastri, S. D. (2010) Element-specific structure of materials with intrinsic disorder by high-energy resonant x-ray diffraction and differential atomic pair-distribution functions: a study of PtPd nanosized catalysts, *Phys. Rev. B*, 81, 165428-1– 165428-7.

[14] Du Plessis, H. E., de Villiers, J. P. R., Kruger, G. J., Steuwer, A., Brunelli, M. (2011) Rietveld and pair distribution function study of Hagg carbide using synchrotron X-ray diffraction, *J. Synchrotron Radiat.*, 18, 266–271.

[15] Playford, H. Y., Modeshia, D. R., Barney, E. R., Hannon, A. C., Wright, C. S., Fisher, J. M., Amieiro-Fonseca, A., Thompsett, D., O'Dell, L. A., Rees, G. J., Smith, M. E., Hanna, J. V., Walton, R. I. (2011) Structural characterization and redox catalytic properties of Cerium(IV) pyrochlore oxides, *Chem. Mater.*, 23, 5464–5473.

[16] Zhao, H., Nenoff, T. M., Jennings, G., Chupas, P. J., Chapman, K. W. (2011) Determining quantitative kinetics and the structural mechanism for particle growth in porous templates, *J. Phys. Chem. Lett.*, 2, 2742–2746.

[17] Coduri, M., Brunelli, M., Scavini, M., Allieta, M., Masala, P., Capogna, L., Fischer, H. E., Ferrero, C. (2012) Rare earth doped ceria: a combined X-ray and neutron pair distribution function study, *Z. Kristallogr.*, 227, 272–279.

[18] Kardash, T., Plyasova, L. M., Kochubey, D. I., Bondareva, V. M., Neder, R. B. (2012) Development of the local and average structure of a V–Mo–Nb oxide catalyst with Mo5O14-like structure during synthesis from nanostructured precursors, *Z. Kristallogr.*, 227, 288–298.

[19] Newton, M. A., Chapman, K. W., Thompsett, D., Chupas, P. J. (2012) Chasing changing nanoparticles with time-resolved pair distribution function methods, *J. Am. Chem. Soc.*, 134, 5036–5039.

[20] Petkov, V., Shastri, S., Wanjala, B., Loukrakpam, R., Luo, J., Zhong, C.-J. (2012) Atomic ordering in nanosized PtxAu1−x (x = 0, 0.51, 1) by resonant X-ray diffraction and differential atomic pair distribution functions, *Z. Kristallogr.*, 227, 262–267.

[21] Zhao, H., Chupas, P. J., Chapman, K. W. (2012) Fast time-resolved pair distribution function studies of supported gold nanoparticle formation, *Z. Kristallogr.*, 227, 268–271.

[22] Chupas, P. J., Chapman, K. W., Halder, G. J. (2011) Elucidating the Structure of Surface Acid Sites on gamma-Al(2)O(3), *J. Am. Chem. Soc.*, 133, 8522–8524.

[23] Zhao, H. Y., Chupas, P. J., Chapman, K. W. (2012) Simultaneous infrared and PDF studies. In preparation.

[24] Warren, B. E. (1990) *X-Ray Diffraction*, Dover, New York.

[25] Egami, T., Billinge, S. J. L. (2004) *Underneath the Bragg Peaks: Structure Analysis of Complex Materials*, Oxford/Pergamon Press, New York.

[26] Gallezot, P. (1984) X-ray techniques in catalysis, in Anderson, J. R., Boudart, M. (eds.), *Catalysis Science and Technology*, Vol. 5, Springer, Berlin, Germany.

[27] Keen, D. (2001) A comparison of various commonly used correlation functions for describing total scattering, *J. Appl. Crystallogr.*, 34, 172–177.

[28] Gilbert, B. (2008) Finite size effects on the real-space pair distribution function of nanoparticles, *J. Appl. Crystallogr.*, 41, 554–562.

[29] Mullen, K., Krayzman, V., Levin, I. (2010) Atomic structure analysis at the nanoscale using the pair distribution function: simulation studies of simple elemental nanoparticles, *J. Appl. Crystallogr.*, 43, 483–490.

[30] Farrow, C. L., Juhás, P., Liu, J. W., Bryndin, D., Božin, E. S., Bloch, J., Proffen, T., Billinge, S. J. L. (2007) PDFfit2 and PDFgui: computer programs for studying nanostructure in crystals, *J. Phys.: Condens. Matter*, 19, 335219.

[31] Proffen, T., Neder, R. B. (1997) DISCUS, a program for diffuse scattering and defect structure simulations, *J. Appl. Crystallogr.*, 30, 171–175.

[32] Tucker, M. G., Dove, M. T., Keen, D. A. (2001) Application of the reverse Monte Carlo method to crystalline materials, *J. Appl. Crystallogr.*, 34, 630–638.

[33] Neder, R. B., Proffen, T. (2008) *Diffuse Scattering and Defect Structure Simulations: A Cook Book Using the Program DISCUS*, Oxford University Press, Oxford.

[34] Page, K., Hood, T. C., Proffen, T., Neder, R. B. (2011) Building and refining complete nanoparticle structures with total scattering data, *J. Appl. Crystallogr.*, 44, 327–336.

[35] Chapman, K. W., Chupas, P. J., Kepert, C. J. (2005) Selective recovery of dynamic guest structure, in a nanoporous Prussian blue through in situ X-ray diffraction: a differential pair distribution function analysis, *J. Am. Chem. Soc.*, 127, 11232–11233.

[36] Besson, C., Finney, E. E., Finke, R. G. (2005) A mechanism for transition-metal nanoparticle self-assembly, *J. Am. Chem. Soc.*, 127, 8179–8184.

6

NEUTRON SCATTERING FOR *IN-SITU* CHARACTERIZATION OF HETEROGENEOUS CATALYSIS

ASHFIA HUQ AND WEI-REN CHEN

6.1 INTRODUCTION

Large penetration depth and sensitivity to lighter elements, especially mobile species such as hydrogen and oxygen, in the presence of heavier elements make neutron scattering an ideal tool to probe heterogeneous catalysts in operational conditions [1, 2]. In order to study materials *in situ*, it is desirable to use a probe that has very weak interaction with the samples in question and can also penetrate through the containment. This is especially true for investigating catalysis, as one is often interested in building a reactor and flow gases through the catalyst to mimic the environment in which it will be used. As such, both X-ray and neutron scattering, which are complementary to each other, can be used in a number of different ways to provide valuable information about these systems. While much work has been accomplished to showcase the possibilities of the type of information that can be obtained using both neutron diffraction and inelastic spectroscopy, the use of neutron scattering in the field of catalysis is still quite limited and underutilized by the community. The principal reasons why the technique has, to date, been barely used to characterize working catalysts is the complexity of the problems often faced in the study of catalysis, weak signals due to lower flux available in

In-situ *Characterization of Heterogeneous Catalysts*, First Edition.
Edited by José A. Rodriguez, Jonathan C. Hanson, and Peter J. Chupas.

neutron sources requiring long beamtimes, and, finally, lack of modeling tools available to interpret the data. However, the recent advances in computation power, newer theoretical modeling tools such as Monte Carlo (MC) and density functional theory, and also the availability of higher flux neutron instruments at various facilities now make the study of the structure and dynamics of heterogeneous catalysts using neutrons a realistic goal and offer compelling prospects. In this chapter, we will introduce some of the relevant neutron scattering techniques along with examples to elucidate their usage and other practical aspects such as material consideration for choice of sample environment, and finally, access to neutron facilities around the world. We will keep our discussion limited to structure, but some reference will be provided for those who are interested in learning more about dynamics using inelastic neutron scattering (INS).

As mentioned earlier, neutrons are highly penetrating and charge neutral. As a result, it is a nondestructive probe and can penetrate through large sample environments. Unlike X-rays, neutrons interact with the nuclei and not the electron cloud of atoms. Hence, the scattering power of neutrons is not dependent on the atomic number Z and neutrons have a different relative sensitivity to different atoms compared to X-rays. Thus, in neutron scattering, heavy and light elements can have comparable scattering power, while elements with very similar atomic numbers can have very different scattering lengths. For example, the scattering cross section of Fe is 11.22 barn, while that of Co is 0.78 barn. This also extends to difference in scattering power with different isotopes of the same element. This has a rather wide range of applicability for catalysis research as it allows the precise determination of hydrogen position in crystal structures via neutron diffraction by simply replacing the hydrogen by deuterium. This is extremely hard to do with X-ray diffraction since hydrogen is a very light element and is hard to detect, especially in the presence of heavier elements. Moreover, if the scattering nucleus contains a mixture of isotopes or has a nonzero nuclear spin, the neutron scattering length consists of a coherent component, which is the average over all spins, and an incoherent part, which gives the deviation from this average value. In other words, coherent scattering describes the interference between waves produced by the scattering of a single neutron from all the neutrons in the nuclei of the sample, while incoherent scattering involves correlations between the position of an atom j at time zero and the position of the same atom in time t, and so the scattered waves from different nuclei no longer interfere. Thus, incoherent scattering provides an excellent tool to study processes involving atomic diffusion [3]. Again, hydrogen is very special in this regard as it has a very large incoherent scattering cross section (σ_{inc}) of 80.26 barn, while its coherent cross section (σ_{coh}) is 1.758. In comparison, for deuterium, $\sigma_{inc} = 2.05$ and $\sigma_{coh} = 5.59$. When these two properties are combined, one immediately understands the significant impact neutron scattering can make in the study of catalysis. Hydrogen is critically important in catalysts used for fossil fuel conversion [4]. Neutrons can easily penetrate steel reaction vessels and

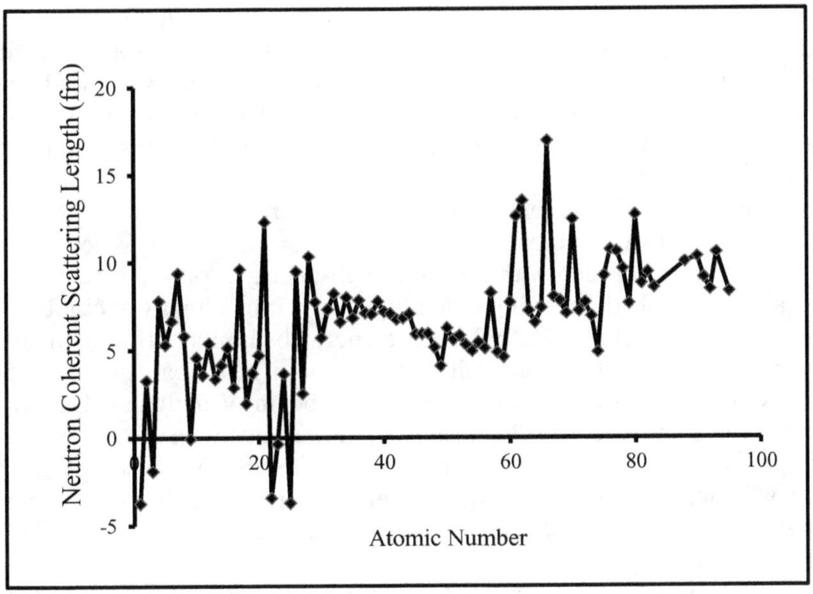

Figure 6.1. Scattering amplitude b shown as a function of atomic weight Z. Data for the plot are obtained from the NIST website.

pressure cells and effectively probe hydrogen atoms and help decipher if hydrogen is present in the form of protons, hydrogen atoms, or hydrides. The scattering lengths and various cross sections of neutrons can be accessed through a website maintained by the National Institute for Standards and Technology (NIST) [5] (Fig. 6.1).

6.2 NEUTRON SOURCES

Neutron scattering requires a source of intense neutrons that can be produced either by nuclear fission at a reactor or by spallation when high-energy protons hit a heavy metal target such as mercury, tungsten, and tantalum. Research reactors induce fission in a critical mass of ^{235}U using thermal neutrons to produce fast neutrons. Their design is significantly different from power reactors, as the purpose of these reactors is to generate neutron beam and not heat. Enriched uranium often has to be used, as the natural abundance of uranium is only around 0.7%. However, due to concern about nuclear weapons proliferation, there is a drive to use lower levels of enrichment. In the United States, two large reactors, namely, the High Flux Isotope Reactor (HFIR) at Oak Ridge National Laboratory and the NIST Center for Neutron Research (NCNR), are maintained by the Department of Energy (DOE) and the

Department of Commerce. Both these facilities maintain a vibrant user program. Other notable research reactors around the world are the Institute Laue Langevin (ILL, France), Forschungs-Neutroenquelle Heinz Maier-Leibnitz (FRM II, Germany), and OPAL (Sydney, Australia). However, this by no means is an extensive list. A research reactor such as ILL houses 28 neutron scattering instruments.

The production of neutrons at a spallation source generally starts with an accelerator that can produce a highly intense proton beam. When a heavy metal target is bombarded with pulses of high-energy proton beam, it triggers an intranuclear cascade, exciting individual nuclei into a highly energetic state. This causes the nuclei to release this extra energy by boiling off some neutrons that can leave the target. Each hit of this proton beam can produce 15–20 neutrons, resulting in an extremely energetic beam of neutrons. Oak Ridge National Laboratory currently hosts the newest and world's most intense pulsed source called Spallation Neutron Source (SNS). When completed, this source will host up to 24 neutron scattering instruments. Other prominent sources include JPARC (Japan), ISIS (United Kingdom), Lansce (United States), and SINQ (Switzerland).

The distribution of neutrons that are produced both at reactor and spallation sources has a generally much higher energy than what is desirable for scattering experiments. As a result, these neutrons are cooled down by bringing them in contact with moderating materials such as hydrogen, methane, or water that have large scattering cross sections. Thus, the neutrons that are used for studying materials have a distribution of thermal or cold neutrons.

At a reactor source, most scattering experiments are carried out using a narrow energy band of neutrons (~constant wavelength). A monochromatic beam is produced just as it is at synchrotron sources, using a highly reflective material such as germanium, copper, or pyrolytic graphite. The incident wavelength can be adjusted by changing the scattering angle (2θ) of the monochromator. Spallation sources, in contrast, generate neutrons in a pulsed fashion, which eliminates the need for the use of monochromators. Scattering measurements take advantage of the time-of-flight (TOF) technique in which neutron velocities can be determined by measuring the time it takes each neutron to travel the distance between the moderator and the detector.

6.3 SAMPLE ENVIRONMENT

The choice of materials for designing a sample environment usually depends on the type of experiments (elastic vs. inelastic), the type of neutron source (constant wavelength or not), and the sample. Aluminum is the metal of choice for cells for most inelastic measurements as Al absorbs neutrons only weakly. This is because the absorption cross section of Al is $\sigma_{abs} = 0.231$ barn for a wavelength of 1.8 Å. Al also has an extremely small $\sigma_{inc} = 0.0082$, which is often the source of background for these measurements. The scattering of vanadium

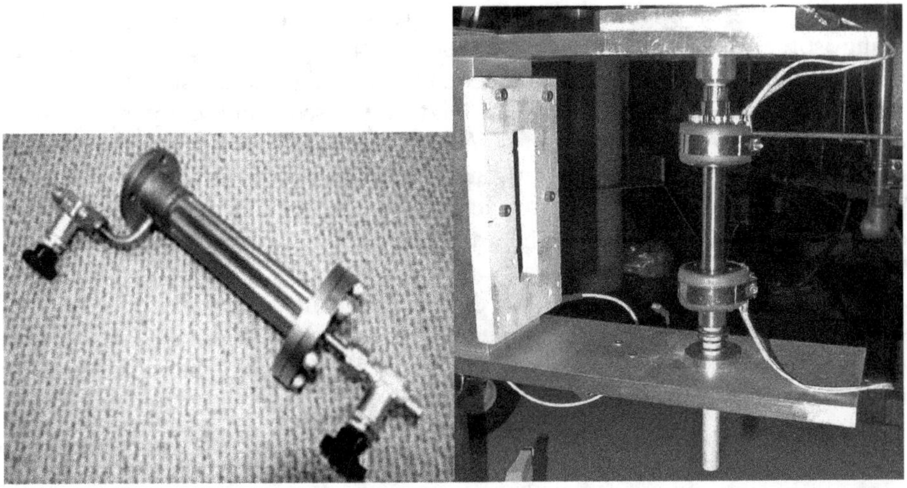

Figure 6.2. Two examples of reactors used for neutron scattering studies. The cell on the left is made of Al for inelastic studies [8] (photo courtesy of Professor Juergen Eckert). The one on the right made of Inconel was used for hydrogen absorption studies in Li_3N [9].

(σ_{coh} = 0.0184) or a TiZr alloy (null scatterer, due to exact matching of the negative scattering length of Ti and the positive scattering length of Zr), in contrast, is almost purely incoherent, making it more ideal for sample containment for diffraction measurements. However, at an elevated temperature, V easily forms oxides or hydrides in the presence of air or hydrogen and so its use is often limited for low temperature studies. However, a thin layer of a noble metal such as gold can be vacuum-deposited inside a V can to keep it from reacting [6]. For experiments requiring hydrogen pressure, Inconel is often the material of choice [7]. However, in this case for diffraction measurements, one has to either exclude the Inconel peaks or make use of a radial collimator to reduce the signal from the vessel (Fig. 6.2).

A frequently used approach for gas flow experiments is to make the sample holder and furnace insert out of quartz tubes, which is easy to work with. Quartz generally does not have any Bragg peaks, but it is not featureless due to the local structure of the SiO_2. This, however, is usually easy to handle in data reduction, by collecting data on the empty reactor and doing background subtraction or modeling the background using a pair distribution-type analysis.

To imitate the actual operational environment of the heterogeneous catalyst and to obtain the needed information to get to the reaction mechanism, it is often desirable to combine multiple probes along with the actual neutron scattering studies. It is obvious that a gas handling system compatible with the furnace is necessary. This can be a flow system or a Sievert system capable of doing isotherms depending on the experiment. If one is studying fuel cell-related catalysts, it is often advantageous to have a good measurement of the partial pressure of oxygen of the gas mixture to ascertain the environment is indeed realistic. A mass spectrometer/residual gas analyzer (RGA) to monitor the output gas is also extremely useful. A system incorporating all these components has been recently built for such studies on the general-purpose powder diffraction instrument POWGEN at the SNS, and a flow diagram is shown in Figure 6.3.

While in most cases absorption of neutrons can be neglected, there are some elements that have rather large absorption cross sections such as Cd and Gd, which are often used very effectively for shielding against parasitic and undesirable scattering.

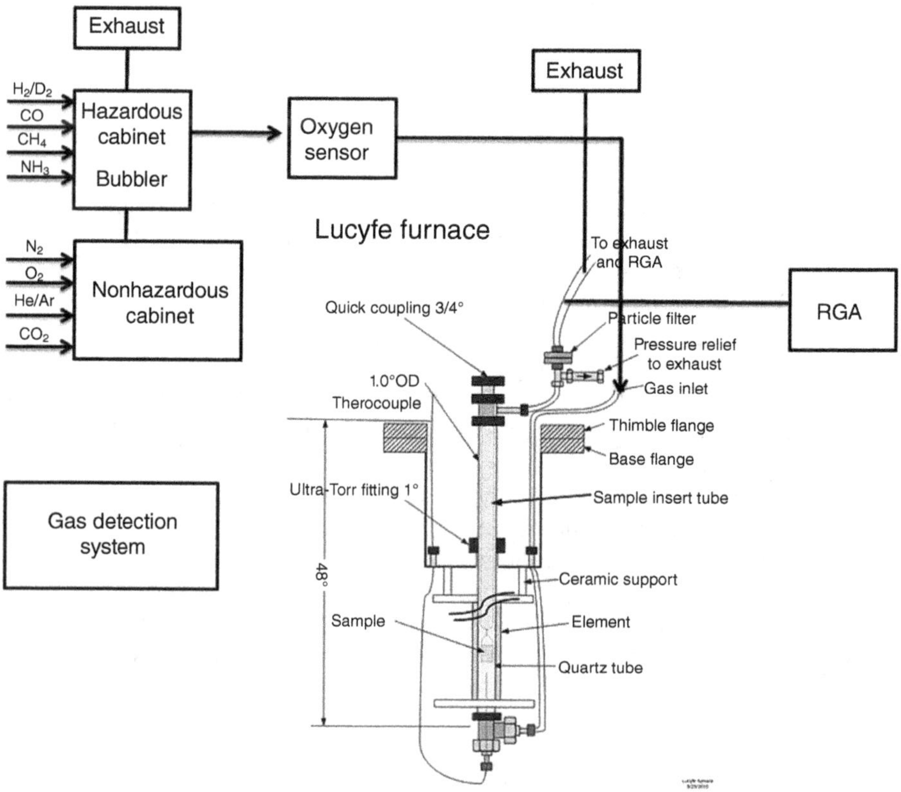

Figure 6.3. A flowchart showing an *in-situ* gas handling reactor insert for a furnace for neutron diffraction measurement recently developed at the POWGEN diffractometer at SNS.

6.4 NEUTRON SCATTERING TECHNIQUES FOR CATALYSIS

There are three important techniques using a neutron beam and a catalyst sample that are of interest, namely,

- *Diffraction*: providing crystallographic structure (average structure, Rietveld method) and also local order such as near-neighbor distance and coordination numbers (pair distribution function [PDF] method)
- *Small-Angle Neutron Scattering (SANS)*: learning pore size distribution
- *Quasielastic and Inelastic Scattering*: providing information about diffusion, rotation, and vibration.

6.4.1 Powder Diffraction

The Nobel Prize winner Francis Crick said, "If you want to understand function, study structure." Knowing accurate structural information is often necessary as a basis for input of theoretical approaches to understand the mechanistic steps involved in the activity and selectivity of the heterogeneous catalyst system. In his 1969 article titled "A Profile Refinement Method for Nuclear and Magnetic Structures" [10], Hugo M. Rietveld described a structure refinement method that, instead of using integrated intensities from individual peaks, applied a least squares method for full powder pattern matching. This revolutionized the field of structural characterization of materials using powder diffraction, as it allowed the user to handle the overlap problem inherent in a powder diffraction experiment and still extract useful quantitative structural information of materials of interest. Traditionally, powder diffraction techniques were used predominantly for phase identification and quantitative phase analysis. Rietveld refinement led to a much wider application of the technique to structure determination of complex oxides, zeolites, and small organic molecules [11]. Recent advances in X-ray and neutron sources and optics deliver higher resolution and flux to the sample. Combined with rapid improvements in computing, this now makes it possible to carry out real-time diffraction measurements that can access processes as fast as seconds, combine stroboscopic techniques to study kinetics of structural changes on a microscopic level on timescales of microseconds [12, 13], and undertake parametric studies [14] of materials under extreme conditions of varying temperatures [15, 16] and pressures [17].

Atomic order of interest in emerging complex functional materials is often limited to the nanometer length scale where crystallography, the basis of Rietveld refinement, is no longer applicable [18]. In these cases, the total scattering approach described in Chapter 5 can be used. This approach treats Bragg and diffuse scattering on an equal footing. In recent years, the analysis of the total scattering data has become an invaluable tool to study nanocrystalline, nanoporous, and disordered crystalline materials. As a sine Fourier transform of the total scattering structure function, $F(Q)$, the atomic PDF, $G(r)$, utilizes both Bragg and diffuse scattering, and hence yields the direct-space information on atomic arrangements on various length scales [19]. With advances on the experimental and modeling software frontiers, and their synergy, the applicability scope of the PDF method is extensive.

Neutron powder diffraction has traditionally found its niche in structural studies of new materials such as high Tc superconductors, ferroelectrics, dielectrics, and other traditional condensed matter physics systems, especially in the determination of magnetic structures. In the recent past, however, due to the availability of high flux instruments at both reactor and spallation sources, the technique is being very successfully used in studies of dynamic processes such as understanding gas absorption phenomenon, for example, in hydrogen storage systems such as complex hydrides and metal organic frameworks (MOFs), oxygen diffusion in solid oxide fuel cells, synthesis and crystallization of solids, solving structures of zeolites, and ion-exchange selectivity processes in porous frameworks. The total scattering measurements have opened up the avenue of looking at the local environment of a species of interest along with the large field of nanostructures. In the following sections, the use of this powerful tool will be elucidated using detailed examples of two different systems that are relevant to the study of heterogeneous catalysis.

6.4.1.1 Example I: Neutron Diffraction of Ammoxidation Catalysts Commercial catalysts used for ammoxidation of propylene to acrylonitrile are usually multicomponent metal (MCM) oxides [20–22]. A mixture of ferric molybdate ($Fe_2Mo_3O_{12}$) and bismuth molybdate (Bi_2MoO_6) form the principal part of these catalysts. It is generally recognized that during catalyst activation ferric molybdate is reduced from Fe^{3+} to Fe^{2+} phases, freeing molybdenum oxide for the formation of the active catalyst phase(s) [23]. Although this reaction is widely believed to occur in a number of catalytic processes, the mechanism of the reaction had never been extensively studied. *In-situ* neutron diffraction studies on two model ammoxidation catalyst systems, namely, $Fe_2Mo_3O_{12}$ and a 1:1 mixture of $Fe_2Mo_3O_{12}$ and Bi_2MoO_6 revealed phase cooperation between the Fe and Bi molybdate phases. The presence of an active catalyst phase containing all three crucial components Fe, Bi, and Mo during the ammoxidation reaction was also observed.

TOF neutron diffraction data were collected at the general-purpose powder diffractometer (GPPD) at the Intense Pulsed Neutron Source (IPNS), Argonne National Laboratory. Approximately 3.5 g of sample was loaded in a special fused silica tube and inserted inside a furnace (similar to Fig. 6.3). In this tube, the sample was held below by a fused silica frit of coarse porosity (40–90 μ) and silica wool was placed on top of the sample to prevent sample movement during gas flow inside the cell. The $Fe_2Mo_3O_{12}$ sample was heated to 450°C, while the Fe/Bi mixed molybdate sample was kept at 430°C to prevent the formation of β $Fe_2Mo_3O_{12}$. The ratio of the flow rates for the reactant gases was O_2:C_3H_6:NH_3 = 2.2:1:1 for acrylonitrile formation. In addition to the reactants, nitrogen gas was added to the feed to help control the temperature of the exothermic reaction. The exhaust end of the silica tube was wrapped with heating tape to prevent the steam generated during the reaction from dripping into the sample. Two thermocouples were placed above and below the sample to monitor and control the reaction temperatures. Diffraction patterns were collected every half hour for ~2 days each, for the two separate systems. The range of data used corresponds to d spacings of 0.5–4.5 Å. Due to the presence of the silica tube in the beam, the background had to be modeled carefully using the diffuse scattering options available in the General Structure Analysis System (GSAS) [24].

Information Extracted from Rietveld Refinements Two critical parameters that can be obtained from Rietveld refinements of the powder data are lattice parameters and scale parameters of the present phases. The calculated intensity in Rietveld refinement for a multiphase sample is given by the following equation:

$$I_c = I_b + I_d + S_h \sum_p S_{ph} Y_{ph},$$

where S_h is an overall scale factor, S_{ph} is the scale factor associated with the individual phases, Y_{ph} is the contribution from h_{th} reflection of the p_{th} phase, and finally, I_b and I_d are contributions from background and diffuse scattering, respectively. Consequently, one can calculate the moles of each present phase by multiplying the scale factor S_h by Z. The change in volume of the unit cell in a constant environment (i.e., no change in temperature or pressure) is often indicative of the change in oxidation state of the compound. For example, Fe^{+2} has a larger ionic radius than Fe^{3+}; as a result, an increase in volume indicates reduction of the Fe in the compound containing the ion.

Reduction of $Fe_2Mo_3O_{12}$ and $Fe_2Mo_3O_{12}/Bi_2MoO_6$ Mixtures The sequential reactions observed during the *in-situ* reduction of $Fe_2Mo_3O_{12}$ with the mixture of propylene, ammonia, oxygen, and nitrogen are

$$\alpha\text{-}Fe_2Mo_3O_{12} \rightarrow \beta\text{-}Fe_2Mo_3O_{12} \quad \text{(with some } Fe^{3+} \text{ reduction within the structure)}, \tag{6.1a}$$

$$\beta\text{-}Fe_2Mo_3O_{12} \rightarrow Fe_2O_3 + 3MoO_3 \quad \text{(with some } Mo^{6+} \text{ reduction within the structure)}, \tag{6.1b}$$

$$Fe_2O_3 + 3MoO_3 \text{ (reduction)} \rightarrow \beta\text{-}FeMoO_4, \tag{6.1c}$$

$$5\beta\text{-}FeMoO_4 \text{ (reduction)} \rightarrow 2MoO_2 + Fe_3O_4 + Fe_2Mo_3O_8. \text{ (Fig. 6.4)} \tag{6.1d}$$

These results are unlike other studies of the reduction of ferric molybdate by different reductants, for example, H_2 [25], which show an initial reduction

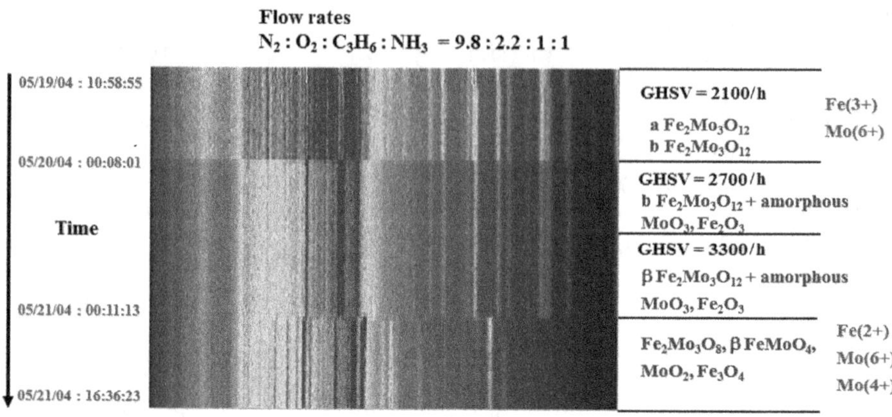

Figure 6.4. Neutron diffraction collected during the *in-situ* reduction measurements of $Fe_2Mo_3O_{12}$over 2 days of beamtime at GPPD. Equation 6.1a–d was derived on the basis of these data. (See color insert.)

of Fe (as indicated by the formation of β-$FeMoO_4$), followed by further reduction (to MoO_2, $Fe_2Mo_3O_8$, etc.). In this study, it was observed that when ammonia and propylene are the reductants, the first reaction is disproportionation (with perhaps some small amount of iron reduction within the β-ferric molybdate phase), followed by molybdenum reduction. Another important difference between this and earlier studies [26] is that phases with mixed Mo oxidation states ($Mo_nO_{(3n-1)}$) are not observed in our experiments. These differences are likely due to the low reactivity of the ferric molybdate to propylene and ammonia. Only after a total of 28 h on stream does one observe any reduced phases. This is in agreement with the previously noted [27] poor reactivity of this material to propylene.

The evolution of phases when a 1:1 molar mixture of α-$Fe_2Mo_3O_{12}$ and Bi_2MoO_6 is used as a catalyst is significantly different from those discussed in the earlier section for α-$Fe_2Mo_3O_{12}$. The addition of Bi_2MoO_6 (which was intimately mixed with α-$Fe_2Mo_3O_{12}$) results in immediate declines in the mole fractions of starting phases forming β-$FeMoO_4$ and $Bi_2Mo_3O_{12}$, which steadily increase with time.

These data are consistent with the following reactions:

$$Fe_2Mo_3O_{12} \rightarrow 2FeMoO_4 + MoO_3 \text{ (reduction, } Fe^{3+} \rightarrow Fe^{2+}\text{)}, \tag{6.2a}$$

$$Bi_2MoO_6 + 2MoO_3 \rightarrow Bi_2Mo_3O_{12}. \tag{6.2b}$$

The reduction of ferric molybdate described earlier and the formation of β $FeMoO_4$ found in these experiments are in agreement with other studies [28].

In contrast to ferric molybdate, we observe no time lag before reactions occur and the formation of beta ferrous molybdate is observed immediately. During the *in-situ* experiment, only four phases are initially observed at 430°C. They are α-$Fe_2Mo_3O_{12}$, β-$FeMoO_4$, Bi_2MoO_6, and $Bi_2Mo_3O_{12}$. As the reaction proceeds, a gradual increase in relative mole percentages of reaction products ($FeMoO_4$ and $Bi_2Mo_3O_{12}$) and a decline in reactants ($Fe_2Mo_3O_{12}$ and Bi_2MoO_6) are observed consistent with Equation 6.2a,b. This analysis ignores any noncrystalline material that is likely to be present.

Of equal interest is the postmortem analysis of the spent catalyst charge. Room temperature diffraction analysis of the catalyst after quenching indicated a mixture of phases different from that at the end of the *in-situ* experiment at 430°C. However, analysis of the quenched room temperature material indicates that significant quantities of $Bi_3FeMo_2O_{12}$ and lesser amounts of $Bi_2Mo_2O_9$ are also present in the spent catalyst. Since these phases are seemingly absent at 430°C, they must have been formed during the quench from higher temperatures. One explanation is that the "$Bi_2Mo_3O_{12}$" phase observed at 430°C is a solid solution of $Bi_2Mo_3O_{12}$ and $Bi_3FeMo_2O_{12}$ with the formula $Bi_{2+x}Fe_xMo_{3-x}O_{12}$.

The possibility of a $Bi_{2+x}Fe_xMo_{3-x}O_{12}$ solid solution at elevated temperatures is reasonable due to the similarities in structures and cell metrics of the two end members. Both $Bi_2Mo_3O_{12}$ [29] and $Bi_3FeMo_2O_{12}$ [30] are ordered structures that are derived from scheelite-type cells. The scheelite structure, formula ABO_4, with tetragonal unit cell $a = 5.5$, $c = 11.2$ Å, can accommodate many eight-coordinate (A) and four-coordinate (B) cations and are well known for converting propylene to acrylonitrile [31]. For both $Bi_2Mo_3O_{12}$ and $Bi_3FeMo_2O_{12}$, the cell sizes have been tripled with respect to the basic scheelite unit cell: $3[A_{2/3}\ A'_{1/3}\ BO_4]$. The *ac* plane of each monoclinic structure ($Bi_2Mo_3O_{12}$: $a = 7.72$ Å, $b = 11.52$ Å, $c = 11.98$ Å, and $\beta = 115.3°$; $Bi_3FeMo_2O_{12}$, $a = 16.91$ Å, $b = 11.65$ Å, $c = 5.25$ Å, and $\beta = 108°$) has an area approximately three times that of the *ab* plane of the scheelite structure, with the *b*-axis of the monoclinic cell coincident with the *c*-axis of the scheelite cell. These tripled unit cells are the result of either A or B cation ordering. Assuming a solid solution of these end members exists at elevated temperatures, during quenching, disproportionation of this phase could proceed according to the following (Fig. 6.5):

$$Bi_{2+x}Fe_xMo_{3-x}O_{12} \rightarrow xBi_3FeMo_2O_{12} + (1-x)Bi_2Mo_3O_{12}. \quad (6.3a)$$

Alternatively, $Bi_{2+x}Fe_xMo_{3-x}O_{12}$ could disproportionate according to Equation 6.3b:

$$Bi_{2+x}Fe_xMo_{3-x}O_{12} \rightarrow x/2Fe_2Mo_3O_{12} + xBi_2Mo_2O_9 + 3x/2Bi_2MoO_6. \quad (6.3b)$$

Since all three Bi_2O_3•MoO_3 compounds (1:3, 1:2, and 1:1) are found in the quenched sample, it is likely that both disproportionation mechanisms are active in the final, nonequilibrated sample.

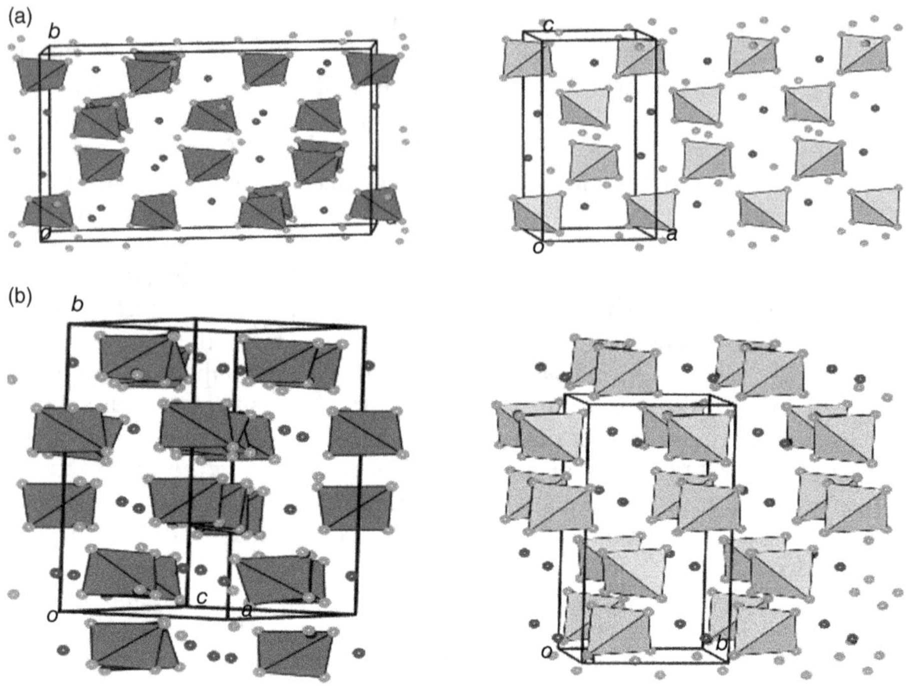

Figure 6.5. (a) Left: structures showing similarities of ordered $Bi_3FeMo_2O_{12}$; right: scheelite structure with tripled *a*-axis. Bi and O are represented by blue and green circles, respectively, along with MoO_4/FeO_4 tetrahedron in red (left) and orange (right). (b) Left: structures showing similarities of ordered $Bi_2Mo_3O_{12}$; right: scheelite structure. In this case, the scheelite lattice is shown with $1.5 \times a$ and $2.2 \times b$.

Effects of Fe and Bi on Phase Cooperation of MCM Oxide Catalyst This comparative study of the MCM shows that the introduction of feed gases over catalyst mixtures can lead to significant changes in the solid-state chemistry of the catalyst. This observation reinforces the belief that oxidation catalysts of this type can have phases present during the reaction that are not represented in the quenched sample and indicates the possibility of the existence of $Bi_{2+x}Fe_xMo_{3-x}O_{12}$. This phase also brings together all the ingredients required for a highly active and selective ammoxidation catalysis in a single phase, namely,

1. the presence of bismuth and molybdenum, elements that have long been known to act in concert in the reaction; the presence of iron, which very early on was found to increase catalyst activity, and
2. cation vacancies in a scheelite-type structure. (Sleight et al. has shown through a series of seminal studies the important role that vacancies play in the selective ammoxidation of propylene [31].)

If the appearance of a Bi/Fe/Mo oxide phase were the only major difference in the two *in-situ* experiments described here, it would be easy to ascribe the enhanced ability of iron containing bismuth molybdate catalysts for the ammoxidation of propylene to this new phase. However, these experiments have served to highlight additional significant differences in the reactivity of iron molybdate phases depending upon the presence or absence of bismuth molybdate, and so the existence of a new Bi/Fe/Mo phase of the formula $Bi_{2+x}Fe_xMo_{3-x}O_{12}$ may not be the only component responsible for the iron effect.

By themselves, both α- and β-ferric molybdates are largely unreactive at the temperatures employed here; the long lag between the introduction of feed gas and the appearance of reaction products is evidence of that. There is, however, a significant increase in ferric molybdate reactivity in the presence of bismuth molybdate; β-ferrous molybdate (β-$FeMoO_4$) is formed immediately when both phases are present. Ferrous molybdate, the initial reaction product of ferric molybdate, also shows a different reactivity depending upon the presence of bismuth molybdate phases. In the absence of bismuth molybdate, ferrous molybdate is immediately reduced (to $Fe_2Mo_3O_8$ and MoO_2), whereas in the presence of bismuth molybdates, these deeper reduction products are largely absent. This suggests that not only does bismuth molybdate enhance the reducibility of ferric molybdate but it also may enhance the ability of ferrous molybdate to be oxidized. In this way, the presence of bismuth molybdate phase (gamma and alpha in this case) mitigates both reduction and oxidation to insure that both Fe^{+2} and Fe^{+3} ions are present at all times during the ammoxidation reaction. This has the additional effect of preventing the formation of molybdenum dioxide (MoO_2), which has been shown to be largely unreactive in propylene ammoxidation.

6.4.1.2 Example II: Zeolites and MOFs Zeolites and MOFs have found extensive application in the field of heterogeneous catalysis and neutron powder diffraction has been an extremely useful tool to elucidate their structures [32–35]. Zeolites have the ability to act as catalysts for chemical reactions that take place within the internal cavities. An important class of reactions is that catalyzed by hydrogen-exchanged zeolites, whose framework-bound protons give rise to very high acidity. This is exploited in many organic reactions, including crude oil cracking, isomerization, and fuel synthesis. Zeolites can also serve as oxidation or reduction catalysts, often after metals have been introduced into the framework. Examples are the use of titanium ZSM-5 in the production of caprolactam and copper zeolites in NO_x decomposition. The catalytic properties of synthetic zeolites originate from their capacity to adsorb a variety of molecular species inside its intricate channels and porous network where the cations reside. To understand the catalytic behavior of the zeolites, it is extremely important to understand the host–guest interaction, guest–guest interaction, and also the cation distribution in the framework. It is also desirable to understand the mechanisms behind the hydrothermal crystallization process of these materials so that newer and better catalysts can be synthesized with better control over pore size and desired surface area.

Reaction Mechanism of Cs Ion Selectivity in Crystalline Silicotitanate Recently, a combined X-ray and neutron powder diffraction study very successfully revealed the mechanism by which hydrogen crystalline silicotitanate (CST) achieves its exceptional ion-exchange selectivity for Cs [36]. CSTs are being considered as candidates for possible exchangers for separation of ^{137}Cs and ^{90}Sr from radioactive nuclear waste solutions. This work was motivated by the need to develop materials to combat the nuclear waste problem. High levels of waste have been generated principally from the nuclear weapons program. If possible, it would be ideal to remove the small quantities of highly radioactive ^{137}Cs, ^{90}Sr, and actinides such as Np, Pu, and U before storing them in underground tanks as the waste tends to cause corrosion in these vessels, causing seepage of the waste solutions. One strategy proposed for removal is developing and synthesizing ion-exchange materials that are highly selective to these species.

In order to make the CST framework sequester Cs ions, it has to first be H-exchanged. The dynamics and exchange process greatly depends on the H positions of the hydroxyl sites and water orientation. To determine the H position, an *in-situ* ion-exchange experiment was conducted in D_2O solution. The measurements were carried out in a specially designed *in-situ* flow-through cell. Diffraction data were collected in 30-min intervals at the medium resolution TOF beamline Polaris at the ISIS spallation neutron source, Rutherford Appleton Laboratory. *Ex-situ* neutron studies were also performed at the BT-1 beamline at the NCNR at the NIST along with in-house *in-situ* X-ray diffraction. Neutron studies revealed that the partially Cs^+ exchanged phase were considerably different in the predeuterated H-CST ($D_{1.6}H_{0.4}Ti_2O_3SiO_4{\cdot}D_{2.66}H_{0.34}O_{1.5}$ and $Cs_{0.38}D_{1.08}H_{0.54}Ti_2O_3SiO_4.\ D_{1.72}H_{0.28}O$). The position of the water protons was revealed by the neutron study, allowing a more detailed picture of the ion-exchange mechanism compared to the *in-situ* X-ray studies. The overall study revealed that the Cs^+ exchange takes place in a two-step process mediated by conformational changes in the framework. This conformational rearrangement in H-CST unlocks the preferred Cs site causing an increase of the overall capacity and selectivity for ion exchange. This study truly highlights the complementarity of the two techniques and the strength neutron scattering brings to solve certain types of problems.

Reaction Mechanism of Cs Ion Selectivity in CST In another study carried out by Walton et al. [37], crystallization of some zeolites (Fig. 6.6) from amorphous aluminosilicate gels were also studied using TOF neutron diffraction. By using different concentrations of NaOD with a starting gel of composition $Al_2O_3{:}2SiO_2{:}n\text{NaOD}{:}20D_2O$ (where n ranges from 2.5 to 4), they were able to determine the optimal conditions for both temperature and NaOD concentration to form the desired hydroxyl sodalite $Na_8[Al_6Si_6O_{24}](OH)_2.nH_2O$. They

Figure 6.6. Large volume Inconel pressure cell for an *in-situ* study of supercritical hydrothermal syntheses using time-resolved neutron diffraction studies [38] (photo courtesy of Professor Dermot O'Hare).

discovered that low concentrations of NaOD results in the formation of Na zeolite A only (in 10 h of heating at 100°C). Medium concentration forms this phase as an intermediate but produces the sodalite phase in 3 h, while the higher concentration takes only 45 min for complete transformation. They also concluded by doing temperature measurements that increasing temperature and NaOD concentration are equivalent in the sodalite formation. By fitting the area under the diffraction peaks, they were able to determine the crystallization curves.

Host–Guest Hydrogen Bonding in Zeolite NaY/Chloroform System The PDF analysis using neutrons can add valuable information to what can be learnt from other similar measurements such as X-ray PDF and extended X-ray absorption fine structure (EXAFS). The PDF method has long been used for the study of materials with no long-range order such as glasses and liquids. However, more recently due to the availability of high-energy X-ray sources such as synchrotron and TOF neutron sources where high-energy neutrons are readily available along with vast improvements in computing, the PDF technique is being widely used to study local orders of polycrystalline solids and also the newly emerging functional nanomaterials. Understanding the host–guest interaction in zeolites is very important for its catalysis applications and PDF measurements can be helpful in shedding light on the local structure at an adsorption site or catalytically active site if structural analysis based on long-range ordered structure is not available. In a study on zeolite NaY/chloroform, Eckert et al. [39] reported how the differential hydrogen PDF obtained from neutron scattering contrast measurements on hydrogenous and deuterated chloroform in NaY provides direct support for H bonding and other facets of host–guest interaction. This technique can clearly be extended to more time-resolved *in-situ* studies as has been carried out for other systems [40].

Small-Angle Scattering In addition to the *microscopic* ordering determined by the chemical composition of material, various industrial applications requires structural modification of catalyst materials at a *mesoscopic* level for fulfilling the designed role of modulating the chemical reaction [41]. It has been common practice in synthetic chemistry to introduce the spatial heterogeneity, such as porosity, at mesoscopic length scale in various catalyst systems to alter the reaction rate via providing an alternative reaction mechanism with a different activation energy. These so-called multiphasic heterogeneous catalysts have received considerable attention in the last few decades because of their fundamental role in a wide context of industrial applications. Understanding the correlation between the material functionality and their mesoscopic structural characteristics has been a subject of numerous scattering investigations [42]. Due to its accessibility of relevant length scale ranging from few angstroms to some microns, the neutron scattering techniques, including the conventional SANS/ultra-small-angle neutron scattering (USANS) [43], and recently available spin-echo small-angle neutron scattering (SESANS) [44], present a complementary experimental means, to the previously mentioned powder diffraction method which probes the structural ordering at atomic level, for investigating the structural properties of heterogeneous catalysts [45].

Small-angle scattering methods both X-ray (small-angle X-ray scattering [SAXS]) and neutron (SANS) can shed valuable insight into the texture of solids such as pore size and distribution and the crystallite nucleation process. Since the relevant length scales are generally in the order of 10- to 1000-Å range, the diffracted intensity of neutrons or X-rays mainly occur at a small-angle range, say, from 0.3 to 10°. In general, the small-angle scattering intensity obtained from these complex multicomponent catalyst systems manifests the sum of the contributions of all constituent components and thus yields only the global structural information of probed systems. It is often desirable to further resolve the spatial arrangements of one certain species. To achieve this goal, it is necessary to compartmentalize the scattering contribution of one specific component from others. In other words, experimentally it needs to be specifically "colored" so that it can be distinguished from that of the rest of the probed samples. One well-known advantage of neutron scattering, in comparison to other elastic small-angle scattering tools such as X-ray and light, which operate at a comparative length scale, is that it allows the detailed structural characterization via the isotopic labeling of the targeted system with generally invariant physical and chemical properties but distinct neutron scattering efficiency. This method referred to as contrast matching is widely used to make part of the system to have no contribution to the scattering, making it possible to study parts of the system whose scattering is unmatched. Taking full advantage of this unique property of neutron scattering, extensive effort has been devoted to develop the SANS data analysis methodologies applicable for determining many spatial morphological and topological features of heterogeneous catalysts. Moreover, the SANS technique also possesses many intrinsic merits: For example, preparation of catalyst samples for SAXS experiment is often mechanically involved for the sake of overcoming the multiple scattering issues since they often contains high-Z elements due to various functional considerations. It has also been known that the requirement of sample transparency has severely limited the applicability of light scattering investigation.

Heterogeneous catalysts find applications in a wide variety of contexts, and significant progresses in materials engineering of catalysts continue to bring issues of fundamental importance to the fore. It is expected that interpretation of a new wealth of experimental data and finding ways to address relevant practical concerns, based on the growing sophistication of the elastic neutron scattering technique and corresponding data analysis, will provide an impetus to further advance the development of this field.

6.5 EXAMPLES OF EXTRACTING STRUCTURAL CHARACTERISTICS FROM SMALL-ANGLE SCATTERING OF CATALYSTS

The heterogeneous catalyst systems have been reduced, in several models of small-angle scattering coherent cross section, as materials consisting of two internally homogeneous phases, matrix and void, with spatial distributions that depend on chemical details of material synthesis [45]. In this approximation, the two most important parameters characterizing general catalyst systems are the specific area and the volume fraction of pores. In comparison to many soft matter systems that can be described as a collection of interacting particles, the structural information contained in the small-angle scattering spectra is much more reduced due to the interconnectivity in each phase. There have been theoretical models, developed for general convolutional two-phase systems, which can be applied, with proper modifications, to small-angle scattering study of heterogeneous catalyst systems [43, 46]. The use of contrast variation SANS and anomalous SAXS (ASAXS) [47, 48] has been applied to varying the scattering power selectively to facilitate the extraction of structural characteristics such as the specific area of voids, connectivity of pores, spontaneous curvature of interface and characteristic lengths, and their evolutions as external thermodynamical parameters.

6.5.1 Inelastic Neutron Scattering

Inelastic neutron spectroscopy (INS) has been used in catalysis research principally due to its sensitivity to the vibrations of hydrogen atoms. As mentioned in the introduction, hydrogen, due to its large incoherent scattering cross section, is far more visible than other elements, whereas Raman and infrared are more sensitive to the motion of heavier atoms (because of the larger number of electrons). INS is not subject to optical selection rules. All vibrations are active and, in principle, measurable. INS observations are not restricted to the center of the Brillouin zone (gamma point) as are the optical techniques. The spectrum obtained from such measurements is readily and accurately modeled. This is because the intensities are proportional to the concentration of elements in the sample and their cross sections; the measured INS intensities relate easily to the associated displacements of the scattering atom; and manipulation of backgrounds is uncomplicated. INS spectrometers cover the whole range of the inter- and intramolecular vibration spectra, 0–500 meV (0–4000/cm). For a more detailed description of this technique and usage in catalysis research, the reader is referred to the book titled *Vibrational Spectroscopy with Neutrons, with Applications in Chemistry, Biology, Materials Science and Catalysis* [3], other books on the theory of INS [49–54], and various review articles that have been written on this topic [55, 56].

6.6 CONCLUSION

In this chapter, we have provided an introduction to the structural characterization of heterogeneous catalysts using neutron powder diffraction and small-angle scattering. References have also been provided for those who want to extend the characterization using INS. Neutron scattering techniques have made a large contribution in the traditional condense matter systems, which was the driving force in the development of instrumentation, theoretical and data analysis tools that have been developed in the latter half of the last century. It is now a very mature technique and is being successfully used in solid-state chemistry research, following structural evolution of ever more complex systems. While the groundwork for catalysis research has been laid, it has not yet been extensively used by a wider community. The various reactor- and spallation-based sources are reaching out to the scientific community to extend their usage for industrial applications, with catalysis research fitting very nicely. These instruments are available for all researchers and up-to-date information about the capabilities can be found on their websites.

ACKNOWLEDGMENTS

The authors acknowledge support by the Scientific User Facilities Division, Office of Basic Energy Sciences, U.S. Department of Energy. A. Huq also acknowledges the sample environment team at the spallation neutron source and collaborators R.G. Teller, J.W. Richardson, and E.R. Maxey for the work on ammoxidation catalysis carried out at Argonne National Laboratory that was supported by the U.S. DOE, Office of Science, Basic Energy Sciences, under contract DE-AC02-06CH11357.

REFERENCES

[1] Thomas, J. M., Thomas, W. J. (eds.) (2005) Neutron scattering: a technique of growing importance in the study of catalysts, in *Principles and Practice of Heterogeneous Catalysis*, VCH, Weinheim, Germany and VCH Publishers Inc., New York, pp. 219–221, 241–242.

[2] Joyner, R. W. (1991) *Catal. Today*, 9, 161.

[3] Mitchell, P. C. H., Parker, S. F., Ramirez-Cuesta, A. J., Tomkinson, J. (2005) *Vibrational Spectroscopy with Neutrons, with Applications in Chemistry, Biology, Materials Science and Catalysis*, World Scientific, Singapore.

[4] Brown, L. F. A. (2001) *Int. J. Hydrogen Energy*, 26, 381.

[5] Neutron scattering lengths and cross sections (http://www.ncnr.nist.gov/resources/n-lengths/, last accessed December 21, 2012).

[6] Turner, J. F. C., Done, R., Dreyer, J., David, W. I. F., Catlow, C. R. A. (1999) *Rev. Sci. Instrum.*, 70, 5.

[7] Bailey, I. F., Done, R., Dreyer, J. W., Gray, E. M. (2004) *High Press. Res.*, 24, 309.

[8] Chinta, S. C., Choudhary, T. V., Daemen, L. L., Eckert, J., Goodman, D. W. (2002) *Angew. Chem. Int. Ed.*, 41, 144.

[9] Huq, A., Richardson, J. W., Maxey, E. R., Chendra, D., Chien, W. (2007) *J. Alloys Comp.*, 436, 256.

[10] Rietveld, H. M. (1969) *J. Appl. Cryst.*, 2, 65.

[11] David, W. I. F., Shankland, K., McCusker, L. L., Baerlocher, C. H. (2002) *Structure Determination from Powder Diffraction Data*, Oxford University Press, Oxford, UK.

[12] Eckold, G., Gibhardt, H., Caspary, D., Elter, P., Elisbihani, K. (2003) *Z. Kristallogr.*, 218, 144.

[13] Jones, J. L., Hoffman, M., Daniels, J. E., Studer, A. J. (2006) *Appl. Phys. Lett.*, 89, 092901.

[14] Stinton, G. W., Evans, J. S. O. (2007) *J. Appl. Cryst.*, 40, 87.

[15] Ali, R., Yashima, M. (2005) *J. Solid State Chem.*, 178, 2867.

[16] David, W. I. F., Evans, J. S. O., Sleight, A. W. (1999) *Europhys. Lett.*, 46, 661.

[17] Salzman, C. G., Radaelli, P. G., Hallbrucker, A., Mayer, E., Finney, J. L. (2006) *Science*, 311, 1758.

[18] Billinge, S. J. L., Levin, I. (2007) *Science*, 316, 561.

[19] Egami, T., Billinge, S. J. L. (2003) *Underneath the Bragg Peaks: Structural Analysis of Complex Materials*, Pergamon Press Elsevier, Oxford, UK.

[20] Kunio, M., Yutaka, S., Kenichi, M., Hirokazu, W. (2002) Method for producing acrylonitrile, catalyst for use therein and the method for preparing same. Patent Number: EP1223164(A1).

[21] Morooka, Y., Ueda, W. (1994) *Adv. Catal.*, 40, 233; Ueda, W., Morooka, Y., Ikawa, T. J. (1981) *J. Catal.*, 70, 409.

[22] Chang, T. S., Guijia, L., Shin, C. H., Lee, Y. K., Yun, S.-S. (2000) *Catal. Lett.*, 68, 229.

[23] Nasu, S., Shimizu, S. (1987) *J. Catal.*, 104, 164; Annenkova, B., Alhazov, T. G., Belenkii, T. G. (1969) *Kinet. Katal.*, 10, 1076.

[24] Larson, A. C., Von Dreele, R. B. (2000) *General Structure Analysis System (GSAS)*, Los Alamos National Laboratory Report LAUR 86-748; Toby, B. H. (2001) *J. Appl. Cryst.*, 34, 210–213.

[25] Rodriguez, J. A., Hanson, J. C., Chaturvedi, S., Maiti, A., Brito, J. L. (2000) *J. Phys. Chem. B*, 104, 8145.

[26] Ge, X., Shen, J. Y., Zhang, H. L. (1996) *Sci. China, Ser. B, Chem. Life Sci. Earth Sci.*, 39, 53.

[27] Trifiro, F., Devecchi, V., Pasquon, I. *J. Catal.*, 15, 8 (1969).

[28] Benaichouba, B., Bussiere, P., Vedrine, J. C. (1991) *Hyp. Inter.*, 69, 739.

[29] Ponceblanc, H., Millet, J. M. M., Coudurier, G., Hermann, J. M., Vedrine, J. C. (1993) *J. Catal.*, 142, 373; Zhang, H. L., Shen, J. Y., Ge, X. (1991) *Hyp. Inter.*, 69, 859.

[30] Van Den Elzen, A. F., Rieck, G. D. (1973) *Acta Crystallogr. B*, 29, 2433.

[31] Jeitschko, W., Sleight, A. W., McCellan, W. R., Weiher, J. F. (1976) *Acta Crystallogr. B*, 32, 1163.

[32] Fitch, A. N., Jobic, H., Renouprez, A. (1986) *J. Phys. Chem.*, 90, 1311.

[33] Morris, R. E., Weigel, S. J., Henson, N. J., Bull, L. M., Janicke, M. T., Chmelka, B. F., Cheetham, A. K. (1994) *J. Am. Chem. Soc.*, 116, 11849.

[34] Yildirim, T., Hartman, M. R. (2005) *Phys. Rev. Lett.*, 95, 215504.

[35] Mulder, F. M., Assfour, B., Huot, J., Dingemans, T. J., Wagemaker, M., Ramirez-Cuesta, A. J. (2010) *J. Phys. Chem. C*, 114, 10648.

[36] Celestian, A. J., Parise, J. B., Smith, R. I., Toby, B. H., Clearfield, A. (2007) *Inorg. Chem.*, 46, 1091.

[37] Walton, R. I., Smith, R. I., O'Hare, D. (2001) *Micropor. Mesopor. Mater.*, 48, 79.

[38] Ok, K. M., O'Hare, D., Smith, R. I., Chowdhury, M., Fikremariam, H. (2010) *Rev. Sci. Instrum.*, 81, 125107.

[39] Eckert, J., Draznieks, C. M., Cheetham, A. K. (2002) *J. Am. Chem. Soc.*, 124, 170.

[40] Jeong, I. K., Darling, T. W., Lee, J. K., Proffen, T., Heffner, R. H., Park, J. S., Hong, K. S., Dmowski, W., Egami, T. (2005) *Phys. Rev. Lett.*, 94, 147602.

[41] Rothenberg, G. (2008) *Catalysis: Concepts and Green Applications*, Wiley-VCH, Weinheim, Germany.

[42] Brumberger, H. (1988) *Makromol. Chem. Macromol. Symp.*, 15, 223.

[43] Linder, P. and Zemb, T. H. (eds.) (2002) *Neutrons, X-rays and Light: Scattering Methods Applied to Soft Condensed Matter*, North-Holland, Amsterdam, The Netherlands.

[44] Andersson, R. A., van Heijkamp, L. F., de Schepper, I. M., Bouwman, W. G. (2008) *J. Appl. Crystallogr.*, 41, 868, and reference therein.

[45] Brumberger, H. (ed.) (1995) *Modern Aspects of Small-Angle Scattering*, Kluwer, Dordrecht, The Netherlands.

[46] Linder, P. and Zemb, T. H. (eds.) (1991) *Neutrons, X-rays and Light Scattering: Introduction to an Investigative Tool for Colloidal and Polymeric Systems*, North-Holland, Amsterdam, The Netherlands.

[47] Brumberger, H., Hagrman, D., Goodismana, J., Finkelstein, K. D. (2005) *J. Appl. Cryst.*, 38, 147.

[48] Brumberger, H., Hagrman, D., Goodismana, J., Finkelstein, K. D. (2005) *J. Appl. Cryst.*, 38, 324.

[49] Turchin, V. F. (1965) *Slow Neutrons*, Israel Program for Scientific Translations Ltd., Jerusalem, Israel.

[50] Gurevich, I. I., Tarasov, L. V. (1968) *Low-Energy Neutron Physics*, North-Holland, Amsterdam, The Netherlands.

[51] Marshall, W., Lovesey, S. W. (1971) *Theory of Thermal Neutron Scattering: The Use of Neutrons for the Investigation of Condensed Matter*, Clarendon Press, Oxford, UK.

[52] Squires, G. L. (1978) *Introduction to the Theory of Thermal Neutron Scattering*, Cambridge University Press, Cambridge, UK.

[53] Lovesey, S. W. (1984) *Theory of Neutron Scattering from Condensed Matter*, Oxford University Press, Oxford, UK.

[54] Izyumov, Y. A., Chernoplekov, N. A. (1994) *Neutron Spectroscopy*, Consultants Bureau, New York.

[55] Albers, P. W., Prescher, G., Seibold, K., Parker, S. F. (1999) *Chem. Eng. Technol.*, 22, 2.

[56] Albers, P. W., Parker, S. F. (2007) *Adv. Catal.*, 51, 99.

7

VISUALIZATION OF SURFACE STRUCTURES OF HETEROGENEOUS CATALYSTS UNDER REACTION CONDITIONS OR DURING CATALYSIS WITH HIGH-PRESSURE SCANNING TUNNELING MICROSCOPY

LUAN NGUYEN, SHIRAN ZHANG, YINGCHUN YE, YUAN ZHU, LEI WANG, AND FRANKLIN (FENG) TAO

7.1 SURFACE STRUCTURES OF HETEROGENEOUS CATALYSTS IN ULTRAHIGH VACUUM

Studies of surface structures of model catalysts have been performed for many decades. The initial structural model of surfaces of model catalysts was a smooth structureless discontinuity proposed in the 1930s [1]. In this model, the surface of single-crystal model catalysts was proposed as a perfect, smooth, and structureless discontinuity on the basis of a jellium model [1]. The available analytical techniques of electron microscopy and field-emission microscopy in the 1960s revealed the existence of surface steps, kinks, and other defects. Therefore, the jellium model was replaced with a heterogeneous rigid model in the 1960s. Figure 7.1 schematically shows the structural heterogeneity of surfaces of catalysts. From the point of view of this model, the atoms of the surface layer are arranged with the same periodicity as those in the bulk. This

In-situ *Characterization of Heterogeneous Catalysts*, First Edition.
Edited by José A. Rodriguez, Jonathan C. Hanson, and Peter J. Chupas.

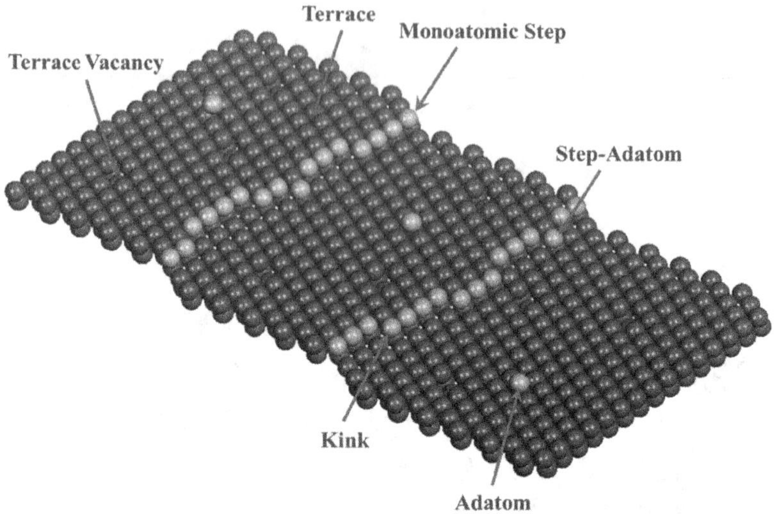

Figure 7.1. Schematic of heterogeneous rigid surface model. Reprinted with permission from Somorjai, G. A., Li, Y. (2010) *Introduction to Surface Chemistry and Catalysis*, Wiley-VCH, New York. Copyright 2010 John Wiley & Sons.

model ignored the surface effect of relaxation and reconstruction toward minimization of the surface free energy. However, it is reasonable to some extent because it can rationalize the kinetics of crystal growth and some surface phenomena such as site-dependent chemisorption and charge distribution. Numerous studies with surface analytical techniques developed in the 1980s, particularly scanning tunneling microscopy (STM), with the capability of visualization of surface structures at the atomic scale, showed that this heterogeneous rigid model is not perfect. The concept of flexibility of catalyst surfaces was thus introduced [2]. A solid piece of evidence is the inward relaxation of the first layer of a clean metal surface to produce a short interlayer distance in contrast to the distance between two adjacent atomic layers in bulk. In fact, extensive studies of surface structures in the last three decades have confirmed the universal phenomenon of surface relaxation. Recently, studies showed that surfaces of model catalysts could be restructured into a reactive environment with a relatively high pressure or during catalysis [3–7]. It illustrated the possibility of restructuring of catalyst surfaces under reaction conditions or during catalysis.

7.2 SURFACE STRUCTURES OF HETEROGENEOUS CATALYSTS UNDER REACTION CONDITIONS

Characterization of the surfaces of heterogeneous catalysts performed under reaction conditions in terms of "reactions on a catalyst surface at ambient or high temperature in the environment of reactant/product gases with ambient or high pressure" is extremely complicated because a catalytic reaction could involve dramatic atomic rearrangement of a surface, restructuring of a subsurface, and mass transfer, driven by both thermodynamic and kinetic factors. Few surface analytical techniques could be applied to study the evolution of surface structures of catalysts during catalysis, though vibrational spectroscopies such as infrared and Raman spectroscopies have been well developed for identification of adsorbates. In fact, most characterizations of catalyst surfaces were performed in high vacuum (HV) or ultrahigh vacuum (UHV) in the last three decades [1]. Here we refer to it as the "vacuum approach." In this approach, reactant molecules are introduced onto the surface of a single-crystal model catalyst, which is exposed to a reactant gas at a certain pressure, usually ~10^{-6}–10^{-9} Torr, within a certain amount of time. Before examining this surface and the adsorbate layer, the reactant gases are completely pumped out of the chamber to restore the original UHV or HV environment. This vacuum approach takes advantage of the easy cleaning processes of the metal surfaces in UHV, the well-known surface structures of single-crystal model catalysts in UHV, and the well-developed vacuum-compatible surface analytical techniques such as Auger electron spectroscopy (AES), X-ray photoelectron spectroscopy (XPS), and low-energy electron diffraction (LEED). By using various surface analytical techniques, molecular adsorption, surface restructuring, and various surface reactions on single-crystal model catalysts were extensively studied in the last four decades [1]. Numerous surface structures and phenomena were revealed and an enormous database of these model studies was created [1]. The insights achieved from those vacuum approaches built the foundation for studying heterogeneous catalysis under reaction conditions or during catalysis [1].

Industrial heterogeneous catalysis is usually carried out on catalyst surfaces at ambient or high temperature in an environment of reactant/product gases with ambient or high pressure. Adsorption and surface structures are the key issues in understanding catalytic mechanisms. However, it has been revealed in some cases that the adsorption and surface structure are pressure-dependent due to both thermodynamic and kinetic factors [3–5]. From the point of view of thermodynamics, it is necessary to consider the entropic contribution to the Gibbs free energy of a catalyst surface and potential restructuring induced by large repulsion of adsorbates at high coverage at high pressure of reactants [5]. The contribution of the former is called pressure factor, $kT \cdot \log P$. It raises a difference of 0.3 eV for a catalytic system at room temperature between UHV (10^{-10} Torr) and atmospheric pressure (760 Torr). Such amount of change in the surface free energy of catalytic systems could largely change the adsorption and surface structure if the enthalpy does not significantly change with the increase of pressure of reactant gases. In contrast, dynamic factors also play an important role since the surface is flexible partially due to the dramatic exchange of the adsorbed molecules with the molecules near the catalyst surface. Thus, a simple extrapolation of the insights of the adsorption and surface structure obtained under UHV or HV conditions is not necessarily applicable to that under ambient or high-pressure conditions since the latter possibly has different energetic pathways. In catalysis studies, this potential difference in surface structure and chemistry *between* UHV, HV, or low pressure *and* an ambient or high pressure is called pressure gap [3, 5, 8, 9]. One example is the catalytic activity of the Ru catalyst for CO oxidation at different pressures. In a low-pressure environment, Ru is inert for CO oxidation. However, it is an extremely active catalyst for CO oxidation under high pressure of oxygen due to the formation of a new phase RuO_2 [9]. To understand catalysis mechanism, it is necessary to study the surface structures and chemistry of catalysts under reaction conditions or during catalysis.

7.3 TECHNIQUES FOR STUDIES OF SURFACE STRUCTURE OF CATALYSTS UNDER REACTION CONDITIONS OR DURING CATALYSIS

The studies of catalysis under a condition of high pressure require surface analytical techniques with the capability of characterizing a catalyst surface when the surface is in an environment of reactant gases at ambient or high pressure. Sum frequency generation (SFG) is one of the vibrational spectroscopies capable of identifying vibrational signatures of adsorbates at high-pressure conditions [1, 10, 11]. In this approach, the catalyst surface is cleaned and then reactant gases are introduced to a high-pressure reactor and kept at a constant pressure during each experiment. The vibrational signatures of adsorbates under reaction conditions obtained from this technique allow us to identify adsorbates on catalyst surfaces and, in some cases, even the reaction intermediates and products.

Heterogeneous catalysis is performed on active sites such as step edges, kinks, vacancies, adatoms, or some ensembles with specific arrangement and composition of atoms. These sites typically have a specific packing of surface atoms and even a different electronic density. Thus, structural information on catalyst surfaces at the atomic level is crucial for understanding a catalytic mechanism. Due to the contribution of the pressure-dependent entropy factor, packing of atoms on catalyst surfaces under reaction conditions or during catalysis could be quite different from that characterized in UHV or HV. With the addition of thermal diffusion at the reaction temperature, it is challenging to predict the surface structure of catalysts under reaction conditions or during catalysis without *operando* studies of surface structures under reaction conditions. The importance in studying the structural evolution of both the adsorbates and catalyst surfaces under reaction conditions or during catalysis has driven the development of new techniques to determine surface structures under these conditions. Environmental transmission electron microscopy (E-TEM) and high-pressure scanning tunneling microscopy (HP-STM) are two important techniques for analyzing structures of higher surface catalysts and low surface ones, respectively. E-TEM [12, 13] offers the capability of *operando* studies of structures of catalysts with high surface area such as metal nanoparticles supported on oxide prepared with impregnation. Unfortunately, it is quite challenging to obtain information on the atomic arrangement of the catalyst surfaces by using E-TEM. Alternatively, HP-STM has the unique capability of examining the structure of adsorbates and surface atoms of catalysts *atom by atom* to address several crucial issues of heterogeneous catalysis including restructuring of catalyst surfaces, rearrangement of surface atoms, ensemble effects on surfaces of alloy catalysts, and binding site and configuration of adsorbates under reaction conditions or during catalysis [3, 4].

Since its birth in the 1980s [14], STM has been one of the main techniques used to perform fundamental studies of heterogeneous catalysis with a vacuum approach. This is because apart from most other surface techniques, it has the unique capability of visualizing surface structures with an atomic resolution and identifying the atomic packing and electron state of any specific location on a catalyst surface. In principle, this technique can be applied in a pressure range from UHV to atmospheric or high pressures since the tunneling process between the sample and the tip only occurs in a very close range of 3–5 Å and the molecular layer between the tip and the adsorbate or surface atoms does not contribute to a tunneling process. The prototype studies of HP-STM are the characterization of the surfaces of model catalysts at a high pressure by filling reactant gases into the STM chamber and holding the reactant gas at a relatively high pressure during STM scanning [15]. Since the early demonstration, progress has been made in the instrumentation of an HP-STM. Studies using these HP-STMs revealed new information on surface structures of catalysts under reaction conditions or during catalysis [12, 16–20].

Frenken et al. built an HP-STM system that combined an STM with a flowing cell with a very small free volume to study catalyst surfaces at a semi-practical condition [18]. In their design, a tip is led through an Au membrane. Figure 7.2 represents the scheme of this HP-STM. This design gives a reasonable resolution of catalyst surfaces under reaction conditions or during catalysis. It has the capability of *in-situ* heating of the sample during scanning, which makes it possible to examine surface structure during catalytic reactions at a condition of high pressure (<1000 Torr). Besenbacher et al. installed their high-resolution Aarhus STM body into a small vacuum chamber through an 8-in. flange [17]. This single-chamber STM reactor can perform catalysis studies from UHV to 1000 Torr with the capability of resolving surface structure at atomic level at room temperature. Figure 7.3 is the schematic of this single-chamber high-resolution HP-STM [17]. Kolmakov and Goodman developed a single-chamber HP-STM by filling reactant gases into a UHV chamber (Fig. 7.4) as well [19]. This HP-STM was used to study catalysis of metal clusters deposited on oxide substrates at a high pressure of reactive gases [19]. In addition, Wintterlin et al. built a high-pressure cell (1.5 L) containing an STM for catalysis studies in a pressure from UHV to 760 Torr (Fig. 7.5) [16]. Thermal isolation of the hot catalysts and warm reactant/product gases from scanning

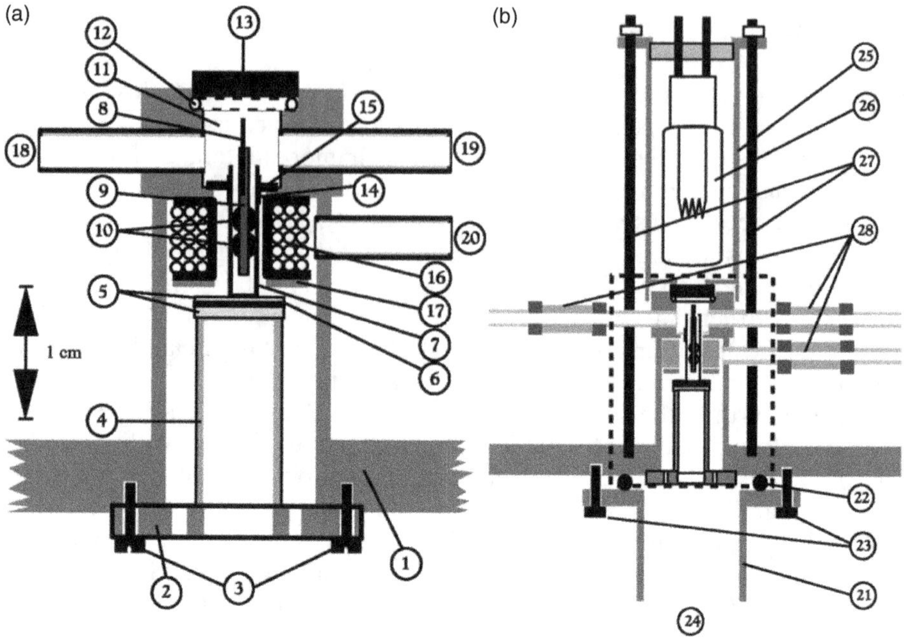

Figure 7.2. (a) Cross section of the STM reactor sample assembly: (1) housing, (2) baseplate, (3) mounting screws for baseplate, (4) piezo element, (5) BN disks, (6) Ta plate, (7) silver tube, (8) tip, (9) Cu capillary, (10) steel balls, (11) reactor, (12) Au ring, (13) sample, (14) isolation ring, (15) Au membrane, (16) solenoid, (17) pole shoe, (18) reactor gas inlet, (19) reactor gas outlet, (20) N_2 outlet. (b) Cross section of the complete reactor STM: (21) lower housing, (22) Viton ring, (23) mounting screws for lower house, (24) electrical feedthroughs and N_2 inlet, (25) sample holder, (26) 150-W halogen light bulb, (27) threaded rods, (28) Al_2O_3 insulators. Reprinted with permission from Frenken, J. W. M. (1998) *Rev. Sci. Instrum.*, 69, 3879–3884. Copyright 1998 American Institute of Physics.

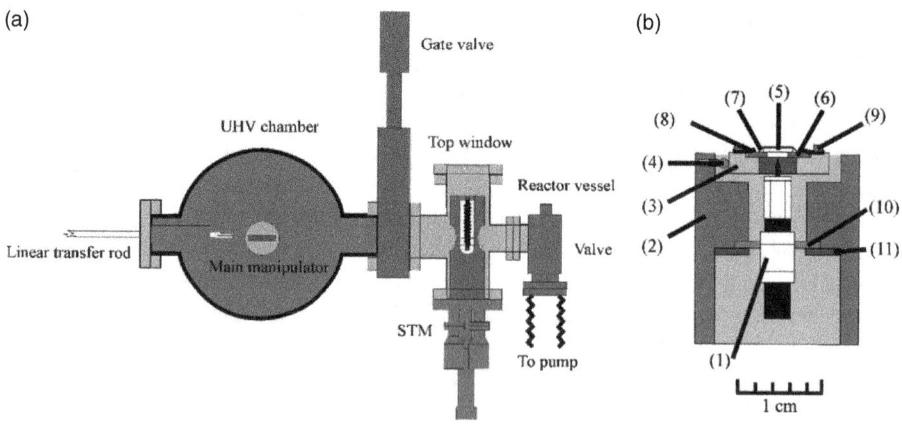

Figure 7.3. (a) Reaction chamber containing the STM, the connection to the UHV system, and the sample transfer mechanism. The manipulator of the main chamber moves perpendicularly to the plane of the figure. (b) High-pressure STM consisting of (1) Inchworm scanner tip assembly, (2) Invar housing, (3) sample holder, (4) quartz balls, (5) sample, (6) Ta support, (7) Ta foil, (8) leaf springs, (9) screws, (10) macor ring, and (11) support ring. Reprinted with permission from Besenbacher, F. (2001) *Rev. Sci. Instrum.*, 72, 3537–3542. Copyright 2001 American Institute of Physics.

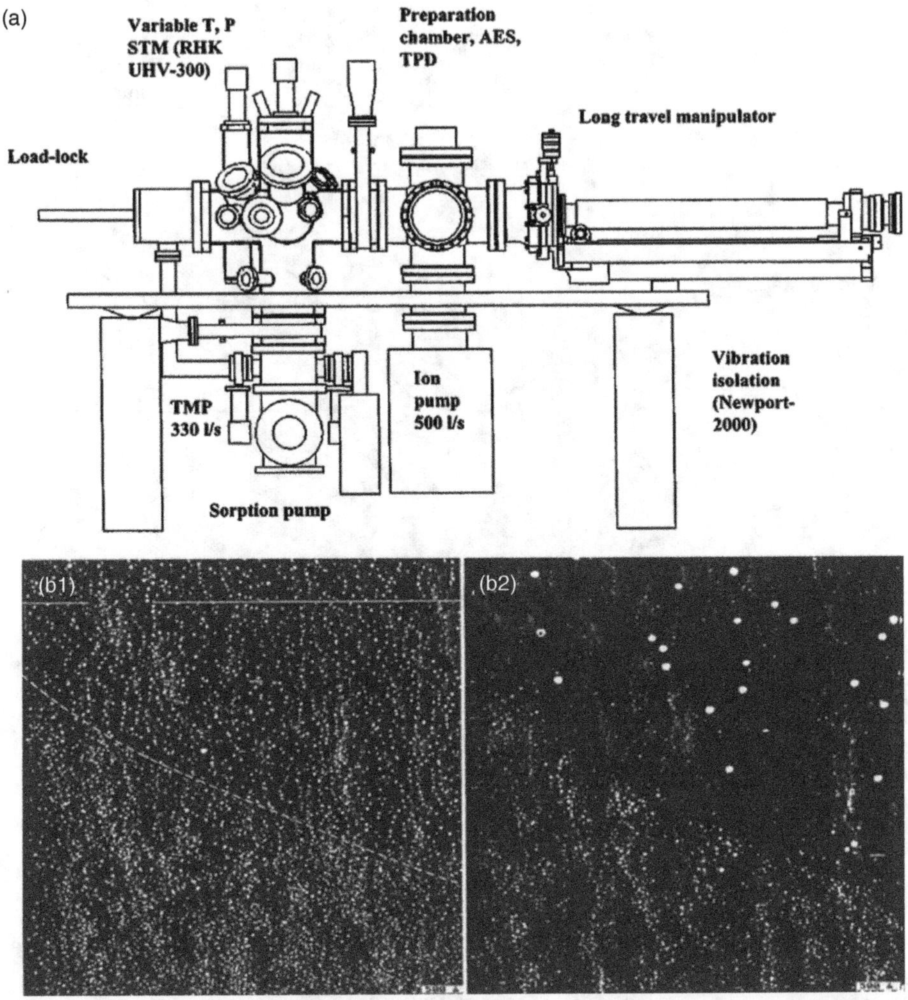

Figure 7.4. (a) A schematic of the experimental setup for *in-situ* elevated pressure STM studies on individual oxide-supported nanoparticles. (b1) The same areas of Ag, Ag–Ag (mixed), and Au clusters on TiO_2 as imaged under UHV conditions ~$0.5 \times 0.5\ \mu m^2$. (b2) The same area in air environment of a 2.5×10^4 Pa. Reprinted with permission from Goodman, D. W. (2003) *Rev. Sci. Instrum.*, 74, 2444–2450. Copyright 2003 American Institute of Physics.

tubes and approaching motor remains challenging in the application of HP-STM to study catalyst surfaces at a temperature higher than room temperature in reactive gases with certain pressure. In terms of the pressure of reactants, the upper pressure is limited to the pressure of the UHV chamber, typically 1–2 bar due to the sealing mechanism of gasket and Del-Seal flange.

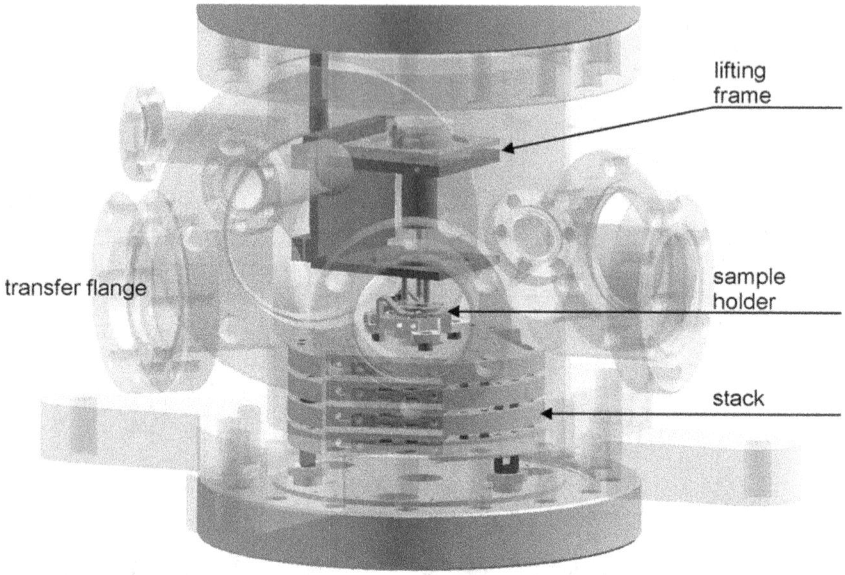

Figure 7.5. High-pressure cell with the STM, viewed from the position of the viewport (DN 40 CF flange in the center). Parts directly visible through the viewport window are the beetle-type STM head and the fitting with the sample holder. The four round plates below the viewport flange form the internal vibration isolation stack. Parts above the viewport belong to the frame used for lifting the STM head during transfer of the sample holder. Reprinted with permission from Rossler, M., Geng, P., Wintterlin, J. (2005) *Rev. Sci. Instrum.*, 76, 023705. Copyright 2005 American Institute of Physics.

Overall, most of these HP-STMs are typically used to study surface structures of catalysts at room temperate in an environment of reactants/products at a pressure up to 1–2 bar.

A new chamber-in-chamber HP-STM was built recently [5]. Figure 7.6 schematically shows this chamber-in-chamber HP-STM. In this design, the whole STM body including scan tube and coarse approach motor, tip, and sample stage was installed into a small cylinder reaction cell with a free volume of 19 mL for gas of reactants or/and products (Fig. 7.6c). Two stainless steel tubes (internal diameter [i.d.] = 1/16 in.) were installed at the top and bottom of the reaction cell for gas introduction and exit, respectively (7 in Fig. 7.6c). The reaction cell was integrated into a detachable scaffold (Fig. 7.6b). A sealable port on the wall of the reactor separates the high-pressure environment of reactant gases in the reactor from the vacuum environment of the UHV chamber and permits the change of sample and tip in UHV (top left in Fig. 7.6c). A combination of a sample transfer arm, wobble stick, and sample load–lock system allows a convenient transfer of samples and tips among the preparation chamber, high-pressure reactor, and ambient environment (Fig. 7.6a).

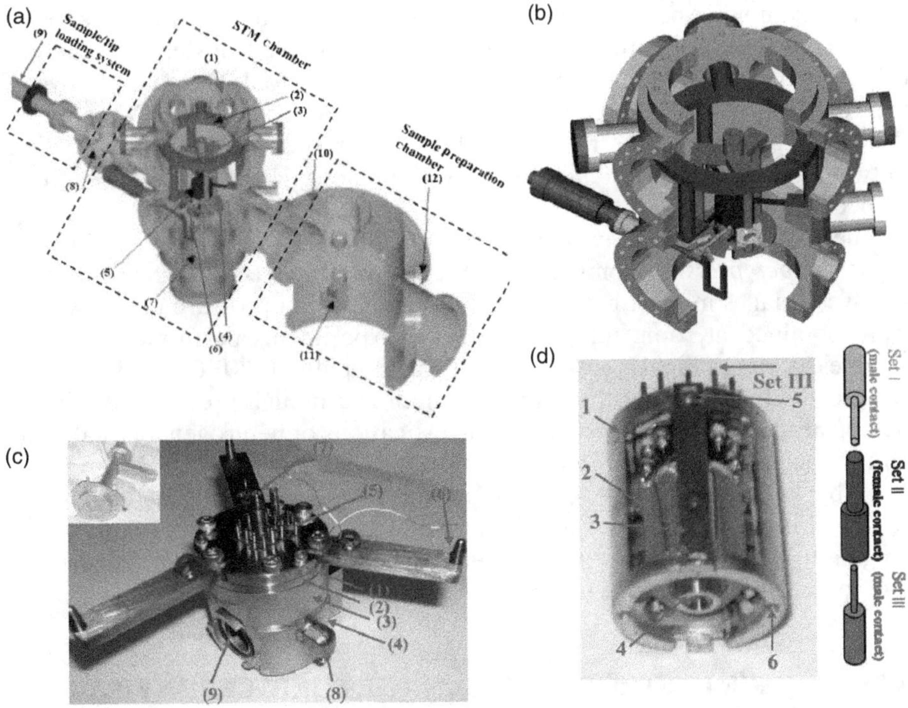

Figure 7.6. (a) CAD drawing of the high-pressure high-temperature STM reactor/ UHV system designed by Tao et al. [20]: (1) view window, (2) mounting framework, (3) docking scaffold, (4) docking disk, (5) high-pressure reactor (STM body housed within), (6) bayonet seal, (7) guide rod of docking scaffold, (8) sample/tip load–lock system, (9) transfer rod, (10) gate valve, (11) four-finger sample stage, and (12) sputtering ion gun. (b) CAD drawing of the STM chamber, scaffold, and high-pressure cell. (c) Photo of the high-pressure reactor (STM body was incanted): (1) cell lid, (2) cell neck, (3) middle stage, (4) bottom stage, (5) pin–socket contacts for wiring of the STM body at the interface of high-vacuum and high-pressure environment, (6) a pin on the lever of the cell lid for mounting, (7) Swagelok fitting welded on the tube of the cell lid for gas introduction, (8) Swagelok fitting welded on the button stage of the reactor for gas exit, (9) port of reactor for sample and tip transfer. (d) Side view of the STM body: (1) wall of the STM body coated with a layer of gold, (2) hexagonal sapphire, (3) CuBe spring plates, (4) receiver of tip holder, (5) screw to adjust the pressure applied to hexagonal sapphire, (6) hole to assemble the sample stage to the STM body. (c) Scheme of pin–socket alignment for wiring. Set 3 is the contacts glued to the holes on top of the STM body. Set 2 is the contacts glued to the holes on top of the cell lid of the reactor. Set 1 is the contacts glued to wires from the docking scaffold. (See color insert.)

While a sample is being scanned, the reaction cell is hung on three springs for the purpose of damping. For the purpose of sample transfer, the reaction cell is fixed to a sample manipulator through the scaffold. The space between the reaction cell and the UHV chamber is an environment of HV or UHV. When the pressure in the reaction is 1 or 10 bar, the pressure in the section between the reaction cell and the UHV chamber is 1×10^{-8} or 5×10^{-7} Torr, respectively.

The feature of this new chamber-in-chamber HP-STM (Fig. 7.6) is the higher *upper pressure* compared to the above single-cell HP-STM. Atom-resolved images in inert gaseous environments with a pressure of a few bars were obtained at room temperature [20]. Experiments performed on inert samples such as highly ordered pyrolytic graphite (HOPG) and reactive samples such as *hex*-Pt(100) both in vacuum and in high-pressure conditions demonstrate the capability of *in-situ* investigations of heterogeneous catalysis and surface chemistry at atomic resolution at a wide pressure range from UHV to 10 bar or even higher. Atomically resolved STM images of metal surfaces can be obtained routinely at room temperature [20]. However, it is quite challenging to obtain atom-resolved images at high temperature. This is partially due to the thermal drift of the tip induced by hot catalysts and warm gases.

7.4 SURFACE RESTRUCTURING IN THE SURFACE LAYER OF A CATALYST

Surface restructuring and flexibility were proposed earlier on the basis of studies of surface science in vacuum. As mentioned earlier, entropy could significantly contribute to surface restructuring and flexibility at a high-pressure condition [5]. Our recent studies showed a surface restructuring in the near surface region (surface layer and subsurface layers) induced by the increased pressure of the reactant gases. It showed that the surface mobility of a catalyst surface is significantly increased at a condition of high pressure [4, 6]. Thus, catalyst surfaces under an environment of reactant gas with high pressure could be considered as a mobile surface in some cases.

Here we take the pressure-dependent adsorption and surface structure of *hex*-Pt(100) under an environment of CO as an example to illustrate the mobility of metal atoms of a catalyst surface [6]. Figure 7.7 shows the image of a clean *hex*-Pt(100) prepared in UHV [6]. The surface structure and the formed nanoclusters at room temperature at different pressures of CO were presented in Figure 7.8 [6]. Clearly, the surface of *hex*-Pt(100) forms nanoclusters under certain pressures of CO at room temperature. More importantly, the size and shape of the formed nanoclusters are pressure-dependent. At a low pressure of 5×10^{-9} Torr (Fig. 7.8a), a portion of the surface is preferentially lifted along the [01-1] direction. However, upon a clean *hex*-Pt(100) exposed to 5×10^{-5} Torr CO, islands with a rectangular shape and a size of 1.5–4.0 nm are formed on the whole surface (Fig. 7.8c). There is no preferential growth

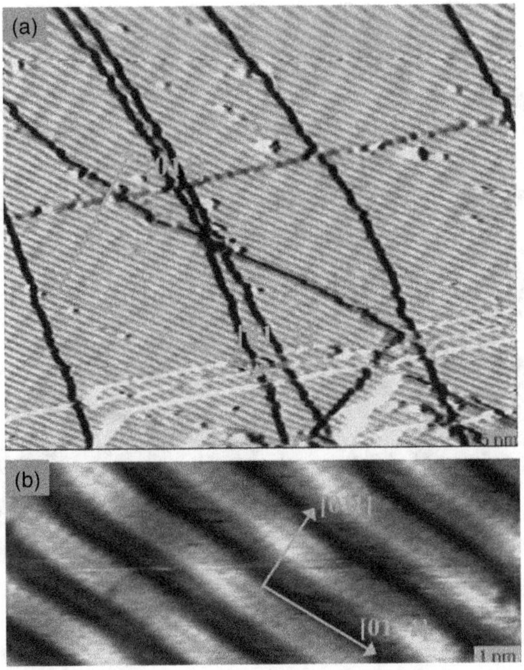

Figure 7.7. STM image of a clean *hex*-Pt(100) surface. (a) Large-scale STM image of a clean *hex*-Pt(100) surface collected at room temperature at 1×10^{-10} Torr. (b) Small-scale atomically resolved STM images of *hex*-Pt(100) collected at room temperature at 1×10^{-10} Torr. Reprinted with permission from Tao, F., Dag, S., Wang, L.-W., Liu, Z., Butcher, D. R., Salmeron, M., Somorjai, G. A. (2009) *Nano Lett.*, 9(5), 2167–2171. Copyright 2009.

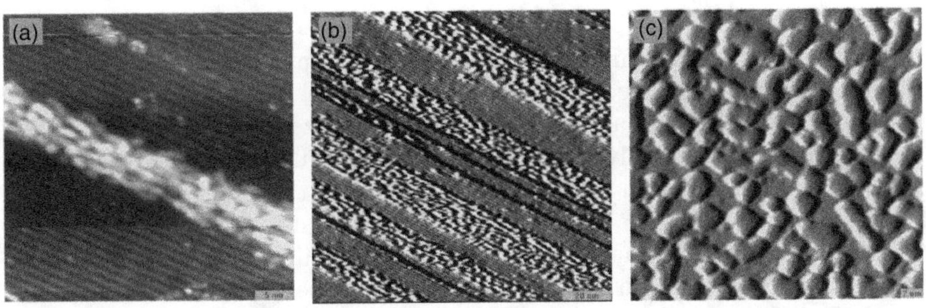

Figure 7.8. STM images of a lifted Pt(100) surface in CO environment with different pressures (a) 5×10^{-9} Torr, (b) 5×10^{-7} Torr, and (c) 5×10^{-5} Torr. *Note*: The listed pressure is the pressure of CO that remained in the reactor during data acquisition. Reprinted with permission from Tao, F., Dag, S., Wang, L.-W., Liu, Z., Butcher, D. R., Salmeron, M., Somorjai, G. A. (2009) *Nano Lett.*, 9(5), 2167–2171. Copyright 2009. (See color insert.)

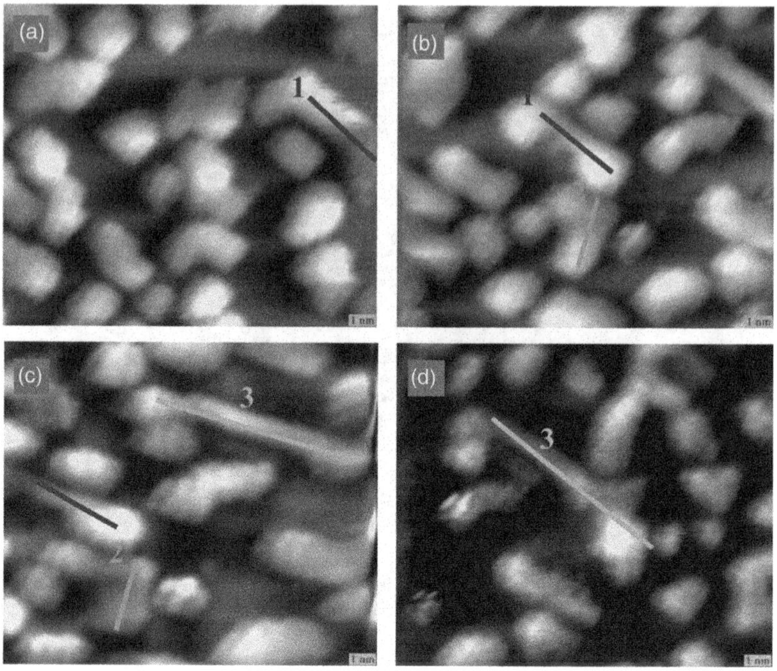

Figure 7.9. STM images of a lifted Pt(100) surface at a condition of 5×10^{-7} Torr CO. (a) $t = t_o$ min, (b) $t = t_o + 10$ min, (c) $t = t_o + 19$ min, and (d) $t = t_o + 28$ min. All the images have the same size of 14.1×14.1 nm. They were obtained with the same tunneling condition and scanning speed. Landmarks 1, 2, and 3 represent the same cluster between (a) and (b), between (b) and (c), and between (c) and (d), respectively. Reprinted with permission from Bowker, M., Davies, P. R. (2010) *Scanning Tunneling Microscopy in Surface Science, Nanoscience and Catalysis*. Copyright 2010 John Wiley & Sons.

along any direction. At a high pressure of 1 Torr CO, nanoclusters with a round or ellipse shape are formed on the whole surface.

The formation of Pt nanoclusters can be visualized with an HP-STM. As described earlier, *hex*-Pt(100) restructures to nanoclusters when it remains in a CO environment with a pressure of $\sim 5 \times 10^{-7}$ Torr. The incubation time for formation of nanoclusters at this pressure range is a few to tens of minutes (Fig. 7.9). Thus, the formation process can be monitored by HP-STM [21]. During this visualization, a landmark that does not change within minutes must be selected [21]. Thus, this landmark is used as a marker to identify how other nanoclusters change as a function of time. Selection of a landmark is necessary since a thermal drift always exists at room temperature. Within about 10 min, the shape of the islands changes and a portion of atoms in adjacent nanoclusters move together and merge to form new nanoclusters in CO with a pressure of 5×10^{-7} Torr, suggesting a surface mobility of the lifted Pt atoms at this pressure. Such a visualization of high-pressure STM can clearly demonstrate the mobility of nanoclusters induced by the presence of a high-pressure reactant gas. Notably, no atomically resolved images can be obtained from a *hex*-Pt(100) surface in an environment of CO with a pressure higher than 10^{-4} Torr upon a number of efforts, though the atomically resolved images were obtained at a pressure of 10^{-10}–10^{-5} Torr CO. It probably results from a higher mobility of Pt atoms on the surface in CO environment of a higher pressure of CO.

7.5 PRESSURE-DEPENDENT SURFACE COVERAGE OF ADSORBATES AND STRUCTURAL EVOLUTION FROM LOW TO HIGH PRESSURE

STM studies at high pressure have been recently performed on Pt(111), Ni(111), and Pt(110) model catalysts [22–24]. On Pt(111) it was found that an increase of the CO pressure from 10^{-9} to 1 Torr caused the coverage to increase from 0.5 to 0.65, and finally to reach a saturation coverage, 0.68 at 100 Torr [23]. Notably, the moiré pattern is formed on Pt(111) at a relatively high pressure (Fig. 7.10). Similar to Pt(111), there is no large-scale change in the surface structure of flat Ni(111) surfaces from low to high pressure [22, 24]. However, this is not a case of stepped surfaces of catalysts as described in the next paragraph.

Industrial catalysts consist of particles with sizes ranging from a few nanometers to a few hundred nanometers. Due to their small size, they contain a large fraction of atoms at step edges and corners, which are thought to be the active sites due to their low coordination number. Stepped surfaces with parallel nanoscale terraces (typically 1–5 nm) are considered the closest model catalysts for industrial catalysts [1]. To model these catalyst systems, stepped surfaces can be prepared by cutting a crystal at small angles relative to a low Miller index direction. This produces a surface with steps and kinks [4]. Recent studies have shown that the coverage of CO on Pt(557) and Pt(332), which consists of 1.2-nm wide (111) oriented terraces separated by monoatomic steps, can be as high as 1.0 [4]. Most importantly, the high coverage of reactant molecules at a pressure higher than 0.1 Torr at room temperature results in a significant restructuring of the catalyst surface, where the terraces break up into triangular-shaped nanoclusters with lateral dimensions of ~2 nm [4]. The formation of this new equilibrium structure on a stepped surface is favored by repulsion resulting from the high density of CO and the facile displacement of step atoms compared to atoms located on a flat surface. The energy cost of the rearrangement of atoms at step is compensated by the decrease in surface free energy through packing a larger number of CO molecules on the same surface area. The accommodation of more CO molecules in terms of the increase of CO coverage is accomplished by fanning out CO molecules adsorbed on the low-coordinated Pt atoms at the edge of the nanoclusters, which decreases their repulsion [4].

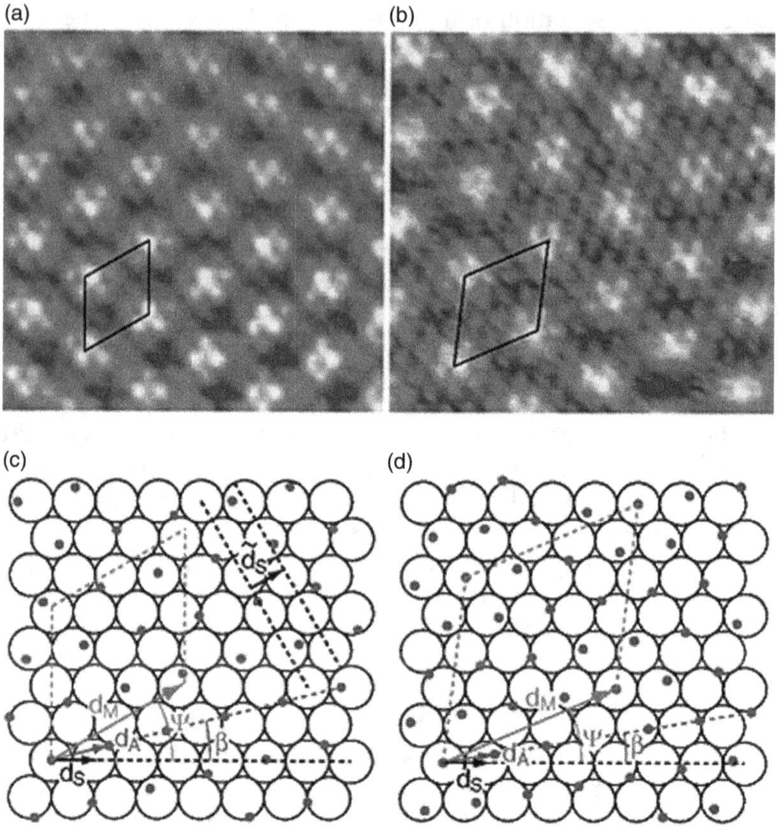

Figure 7.10. STM images of 55×55 Å^2 and corresponding ball models showing moiré superstructures of CO on Pt(111) at room temperature. (a) Incommensurate structure at $p = 10^{-2}$ Torr, $I_t = 1.06$ nA, $V_t = 8.2$ mV; (b) $p = 720$ Torr, $I_t = 1.27$ nA, $V_t = 4.9$ mV. The images (a) and (b) are aligned so that the bulk [110] direction is oriented along the x-axis. The image treatment comprised frequency filtering and interpolation. In the ball models in (c) and (d), the open circles represent Pt atoms and the solid dots CO molecules. The rotational angles β between the substrate and adsorbate lattices and ψ between the substrate and moiré lattices are indicated. The moiré unit cell with lattice constant d_M is also shown. (c) Incommensurate structure at 10^{-2} Torr, ψ = 30°, β = 10.4°, $d_M/d_S = 3.6$; d_S is the real space equivalent of the reciprocal vector q_S observed in the fast Fourier transform. d_S is rotated by 30° from d_S, and the length relationship is $d_S = \sqrt{3}d_s/2$. (d) Commensurate ($\sqrt{19} \times \sqrt{19}$) R23.4°-13CO structure at 720 Torr, ψ = 23.4°, β = 9.5°, $d_M/d_S = 4.4$. Reprinted with permission from Besenbacher, F. (2004) J. *Phys. Chem. B*, 108(38), 14497–14502. Copyright 2004.

7.6 TEMPERATURE-DRIVEN SURFACE RESTRUCTURING AT A RELATIVELY HIGH PRESSURE OF REACTANTS AND THE CHALLENGE IN OBTAINING ATOMICALLY RESOLVED IMAGES AT HIGH TEMPERATURE AND HIGH PRESSURE

As temperature is certainly a factor in increasing the mobility of catalyst atoms at surfaces, restructuring could be much more dramatic at a higher temperature of catalysts in an environment of reactant gases. Most catalytic reactions are performed at a temperature higher than room temperature. Achieving high-resolution images of catalysts at high temperature in reactant gases is quite challenging since there were several important technical issues to overcome.

One example is that the small Pt nanoclusters terminated with CO molecules formed in 1 Torr CO at room temperature aggregate into large clusters when the *hex*-Pt(100) catalyst is heated to 420 K in 1 Torr of CO [21]. The surface formed at this pressure of CO at room temperature consists of Pt nanoclusters with a size of 2–3 nm. When this surface is heated to 150°C in 1 Torr CO, the lifted surface structure significantly changes. The nanoclusters formed at room temperature largely aggregate together at 150°C, forming clusters with a size of 3–7 nm. This aggregation illustrates again the high flexibility and mobility of a catalyst surface. More importantly, it clearly demonstrates the necessity of visualizing surface structures of catalysts under reaction conditions at high temperature and high pressure, though it is challenging.

7.7 SUMMARY

Heterogeneous catalysis is a chemical process performed on specific catalytic sites on a catalyst surface at a certain temperature in an environment of reactant gases with certain pressure. Surface structures of catalysts at the atomic level under reaction conditions and during catalysis are likely different from those examined before or after the reaction or catalysis for some reactions, which drives the *operando* studies of catalysts at atomic scale with the purpose of building a direct correlation between surface structures under reaction conditions and their catalytic performance (activity, selectivity, and deactivation). The instrumentation of high-pressure STM has allowed us to visualize surface structures of catalysts at a relatively high pressure, particularly during catalysis. Recent studies have shown the distinctly different surface structures of catalysts under reaction conditions in contrast to those before or after reactions. Overall, these studies of the surface structure of catalysts under reaction conditions clearly show the possibility of being restructured for catalyst surfaces under reaction conditions. To explore catalytic mechanisms at atomic scale, it is necessary to perform *operando* studies of catalyst surfaces at the atomic level. A great challenge from the point of view of technique is how to minimize the thermal drift of STM tip since it is the bottleneck to obtain atom-resolved images at a temperature higher than room temperature.

ACKNOWLEDGMENT

This work was supported by the Chemical Sciences, Geosciences and Biosciences Division, Office of Basic Energy Sciences, Office of Science, U.S. Department of Energy.

REFERENCES

[1] Somorjai, G. A., Li, Y. (2010) *Introduction to Surface Chemistry and Catalysis*, Wiley-VCH, New York.

[2] Somorjai, G. A. (1991) The flexible surface—correlation between reactivity and restructuring ability, *Langmuir*, 7, 3176–3182.

[3] Tao, F., Salmeron, M. (2011) *In situ* studies of chemistry and structure of materials in reactive environments, *Science*, 331, 171–174.

[4] Tao, F., Dag, S., Wang, L.-W., Liu, Z., Butcher, D. R., Bluhm, H., Salmeron, M., Somorjai, G. A. (2010) Break-up of stepped platinum catalyst surfaces by high CO coverage, *Science*, 327, 850–853.

[5] Tao, F., Grass, M. E., Zhang, Y., Butcher, D. R., Renzas, J. R., Liu, Z., Chung, J. Y., Mun, B. S., Salmeron, M., Somorjai, G. A. (2008) Reaction-driven restructuring of Rh-Pd and Pt-Pd core–shell nanoparticles, *Science*, 322, 932–934.

[6] Tao, F., Dag, S., Wang, L.-W., Liu, Z., Butcher, D. R., Salmeron, M., Somorjai, G. A. (2009) Restructuring of hex-Pt(100) under CO gas environments: formation of 2-D nanoclusters, *Nano Lett.*, 9, 2167–2171.

[7] Zhu, Z., Tao, F., Zheng, F., Chang, R., Li, Y., Heinke, L., Liu, Z., Salmeron, M., Somorjai, G. A. (2012) Formation of nanometer-sized surface platinum oxide clusters on a stepped Pt(557) single crystal surface induced by oxygen: a high-pressure STM and ambient-pressure XPS study, *Nano Lett.*, 12, 1491–1497.

[8] Vang, R. T., Laegsgaard, E., Besenbacher, F. (2007) Bridging the pressure gap in model systems for heterogeneous catalysis with high-pressure scanning tunneling microscopy, *Phys. Chem. Chem. Phys.*, 9, 3460–3469.

[9] Over, H., Muhler, M. (2003) Catalytic CO oxidation over ruthenium—bridging the pressure gap, *Prog. Surf. Sci.*, 72, 3–17.

[10] Somorjai, G. A., McCrea, K. R. (2000) Sum frequency generation: surface vibrational spectroscopy studies of catalytic reactions on metal single-crystal surfaces, *Adv. Catal.*, 45, 385–438.

[11] Somorjai, G. A., York, R. L., Butcher, D., Park, J. Y. (2007) The evolution of model catalytic systems; studies of structure, bonding and dynamics from single crystal metal surfaces to nanoparticles, and from low pressure (<10(−3) Torr) to high pressure (>10(−3) Torr) to liquid interfaces, *Phys. Chem. Chem. Phys.*, 9, 3500–3513.

[12] Thiberge, S., Zik, O., Moses, E. (2004) An apparatus for imaging liquids, cells, and other wet samples in the scanning electron microscopy, *Rev. Sci. Instrum.*, 75, 2280–2289.

[13] Thiel, B. L., Toth, M., Schroemges, R. P. M., Scholtz, J. J., van Veen, G., Knowles, W. R. (2006) Two-stage gas amplifier for ultrahigh resolution low vacuum scanning electron microscopy, *Rev. Sci. Instrum.*, 77, 033705.

[14] Binnig, G., Rohrer, H. (1983) Scanning tunneling microscopy, an atomic probe, *Scan. Electron Microsc.*, Pt 3, 1079–1082.

[15] Mcintyre, B. J., Salmeron, M. B., Somorjai, G. A. (1992) A scanning tunneling microscope that operates at high-pressures and high-temperatures (430-K) and during catalytic reactions, *Catal. Lett.*, 14, 263–269.

[16] Rossler, M., Geng, P., Wintterlin, J. (2005) A high-pressure scanning tunneling microscope for studying heterogeneous catalysis, *Rev. Sci. Instrum.*, 76, 023705.

[17] Laegsgaard, E., Osterlund, L., Thostrup, P., Rasmussen, P. B., Stensgaard, I., Besenbacher, F. (2001) A high-pressure scanning tunneling microscope, *Rev. Sci. Instrum.*, 72, 3537–3542.

[18] Rasmussen, P. B., Hendriksen, B. L. M., Zeijlemaker, H., Ficke, H. G., Frenken, J. W. M. (1998) The "reactor STM": a scanning tunneling microscope for investigation of catalytic surfaces at semi-industrial reaction conditions, *Rev. Sci. Instrum.*, 69, 3879–3884.

[19] Kolmakov, A., Goodman, D. W. (2003) *In situ* scanning tunneling microscopy of individual supported metal clusters at reactive gas pressures from 10(–8) to 10(4) Pa, *Rev. Sci. Instrum.*, 74, 2444–2450.

[20] Tao, F., Tang, D., Salmeron, M., Somorjai, G. A. (2008) A new scanning tunneling microscope reactor used for high-pressure and high-temperature catalysis studies, *Rev. Sci. Instrum.*, 79, 084101.

[21] Somorjai, G. A., Tao, F., Butcher, D. R. (2009) Surface mobility of atoms and molecules studies with high-pressure scanning tunneling microscopy, in Bowker, M., Davies, P. R. (eds.), *Scanning Tunneling Microscopy in Surface Science, Nanoscience and Catalysis*, Vol. 1, John Wiley & Sons, Chichester, UK, p. 197.

[22] Thostrup, P., Christoffersen, E., Lorensen, H. T., Jacobsen, K. W., Besenbacher, F., Norskov, J. K. (2001) Adsorption-induced step formation, *Phys. Rev. Lett.*, 87, 04990.

[23] Longwitz, S. R., Schnadt, J., Vestergaard, E. K., Vang, R. T., Laegsgaard, E., Stensgaard, I., Brune, H., Besenbacher, F. (2004) High-coverage structures of carbon monoxide adsorbed on Pt(111) studied by high-pressure scanning tunneling microscopy, *J. Phys. Chem. B*, 108, 14497–14502.

[24] Quiros, C., Robach, O., Isern, H., Ordejon, P., Ferrer, S. (2003) Compressibility of CO adsorbed on Ni(111) from 10(–6) mbar to 1.2 bar ambient CO pressures investigated with X-ray diffraction, *Surf. Sci.*, 522, 161–166.

8

IN-SITU INFRARED SPECTROSCOPY ON MODEL CATALYSTS

KUMUDU MUDIYANSELAGE AND DARÍO J. STACCHIOLA

8.1 INTRODUCTION

Infrared (IR) spectroscopy is one of the most widely used tools in surface science and catalysis research. It is a very effective spectroscopic method for the characterization of reactants and intermediates adsorbed on catalyst surfaces and products in the gas phase. The most common IR measurement modes for the characterization of powder samples are transmission–absorption IR spectroscopy, diffuse reflectance infrared Fourier transform (DRIFT) spectroscopy, and attenuated total reflection infrared (ATR-IR) spectroscopy (see Chapter 9 on IR spectroscopy for powder catalysts), and for the study of planar samples, ATR-IR and infrared reflection–absorption spectroscopy (IRRAS) are used [1, 2].

In order to resolve catalytic reaction mechanisms at the molecular level, studies need to be performed on well-defined model catalytic systems under elevated pressures and temperatures. The main focus of this chapter is to showcase the characterization of adsorbed species on single crystal-based model systems under elevated pressure conditions utilizing IRRAS and polarization–modulation infrared reflection–absorption spectroscopy (PM-IRRAS). There are three different acronyms commonly used in the literature for infrared reflection absorption spectroscopy: IRAS, RAIRS, and IRRAS, with an

In-situ *Characterization of Heterogeneous Catalysts*, First Edition.
Edited by José A. Rodriguez, Jonathan C. Hanson, and Peter J. Chupas.

approximately equal number of ~500 publications each, with the latter being slightly the most popular. For polarization modulation infrared reflection absorption spectroscopy, only two acronyms have been extensively used: PM-IRAS and PM-IRRAS, where less than 50 publications have used the first denomination for gas/solid systems while ~300 publications have used the latter for liquid/solid systems. One conflict in the literature with the acronym IRAS is that it coincides with the more popular infrared astronomical satellite (IRAS) nomenclature, which has thousands of citations. We therefore will use IRRAS and PM-IRRAS in this chapter and hope that over time, more and more practitioners are persuaded to adopt a unified acronym. First, we will introduce IRRAS and its applications on metal and oxide model samples under ultra-high-vacuum (UHV) conditions. Then some selected studies performed under elevated pressures utilizing IRRAS and PM-IRRAS will be described. The development, current status, and limitations of PM-IRRAS and the potential future applications of it in combination with other techniques for the characterization of model catalytic systems will then be briefly discussed.

8.2 IRRAS APPLICATIONS ON CRYSTALLINE PLANAR SAMPLES UNDER UHV CONDITIONS

8.2.1 Metal Single Crystals

The use of IR spectroscopy for the study of adsorbed species on metals was already well established by the middle of the 1900s [3]. Until then, all the studies had been carried out by IR transmission mode using metal catalysts supported on high surface area oxides to increase the number of oscillators. However, the absorbance of the oxide support and its influence on the morphology and properties of the metal catalysts prevented a clear molecular study of the adsorbate–metal interactions. In the mid-1960s, Greenler showed that submonolayer concentrations of adsorbed species on bulk metals could be detected by using a single reflection of IR light at grazing incidence from the surface [4] (Fig. 8.1). This approach, IRRAS, allows studying simplified systems and obtaining detailed molecular information about the chemical identity, geometry, and adsorption site of adsorbed species on surfaces. There has been a large number of studies applying IRRAS to identify adsorbed species on model catalytic systems under UHV conditions [5]. We will briefly describe here the fundamentals of IRRAS (more detailed descriptions can be found in the literature [5–8]).

For adsorbates on a metal surface, the detection of vibrational modes by IRRAS is governed by the selection rule for electric dipole transitions with the additional "surface selection rules," which establish that only vibrations with a dynamic dipole moment perpendicular to the surface can be excited. The theory of reflection at metal surfaces was earlier studied by Francis and

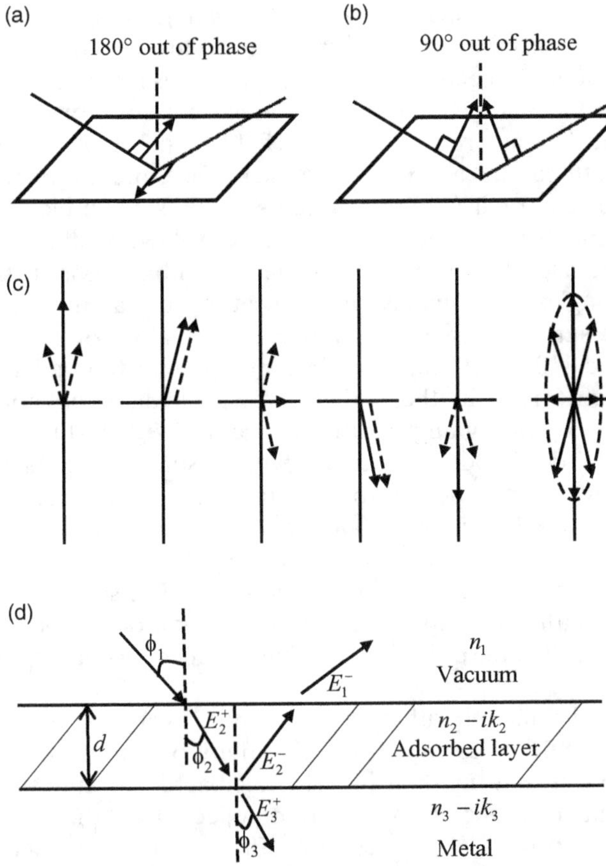

Figure 8.1. Greenler's original description of IRRAS.

Ellison [9] and then developed by Greenler [4], who showed that the optimum angle of incidence depends on the metal used and is always close to grazing. The electric field vector of IR light can be resolved into two components, in the incident plane E_s (s-polarized) and normal to the incident plane E_p (p-polarized). Figure 8.1a shows the geometry of the electric vectors near the surface for the incident and reflected waves of the s-polarized light. On reflection, the phase of the s-polarized component changes by nearly 180° at all angles of incidence, and since the reflection coefficient of metals is close to unity at IR frequencies, the resultant field is close to zero. Therefore, the interaction of the surface electric field E_s with the adsorbates is weak. In addition, the screening of dipoles parallel to the surface by the metal electrons further contributes to cancel the signal from their corresponding vibrational modes.

The p-polarized radiation also undergoes a phase change on reflection and strongly depends on the incident angle. Figure 8.1b shows the geometry of the electric vectors of the p-polarized light at a high angle of incidence where the phase shift upon reflection is 90°. As shown in Figure 8.1c, the vector addition of these incident and reflected vectors at the surfaces gives rise to an elliptical standing wave, which has a component of electric vector normal to the metal surface. The incident and reflected electric fields of p-polarized radiation have components both parallel and normal to the surface, and the vector sum gives parallel ($E_p^{\parallel}$) and normal ($E_p^{\perp}$) components of the surface electric field. The parallel components of $E_p^{i\parallel}$ and $E_p^{r\parallel}$ combine to give a very small resultant field at all incident angles. The normal components combine constructively, but at low incident angles, $E_p^{\perp}$ remains small because only a small proportion is resolved in the surface normal direction. However, at grazing incidence, $E_p^{\perp}$ is almost doubled [10, 11]. Therefore, incident p-polarized radiation gives a significant electric field at the surface only in a direction normal to the surface at close to grazing angles of incidence. The $E_p^{\perp}$ can therefore interact with any vibrational modes associated with adsorbed species that have a dipole derivative perpendicular to the surface, but the sensitivity to any parallel modes of vibrations is far lower. This constitutes an important surface selection rule in IRRAS. The image dipole created perpendicular to the surface further enhances IR absorption bands. The surface selection rule in conjunction with group theory analysis of vibrational modes provides a powerful tool for the interpretation of infrared reflection–absorption (IRRA) spectra.

Greenler used a model with a thin adsorbed layer on a metal surface as shown in Figure 8.1d to calculate the reflectivity of a covered metal surface using a boundary value method [4]. He calculated the intensity of the IR absorption band resulting from radiation reflected by the surface for (1) various optical constants (k-extinction coefficients and n-refractive indices), (2) various thicknesses of the adsorbed layer, (3) various angles of incidence, and (4) both states of polarization of the incident radiation. These calculations showed that the absorption factor, $A = (R_0 - R)/R_0$ (R_0 is the value of the reflectance for $k_2 = 0$), for p-polarized IR light has a maximum value at an incident angle of 88°, where the absorption is 5000 times greater than that at 0°. The values of the absorption factor for s-polarized IR light are smaller than that for p-polarized light by a factor of 10^5. The absorption factor also varies linearly with film thickness and the extinction coefficient (k_2) of the adsorbed layer. When compared with transmission through a similar film at normal incidence, a reflection method at the optimum conditions is more sensitive by a factor of 25 [4]. In addition, an increase of incident angle of the IR radiation enhances the absorption due to an increase in the number of adsorbed species sampled in the area irradiated. This area varies as sec θ, where θ is the angle of the incidence of IR radiation, which must be close to grazing [10, 11]. The optimum angle of incidence is close to 88° in the wave number region of interest; $\theta_{opt} = 88.5°$ at 500 cm^{-1} and 86.7° at 2100 cm^{-1} for the Pt surface [12]. The total absorption intensity is proportional to $E^2 \sec \theta$, where E is the amplitude of the resultant electric field component. The intensity of IRRA spectra is described by the intensity of the reflected beam, that is, reflectivity. The change in reflectivity due to the adsorbed molecules is generally defined by $\Delta R/R = R - R_0/R_0$, where R and R_0 are the reflectivity of the surface with and without adsorbed molecules, respectively. Since changes in $\Delta R/R$ are due to the absorption of IR radiation by adsorbed molecules, IRRA spectra will be similar in appearance to a conventional "peak-down" absorption spectra.

During the decades following the pioneer work by Greenler [4] in the 1960s, thousands of studies on metal thin films and single crystals helped to advance significantly our understanding of the interaction of molecules with metals under UHV conditions [6–8, 13, 14]. The commercialization of Fourier transform IR spectrometers also greatly enhanced the application of IRRAS (Fourier transform infrared reflection–absorption spectroscopy [FT-IRRAS]) to study surface reactions on model systems. The main advantage of FT-IRRAS is that a large spectral range can be covered within a very short time, a critical aspect for the small signals normally obtained from monolayer coverages of adsorbates on ~1 cm^2 area samples.

8.2.2 Oxide Single Crystals

A similar setup to the one utilized in IRRAS for metal single crystals has been applied to the study of oxide single crystals [15–18], even though less extensively. A strong signal attenuation of adsorbate vibrational bands (one to two orders of intensity loss with respect to metal surfaces) results from the optical properties of oxides [15]. The selection rules of IRRAS at oxide surfaces are different from those of metal surfaces [18]. On metal surfaces, the excitation of adsorbate vibrations by s-polarized light is negligible, whereas for oxide surfaces, the situation is more complicated because both s-polarized and p-polarized light interacts with adsorbate vibrations. In contrast to metals, the excitation of adsorbate vibrations does not always lead to a reduction in reflectivity but can also lead to an enhancement of reflectivity on oxide surfaces. For s-polarized light, vibrational absorption bands produce always "negative" peaks (enhancement of reflectivity, $\Delta R_s = R - R_0 > 0$; note that here the change of reflectivity [ΔR] is defined as $\Delta R = R_o - R_a$, hence resulting in a negative peak, where R_o and R_a denote the reflectivities of a clean substrate and an adsorbate-covered substrate). For p-polarized light, the bands can be positive or negative depending on the incident angle of the IR beam and the refractive index, n, of the oxide [18]. In addition, the total intensities of p-polarized bands are one or two orders of magnitudes weaker than those of metals due to the absence of image dipoles in dielectric substrates [19]. Only a few numbers of studies by IRRAS have been reported on oxides, mostly related to TiO_2 [15–18]. Xu et al. applied IRRAS to study the adsorption of NO on single-crystalline rutile $TiO_2(110)$ surfaces at 100 K and the subsequent formation of N_2O via a NO dimer. This study showed that the parallel absorption vibrational bands always increase the reflectivity, giving negative peaks, whereas p-polarized bands are negative on oxidized $TiO_2(110)$ and positive on reduced $TiO_2(110)$ [18]. IRRAS can therefore be used to investigate the electronic structure changes on flat, single-crystal oxide surfaces by monitoring the changes in the sign of p-polarized vibrational bands [18]. Hayden et al. studied the reaction of hydrogen with $Rh(CO)_2$ supported on $TiO_2(110)$ and the adsorption of formate species on $TiO_2(110)$ with IRRAS [16, 17]. The spectrum of $Rh(CO)_2$ on $TiO_2(110)$ shows two bands for the monodispersed geminal dicarbonyl, with $\nu_{sym}(C–O) = 2112\ cm^{-1}$ observed as a transmission (negative) band, and $\nu_{asym}(C–O) = 2028\ cm^{-1}$ observed as an absorption (positive) band [15]. Since the p-polarized light has components parallel and perpendicular to the surface, there is a coupling of ν_{sym} with the perpendicular component (ΔR is negative, i.e., an increase in reflectivity following adsorption) and a coupling of ν_{asym} (C–O) with the parallel component (ΔR is positive, i.e., a decrease in reflectivity following adsorption; note that the mode polarized parallel to the surface can also couple to the s-polarized radiation). The IRRAS technique not only has the advantages of intrinsic resolution for surface vibrational spectroscopy but, when carried out on a surface with a dielectric response of an insulator or a semiconductor, it can also differentiate vibrational modes polarized both parallel and perpendicular to the surface normal. In addition, the azimuthal dependence of the parallel mode can be used to assess the orientation of the adsorbate [17].

Hayden et al. have studied the adsorption of formate derived from the dissociation of formic acid on $TiO_2(110)$ by using IRRAS [17]. They performed p-polarized IRRAS measurements of the adsorbed formate as a function of azimuthal angle (Φ). All the IRRA spectra showed two absorption features for the adsorbed formates. One is for an absorption band resulting from the coupling of the antisymmetric ($\nu_{asym}(OCO)$) vibrational mode of formate to the p-polarized light parallel to the surface plane. The second feature is a transmission band resulting from the coupling of the symmetric ($\nu_{sym}(OCO)$) vibrational mode of formate to the normal component of the p-polarized light. The azimuthal-dependent measurements revealed the presence of two formate species through the coupling of both the parallel and perpendicular components of the radiation to the symmetric ($\nu_{sym}(OCO)$) and antisymmetric ($\nu_{asym}(OCO)$) vibrational modes of the molecule.

Xu et al. recently studied the photocatalytic activity of TiO_2 anatase(101) and rutile(110) single crystals for CO photo-oxidation using IRRAS [20]. Exposure of anatase(101) to CO at 100 K led to the appearance of a negative IR peak at 2180 cm^{-1} for adsorbed CO on fivefold coordinated Ti^{4+} cations. When this CO adsorbed surface was further exposed to both molecular oxygen gas and UV photons with an energy of 3.40 eV, the photoinduced oxidation of CO occurred as confirmed by the significant decrease of the CO band intensity and the simultaneous appearance of a positive IR peak at 2340 cm^{-1} for physisorbed CO_2. The two peaks showed different sign (negative and positive) results, which was attributed to the orientation of the adsorbed CO and CO_2 relative to the substrate as described in the previous section. Based on the IRRAS data, the photo reaction cross-section Q was determined and it showed that the anatase(101) surface has higher activity, by an order of magnitude, for CO photo-oxidation to CO_2 than the rutile(110).

8.2.3 Oxide Ultrathin Films

An alternative approach to the study of model oxides is to use well-ordered metal oxide ultrathin films, grown on metal single-crystal substrates [21, 22]. The IR response from these systems is dictated by the dielectric properties of the metal since the oxide films are thin with respect to IR wavelengths [15]. IRRAS has been extensively applied to study changes in the termination of atomically characterized ultrathin oxides and the interaction of adsorbates with model systems where metal nanoclusters are deposited on top of these well-ordered oxide substrates [23]. One of the main advantages of using conducting single crystal-based model systems is that a variety of complementary UHV surface science techniques that make use of electrons, such as low-energy electron diffraction (LEED), Auger electron spectroscopy (AES), X-ray photoelectron spectroscopy (XPS), and scanning tunneling microscopy (STM), can also be applied to obtain structural information as described in the following example.

Vanadium oxides have been widely investigated in connection to their selective oxidation of alcohols and hydrocarbons (see Chapter 9 for a detailed discussion of studies on powder samples). The morphology, electronic structure, and vibrational properties of vanadia nanoclusters deposited on various ultrathin oxides, from the least active SiO_2 to the most active CeO_2, were studied by STM, XPS, and IRRAS [24, 25]. As an example, Figure 8.2 summarizes the key observations obtained for vanadia supported on clean silica films, which were grown on Mo(112) substrates. Vanadium deposition on top of an ultrathin silica film, in the presence of 2×10^{-7} mbar of O_2 at 300 K, results in the formation of three-dimensional nanoparticles, which are randomly distributed on the surface (Fig. 8.2a,b). At the oxygen pressures used, vanadium oxidation and clustering occurs only on the surface and not in the gas phase. At low coverages, the morphology of the silica substrate, forming wide terraces separated by monatomic steps of approximately 0.12 nm in height, can also be recognized. Comparing these STM images with those observed for Au deposition on similar thin silica films, where particles are mostly found on the step edges (not shown here), it can be concluded that the V adatoms quite strongly interact with the silica film and therefore exhibit a low diffusivity. As a result, the particle density increases, while the mean particle size, which ranges between 0.4 and 1.2 nm, stays nearly the same at increasing coverages up to 1 monolayer (ML).

The O1s and V2p regions in the X-ray photoelectron (XP) spectra are shown in Figure 8.2c. The bottom curve in this figure shows the spectrum for the silica film, which exhibits two O1s lines centered at 532.5 and 531.2 eV. These states are associated with the O atoms in the topmost layer, formed by sharing three oxygen atoms of $[SiO_4]$ tetrahedra, and to the O atoms linking

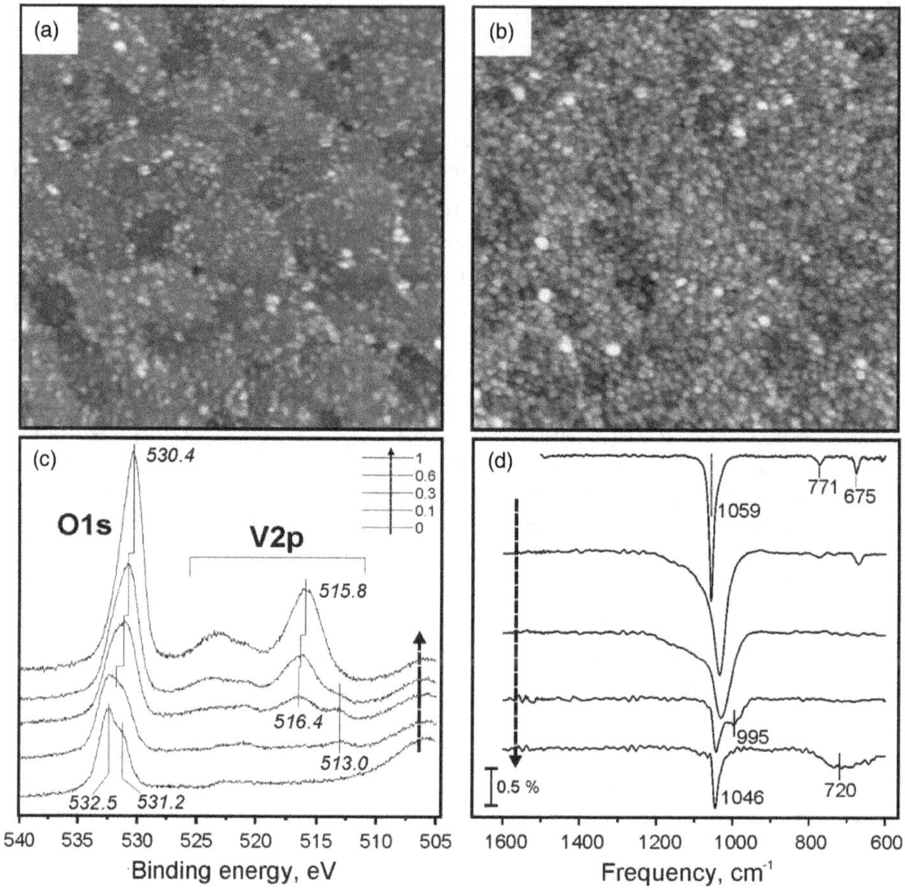

Figure 8.2. STM images (100×100 nm^2) of (a) 0.12 ML and (b) 0.4 ML vanadium oxide particles formed by deposition on the silica film in 2×10^{-7} mbar O_2 at 300 K (tunneling parameters: $V_S = 3$ V, $I = 0.2$ nA). The (c) XP and (d) IRRA spectra as a function of V coverage are indicated. All measurements were performed at 300 K. Reprinted with permission from Kaya, S., et al. (2007) Ice-assisted preparation of silica-supported vanadium oxide particles, *The Journal of Physical Chemistry C*, 2007 111(14), 5337–5344. Copyright 2007 American Chemical Society.

Si and the Mo substrate atoms, respectively [26, 27]. Upon deposition of vanadia, the O1s signal grows in intensity until only a symmetric peak at 530.4 eV is observed, which is typical for the binding energies (BEs) of oxygen in many transition metal oxides. Regarding the valence state of vanadium, a broad signal centered at 513 eV is observed at low and mid coverages, which is characteristic for the metallic state. In addition, initially fully oxidized Si^{4+} atoms were found partially reduced after reactive vanadia deposition (not shown here). This again indicates a strong interaction of the V atoms with the silica film proposed above on the basis of STM images. At increasing coverage, a high BE peak emerges and dominates the spectra. This peak slightly shifts from 516.4 eV at 0.3 ML to 515.6 eV at 1 ML probably because of vanishing of the well-known final state effect, leading to higher BE values for the smallest particles. The last spectrum is in fact very similar to that observed on well-ordered V_2O_3 films, thus suggesting that the oxide particles formed on the silica films under these conditions are of V_2O_3 in nature.

Figure 8.2d shows IRRA spectra for the vanadia/silica surface as a function of V coverage. On the clean silica film, a very intense peak at 1059 cm^{-1} corresponds to the Si–O–Mo asymmetric stretching, and the two weak absorption bands at 771 and 675 cm^{-1} are assigned to the combination of Si–O–Si stretching and bending modes, respectively [27]. Vanadia deposition leads to a broadening and shift to lower frequencies of the main silica phonon with increasing coverage. Concomitantly, a band at 1046 cm^{-1} grows up, which is assigned to the stretching vibration of V=O species. A broad signal centered at 720 cm^{-1} that emerges at the highest vanadia coverage studied can be assigned to the V–O–V stretching vibrations in the bulk vanadia [28]. Therefore, a combined STM, XPS, and IRRAS study shows that vanadium deposited in oxygen ambient strongly interacts with the ultrathin silica film and forms V_2O_3-like nanoparticles exposing V=O surface species. Further studies of deposition of vanadia clusters on reducible ceria thin films [25], where isolated vanadia species were stabilized and imaged with atomic resolution by STM, showed a blue shift of the vanadyl frequency from 1006 to 1046 cm^{-1} described earlier, upon increasing vanadia cluster size from mono- to polyvanadates. These detailed studies on atomically characterized model systems, in combination with density functional theory (DFT) calculations, helped to revised previous assignments for vanadia/ceria powder catalysts, where in the absence of a full atomic description of the system the opposite assignment had been carried out with the lower Raman frequency (1017 cm^{-1}) assigned to polymeric species and the higher (1034 cm^{-1}) to isolated species.

8.3 *IN-SITU* IRRAS STUDIES ON MODEL SYSTEMS AT ELEVATED PRESSURES

The investigation of reactions under atmospheric pressure conditions is essential to fully understand the reaction mechanisms in practical catalytic systems, since the appearance of new catalyst phases or the existence of coadsorbed species can profoundly alter the activity or selectivity of a catalyst. Since most surface science techniques such as XPS, STM, LEED, and AES make use of electrons, they had traditionally been limited to UHV environments [29]. Therefore, a significant emphasis has been devoted to increasing the pressure range of operation of traditional surface science techniques in order to apply these tools for investigating the reactions under elevated pressures. Currently, STM and XPS measurements can be carried out at ~millibar pressures, 10 orders of magnitude above UHV conditions (see Chapter 9 and Chapter 15).

IRRAS makes use of photons, which, in contrast to electrons, can readily be used under atmospheric pressure conditions to characterize surface species. Model systems can also be studied at higher local partial pressures by using molecular beams while following *in-situ* surface species by IRRAS [30]. Although there are difficulties in using IRRAS to characterize adsorbed species at pressures above 1 mbar due to the interference from gas-phase absorption, there are many systems where it can be applied without further modifications, as described next. If gas-phase species do not absorb in the mid-IR region (e.g., hydrogen or oxygen) or their absorption does not interfere with that of adsorbed species, then IRRAS can be applied without further modifications to obtain information of adsorbed species under elevated pressure conditions, as shown in the following examples. Another advantage of IR spectroscopy is that, under certain circumstances, results from well-controlled single crystal studies can be directly compared to those obtained on complex high surface area powder catalysts, as described for the case of vanadia catalysts, and thus can help in the identification of adsorbed species. In the next sections, we will first provide an example for a model system study with a molecular beam/high-pressure cell apparatus, and then describe examples of adsorption/reaction studies on palladium followed *in situ* by IRRAS under elevated pressures.

8.3.1 Model NO_x Storage Materials at Elevated NO_2 Pressures

Libuda et al. investigated the reactions of NO_2 with barium aluminate-like nanoparticles supported on ultrathin Al_2O_3/NiAl(110) by IRRAS at NO_2 partial pressures up to 1.75 mbar [31]. These experiments were performed in a combined UHV/IRRAS high-pressure cell and molecular beam apparatus. Molecular beams can be used to achieve high local partial pressures at the surface while maintaining a lower pressure in the system [30].

Exposure of $BaAl_2O_4/Al_2O_3$/NiAl(110) to NO_2 in the millibar regime leads to rapid formation of surface nitrites and nitrates. The vibrational spectra of these surface species formed at elevated pressures are similar to those for species formed under UHV conditions, but the formation of bulk nitrates is observed only at elevated pressures. The results of this study showed that model studies under UHV conditions mainly provide information on the initial reaction mechanism, whereas the observation of actual bulk nitrates requires experiments at elevated temperatures and pressures, indicating the importance of high-pressure experiments. The gas-phase NO_2 absorption occurs at ~1600 cm^{-1} with weak features at 1250 and 1750 cm^{-1}, whereas absorption bands of adsorbed species appear at ~1245, 1356, 1380, 1420, and 1455 cm^{-1}. In this case study, the absorption of gas phase does not interfere with that of adsorbed species.

8.3.2 Palladium-Catalyzed Cyclotrimerization of Acetylene

The catalytic process to form benzene from acetylene on Pd-based catalysts is an ideal system to study a reaction mechanism at the molecular level because benzene is formed both in UHV and elevated pressure conditions [32], and results can be compared directly between the model systems and industrial catalysts.

Early on, it had been reported in the literature that exposing a powder Pd/Al_2O_3 catalyst at 300 K to acetylene has led to the formation of ethylidyne, but exposing a Pd(111) single crystal to the same reactant has led to the formation of vinylidene surface species. In the attempt to resolve this apparent contradiction, the formation of benzene from the reaction of gas-phase acetylene was studied at elevated pressure on an alumina-supported palladium (5% Pd/Al_2O_3) catalyst with ^{13}C-labeled vinylidene [33], where the powder catalyst was originally cleaned and left under high vacuum for a day before the experiment. The presence of vinylidene on Pd/Al_2O_3 was confirmed by comparing the IRRA spectra of the species formed on the supported sample with those found on a Pd(111) single crystal as shown in Figure 8.3. The single IR feature observed on Pd(111) at ~1268 cm^{-1} in Figure 8.3 was assigned to the CH_2 bending mode of vinylidene. The corresponding spectrum for alumina-supported palladium saturated with acetylene is shown in Figure 8.3a (^{13}C-labeled acetylene). The spectra on Pd(111) and supported Pd are identical, indicating the formation of vinylidene on alumina-supported palladium. For comparison, Figure 8.3d,e displays the IR spectra collected following the exposure of alumina-supported Pd and Pd(111) to ethylene, respectively. These spectra show characteristic peaks for ethylidyne and confirm that acetylene adsorption on alumina-supported Pd does not form ethylidyne.

The problem with the early results on powder samples was that after the cycles of oxidation and reduction with hydrogen used for the cleaning of the catalysts, hydrogen was likely left trapped in the catalyst even after leaving the samples under high vacuum for several hours, and the reaction of that hydrogen with acetylene led to the formation of ethylidyne, as it will be shown to be the case.

The concentration of vinylidene-labeled species on powder samples was followed using magic-angle solid state nuclear magnetic resonance (NMR). At high pressures (~513 mbar), gas-phase acetylene reacts with adsorbed vinylidene at the same rate at which benzene is formed. The resulting benzene was analyzed by mass spectroscopy and was found to incorporate two ^{13}C atoms, indicating that it is formed by the reaction between gas-phase (^{12}C-labeled) acetylene and adsorbed vinylidene ($^{13}CH_2{=}^{13}C{=}$) to produce C4 intermediates that react with acetylene to yield benzene (Scheme 8.1). This proposed reaction pathway differs from that elucidated previously from UHV studies on clean Pd(111), where it has been found that benzene synthesis also proceeds via a C4 intermediate, but under UHV, the intermediate is formed from two adsorbed acetylene molecules [32].

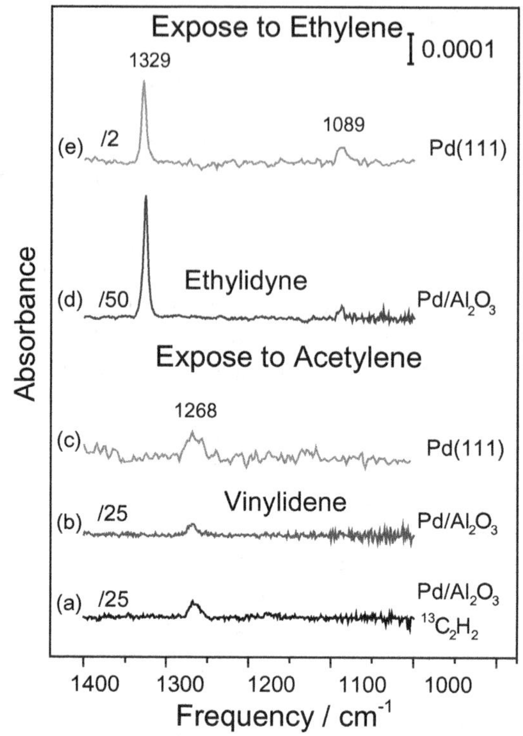

Figure 8.3. Infrared spectra at 300 K of (a) $^{13}C_2H_2$ adsorbed on 5% Pd/Al_2O_3, (b) $^{12}C_2H_2$ adsorbed on 5% Pd/Al_2O_3, (c) acetylene adsorbed on a Pd(111) single crystal, (d) ethylene adsorbed on 5% Pd/Al_2O_3, and (e) ethylene adsorbed on Pd(111). Reprinted with kind permission from Kaltchev, M., et al. (1999) On the reaction pathway for the formation of benzene from acetylene catalyzed by palladium, *Catalysis Letters*, 60(1–2), 11–14. Copyright 1999 Springer Science+Business Media.

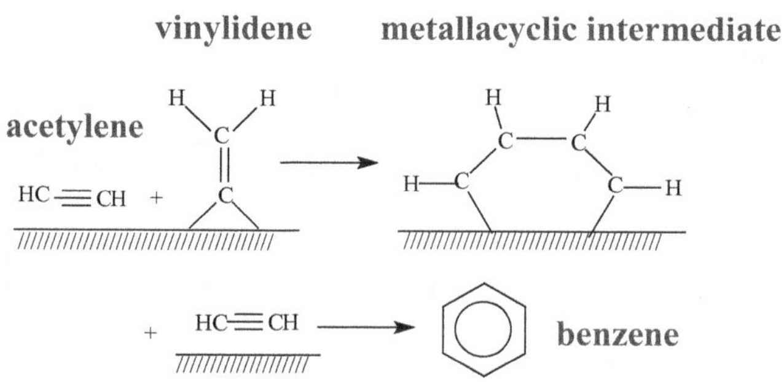

Scheme 8.1. The reaction between gas-phase acetylene and adsorbed vinylidene to produce C4 intermediates, which react with acetylene to yield benzene.

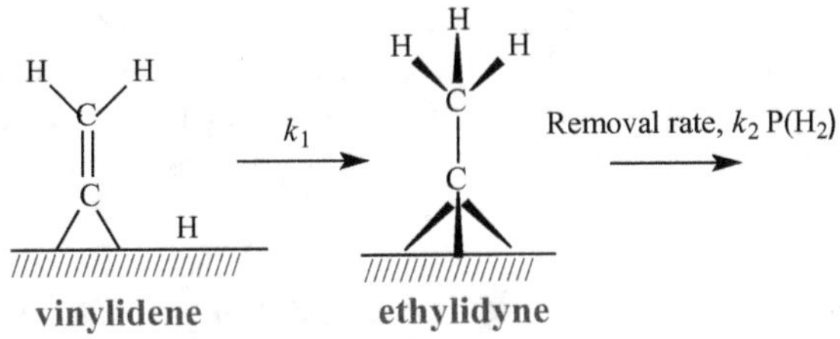

Scheme 8.2. The reaction of vinylidene formed from acetylene with hydrogen at elevated pressures to form ethylidyne species.

The addition of hydrogen to the feed of acetylene increases the rate of formation of benzene. To understand this effect, a mechanistic study was carried out on Pd(111) single crystals. It was found that benzene is synthesized on clean Pd(111) by an initial fast reaction between two adsorbed acetylenes to form a tilted C_4H_4 metallacyclic intermediate, which in turn reacts with a third acetylene molecule. A relatively unreactive vinylidene layer covers the surface under catalytic conditions, and at UHV conditions, it completely suppresses the formation of benzene. As shown previously, at elevated pressures, vinylidene eventually reacts with acetylene to form benzene, as shown in Scheme 8.1. In addition, vinylidene formed from acetylene reacts with hydrogen at elevated pressures to form ethylidyne species (Scheme 8.2), as observed by IRRAS on a sample saturated initially with vinylidene. The ethylidyne in turn can be titrated away by hydrogen, and its removal rate was determined by monitoring the concentration of ethylidyne by IR spectroscopy. These experiments were performed for both Pd(111) and Pd/Al_2O_3, where the reaction rates were identical for both samples, indicating the transferability of the information obtained from single crystal studies to powder systems. The conversion of vinylidene species, which crowd the surface with a saturation coverage of 1 ML, into ethylidyne (saturation coverage 0.25 ML) opens up the surface and helps to increase the rate of formation of benzene.

This is an example showing the application of IR spectroscopy to follow the kinetics of relatively slow surface reactions. Fast reactions in the millisecond or nanosecond timescale could also be followed by IR spectroscopy using rapid and step scan methods, respectively. In the reaction of vinylidene with 100 mbar of hydrogen, the coverage of ethylidyne initially increases and subsequently decreases, suggesting a sequential reaction as shown in Scheme 8.2. Although these vinylidene–ethylidyne removal processes were monitored by IR spectroscopy at high pressure of hydrogen, PM-IRRAS is not necessary because gas-phase H_2 does not absorb IR radiation in the mid-IR region. As long as gas-phase species do not absorb IR radiation, or their absorption does not interfere with that of surface species, we can identify and characterize the surface species under high-pressure conditions by regular IRRAS. Under UHV conditions, acetylene does not adsorb on top of a vinylidene layer or react with it, and also vinylidene formed from acetylene and ethylidyne formed from ethylene do not react with hydrogen. However, these reactions take place only under millibar pressure conditions, indicating the importance of *in-situ* investigations of the surface reaction under high-pressure conditions.

8.3.3 Adsorption of Carbon Monoxide on Ethylidyne-Covered Pd(111)

Carbon monoxide is commonly used as a probe molecule on powder and single-crystal samples due to its very intense IR signal and sensitivity to changes in local environments and adsorption sites. The explanation for the increase in the rate of formation of benzene on Pd catalysts due to the addition of hydrogen to the gas feed is the conversion of crowded vinylidene overlayers into open ethylidyne overlayers. To prove that open ethylidyne overlayers allow the adsorption of extra molecules on the surface, IRRAS was applied to study the adsorption of CO on ethylidyne-covered/Pd(111) [34]. The saturation coverage of ethylidyne on Pd(111) is 0.25 ML, whereas the CO coverage achieved on ethylidyne-covered/Pd(111) is also 0.25 ML. The saturation coverage of CO on clean Pd(111) is 0.55 ML. At lower coverages (0.33 ML), CO adsorbs on the more stable face-centered cubic (fcc) sites, while at saturation coverage, it occupies both fcc and hexagonal close-packed (hcp) sites. The proposed structures for CO on clean Pd(111) are displayed in Figure 8.4b,c. Ethylidyne adsorbs on fcc sites as shown in Figure 8.4a. IRRAS results reveal that carbon monoxide adsorbs on less stable hcp sites (~1907 cm^{-1}) on ethylidyne-covered/Pd(111) under UHV conditions, as shown in Figure 8.4d. Atop CO, sites become occupied (~2080 cm^{-1}) when the sample is pressurized to 0.01 mbar. Therefore, *in-situ* investigation of this system can be performed by utilizing regular IRRAS. In contrast, CO does not adsorb at all on vinylidene-covered/Pd(111) surfaces under similar conditions. However, after pressurizing a vinylidene-saturated surface with CO, an IR peak is detected at ~1800 cm^{-1} for adsorbed CO species [32]. Therefore, in this system, *in-situ* IR spectroscopy should be performed under elevated pressure conditions by utilizing PM-IRRAS since at pressures above a few millibars, the gas-phase absorption does not allow clear identification of atop CO species. The results obtained on the adsorption of CO on ethylidyne- or vinylidene-covered surfaces confirm the hypothesis that transforming the latter into ethylidyne by exposure to hydrogen opens the surface for more benzene formation.

8.3.4 Vinyl Acetate Formation on Oxygen-Covered Pd(111)

Complex reactions can be investigated using IRRAS at intermediate pressures (~1×10^{-7} to 1×10^{-2} mbar) under *operando* conditions. The acetoxylation of ethylene to make vinyl acetate monomer (VAM) is an established industrial reaction. Two basic mechanisms had been proposed for the palladium-catalyzed synthesis of the VAM from acetic acid, ethylene, and oxygen. The first, suggested

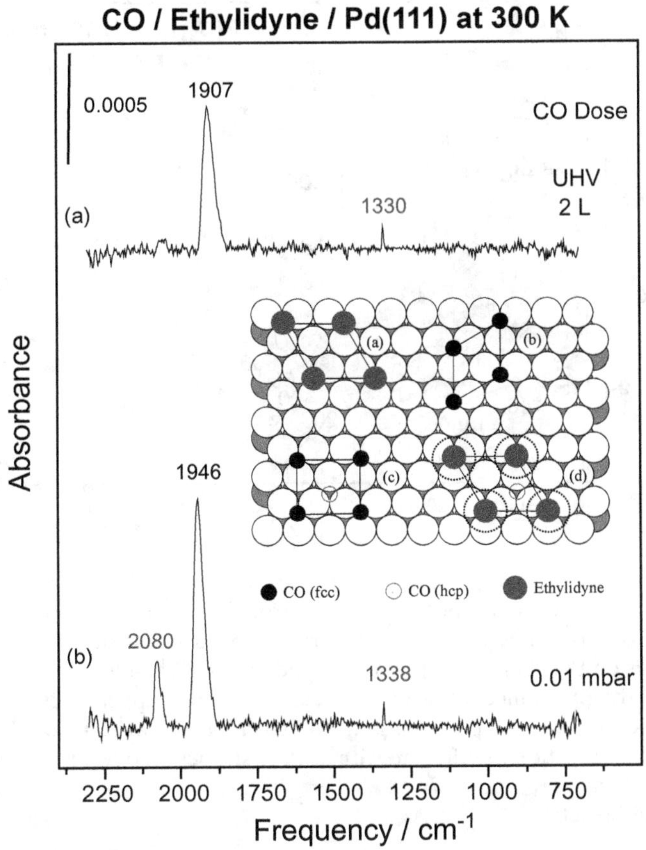

Figure 8.4. IRRAS of (a) 2 L of CO on ethylidyne-covered Pd(111), (b) under a pressure of 0.01 mbar of CO on ethylidyne-covered Pd(111). Reprinted with permission from Stacchiola, D., et al. (2000) The adsorption and structure of carbon monoxide on ethylidyne-covered Pd(111), *Surface Science*, 470(1–2), L32–L38. Copyright (2000) Elsevier.

by Samanos et al., involves the coupling of ethylene directly with chemisorbed acetate [35]. The resulting acetoxyethyl–palladium intermediate then undergoes β-hydride elimination to form vinyl acetate. Alternatively, ethylene could first dehydrogenate to form a vinyl–palladium intermediate, which then couples with an acetate species adsorbed on the surface to form the VAM directly as proposed by Nakamura and Yasui [36]. For decades, studies on powder catalysts could not resolve which one of the two proposed mechanisms operated on Pd catalysts. Stacchiola et al. reported the observation of the reaction between gas-phase ethylene at intermediate pressures ($\sim 1 \times 10^{-4}$ mbar) and acetate species preadsorbed on oxygen-covered Pd(111) [37]. To carry out the model reactions, first acetate overlayers adsorbed on oxygen-covered

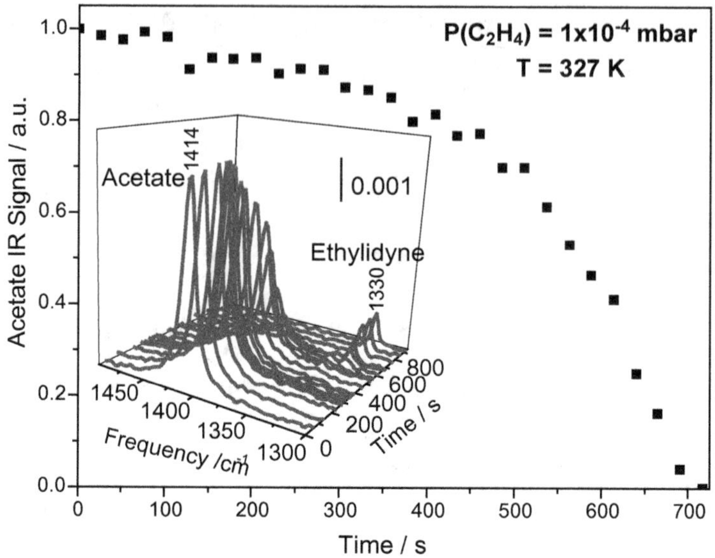

Figure 8.5. Plot of the integrated intensity of the 1414-cm^{-1} feature due to acetate species, as a function of time. The line plotted through this data is the result of a kinetic fit. Shown also is the time sequence of infrared spectra of a saturated overlayer of acetate species on Pd(111)- (2×2)O pressurized by 1×10^{-4} mbar of ethylene at 327 K, showing the disappearance of acetate species and the appearance of ethylidyne (1330 cm^{-1}). Reprinted with permission from Stacchiola, D., et al. (2004) Vinyl acetate formation by the reaction of ethylene with acetate species on oxygen-covered Pd(111), *Journal of the American Chemical Society*, 126(47), 15384–15385. Copyright 2004 American Chemical Society.

Pd(111) surfaces were prepared, and then the system was pressurized with ethylene. Surface species were monitored with IRRAS as a function of time under an ethylene environment, as shown in Figure 8.5. Under ethylene pressures of ~1×10^{-4} mbar, no gas-phase ethylene was detected. At time $t = 0$ (prior to ethylene exposure), the spectrum showed an intense feature at 1414 cm^{-1}, which was assigned to the symmetric acetate OCO stretching mode. This intense feature attenuates following ethylene exposure due to the removal of acetate by ethylene producing vinyl acetate according to the following overall reaction:

$$CH_3COOH + C_2H_4 + \frac{1}{2}O_2 \rightarrow CH_3COOC_2H_3 + H_2O.$$

As the surface becomes depleted of acetate species, another feature grows at 1330 cm^{-1}, which is assigned to ethylidyne. Using partially deuterated ethylene to identify isotope effects related with potential intermediates, it was possible to determine which one of the two possible mechanisms reported on the literature was the main pathway [38]. The experimental results provided clear evidence that the reaction of ethylene with the acetate species proceeds through the pathway proposed by Samanos, that is, by insertion of ethylene into an adsorbed acetate species and subsequent β-hydride elimination to form vinyl acetate. The studies described earlier show the applicability of IRRAS at intermediate pressures of 10^{-4} mbar (~5 orders of magnitude higher than UHV) without the interference from gas-phase reactants, which absorb in the mid-IR region. In general, IRRAS can be applied to investigate systems under *operando* conditions, at elevated gas pressures up to millibar levels without interference of gas-phase absorption.

8.4 PM-IRRAS

The two vibrational techniques used for model studies at pressures above 1 mbar are sum frequency generation (SFG) and PM-IRRAS. They have been applied to study the reactions on surfaces and interfaces of model systems under atmospheric pressure conditions. SFG is a powerful technique for investigating the catalytic reactions on smooth surfaces such as those of a single crystal or an ordered material [39–41]. It is a nonlinear optical phenomenon where two incident photons are absorbed, and a photon with a frequency equal to the sum of the frequencies of the incident photons is emitted [42]. SFG selection rules require that a vibrational mode must have a change in both dipole and polarizability in order for the process to occur. These selection rules are met only at surfaces and interfaces. SFG is a surface-specific vibrational spectroscopy with submonolayer sensitivity. In SFG, a visible laser beam at a fixed wavelength is overlapped with a tunable IR laser beam. Vibrational spectra can be acquired by scanning the IR beam over the vibrational region of interest. No gas-phase subtraction or background spectra are needed due to the inherent surface sensitivity of the technique. This makes SFG spectroscopy useful in studying heterogeneous catalysis because it is not limited by the presence of gas-phase species at high pressures.

SFG has been used for the molecular level investigations of adsorbates under both UHV and elevated pressure conditions to characterize several important catalytic reactions [40, 43–47]. Reflective substrate materials with well-defined geometry and total internal reflection can be used to enhance the SFG signal [48]. The total internal reflection SFG (TIR-SFG) expands the application of SFG beyond flat model catalysts to powdered samples. The potential application of SFG to characterize powdered catalysts under practical conditions could make it a powerful technique to investigate the catalytic processes under industrial conditions. More detailed descriptions of SFG can be found in the literature [39–41, 49–52]. As other techniques, SFG also has some limitations and drawbacks such as applicability only to nonporous materials, relatively long spectral acquisition times, and limited spectral range (1600–4000 cm^{-1}).

In addition, the SFG technique is relatively more difficult to set up as compared to IRRAS.

The initial studies related to the use of IR spectroscopy in the investigation of adsorbates on planar metal surfaces under elevated pressures were conducted in the 1970s [53]. By taking advantage of the fact that only the interaction of p-polarized (normal to the surface) light with molecules on the surface can be detected, the contributions from molecules in the gas phase, which absorb equally s- and p-polarized light, can be discriminated from those adsorbed onto the metal surface by switching between p- and s-polarized light. The difference between the two spectra removes the gas-phase contribution and gives the spectrum for the surface species. Therefore, this technique makes it possible to obtain in parallel *in-situ* vibrational spectra of adsorbed surface species under high pressure of reactants, as well as the composition of the gas-phase reactants and products.

In PM-IRRAS, the differential reflectivity or differential reflection spectrum $\Delta R/R$ is defined as

$$\frac{\Delta R}{R} = \frac{I_p - I_s}{I_p + I_s},$$

where I_p is the IR spectrum obtained with p-polarized light and I_s is the IR spectrum obtained with the s-polarized light.

Double-modulation spectroscopy is a general method for obtaining the differential absorption spectrum of a sample [2]. In a PM-IRRAS setup with a Michelson interferometer, there is an intensity modulation (due to the interferometer) and a polarization modulation. The switching of the polarization can be achieved by rotating a polarizer in the optical path to alternate spectra taken in s- or p-polarized light [53] or by using a fast (usually ~50 kHz) photoelastic modulator (PEM) in front of a polarizer to obtain s − p and s + p spectra almost simultaneously [54]. Due to differences in the reflectivity of s- and p-polarized light on the optical elements, in the first case a second polarizer is also needed in order to obtain a defined polarized beam right before the modulation [55, 56]. The main advantage of the rotating polarizer method is that a PM-IRRAS spectrum can be obtained over the whole spectral range. The main disadvantage is the difficulty with the precise alignment of the optical components. The main disadvantage of the PEM method is that the polarization is only modulated by 90° at a single frequency. In practice, this means that the background signal is effectively canceled out over a relatively narrow frequency range. However, the spectral range of interest is often well known from IRRA spectra carried out in UHV conditions, so that this is not necessarily a major problem [57]. In order to follow the rapid variations of the modulation amplitude with a lock-in amplifier the moving mirror in the Michelson interferometer is required to be run at a slow speed [58]. The main advantage of the PEM method is the almost simultaneous acquisition of the spectra, which eliminates changes due to mechanical or environmental

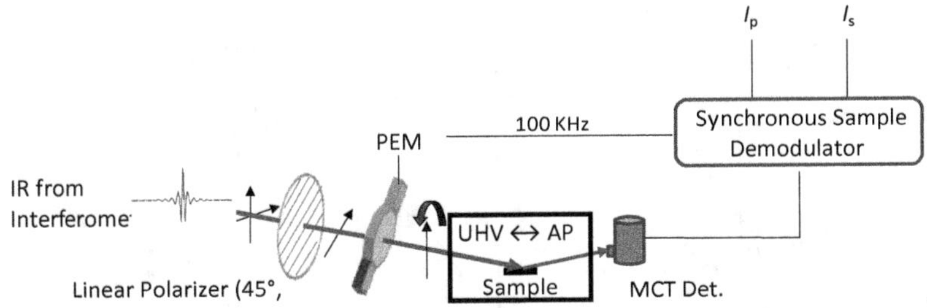

Figure 8.6. Schematic diagram of a PM-IRRAS setup. AP, ambient pressure; MCT, mercury cadmium telluride.

instabilities. A schematic diagram of the setup used for the PM-IRRAS experiment is shown in Figure 8.6 [57]. IR light from the Fourier transform infrared (FTIR) spectrometer passes through a polarizer to select its p-polarized component, and then it is focused into the sample at a grazing incidence angle of ~80°. The reflected IR beam from the sample is focused to the detector. A PEM is placed in the incident light path just before the entrance window to the sample cell at an angle of 45° to the incident p-polarized light, allowing the incident light to be modulated between p and s polarization. Corn et al. [58] have described a synchronous sampling demodulator circuit, where the composite output of the detector is synchronously sampled to I_p and I_s and then recombined to output a derived $I_p - I_s$ signal as well as an average composite signal, $I_p + I_s$, before being fed to the IR spectrometer's two analog-to-digital converters (ADCs) [58].

A detailed description of the signal processing in PM-IRRAS has been provided by Buffeteau et al. [55]. The different steps of a PM-IRRAS signal analysis are shown in Figure 8.7a–f [55]. The detected intensity I_d is the sum of two interferograms, $(R_p + R_s)$ and $J_2(\varphi_0)(R_p - R_s)$, where $J_2(\varphi_0)$ is a second-order Bessel function coming from the wavelength dependence introduced by the modulator. As only the second interferogram has a double modulation, it is easy to split the two signals into two different channels, with the use of an electronic filter with high-pass and low-pass outputs. After the demodulation of the $J_2(\varphi_0)(R_p - R_s)$ interferogram and processing (high-pass/low-pass filtering, amplification, simultaneous sampling, and ADC conversion), two interferograms are obtained as shown in Figure 8.7a,b. Figure 8.7c shows the results of the Fourier transformation of the two interferograms after a large number of scans. The ratio of the two single beams in Figure 8.7c is shown in Figure 8.7d. This ratio, $J_2(\varphi_0)(R_p - R_s)/(R_p + R_s)$, is the expected theoretical differential reflectivity spectrum, $\Delta R/R$. Spectra for a rough qualitative visualization of the surface absorption bands can be obtained by performing a baseline correction using the facilities of the FTIR spectrometer as shown in Figure 8.7e,f. More accurate methods need to be applied for a correct quantitative analysis of the spectra [55].

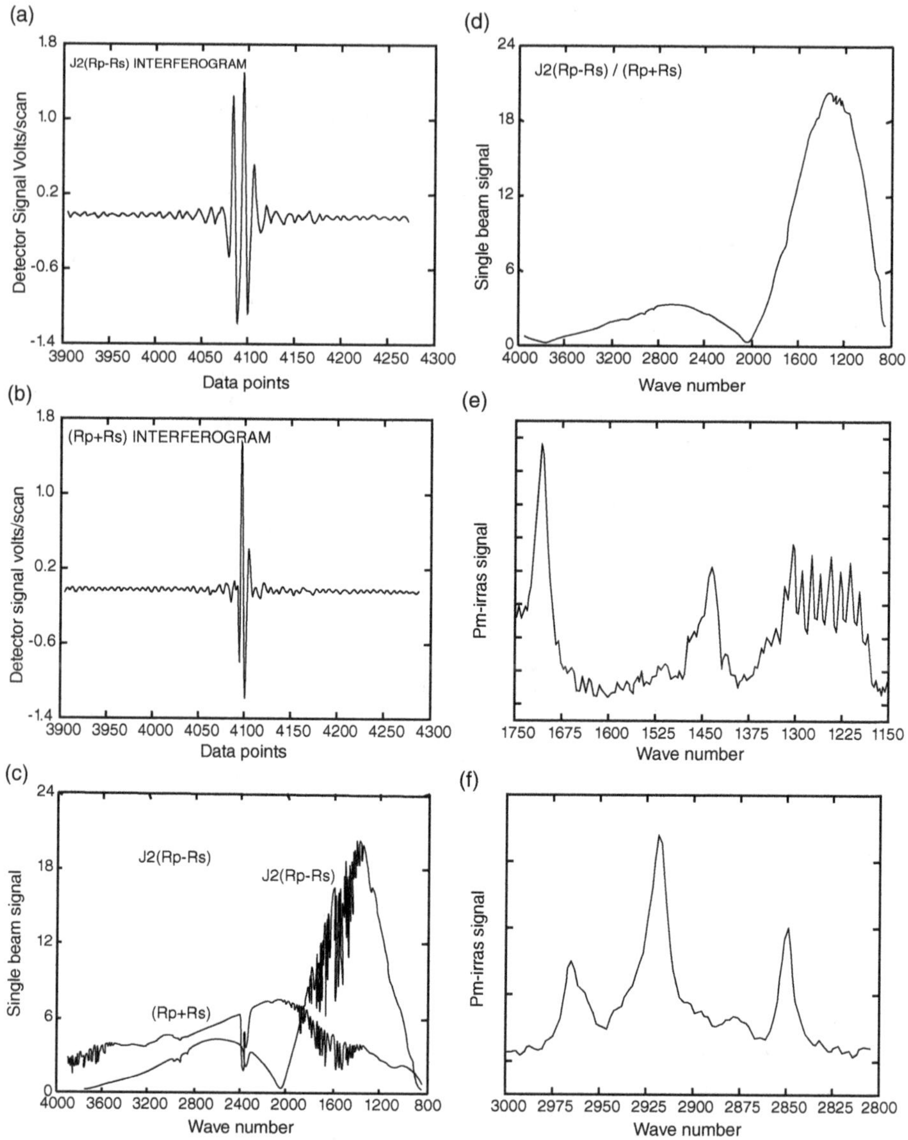

Figure 8.7. PM-IRRAS signal processing. Reprinted with permission from Buffeteau, T., et al. (1991) Polarization modulation FTIR spectroscopy of surfaces and ultra-thin films—experimental procedure and quantitative-analysis, *Applied Spectroscopy*, 45(3), 380–389. Copyright 1991 The Optical Society.

TABLE 8.1. Systems Studied by PM-IRRAS

Reactants	Surface	Reference
CO	Co(0001)	Beitel et al. [64]
CO	Co(0001)	Oosterbeek [65]
CO	Pd(111)	Ozensoy and Goodman [66]
CO	Pd(111)	Stacchiola and Tysoe [57]
CO	Zn/Pd(111)	Rameshan et al. [67]
CO	Pt(111)	Sterrer et al. [68]
CO	Au(110)	Jugnet and Bertolini [69]
CO/H_2	Co(0001)	Beitel et al. [70]
CO/O_2	TiO_x/Pt(111); Au/TiO_x/Pt(111)	Behm and Granozzi [71]
CO/O_2	Pt-group single crystals	Gao and Goodman [72]
NO	Pt-BaO/Al	Maeda and Baiker [73]
CO + NO	Pd(111)	Hess and Goodman [62]
		Ozensoy et al. [63]
CH_3OH	Pd(111)	Rupprechter and Freund [59]
	Zn/Pd(111)	Rameshan et al. [67]
1,3-Butadiene	$Pt_3Sn(111)$	Jugnet and Bertolini [74]

Table 8.1 summarizes most of the systems studied by PM-IRRAS to date on single crystal-based model surfaces. Most PM-IRRAS studies have monitored the IR peaks of CO and NO due to their strong IR absorption. However, PM-IRRAS has also been used to identify adsorbed formaldehyde (ρ_{CH2} of formaldehyde in two different adsorption geometries at 1305 and 1255 cm^{-1}), formyl CHO (CH bending or ν_{CO} at 1200 cm^{-1}) and minute amounts of methoxy species (around 2900 cm^{-1}) in the decomposition and oxidation of methanol on Pd(111) [59]. These results indicate the applicability of PM-IRRAS to a variety of systems, not limited only to species with very strong IR absorption such as CO and NO.

8.4.1 Adsorption of CO at Ambient Pressure on Pd(111)

PM-IRRAS has been used to study the adsorption of CO on Pd(111) up to ~200 mbar [57]. The data collected at a sample temperature of 300 K with various pressures of CO are summarized in Figure 8.8. The bottom IRRA spectrum for CO on Pd(111) was obtained following saturation at 300 K in UHV (Θ[CO] = 0.5 ML). It shows a feature at 1925 cm^{-1}, which is assigned to CO in both fcc and hcp sites where the resulting frequency arises from a combination of chemical and dipole–dipole coupling interactions [60, 61]. This feature shifts to higher frequencies (~1950 cm^{-1}) as the pressure is increased to 1 mbar. The gas-phase CO signal is now evidently displaying the typical *P* and *R* branches. A small peak at 2090 cm^{-1} due to CO adsorbed on the atop site overlaps with the gas-phase features. As the pressure is further increased to 100 mbar, the intensity of the feature at ~1950 cm^{-1} decreases and another

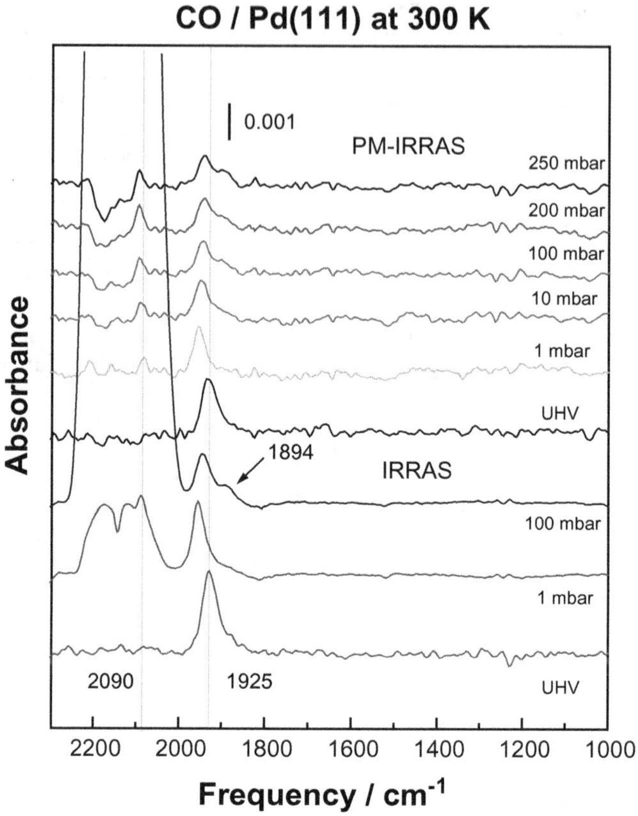

Figure 8.8. IRRAS of CO on Pd(111) after dosing 10 L at 80 K (marked UHV) and with CO pressures of 1 and 100 mbar. A series of PM-IRRAS of 10 L of CO adsorbed on Pd(111) at 80 K (marked UHV) and PM-IRRAS spectra of a Pd(111) surface pressurized to 1, 10, 100, 200, and 250 mbar of CO, where pressures are marked adjacent to the corresponding spectra. Reprinted with permission from Stacchiola, D., et al. (2002) Photoelastic modulation-reflection absorption infrared spectroscopy of CO on Pd(111), *Journal of Vacuum Science & Technology. A, Vacuum, Surfaces, and Films*, 20(6). 2101–2105. Copyright 2002 American Institute of Physics.

peak appears at ~1894 cm^{-1}. This has been assigned to CO adsorbed in a combination of fcc and hcp hollow and atop sites at $\Theta(CO) = 0.75$ ML [61]. Now, however, the gas-phase CO signal is extremely large and completely obscures any signals that may be due to CO adsorbed in the atop site. These results indicate that CO adsorption on Pd(111) at high pressures cannot be studied using IRRAS without interference from the gas-phase absorption. Therefore, PM-IRRAS needs to be applied to this system.

A series of PM-IRRA spectra collected as a function of CO pressure is also shown in Figure 8.8. The spectrum marked UHV was collected for a monolayer of CO adsorbed at a sample temperature of 300 K. This exactly reproduces the IRRA spectrum. The signal due to atop CO, which is coincident with the gas-phase signal, is clearly evident at ~2090 cm^{-1} at a pressure of 1 mbar, which is also seen in the IRRA spectrum. Coincidentally, the feature at 1925 cm^{-1} for the monolayer shifts to ~1950 cm^{-1} due to the coupling between CO on threefold hollow and atop sites. As the CO pressure increases to 10 mbar, the atop feature shifts slightly to ~2099 cm^{-1} as seen previously in the SFG data [45] and its intensity increases. The mode at ~1950 cm^{-1} decreases further in intensity, where both the shifts in frequency and relative intensity changes are consistent with an increase in the atop site population. In addition, a small shoulder appears at ~1890 cm^{-1}, also detected by SFG [45] and assigned as above to the high-coverage, $\Theta(CO) = 0.75$ ML, structure. This feature is also shown in the IRRA spectrum collected at 100 mbar. As the pressure is further increased to 200 mbar, intensities of both features at 2099 and 1948 cm^{-1} are increased. A small amount of gas-phase signal is evident at pressures of 100 and 200 mbar, although these are sufficiently small that they do not unduly obscure the background. Proper alignment of the optics allows cancellation of the gas-phase signal up to atmospheric pressure.

8.4.2 CO + NO Reaction on Pd(111)

PM-IRRAS has been used by Hess et al. to study the reaction of CO + NO on Pd(111) at pressures up to 300 mbar [62]. They observed the formation of isocyanate only at elevated pressures, indicating the importance of *in-situ* IR spectroscopic experiments under high-pressure conditions to obtain detailed understanding of the surface reactions. The formation of isocyanate indicates the dissociation of NO, which is a crucial step in the CO + NO reaction, and provides important information on the mechanism of the overall reaction, although isocyanate acts as a spectator in this process. They also followed the kinetics of the CO + NO reaction to form CO_2, N_2O, and N_2 by monitoring the evolution of CO_2 and N_2O into the gas phase, and determined the apparent activation energy of 54 ± 21 kJ/mol. Figure 8.9 shows PM-IRRA spectra recorded in the presence of a CO + NO mixture at pressure ratios of $P_{CO}/P_{NO} = 2$ (left) and 1.5 (right) and total pressures of 180 and 300 mbar, respectively. After the samples were heated at 650 K, isocyanate was formed, and it was stable during the cooling process back to 300 K as well as after reducing the pressure to 1×10^{-7} mbar at room temperature. As shown in Figure 8.9, an increase in the CO/NO pressure ratio favored the formation of isocyanate.

8.4.3 Decomposition and Oxidation of Methanol on Pd(111)

Rupprechter et al. studied the decomposition and oxidation of methanol on Pd(111) at millibar pressures by *in-situ* PM-IRRAS [59]. Dehydrogenation of methanol at 300 K leads to the formation of adsorbed CO, CH_2O, CHO, and

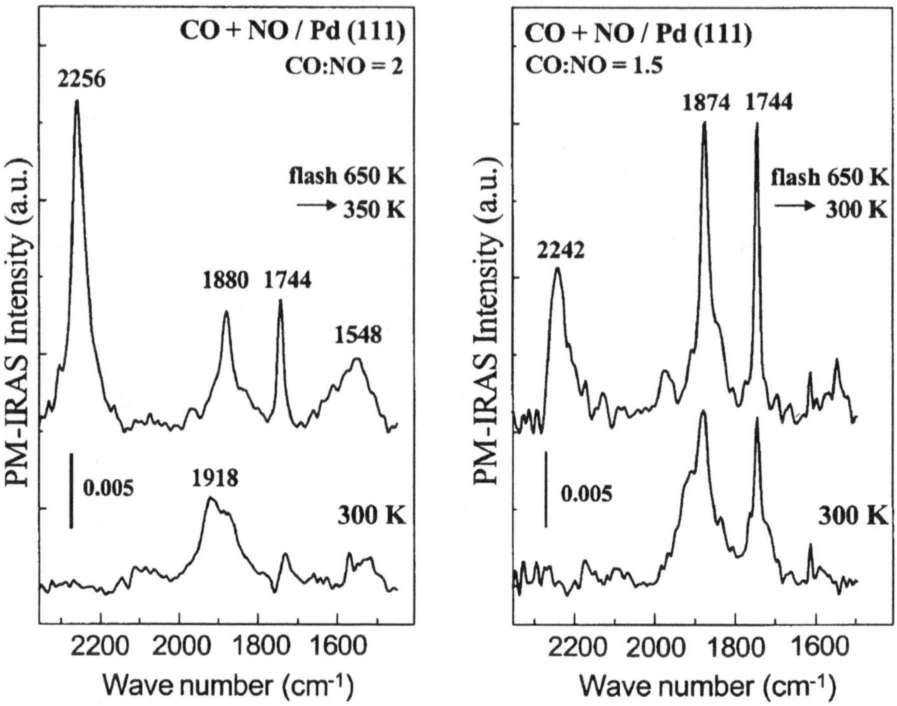

Figure 8.9. PM-IRRA spectra of Pd(111) in the presence of a CO + NO mixture at pressure ratios of P(CO)/P(NO) = 1.5 (right) and 2 (left) and total pressures of 240 and 180 mbar, respectively. The upper spectra were obtained after heating the catalyst to 650 K. Reprinted with permission from Ozensoy, E., et al. (2002) Isocyanate formation in the catalytic reaction of CO + NO on Pd(111): an in situ infrared spectroscopic study at elevated pressures, *Journal of the American Chemical Society*, 124(29), 8524–8525. Copyright 2002 American Chemical Society.

CH_3O, as shown in Figure 8.10. In addition, carbonaceous overlayers (CH_x) were detected by postreaction XPS, indicating the importance of combining techniques to gain a detailed understanding of the surface reaction mechanisms. Although the species formed by the dehydrogenation of methanol were observed under UHV conditions, they were unstable at 300 K. In addition to PM-IRRAS and XPS, online gas chromatography (GC) was also utilized to analyze and quantify the gas-phase contribution to obtain turnover frequencies. All the adsorbed species were stable up to 350 K after introducing O_2. The catalytic activity was observed to start at 400 K, where formation of CH_2O, CO_2, and H_2O was identified by GC and PM-IRRAS (p + s), as shown in Figure 8.10. PM-IRRAS detected only CO on the surface, but CH_2O and CHO were not observed most likely due to the surface concentrations of them being below the detection limit under reaction conditions. Based on the observed

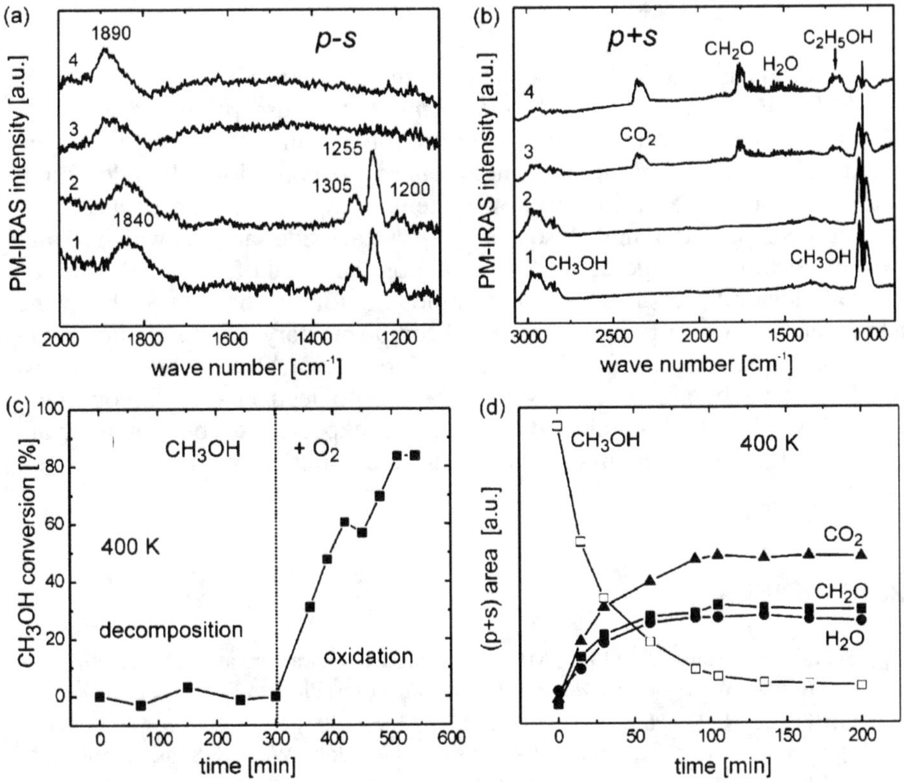

Figure 8.10. (a) Surface (p − s) and (b) gas-phase ($p + s$) PM-IRRAS spectra measured during CH_3OH oxidation at 300–400 K (5 mbar CH_3OH added at 300 K [trace 1] prior to 5 mbar O_2 [trace 2]). Spectra 3 and 4 were acquired during the oxidation reaction at 400 K, after 15 and 60 min, respectively. Methanol conversion and formation of CO_2, H_2O, and CH_2O, as measured by gas chromatography and PM-IRRAS (p + s), are displayed in (c) and (d). Reprinted with permission from Borasio, M., et al. (2005) In situ studies of methanol decomposition and oxidation on Pd(111) by PM-IRAS and XPS spectroscopy, *The Journal of Physical Chemistry C*, 109(38), 17791–17794. Copyright 2005 American Chemical Society.

results, a reaction mechanism was proposed in which methanol oxidation proceeds via dehydrogenation to CH_2O, which either desorbs or further dehydrogenates to CO, which is subsequently oxidized to CO_2. During the various dehydrogenation processes, hydrogen forms and reacts with oxygen to produce H_2O. Based on the IRRAS and XPS results, the authors concluded that CH_x may favor the formation of CH_2O by hindering its dehydrogenation to CO and metallic Pd. This study clearly shows the importance of the application of combined *in-situ* spectroscopic techniques under elevated pressure conditions to characterize catalytic reactions.

8.5 CONCLUSIONS

A number of case studies have been presented earlier to illustrate how IRRAS and PM-IRRAS can be applied to study *in-situ* the adsorption and coadsorption of gas-phase molecules on single crystal-based model systems under elevated pressures, and reactions under *operando* conditions. In addition to extending these applications to new systems using similar approaches, more studies are expected to make use of single-crystal oxide samples, which require two orders of magnitude better S/N detection. The goal of *in-situ* IR spectroscopy at elevated pressures is to determine reaction mechanisms of heterogeneous catalysis, which requires additional complementary structural information by ambient-pressure scanning tunneling microscopy (AP-STM) and compositional analysis by ambient-pressure X-ray photoelectron spectroscopy (AP-XPS), and therefore combined studies are expected to be pursued, and ultimately the development of setups with a combination of these complementary techniques.

REFERENCES

[1] Suëtaka, W., Yates, J. T. (1995) Methods of surface characterization, *Surface Infrared and Raman Spectroscopy: Methods and Applications*, Plenum Press, New York, p. xiii.

[2] Griffiths, P. R., De Haseth, J. A., Books24x7 Inc. (2007) *Fourier Transform Infrared Spectrometry*, Vol. 171, Chemical Analysis, 2nd ed., John Wiley & Sons, Hoboken, NJ.

[3] Eischens, R. P., Pliskin, W. A. (1958) Advances in catalysis and related subjects, *Advances in Catalysis*, 10, 1–56.

[4] Greenler, R. G. (1966) Infrared study of adsorbed molecules on metal surfaces by reflection techniques, *The Journal of Chemical Physics*, 44, 310.

[5] Wilson, E. L., Brown, W. A. (2010) Low pressure RAIRS studies of model catalytic systems, *The Journal of Physical Chemistry C*, 114(15), 6879–6893.

[6] Hoffmann, F. M. (1983) Infrared reflection-absorption spectroscopy of adsorbed molecules, *Surface Science Reports*, 3(2–3), 107–192.

[7] Chabal, Y. J. (1988) Surface infrared-spectroscopy, *Surface Science Reports*, 8(5–7), 211–357.

[8] Hollins, P. (1992) The influence of surface-defects on the infrared-spectra of adsorbed species, *Surface Science Reports*, 16(2), 51–94.

[9] Francis, S. A., Ellison, A. H. (1959) Infrared spectra of monolayers on metal mirrors, *Journal of the Optical Society of America*, 49(2), 131–138.

[10] Rivière, J. C. (1990) *Surface Analytical Techniques*, Monographs on the Physics and Chemistry of Materials, Oxford University Press, New York, p. 363.

[11] Yates, J. T., Jr., Madey, E. T. (1987) Methods of surface characterization, in *Vibrational Spectroscopy of Molecules on Surfaces*, Plenum Press, New York, p. 269.

[12] Greenler, R. G. (1975) Design of a reflection—absorption experiment for studying the IR spectrum of molecules adsorbed on a metal surface, *Journal of Vacuum Science and Technology*, 12(6), 1410–1417.

[13] Sheppard, N., DelaCruz, C. (1996) Vibrational spectra of hydrocarbons adsorbed on metals .1. Introductory principles, ethylene, and the higher acyclic alkenes, *Advances in Catalysis*, 41, 1–112.

[14] Sheppard, N., De la Cruz, C. (1998) Vibrational spectra of hydrocarbons adsorbed on metals—part ii. Adsorbed acyclic alkynes and alkanes, cyclic hydrocarbons including aromatics, and surface hydrocarbon groups derived from the decomposition of alkyl halides, etc., *Advances in Catalysis*, 42, 181–313.

[15] Hayden, B. E. (2001) Vibrational spectroscopy at oxide surfaces, Chapter 13, in Woodruff, D. P. (ed.), *The Chemical Physics of Solid Surfaces*, Elsevier, Amsterdam, The Netherlands, pp. 514–549.

[16] Hayden, B. E., King, A., Newton, M. A. (1998) The reaction of hydrogen with TiO_2(110) supported rhodium gem-dicarbonyl, *Surface Science*, 397(1–3), 306–313.

[17] Hayden, B. E., King, A., Newton, M. A. (1999) Fourier transform reflection–absorption IR spectroscopy study of formate adsorption on TiO_2(110), *The Journal of Physical Chemistry B*, 103(1), 203–208.

[18] Xu, M., Gao, Y., Wang, Y., Woll, C. (2010) Monitoring electronic structure changes of TiO_2(110) via sign reversal of adsorbate vibrational bands, *Physical Chemistry Chemical Physics*, 12(15), 3649–3652.

[19] Kattner, J., Hoffmann, H. (2002) External reflection infrared spectroscopy of thin films on dielectric substrates, in Chalmers J. M., Griffiths P. R. (eds.), *Handbook of Vibrational Spectroscopy*, Vol. 2, John Wiley & Sons, Chichester, UK,

[20] Xu, M. C., Gao, Y. K., Moreno, E. M., Kunst, M., Muhler, M., Wang, Y. M., Idriss, H., Woll, C. (2011) Photocatalytic activity of bulk TiO_2 anatase and rutile single crystals using infrared absorption spectroscopy, *Physical Review Letters*, 106(13), 138302-1–138302-4.

[21] Freund, H. J., Pacchioni, G. (2008) Oxide ultra-thin films on metals: new materials for the design of supported metal catalysts, *Chemical Society Reviews*, 37(10), 2224–2242.

[22] Freund, H.-J., Goodman, D. W. (2008) *Handbook of Heterogeneous Catalysis*, Vol. 3, Wiley-VCH Verlag, Weinheim, Germany.

[23] Freund, H. J., Goodman, D. W. (2008) Ultrathin oxide films, in Ertl, G., Knözinger, H., Schüth, F., Weitkamp, J. (eds.), *Handbook of Heterogeneous Catalysis*, 2nd ed., Wiley-VCH Verlag, Weinheim, Germany, pp. 1309–1338.

[24] Kaya, S., Sun, Y. N., Weissenrieder, J., Stacchiola, D., Shaikhutdinov, S., Freund, H. J. (2007) Ice-assisted preparation of silica-supported vanadium oxide particles, *The Journal of Physical Chemistry C*, 111(14), 5337–5344.

[25] Baron, M., Abbott, H., Bondarchuk, O., Stacchiola, D., Uhl, A., Shaikhutdinov, S., Freund, H. J., Popa, C., Ganduglia-Pirovano, M. V., Sauer, J. (2009) Resolving the atomic structure of vanadia monolayer catalysts: monomers, trimers, and oligomers on ceria, *Angewandte Chemie-International Edition*, 48(43), 8006–8009.

[26] Weissenrieder, J., Kaya, S., Lu, J. L., Gao, H. J., Shaikhutdinov, S., Freund, H. J., Sierka, M., Todorova, T. K., Sauer, J. (2005) Atomic structure of a thin silica film on a Mo(112) substrate: a two-dimensional network of SiO_4 tetrahedra, *Physical Review Letters*, 95(7), 076103-1–076103-4.

[27] Todorova, T. K., Sierka, M., Sauer, J., Kaya, S., Weissenrieder, J., Lu, J. L., Gao, H. J., Shaikhutdinov, S., Freund, H. J. (2006) Atomic structure of a thin silica film on a Mo(112) substrate: a combined experimental and theoretical study, *Physical Review B*, 73, 165414-1–165414-9.

[28] Sock, M., Surnev, S., Ramsey, M. G., Netzer, F. P. (2001) Adsorption and reaction of CO on vanadium oxide-Pd(111) "inverse" model catalysts: an HREELS study, *Topics in Catalysis*, 14(1–4), 15–23.

[29] Somorjai, G. A., Li, Y. (2010) *Introduction to Surface Chemistry and Catalysis*, 2nd ed., John Wiley & Sons, Hoboken, NJ, p. xii.

[30] Libuda, J., Freund, H. J. (2005) Molecular beam experiments on model catalysts, *Surface Science Reports*, 57(7–8), 157–298.

[31] Desikusumastuti, A., Schernich, S., Happel, M., Sobota, M., Laurin, M., Libuda, J. (2009) Model NO_x storage materials at realistic NO_2 pressures, *ChemCatChem*, 1(2), 318–325.

[32] Stacchiola, D., Molero, H., Tysoe, W. T. (2001) Palladium-catalyzed cyclotrimerization and hydrogenation: from ultrahigh vacuum to high-pressure catalysis, *Catalysis Today*, 65(1), 3–11.

[33] Kaltchev, M., Stacchiola, D., Molero, H., Wu, G., Blumenfeld, A., Tysoe, W. T. (1999) On the reaction pathway for the formation of benzene from acetylene catalyzed by palladium, *Catalysis Letters*, 60(1–2), 11–14.

[34] Stacchiola, D., Kaltchev, M., Wu, G., Tysoe, W. T. (2000) The adsorption and structure of carbon monoxide on ethylidyne-covered Pd(111), *Surface Science*, 470(1–2), L32–L38.

[35] Samanos, B., Boutry, P., Montarna, R. (1971) Mechanism of vinyl acetate formation by gas-phase catalytic ethylene acetoxidation, *Journal of Catalysis*, 23(1), 19–30.

[36] Nakamura, S., Yasui, T. (1970) Mechanism of palladium-catalyzed synthesis of vinyl acetate from ethylene in a heterogeneous gas reaction, *Journal of Catalysis*, 17(3), 366–374.

[37] Stacchiola, D., Calaza, F., Burkholder, L., Tysoe, W. T. (2004) Vinyl acetate formation by the reaction of ethylene with acetate species on oxygen-covered Pd(111), *Journal of the American Chemical Society*, 126(47), 15384–15385.

[38] Stacchiola, D., Calaza, F., Burkholder, L., Schwabacher, A. W., Neurock, M., Tysoe, W. T. (2005) Elucidation of the reaction mechanism for the palladium-catalyzed synthesis of vinyl acetate, *Angewandte Chemie-International Edition*, 44(29), 4572–4574.

[39] Rupprechter, G. (2007) Sum frequency generation and polarization-modulation infrared reflection absorption spectroscopy of functioning model catalysts from ultrahigh vacuum to ambient pressure, *Advances in Catalysis*, 51, 133–263.

[40] Somorjai, G. A., McCrea, K. R. (2000) Sum frequency generation: surface vibrational spectroscopy studies of catalytic reactions on metal single-crystal surfaces, *Advances in Catalysis*, 45, 385–438.

[41] Shen, Y. R. (1989) Surface-properties probed by 2nd-harmonic and sum-frequency generation, *Nature*, 337(6207), 519–525.

[42] Rupprechter, G. (2007) Sum frequency laser spectroscopy during chemical reactions on surfaces, *MRS Bulletin*, 32, 1031–1037.

[43] Dellwig, T., Rupprechter, G., Unterhalt, H., Freund, H. J. (2000) Bridging the pressure and materials gaps: high pressure sum frequency generation study on supported Pd nanoparticles, *Physical Review Letters*, 85(4), 776–779.

[44] Baldelli, S., Eppler, A. S., Anderson, E., Shen, Y. R., Somorjai, G. A. (2000) Surface enhanced sum frequency generation of carbon monoxide adsorbed on platinum nanoparticle arrays, *The Journal of Chemical Physics*, 113(13), 5432–5438.

[45] Unterhalt, H., Rupprechter, G., Freund, H. J. (2002) Vibrational sum frequency spectroscopy on Pd(111) and supported Pd nanoparticles: CO adsorption from ultrahigh vacuum to atmospheric pressure, *The Journal of Physical Chemistry B*, 106(2), 356–367.

[46] Rupprechter, G., Unterhalt, H., Morkel, M., Galletto, P., Hu, L. J., Freund, H. J. (2002) Sum frequency generation vibrational spectroscopy at solid-gas interfaces: CO adsorption on Pd model catalysts at ambient pressure, *Surface Science*, 502, 109–122.

[47] Rupprechter, G., Morkel, M., Freund, H. J. (2005) Finite size effects on supported Pd nanoparticles: interaction of hydrogen with CO and c/sub 2/h/sub 4, *Surface Science*, 588(1–3), L209–L219.

[48] Yeganeh, M. S., Dougal, S. M., Silbernagel, B. G. (2005) Sum frequency generation studies of surfaces of high-surface-area powdered materials, *Langmuir*, 22(2), 637–641.

[49] Shen, Y. R. (1994) Surfaces probed by nonlinear optics, *Surface Science*, 299(1–3), 551–562.

[50] Somorjai, G. A., Rupprechter, G. (1999) Molecular studies of catalytic reactions on crystal surfaces at high pressures and high temperatures by infrared-visible sum frequency generation (SFG) surface vibrational spectroscopy, *The Journal of Physical Chemistry B*, 103(10), 1623–1638.

[51] Williams, C. T., Beattie, D. A. (2002) Probing buried interfaces with non-linear optical spectroscopy, *Surface Science*, 500(1–3), 545–576.

[52] Buck, M., Himmelhaus, M. (2001) Vibrational spectroscopy of interfaces by infrared-visible sum frequency generation, *Journal of Vacuum Science & Technology. A, Vacuum, Surfaces, and Films*, 19(6), 2717–2736.

[53] Bradshaw, A. M., Hoffmann, F. (1975) IR reflection-absorption spectroscopy of CO adsorbed on palladium, *Surface Science*, 52(2), 449–454.

[54] Golden, W. G., Dunn, D. S., Overend, J. (1981) A method for measuring infrared reflection—absorption spectra of molecules adsorbed on low-area surfaces at monolayer and submonolayer concentrations, *Journal of Catalysis*, 71(2), 395–404.

[55] Buffeteau, T., Desbat, B., Turlet, J. M. (1991) Polarization modulation FT-IR spectroscopy of surfaces and ultra-thin films—experimental procedure and quantitative-analysis, *Applied Spectroscopy*, 45(3), 380–389.

[56] Schennach, R., Hirschmugl, C., Gilli, E., Tysoe, W. T. (2009) A new method for performing polarization modulation infrared reflection-adsorption spectroscopy of surfaces, *Applied Spectroscopy*, 63(3), 369–372.

[57] Stacchiola, D., Thompson, A. W., Kaltchev, M., Tysoe, W. T. (2002) Photoelastic modulation-reflection absorption infrared spectroscopy of CO on Pd(111), *Journal of Vacuum Science & Technology. A, Vacuum, Surfaces, and Films*, 20(6), 2101–2105.

[58] Green, M. J., Barner, B. J., Corn, R. M. (1991) Real-time sampling electronics for double modulation experiments with Fourier-transform infrared spectrometers, *Review of Scientific Instruments*, 62(6), 1426–1430.

[59] Borasio, M., Fuente, O. R., Rupprechter, G., Freund, H. J. (2005) In situ studies of methanol decomposition and oxidation on Pd(111) by PM-IRAS and XPS spectroscopy, *The Journal of Physical Chemistry B*, 109(38), 17791–17794.

[60] Gießel, T., Schaff, O., Hirschmugl, C. J., Fernandez, V., Schindler, K. M., Theobald, A., Bao, S., Lindsay, R., Berndt, W., Bradshaw, A. M., Baddeley, C., Lee, A. F., Lambert, R. M., Woodruff, D. P. (1998) A photoelectron diffraction study of ordered structures in the chemisorption system Pd{111}-CO, *Surface Science*, 406(1–3), 90–102.

[61] Loffreda, D., Simon, D., Sautet, P. (1999) Dependence of stretching frequency on surface coverage and adsorbate-adsorbate interactions: a density-functional theory approach of CO on Pd(111), *Surface Science*, 425(1), 68–80.

[62] Hess, C., Ozensoy, E., Goodman, D. W. (2003) Combined in situ infrared and kinetic study of the catalytic CO + NO reaction on Pd(111) at pressures up to 240 mbar, *The Journal of Physical Chemistry B*, 107(12), 2759–2764.

[63] Ozensoy, E., Hess, C., Goodman, D. W. (2002) Isocyanate formation in the catalytic reaction of CO + NO on Pd(111): an in situ infrared spectroscopic study at elevated pressures, *Journal of the American Chemical Society*, 124(29), 8524–8525.

[64] Beitel, G. A., de Groot, C. P. M., Oosterbeek, H., Wilson, J. H. (1997) A combined in-situ PM-RAIRS and kinetic study of single-crystal cobalt catalysts under synthesis gas at pressures up to 300 mbar, *The Journal of Physical Chemistry B*, 101(20), 4035–4043.

[65] Oosterbeek, H. (2007) Bridging the pressure and material gap in heterogeneous catalysis: cobalt Fischer-Tropsch catalysts from surface science to industrial application, *Physical Chemistry Chemical Physics*, 9(27), 3570–3576.

[66] Ozensoy, E., Meier, D. C., Goodman, D. W. (2002) Polarization modulation infrared reflection absorption spectroscopy at elevated pressures: CO adsorption on Pd(111) at atmospheric pressures, *The Journal of Physical Chemistry B*, 106(36), 9367–9371.

[67] Rameshan, C., Weilach, C., Stadlmayr, W., Penner, S., Lorenz, H., Havecker, M., Blume, R., Rocha, T., Teschner, D., Knop-Gericke, A., Schlogl, R., Zemlyanov, D., Memmel, N., Rupprechter, G., Klotzer, B. (2010) Steam reforming of methanol on PdZn near-surface alloys on Pd(111) and Pd foil studied by in-situ XPS, LEIS and PM-IRAS, *Journal of Catalysis*, 276(1), 101–113.

[68] Carrasco, E., Aumer, A., Brown, M. A., Dowler, R., Palacio, I., Song, S., Sterrer, M. (2010) Infrared spectra of high coverage co adsorption structures on Pt(111), *Surface Science*, 604(15–16), 1320–1325.

[69] Jugnet, Y., Cadete Santos Aires, F. J., Deranlot, C., Piccolo, L., Bertolini, J. C. (2002) CO chemisorption on Au(1 1 0) investigated under elevated pressures by polarized reflection absorption infrared spectroscopy and scanning tunneling microscopy, *Surface Science*, 521(1–2), L639–L644.

[70] Beitel, G. A., Laskov, A., Oosterbeek, H., Kuipers, E. W. (1996) Polarization modulation infrared reflection absorption spectroscopy of CO adsorption on CO(0001) under a high-pressure regime, *The Journal of Physical Chemistry*, 100(30), 12494–12502.

[71] Artiglia, L., Diemant, T., Hartmann, H., Bansmann, J., Behm, R. J., Gavioli, L., Cavaliere, E., Granozzi, G. (2010) Stability and chemisorption properties of ultrathin TiOx/Pt(111) films and Au/TiOx/Pt(111) model catalysts in reactive atmospheres, *Physical Chemistry Chemical Physics*, 12(25), 6864–6874.

[72] Gao, F., Goodman, D. W. (2010) Reaction kinetics and polarization modulation infrared reflection absorption spectroscopy investigations of CO oxidation over planar Pt-group model catalysts†, *Langmuir*, 26(21), 16540–16551.

[73] Maeda, N., Urakawa, A., Baiker, A. (2010) Surface reactions on Pt during NO_x storage-reduction studied by polarization-modulation infrared reflection-absorption spectroscopy, *Journal of Physical Chemistry Letters*, 1(1), 54–58.

[74] Jugnet, Y., Sedrati, R., Bertolini, J. C. (2005) Selective hydrogenation of 1,3-butadiene on $Pt_3Sn(111)$ alloys: comparison to Pt(111), *Journal of Catalysis*, 229(1), 252–258.

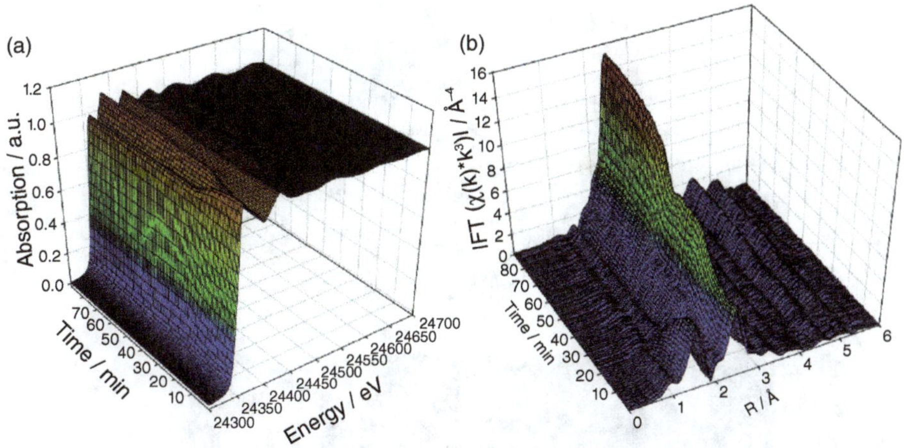

Figure 1.8. (a) Normalized XANES spectra of the 5 wt % Pd/Al_2O_3 catalyst as a function of time. (b) Fourier-transformed EXAFS spectra of the 5 wt % Pd/Al_2O_3 catalyst as a function of time. Reprinted with permission from Reimann, S., Stötzel, J., Frahm, R., Kleist, W., Grunwaldt, J.-D., Baiker, A. (2011) *Journal of the American Chemical Society*, 133, 3921–3930. Copyright 2011 American Chemical Society.

In-situ *Characterization of Heterogeneous Catalysts*, First Edition.
Edited by José A. Rodriguez, Jonathan C. Hanson, and Peter J. Chupas.

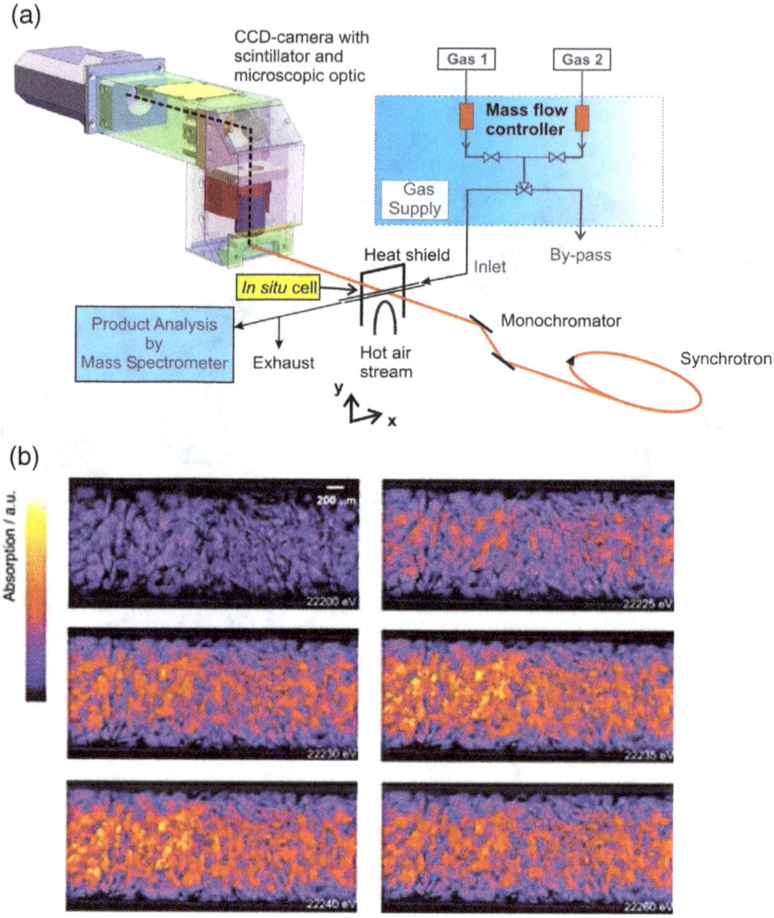

Figure 2.5. (a) Experimental setup for *in-operando* imaging of the catalyst bed during partial oxidation of methane. (b) X-ray absorption in the catalyst bed as a function of X-ray energy (cf. Reference [36] for details). Reprinted with permission from Grunwaldt, J.-D., Hannemann, S., Schroer, C. G., Baiker, A. (2006) 2D-mapping of the catalyst structure inside a catalytic microreactor at work: partial oxidation of methane over Rh/Al_2O_3, *J. Phys. Chem. B*, 110, 8674–8680. Copyright 2006 by American Chemical Society.

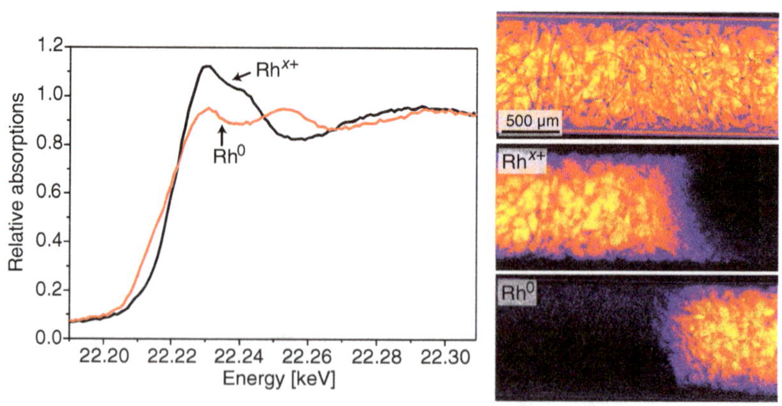

Figure 2.6. Left: Absorption spectra at the Rh K-edge extracted from the stack of absorption images at the far left (Rh^{x+}) and the far right (Rh^0). (See text for full caption.)

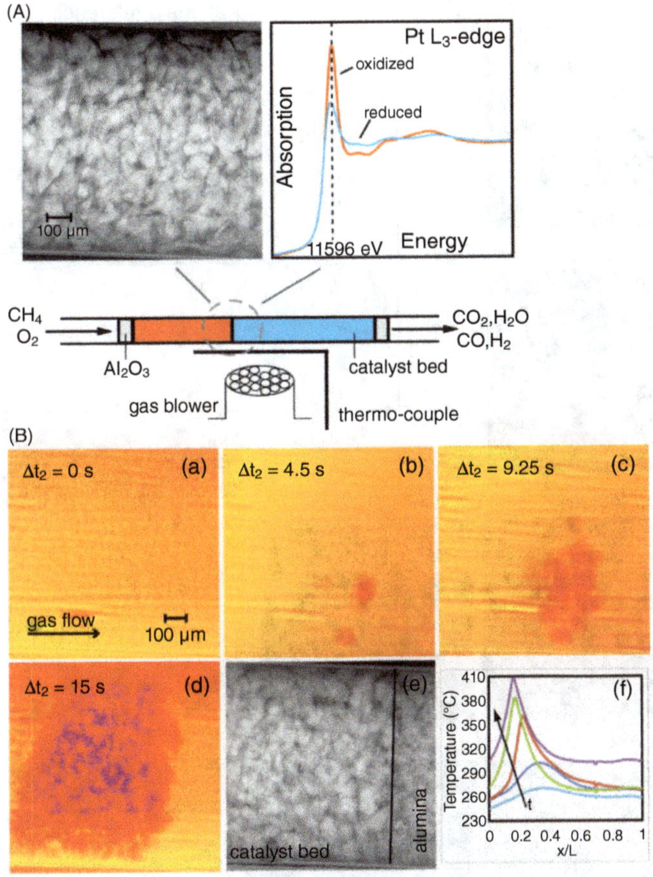

Figure 2.7. (A) Schematic sketch of the reactor capillary for partial oxidation of methane over a $Rh/Pt/Al_2O_3$ catalyst inside a heated reactor capillary together with a transmission image through the capillary and absorption spectra of reduced and oxidized Pt at the Pt L_3-edge. (B) (a–d) Time series of the oxidation state of Pt (yellow oxidized, orange-red reduced) during the ignition of the partial oxidation reaction. (e) Transmission image at the end of the catalyst bed. The black line delineates the end of the catalyst bed. (f) Evolution of the temperature along the capillary during ignition [41]. Reprinted with permission from Kimmerle, B., Grunwaldt, J.-D., Baiker, A., Glatzel, P., Boye, P., Stephan, S., Schroer, C. G. (2009) Visualizing a catalyst at work during the ignition of the catalytic partial oxidation of methane, *J. Phys. Chem.*, C, 113, 3037–3040. Copyright 2009 American Chemical Society.

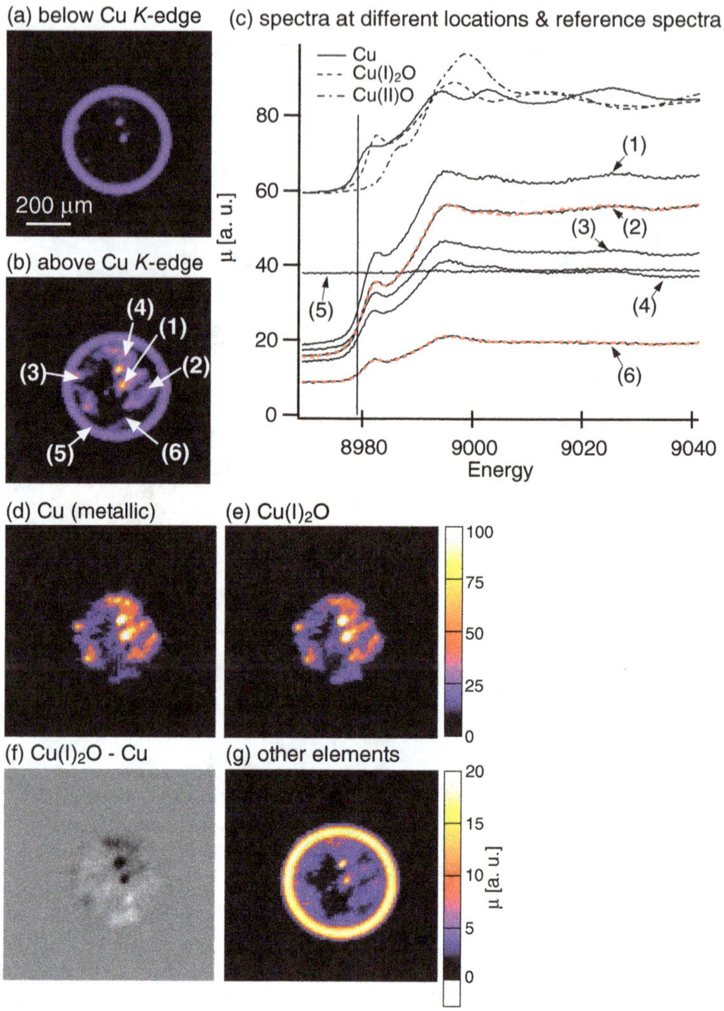

Figure 2.11. XANES tomogram of a Cu/ZnO catalyst mixed with inert material inside a quartz microreactor taken after several reduction/reoxidation cycles between 200 and 400°C [62]. Panels (a) and (b) show the tomographic reconstruction of the sample at an energy below and above the absorption edge of copper, respectively. (c) Absorption spectra reconstructed on the tomographic slice through the sample at different locations depicted in (b) and reference spectra of copper in different oxidation states. The reference spectra are fitted to the measured spectra at each location on the reconstructed slice. In this way, the distribution of copper in different oxidation states can be obtained, as shown in panels (d) and (e), together with the distribution of other elements (g). Panel (f) shows the difference between monovalent and metallic copper. The concentration of bivalent copper lies below the detection limit. Reprinted with permission from Schroer, C. G., Kuhlmann, M., Günzler, T. F., Lengeler, B., Richwin, M., Griesebock, B., Lützenkirchen-Hecht, D., Frahm, R., Ziegler, E., Mashayekhi, A., Haeffner, D., Grunwaldt, J.-D., Baiker, A. (2003) Mapping the chemical states of an element inside a sample using tomographic x-ray absorption spectroscopy, *Appl. Phys. Lett.*, 82, 3360–3362. Copyright 2003 by American Institute of Physics.

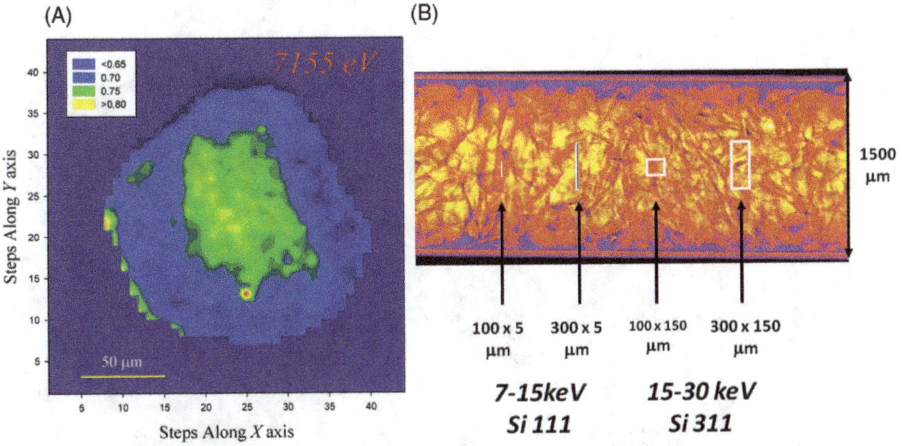

Figure 3.3. A comparison of sample uniformity for two types of sample commonly studied using dispersive EXAFS. (See text for full caption.)

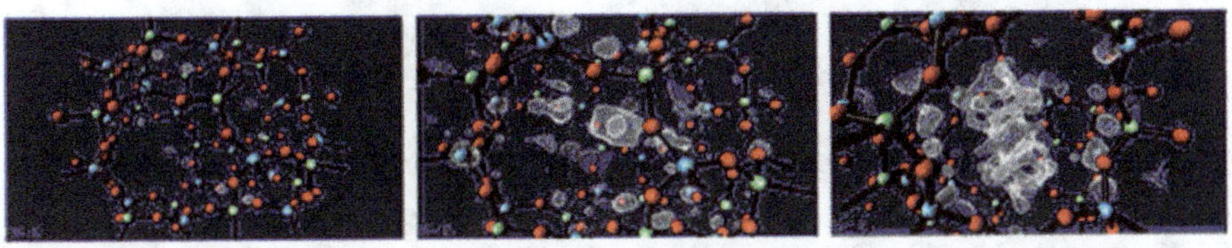

Figure 4.16. Images of the residual electron density in the SAPO-34 cage, determined from difference Fourier maps. From left to right: 440°C before reaction, and 4 and 20 min into reaction. Red spheres = oxygen, green = phosphorus, and blue = aluminum. The white contours represent residual electron density [28]. Copyright © 2009, Elsevier.

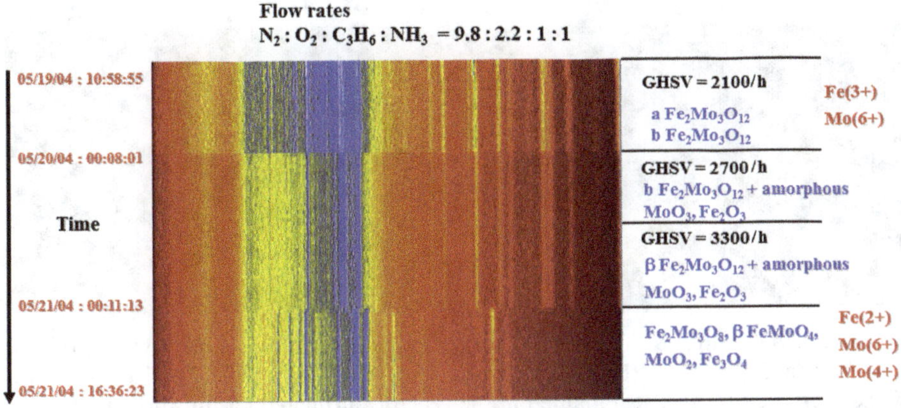

Figure 6.4. Neutron diffraction collected during the *in-situ* reduction measurements of $Fe_2Mo_3O_{12}$over 2 days of beamtime at GPPD. Equation 6.1a–d was derived on the basis of these data.

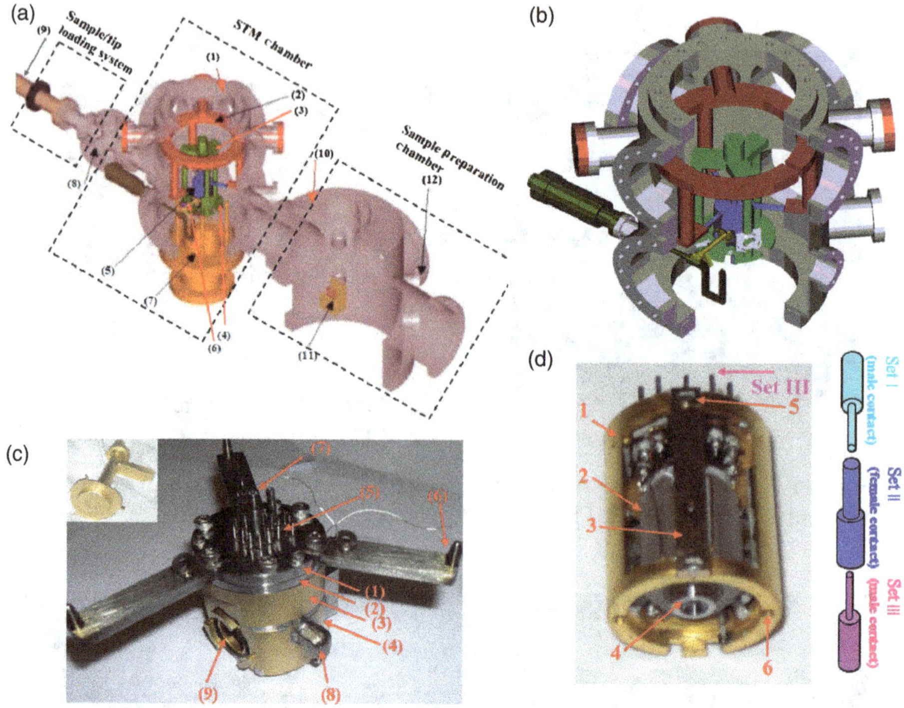

Figure 7.6. (a) CAD drawing of the high-pressure high-temperature STM reactor/UHV system designed by Tao et al. [20]. (b) CAD drawing of the STM chamber, scaffold, and high-pressure cell. (c) Photo of the high-pressure reactor. (d) Side view of the STM body. (See text for full caption.)

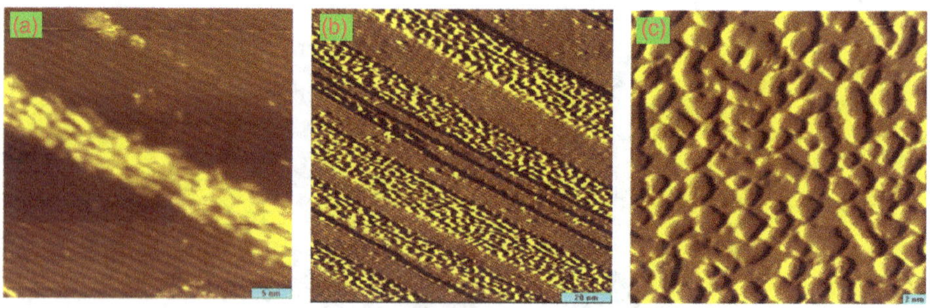

Figure 7.8. STM images of a lifted Pt(100) surface in CO environment with different pressures (a) 5×10^{-9} Torr, (b) 5×10^{-7} Torr, and (c) 5×10^{-5} Torr. *Note*: The listed pressure is the pressure of CO that remained in the reactor during data acquisition. Reprinted with permission from Tao, F., Dag, S., Wang, L.-W., Liu, Z., Butcher, D. R., Salmeron, M., Somorjai, G. A. (2009) *Nano Lett.*, 9(5), 2167–2171. Copyright 2009.

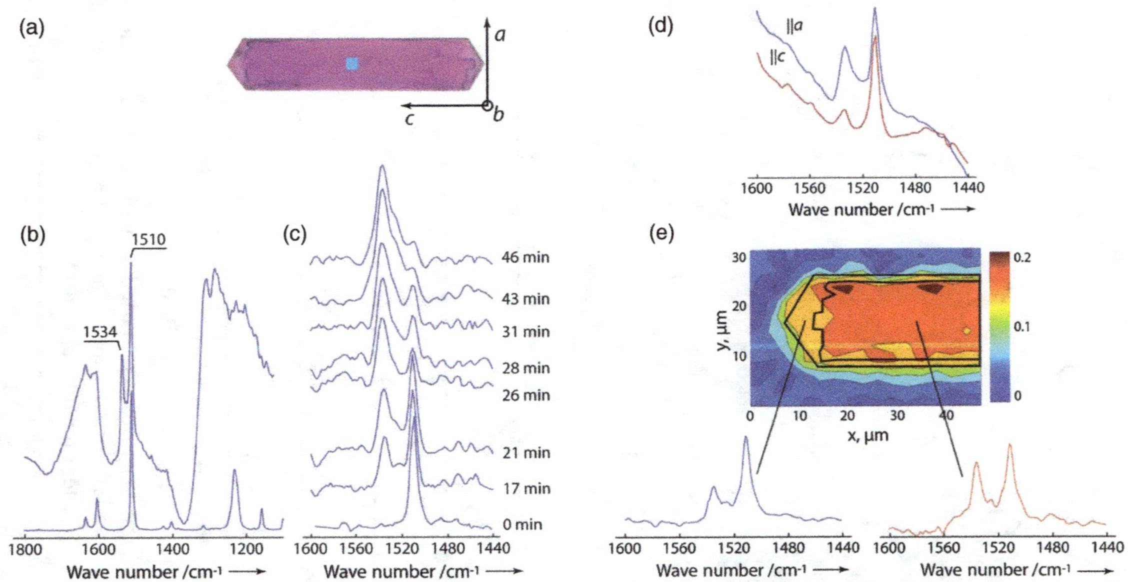

Figure 9.6. (a) Optical microphotograph of the H-ZSM-5 crystal after reaction with 4-fluorostyrene with the square probed area marked. (b) IR spectrum of an individual H-ZSM-5 crystal contacted with 4-fluorostyrene (top) and the IR spectrum of liquid 4-fluorostyrene (bottom). (c) Region of the IR spectra (1440–1600 cm^{-1}) taken *in situ* during the 4-fluorostyrene oligomerization reaction; the spectra are background corrected. (d) *In-situ* IR spectra taken with two different light polarizations. (e) Intensity of the IR band at 1534 cm^{-1} mapped over the crystal after reaction and the IR spectra taken form the edge and the body of the crystal, demonstrating differences in the intensity ratio of the bands.

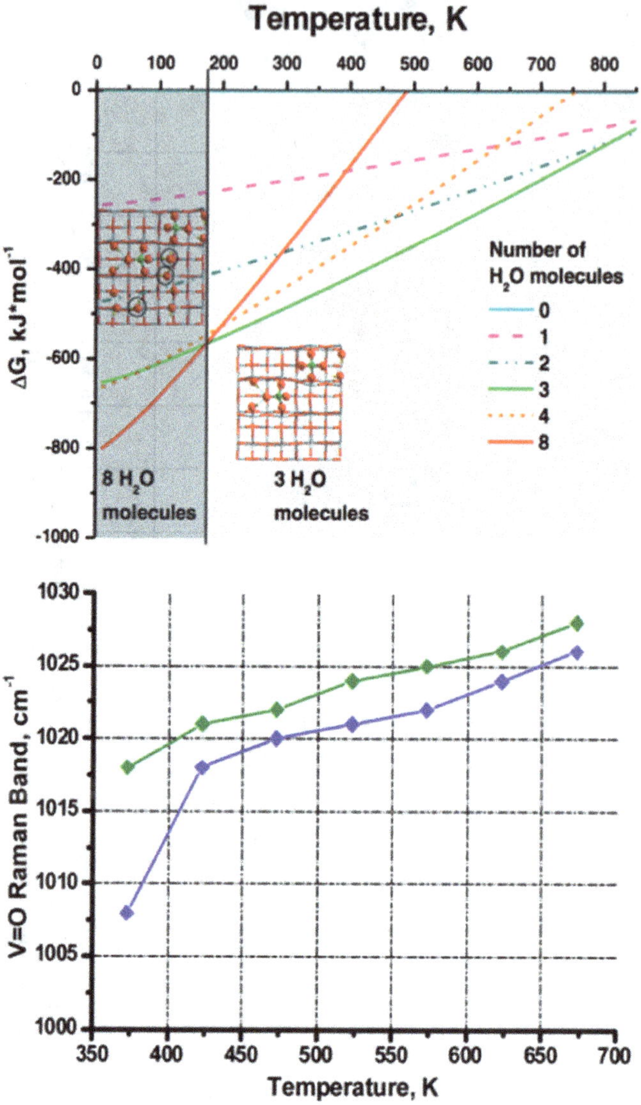

Figure 10.1. Top: Free energy ΔG of the supported vanadium oxide species on the titania surface as a function of temperature for a water partial pressure equivalent to the ambient air (1500 Pa). (See text for full caption.)

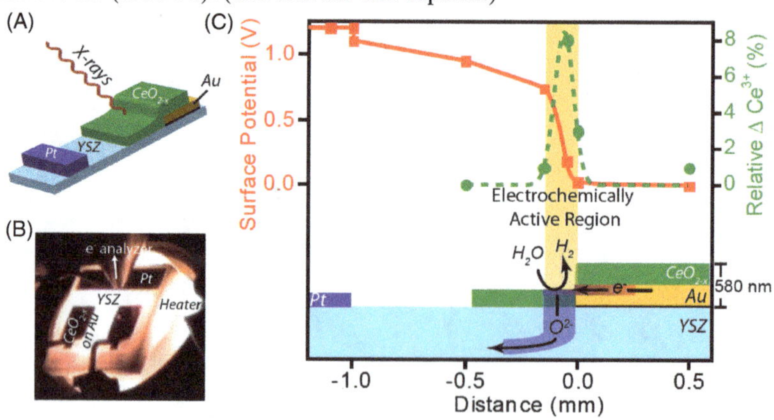

Figure 12.6. Electrochemical activity spreads over 150 μm of the ceria anode. (See text for full caption.)

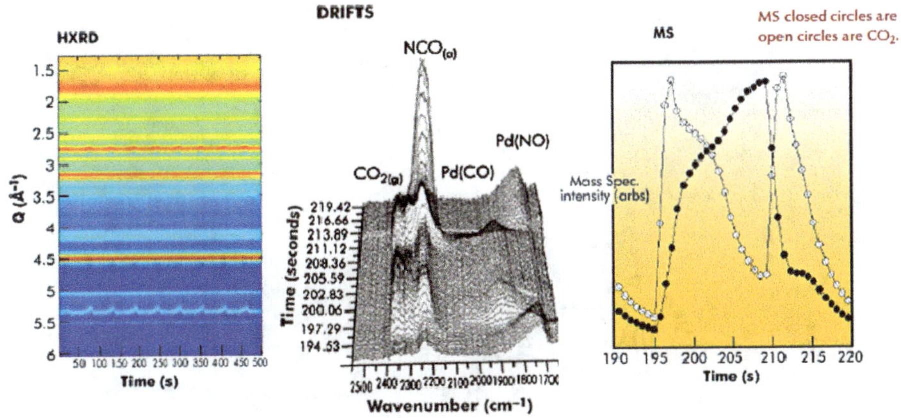

Figure 14.5. Results obtained through synchronously coupling HXRD, DRIFTS and MS using the reactor/optics design shown in Figure 14.2B and C [49]. (See text for full caption.)

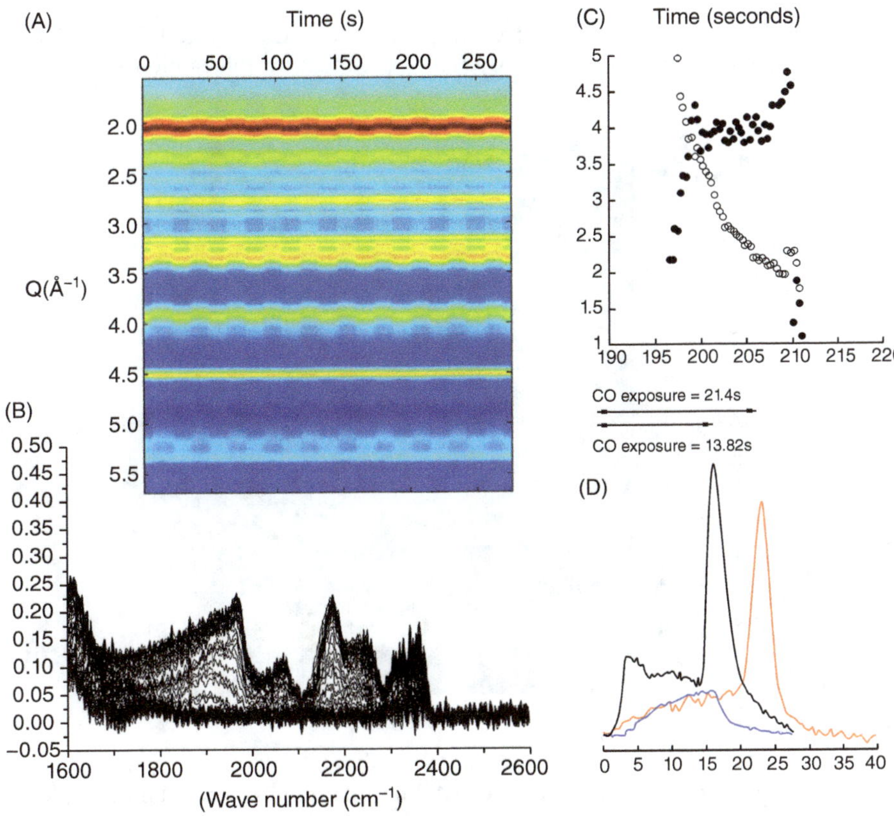

Figure 14.7. (A) HXRD obtained during redox (CO/NO) cycling of 4Pd33ZCA catalysts sample at 673 K [63]. (See text for full caption.)

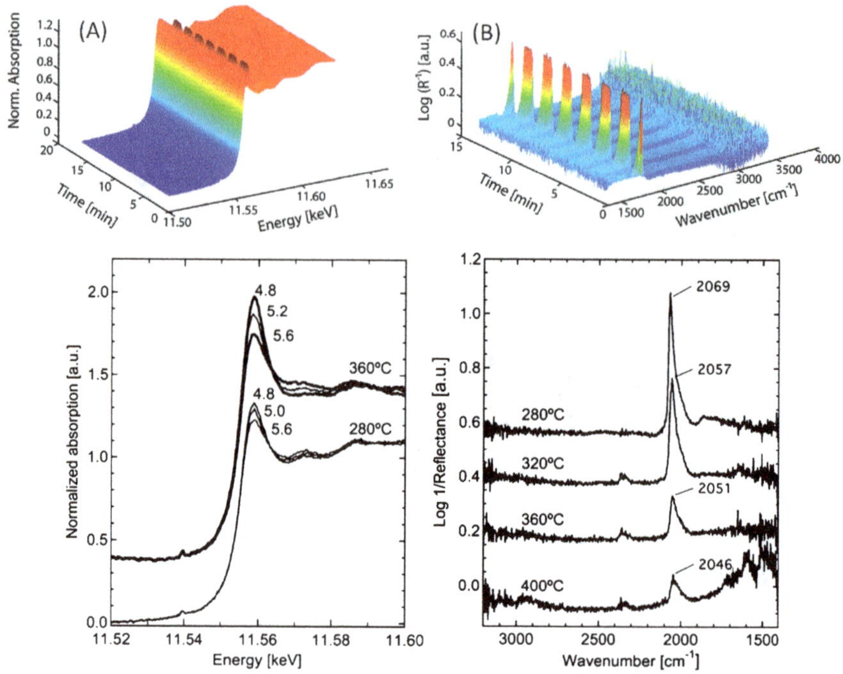

Figure 14.8. Synchronous time-resolved dispersive XANES (A) and DRIFTS (B) during CH_4 oxidation over 4 wt% Pt/Al_2O_3 [65]. (See text for full caption.)

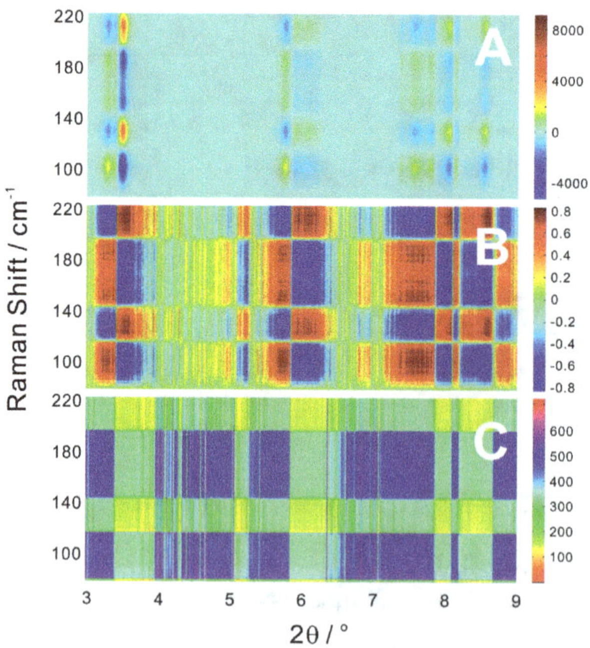

Figure 15.7. 2D XRD-Raman covariance (A), correlation (B), and sum of in-phase angle (C) plots from the spin crossover transition of [Fe(Htrz)$_2$(trz)]BF_4. (See text for full caption.)

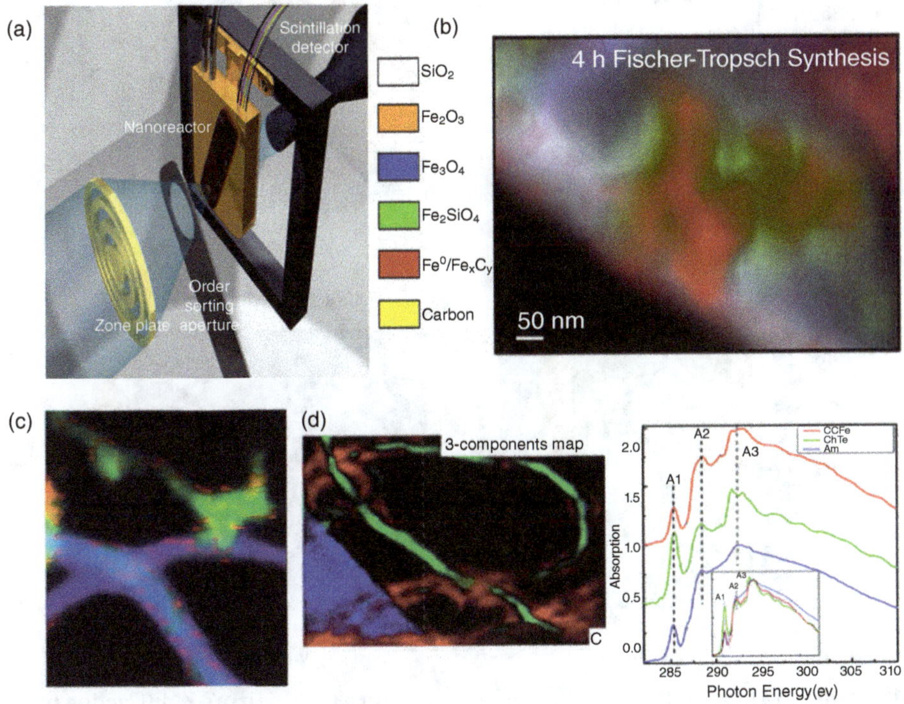

Figure 16.3. (a) Setup of an STXM microscope with, from left to right, the zone plate lens, order-sorting aperture, the nanoreactor, and the scintillation detector. (See text for full caption.)

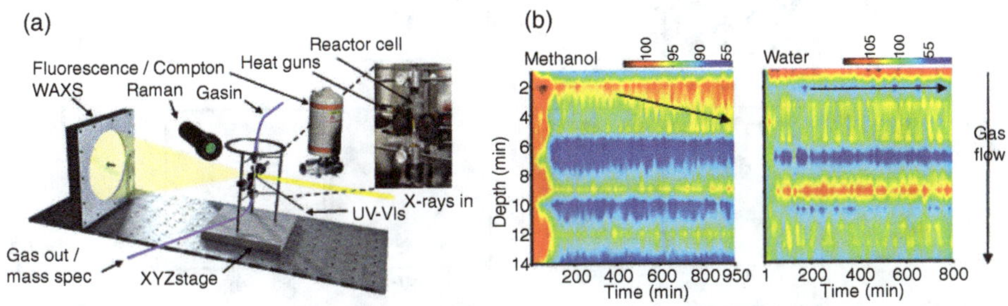

Figure 16.6. (a) A schematic of the novel multitechnique high-energy setup for spatiotemporal characterization studies on catalytic solids. The SR techniques WAXS, fluorescence, and Compton scattering were utilized, while in this experiment complementary Raman, UV-Vis, and online mass spectrometry were also used. The reactor consists of a large (5.5 mm ∅, 0.55 g) quartz reactor heated by air guns in an "open" architecture. (b) Spatiotemporal Compton scattering intensity plots through the depth of the reactor under both methanol and water vapor conditions over extended time periods. Reprinted from Reference 27, copyright John Wiley & Sons 2009.

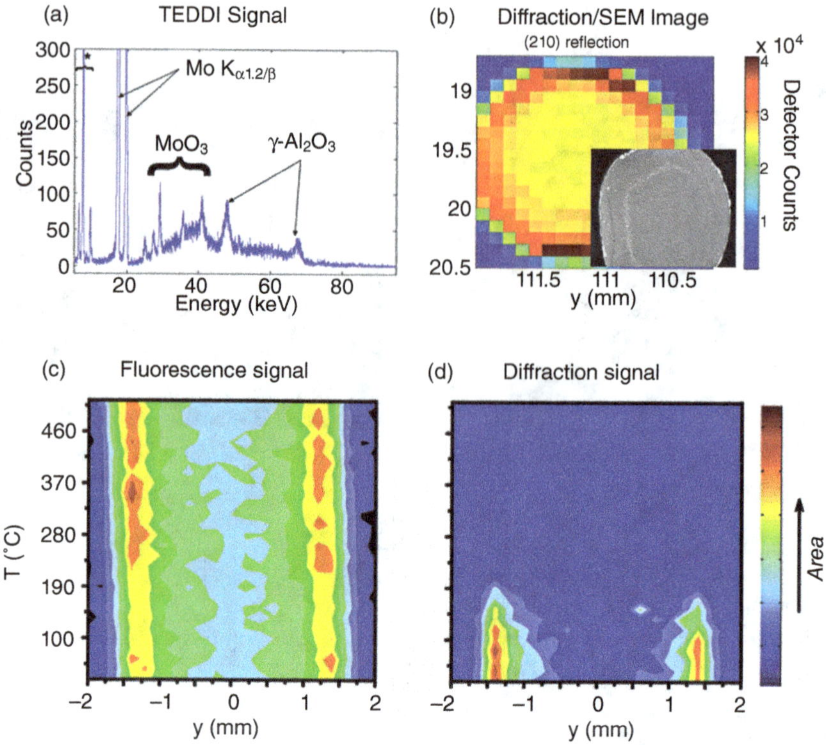

Figure 16.9. (a) TEDDI detector signal together with a typical (b) Y–Z diffraction map created from the (210) Bragg peak from the MoO_3 phase from a Mo/Al_2O_3 catalyst body after calcination at 500°C. (See text for full caption.)

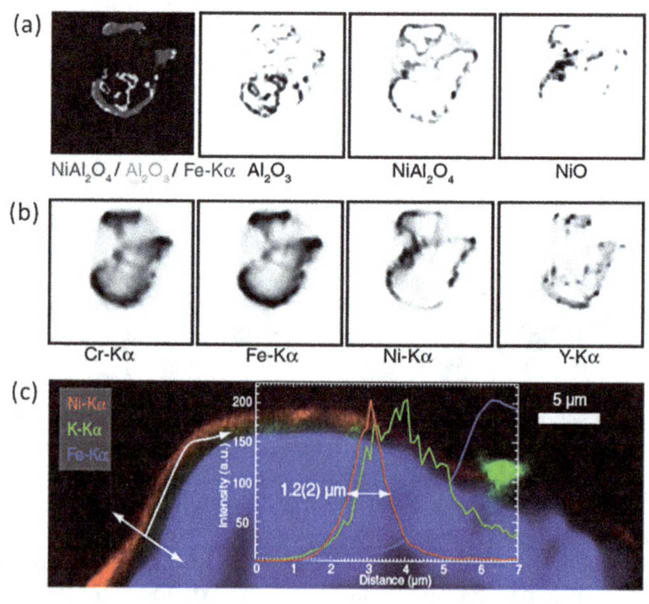

Figure 16.10. (a) Diffraction phase maps created from diagnostic diffraction peaks obtained by XRF/XRPD tomography measurements. (See text for full caption.)

9

INFRARED SPECTROSCOPY ON POWDER CATALYSTS

Eli Stavitski

9.1 INTRODUCTION

It has been widely accepted that rational design of catalytic materials is hardly possible without thorough and detailed understanding of the chemical transformations taking place during the catalyst life cycle. To achieve this level of knowledge, the complex, multilayered processes of catalyst preparation, activation, action, and regeneration should be monitored as they occur. Traditional spectroscopic approaches, namely, probing the catalyst before and after the process, often away from the reactor and at ambient conditions, termed *ex situ*, may offer only partial answers; therefore, considerable effort has been channeled into the development of spectroscopic tools capable of "live broadcasting" chemical information as the process unfolds. It would not be an exaggeration to say that almost every conceivable characterization tool, including, but not limited to, optical and vibrational spectroscopy, magnetic resonance, X-ray absorption, and photoelectron spectroscopy, has found its way to the field of catalysis to help realize the rational design approach which will eventually substitute the current modus operandi which is mostly driven by serendipity.

In the lineup of *in-situ* methods applied to tackle problems relevant to heterogeneous catalysis, infrared (IR) spectroscopy stands out for a number

In-situ *Characterization of Heterogeneous Catalysts*, First Edition.
Edited by José A. Rodriguez, Jonathan C. Hanson, and Peter J. Chupas.

of reasons. First, it has perhaps the longest history among the *in-situ* methods [1]; the first spectroscopic measurement of a catalyst that can be termed *in situ* has been performed using the IR technique [1, 2]. Since this pioneering work, more than five decades of continuous development have made IR spectroscopy one of the most ubiquitous methods in catalysis research. The method is unequivocally embraced by the community due to the combination of its extraordinary chemical sensitivity, relatively straightforward methodology, and affordable instrumentation. It is uncommon today to find a report on catalyst characterization that does not include IR spectroscopy results in one way or another. The wealth of information that can be obtained with *in-situ* IR spectroscopy is summarized in a number of comprehensive reviews in recent years [3–7]. For more information on specific topics, the reader is referred to the overview of *in-situ* methods covering active sites [7], surface species [6], solid–liquid interfaces [8], and IR/Raman imaging [9].

The goal of this chapter is to provide the reader with a general background on IR experimental aspects, briefly discussing the suitability of different methodologies for practical applications. Then we will focus on a number of case studies, which demonstrate the strength of the *in-situ* IR spectroscopy applications. We will also briefly touch upon very recent developments in the field, such as fast time-resolved studies and *in-situ* IR spectral imaging.

9.2 EXPERIMENTAL APPROACHES

IR spectroscopy probes the transitions between vibrational states of the absorber over an energy range of $0.496 > E > 0.0496$ eV (4000–400 cm^{-1}, commonly known as medium/mid infrared [MIR]) and $0.0496 > E > 1.24 \times 10^{-3}$ eV (400–10 cm^{-1}, known as far infrared [FIR]). Thus, the technique is able to probe the chemical and geometric structures of (adsorbed) molecules (molecular vibrational/rotational modes) and solids (lattice vibrations/acoustic modes or phonons). Therefore, in principle, both the chemical transformation in the catalyst itself and the reactant molecules can be sensed when the catalytic process is interrogated. Analysis of the spectra taken during the reaction can provide a number of important insights. For example, interaction of the adsorbate with catalyst surface sites perturbs the spectra of both; spectral changes can be interpreted to determine the strength of the interaction (see, e.g., the following discussion on probe molecules). The *in-situ* approach allows detecting transient species that only form under reaction conditions. Finally, accumulation of residues on the surface leading to degradation of catalytic performance can also be monitored directly.

First, IR spectrometers employed dispersive elements, for example, diffraction grids, to produce the spectra, which, when used with a single detection element, resulted in a substantial measurement time. In the early 1980s, the arrival of Fourier transform infrared (FTIR) spectrometers has revolutionized the field. These instruments allowed for short recording times (seconds to minutes) and high resolution (better than 0.001 cm^{-1}), adjustable to obtain the desired signal-to-noise ratio. Nowadays pricewise, the FTIR spectrometer is comparatively inexpensive and, as such, is one of the workhorses in both academic and industrial heterogeneous catalysis laboratories.

In general, there are three commonly employed measurement modes to probe catalytic reactions on powders under *in-situ* conditions (for a discussion of IR spectroscopic methods for studying surfaces of model catalysts, such as infrared reflection–absorption spectroscopy [IRRAS] and polarization–modulation infrared reflection–absorption spectroscopy [PM-IRRAS], see Chapter 8). Selection of the most appropriate method has to be made based, among other factors, on the desired local (surface/bulk) sensitivity as discussed by Baiker et al. [10]. The most traditional way of measuring an IR spectrum is a transmission measurement mode. The sample is prepared in the form of a self-supported pellet, such as potassium bromide, and then placed in the reaction cell equipped with the pair of IR transparent windows (CaF_2, KBr, and ZnSe are most common materials), heater, and gas injection system. When designing the experiment, one should take into account the IR transmission of the catalyst material since it is difficult to fabricate a sustainable pellet with a low amount of material. Moreover, depending on the fraction of the IR light absorbed by the material, the methods can be surface or surface/bulk sensitive. Another consideration is the gas flow profile in the cell: A dense pellet will prevent the gas transport through the bed.

A popular alternative to transmission measurements is diffuse reflectance infrared Fourier transform spectroscopy (DRIFTS) spectroscopy, which measures both light scattering and absorption phenomena. The DRIFT spectra contain mostly surface information because all reflected light arriving at the detector is scattered on the surface. It is particularly attractive because it does not employ complicated sample preparation approaches and therefore is often very close to realistic reaction conditions. On the other hand, the interpretation of DRIFT spectra is based on the phenomenological theory of Kubelka and Munk, and some precautions have to be taken when comparing transmission IR and DRIFT data. Commercial DRIFTS cells can accommodate catalytic beds a few millimeters thick; however, the penetration depth of the IR radiation may vary for different catalysts. Furthermore, to reduce specular reflectance from the sample, the powder should be finely ground, hindering the gas flow through the cell.

When designing the IR cell for transient studies, whether for transmission of DRIFTS measurements, one should be aware of the dead volume of the reactor. Even when the recording time of the spectra is short (fraction of a second is easily achievable with modern commercial instruments), slow mixing of gases will result in loss in the time resolution of the experiments. Details of the design and application of the ultra-low-volume cells for catalytic studies, allowing for the simultaneous IR measurements and gas analysis, can be found in Reference 11.

An approach based on total reflection of IR radiation is popular for exploring solid–liquid interfaces [8]. In an attenuated total reflection (ATR) measurement, the IR beam traverses through a crystal with high refractive index, for example, Ge, ZnSe, or diamond, and hits the surface at an angle greater than the critical angle. Under such conditions, an evanescent wave orthogonal to the interface is formed, which decays exponentially with distance. Absorption of the sample, which is in close contact with the crystal surface, can be measured in this manner; note that only a thin layer of the sample can be measured with the depth defined by the incidence angle, wavelength, and the refractive indices of both the crystal and the sample. For *in-situ* ATR measurements, a thin layer of the sample is deposited onto the crystal with the thickness less than the penetration depth of the evanescent field. Thin metal films are prepared by chemical or physical deposition, whereas powder samples could be deposited from slurry with subsequent drying. Even though there might be no strong adhesion between powder and the ATR crystal, the layer is often sufficiently stable to withstand shear tension induced by the solvent flow.

Since the ATR crystal comes into direct contact with the components of the reaction mixture, including catalyst particles, this imposes limitations on the materials used in terms of their chemical, mechanical, and thermal stability. A number of reaction cell designs, including batch and continuous flow systems, have been proposed for *in-situ* ATR experiments [12]. Immersion setups, where IR fiber optics-based elements allow probing of the content of regular chemical reactors or autoclaves [13] and ATR-based cells for monitoring multiphase reactors at pressures up to 200 bar, have been also reported [14].

In *in-situ* spectroscopy, the quantitative measurements are essential for a reliable kinetic interpretation of catalytic data. Extinction coefficients for the adsorbed molecules may deviate greatly from those for gas-phase species. The surface structure, for example, flat planes versus confined space in the zeolite pores, can affect the extinction coefficient even further. One of the potential approaches to accurately quantify IR spectroscopic response is by combining a microbalance and a spectroscopic cell (gravimetric analysis combined with infrared [AGIR] technique) [15]. By applying this method, the changes in the weight of the sample due to accumulation and desorption of reactants can be correlated with the changes in IR band intensities. With a coupled online analysis of the gases evolved from the reactor, a true *operando* approach can be realized, assessing catalytic activity and selectivity in real time.

9.3 MOLECULAR PROBE ACID/BASE PROPERTIES OF MOLECULAR SIEVES

Many industrial-scale chemical processes are based on acid/base reaction catalyzed by zeolite materials. Similar to other solid acids, zeolites possess both Brønsted and Lewis acid sites, which are typically hydroxyl groups and

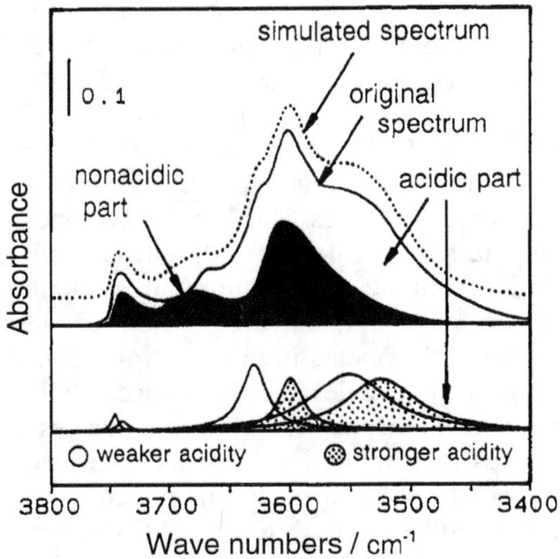

Figure 9.1. The hydroxyl region of the IR spectrum of ultrastable zeolite Y and its deconvolution into individual components. Reprinted with permission from Makarova, M. A., Dwyer, J. (1993) *J. Phys. Chem.*, 97, 6337. Copyright (1993) American Chemical Society.

coordinatively unsaturated cations, respectively. Depending on the chemical content, crystalline structure, postsynthetic treatments and, last but not the least, the state of hydroxylation under the reaction conditions, activity could be determined predominantrly by Lewis or Brønsted acidity, or both. In order to correlate the acidic properties of zeolites and their catalytic properties, it is essential to obtain quantitative information on the number, nature, location, and strength of the acidic sites.

IR spectroscopy has become an essential tool to investigate intermolecular interactions, for example, hydrogen and coordination bonding (Fig. 9.1 and References [16, 17]). It has been shown that the most valuable information on the zeolite acidic properties can be obtained using adsorbed probe molecules. Perturbations introduced to the IR spectra of both the surface groups and probe molecules upon the interaction, such as intensity variations and frequency shifts, can be translated into the properties of relevant acid sites. The extensive criteria for a selection of suitable probe molecules have been formulated in great detail (e.g., see References [17–19]). In general, small and weakly interacting molecules are recommended for probing surface properties of acidic and basic solids. Other criteria include (1) selectivity of interaction with specific surface sites; (2) low reactivity, even at catalytically relevant conditions; (3) detectable magnitude of spectral response upon interaction; (4) high and experimentally measurable extinction coefficients; and (5) whenever possible, the reactant itself shall be used as a probe molecule.

Zeolite hydroxyl groups typically exhibit two or more major bands in a 3800–3400 cm^{-1} spectral region. Surface silanol groups give rise to bands centered at 3710–3760 cm^{-1} (Fig. 9.1). External and internal silanols can be differentiated in the deconvoluted spectra [20]. Second, the IR band at ~3600 cm^{-1} is usually attributed to the O–H stretching mode of the Si(OH)Al bridged group. The latter is the most important chemical entity as it is typically associated with strong Brønsted acid sites. This band may appear asymmetric due to heterogeneity of the sites located at different crystallographic positions within the zeolite unit cell. In the cases when zeolite crystallites are defect rich, a broad band at 3460 cm^{-1}, assigned to hydrogen-bonded silanol groups located in the interior of the crystal, appears in the spectrum [21].

Several classes of probe molecules have been successfully used to characterize acidic properties of zeolites, namely, diatomic molecules (CO, N_2), nitrogen-containing organic bases (pyridine and its derivatives, nitriles) and hydrocarbons (ethene, benzene). CO forms hydrogen-bonded complexes with hydroxyl groups, which can be identified in both C–O and O–H spectral regions. Upon the H-band formation, the O–H stretching band is redshifted and broadened, with the magnitude of shift interrelated to the acidity of the site. At the same time, the C–O band is blue shifted compared to the gas-phase CO frequency. Pyridine, first proposed as a reporter molecule in as early as 1963 [22], leads to the similar effect on the O–H bands, whereas ring vibrations in the region of 1600–1400 cm^{-1} allow identification of pyridine bound to different surface sites. An especially advantageous approach involves using two probes: One is weakly basic, such as CO, to access the acidic strength of the sites and another is a strong base, for example, pyridine, to obtain information of their nature and location within the zeolite framework. Several comprehensive reviews covering investigation of zeolite acidic properties with probe molecules have been published in the 1990s [18, 19, 23].

When several acid sites differing in their nature and strength coexist in a zeolite material, they can be distinguished using a combination of two or more molecular probes. For example, two types of Brønsted sites were identified in mordenite, that is, hydroxyl groups located in the main channels and in the side pockets [24, 25]. The former are more acidic, as evidenced by the redshift upon CO adsorption, and are accessible to pyridine. Buzzoni et al. compared the acidity of several acidic zeolites using pyridine [26]. It was shown that for all the materials studied, pyridine reacts quantitatively with the available acid sites to form pyridinium species. The formation of pyridine dimers was also detected. Unlike zeolites featuring a three-dimensional pore system (ZSM-5 and β), mordenite showed a pore blockage upon pyridine adsorption. Moreover, the acid sites in the small side pockets of the mordenite channels were also shown to be inaccessible. This conclusion is confirmed by another study where only the partial protonation of pyridine on mordenite was observed [27]. It was inferred that pyridine could only slightly enter the side pockets, making protonation improbable due to sterical hindrances.

Pyridine derivatives are claimed to be advantageous over pyridine for probing weak Brønsted acid sites due to their stronger basicity and its weaker affinity for Lewis sites due to steric hindrances induced by the methyl groups [28]. In addition, several IR bands in the spectra of absorbed methylpyridines are sensitive to the acid site strength [28]. An interesting application of bulky pyridine derivatives is the assessment of acid sites with different accessibility. We will illustrate this point with a few examples. One of the important aspects of zeolite catalysis is to overcome the transport limitations that stem from the exclusively microporous character of these materials by introducing controlled mesoporosity. Such a treatment would unavoidably affect the acidic properties. A methodology for the characterization and quantification of the acidic sites with different accessibilities has been developed and applied for a series of dealuminated mordenites. These materials were probed with IR spectroscopy of adsorbed alkyl pyridines of increasing molecular dimensions, followed by the characterization of the nonaccessible sites using subsequently adsorbed CO [29], allowing for a step-by-step characterization of the nature and strength of sites with different accessibilities. It was shown that upon dealumination of the small-port mordenite samples, the acid sites in the side pockets become accessible for pyridine due to the partial destruction of side pockets upon dealumination. Subsequently, the formation of secondary mesoporosity systems due to further dealumination makes the zeolite crystals completely accessible to relatively bulky molecules, such as lutidine, collidine, and 2,6-di-tert-butylpyridine. Mordenite dealumination has been shown to cause destruction of Brønsted sites and the formation of new hydroxyl groups with lower acidity [30]. Combination of several probe molecules has been applied to desilicated and dealuminated zeolites [21, 31], hierarchical zeolites [32], partially dissolved and recrystallized mesoporous materials [33], and silicoaluminophosphates [34].

9.4 WATER–GAS SHIFT REACTION

Recently, the water–gas shift (WGS) reaction, $CO + H_2O \rightarrow CO_2 + H_2$ has attracted significant interest as a possible route toward hydrogen production in mobile applications, such as fuel cells [35–37]. Whereas current commercial technology ultimately involves multiple reaction steps, a number of effective single-stage WSG catalysts have been proposed, typically based on precious metals (Au, Pd, Pt) on oxide supports (titania, zirconia, ceria, or combinations thereof). It has been established that both metal and support actively partake in the catalytic process. Two reaction pathways were proposed: (1) formate formation from surface hydroxyl groups and CO and its subsequent decomposition and (2) ox–red mechanism where CO adsorbs on the metal and oxidizes at the metal–support interface, followed by reoxidation of the surface to produce hydrogen. Formation of the surface formate groups in WSG process

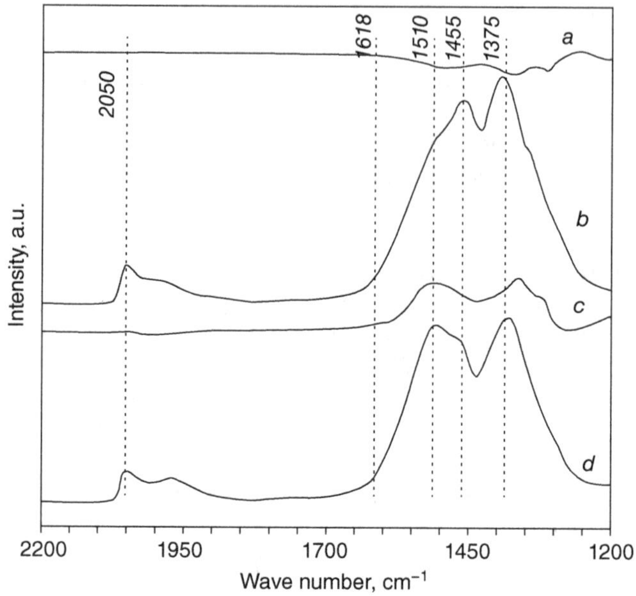

Figure 9.2. *In-situ* FTIR spectra at 300°C of Pt/CeO_2 catalyst; reduced in 10 vol% H_2/He flow (a), after CO pulses (b), after H_2O pulses (c), and *in-situ* WGS conditions (d) showing the formation of surface formates and carbonates. Reprinted with permission from Azzam, K. G., Babich, I. V., Seshan, K., Lefferts, L. *J. Catal.*, 251, 153. Copyright 2007 Elsevier.

on Pt/ceria was observed with *in-situ* DRIFTS [38]. Specifically, 1300 and 1580 cm^{-1} OCO vibrations and C–H stretching vibrations (2845 and 2950 cm^{-1}) were assigned to a bidentate formate. Systematic decrease in the surface formate band intensity under WGS conditions with increasing Pt loading was interpreted as an evidence for a faster decomposition of surface formats, and therefore their direct involvement in the reaction [38]. Lefferts et al. [39] observed that formation of hydrogen on Pt/CeO_2 is suppressed when formats are decomposed with N_2O without regenerating the surface hydroxyl groups, substantiating the formate route (Fig. 9.2). In contrast to the ceria support, no carbon species were detected on the surface of Pt/TiO_2 and Pt/ZrO_2 at WGS conditions, pointing out that the process involves red–ox regeneration step, that is, replenishment of surface hydroxyls via reaction with water molecules.

Isotopic exchange has been demonstrated as a powerful method for *in-situ* WSG studies [40]. Spectra interpretation can be significantly simplified when $^{12}CO/^{13}CO$ combination is used; moreover, fast switching between reactant feeds makes possible transient kinetic analysis of the products [41]. Based on the isotope exchange DRIFTS/kinetics study, Meunier et al. argue that formate species on the Au/ceria catalyst, observed in IR spectra, play a minor role in the overall process since their decomposition rate was found to be approximately 60 times slower than the formation of CO_2. Even in the case where isotope exchange in the formates was as rapid as in the CO_2 product, a fully quantitative analysis of the DRIFTS data shows that only 10% of the formate species account for the product formation [42]. In another study, metal–carbonyl bands due to adsorption of CO on Au surface can also be distinguished from the electronic transitions of partially reduced ceria [40].

9.5 CARBON MONOXIDE OXIDATION

While bulk noble metals are rather inert in the oxidation of carbon monoxide, nanoparticulate metals exhibit unusually high activity even at low temperatures. *In-situ* IR spectroscopy is instrumental in helping to gain insight into the different factors controlling the nanoparticle activity. The particular application capitalizes on the fact that the CO molecule is an extremely sensitive surface probe. The mode of interaction with the surface sites, that is, polarization of the molecule and σ- and/or π-back donation leads to a shift of the vibration frequency with respect to that of gaseous CO (2143 cm^{-1}). Since more than one factor determines the magnitude of the spectral shift, there is no simple correlation between the band position and the strength of the CO interaction with the surface.

Grunwaldt et al. characterized different active sites in the Au/TiO_2 and Au/ZrO_2 catalysts [43]. Upon adsorption of CO on the catalysts, two bands located at 2110–2120 and 2130 cm^{-1} were observed due to CO adsorbed on low-coordinated step/kink defect sites and positive polarized gold species, respectively. Au/TiO_2 exhibits a larger fraction of the former sites compared with Au/ZrO_2, which correlates with the CO oxidation performance. In the reaction conditions, the defect-related band is not observed, leading authors to conclude that the defect sites are responsible for the catalytic activity.

Structural changes of the Ru/MgO_2 and Ru/SiO_2 catalysts during the oxidation of CO were observed with DRIFTS under reaction conditions [44]. Under excess of oxygen, the spectra were dominated by two bands, at 2130 and 2070 cm^{-1}, and a weak shoulder at 2000 cm^{-1} (Fig. 9.3). These features were assigned to CO adsorbed on a RuO_2 surface layer, which is comparable to the RuO_2(110) single-crystal surface. In the reducing conditions, bands due to absorption on reduced and oxidized metal sites can be identified, at 2050 and 2130/2070 cm^{-1}, respectively. Based on these results, catalytic activity was correlated with the presence of partially reduced RuO_2 layer supported on metal nanoparticles. Further reduction of this layer, as well as the oxidation of nanoparticles to bulk RuO_2, led to the deactivation.

Another example of excellent sensitivity of the metal nanoparticle surface composition is the study of Pt–Ru catalysts, which shows high activity in the preferential oxidation of CO in hydrogen-rich feeds [45]. IR spectra of the CO

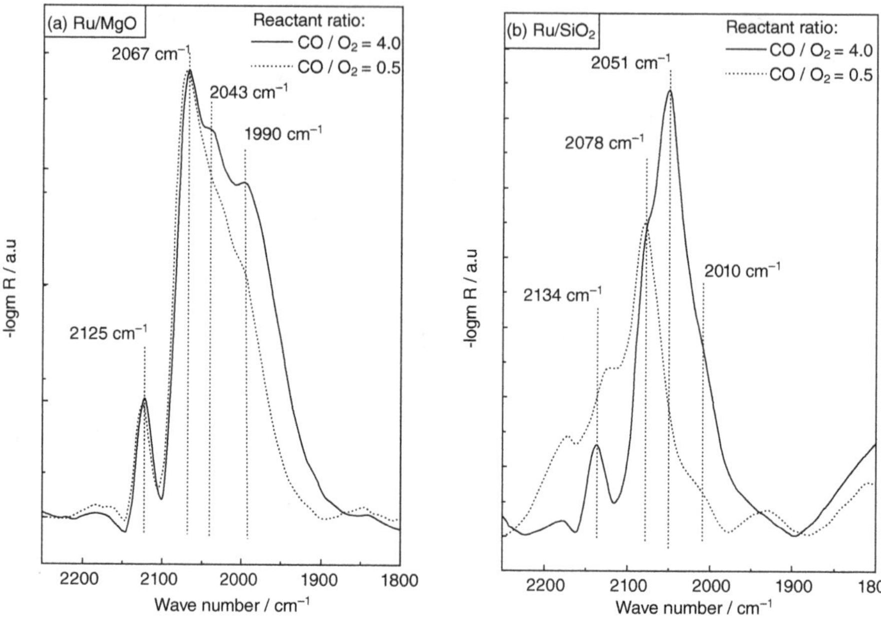

Figure 9.3. DRIFT spectra obtained during CO oxidation on Ru/MgO and (b) Ru/SiO_2 at 423 K in oxidative (dotted traces) and reducing (solid traces). Reprinted with permission from Assmann, J., Narkhede, V., Khodeir, L., Löffler, E., Hinrichsen, O., Birkner, A., Over, H., Muhler, M. (2004) *J. Phys. Chem. B*, 108, 14634. Copyright (2004) American Chemical Society.

adsorbed on the catalyst show a single band at 2061 cm^{-1}, as opposed to that of physical mixture of Pt and Ru nanoparticles, where two distinct bands can be discerned. Band position is slightly shifted compared to that found for monometallic Pt nanoparticles. These finding are consistent with the core–shell nanoparticle structure in which Ru core is coated with 1- to 2-monolayer-thick Pt.

9.6 METHANOL-TO-HYDROCARBON CONVERSION

Methanol can be produced from natural gas, coal, or biomass and further converted into fuel-range hydrocarbons or, alternatively, to polymer-grade olefins. The latter process, referred to as methanol-to-hydrocarbon (MTH) or methanol-to-olefin (MTO) conversion, utilizes acidic zeolites and alumino-phosphates, such as ZSM-5 and H-SAPO-43, as catalysts. FTIR spectroscopy assisted in identifying surface species formed during the MTO reaction. Bilbao et al. observed adsorption of methanol on hydroxyl groups of SAPO-34 in the form of methylcarboxonium ion, $CH_3OH_2^+$ [46]. These adsorbed species transform into surface methoxyl groups, $SiOCH_3Al$, which are characterized by the IR bands at 2868 and 1980 cm^{-1}. With the increase of the reaction temperature to 400°C, methoxy groups could not be detected, indicating their high reactivity. The authors proposed that the reaction of the surface methoxy groups with methanol occurs, resulting in the dimethyl ether (DME) and, subsequently, C–C bond formation. However, direct involvement of methoxy species and DME has been challenged based on NMR studies that detected formation of the aromatic hydrocarbons during the reaction. It is now commonly accepted that the MTH reaction proceeds via "hydrocarbon pool" mechanism, in which relatively large aromatic intermediates, confined in zeolite pores, are methylated by methanol/DME. Formation of end products occurs through the elimination of the alkyl side chains. Bordiga et al. reported the formation of entrapped hydrocarbon species in H-ZSM-5 (bands at 1690, 1621, and 1510 cm^{-1}) [47]. The aromatic species are highly methylated as concluded from the absence of aromatic C–H stretching bands. Bands at 1690 cm^{-1} are associated with C=C modes of polyaromatic species, while the weak absorption at 1620 cm^{-1} was assigned to carbonaceous hydrogen-deficient compounds ("coke"). Park and Seo systematically studied the composition of the catalytic scaffold formed in the course of MTO process on zeolites with different pore sizes and geometries [48]. The IR results show a clear correlation between the dimensions of the channel network and the degree of condensation of aromatic rings. In more detail, large pore zeolites, for example, faujasite, show formation of intense bands at 1300–1370 and 1570 cm^{-1} during the reaction, assigned to coke-like deposits, whereas in the case of MFI and chabazite these bands remain weak. These spectral features are associated with polyaromatic compounds, which are responsible for rapid deactivation of faujasite. On the other hand, bands at 1465 and 1590 cm^{-1} remain constant for MFA and chabazite, indicating steady concentration of methylbenzenes, which are essential intermediates in the MTO process [48].

9.7 NO_X STORAGE AND REDUCTION

Vanadia/titania (V_2O_5/TiO_2) catalysts are commonly used as selective catalytic reduction (SCR) catalysts for the removal of NO_x from automotive and industrial emissions by reduction with ammonia [49]. The catalyst is very active, with NO_x conversions >95% at space velocities below 10,000 N m^3/m^3 h at NH_3/NO_x ratios ~1. Time-resolved *in-situ* IR spectroscopy helps unraveling the details of the process. Despite the difficulty in differentiating between active intermediates and spectator species, *in-situ* IR studies combined with on line catalyst activity measurements (mass spectrometry), revealed that a correlation between the NO_x conversion and the concentration of Brønsted acid sites, associated with V–OH groups, exists. Moreover, V=O group also participate in the reaction. Absorption of NH_3 results in a redshift of the V=O overtone band (2040 cm^{-1}), indicating the weakening of the bond. It was concluded from

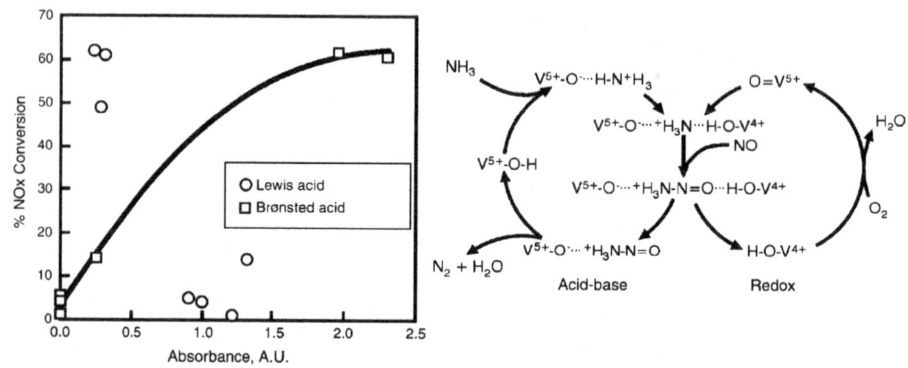

Figure 9.4. Left: NO_x conversion versus absorbance of NH_4^+ band at 1420 cm^{-1} and coordinated NH_3 (~1605 cm^{-1}). Right: proposed SCR catalytic cycle over V_2O_5/TiO_2. Reprinted with permission from Topsoe, N. Y. (2006) *Catal. Today*, 113, 58. Copyright 2006 Elsevier.

these studies that the catalytic cycle consisted of both acid and reduction/oxidation (redox) reactions and involved both surface V–OH (Brønsted sites) and V=O species. A formulated catalytic cycle for the $DeNO_x$ reaction is shown in Figure 9.4 [50]. Based on this proposal, microkinetic modeling could describe a large variety of industrial data satisfactorily, and this has helped both catalyst selection and process optimization [50].

NO_x catalytic reduction becomes challenging in the case of lean gasoline or diesel engines. Both engine types offer an improvement in terms of fuel economy; however, high concentration of oxygen in the exhaust makes it difficult to effectively utilize NO_x reductants. Supported Pt- and Pd-based catalysts were found to be very active in NO_x reduction with hydrocarbons and hydrogen in lean conditions. DRIFTS was employed to elucidate the mechanism of the reaction on Pt/Al_2O_3 and Pd/Al_2O_3 [51]. With H_2 as a reductant, formation of the hydrated nitrite species (1411 and 1335 cm^{-1}) was observed. In the case of $NO–CO–O_2$ reaction mixture, the spectra exhibit features due to bicarbonate and Pt-bound carbonyl species. In the case of Pd/Al_2O_3 exposed to $NO–CO–O_2$, more complicated spectra were obtained, with contributions from isocyanate species adsorbed on alumina, nitrate, nitrite carbonate, and bicarbonate species. With both reductants present in the gas feed, the carbonyl bands due to CO adsorbed on Pd in different configurations were redshifted, possibly due to electron donation from chemisorbed H_2 to Pd and, consequently, an increase in π-donation to CO. In addition, formate species were detected. Based on time-resolved measurements, the authors conclude that formates, formed through the reaction of the Pd-bound CO with hydroxyl group of the support are spectator species, whereas isocyanate groups are hydrolyzed under reaction conditions to yield NH_x species which are very active toward NO_x reduction.

Another potential strategy for nitric oxide abatement is a cyclic process of NO_x storage on alkaline metal oxides followed by rich periods when it is released on the noble metal component of the catalyst. Analysis of the surface chemistry on a model $Pt/BaO/Al_2O_3$ catalyst has been widely used to shed light onto the mechanism of the process [52, 53]. It was established that NO, which is the most abundant component of NO_x, can be adsorbed on the BaO in the form of nitrites, which can be identified from their band at 1210 cm^{-1}, which is slowly oxidized to nitrates (1320 and 1420 cm^{-1}) [52]. A comparison of the catalyst and BaO/Al_2O_3 support upon exposure to NO/O_2 mixtures shows that Pt is essential to promote nitrite to nitrate oxidation through a spillover process. Two different nitrate species, namely, bidentate and monodentate, can be identified, which are formed on Al and Ba sites, respectively [53]. The former is not bound strongly and does not play a role in the overall process, whereas the latter is retained even at high temperatures. Both nitrite and nitrate species associated with the Ba component partially decompose upon exposure to hydrocarbons. In the absence of Pt, nitrite groups formed exclusively are not reduced by hydrocarbons, suggesting activation of the reducing agent during the fuel-rich cycle.

9.8 HYDRODESULFURIZATION CATALYSTS

Hydrodesulfurization (HDS), or hydrotreating, is the process of sulfur removal from fossil fuels in order to reduce poisoning of the noble metal catalysts and SO_x emissions. HDS catalysts are typically based on MoS_2 supported on alumina promoted by either Co or Ni. IR spectra of the catalyst, with increasing amounts of Mo present [54], show the gradual disappearance of the alumina hydroxyl bands, indicating that Mo interacts strongly with the Al–OH groups, with the preferential interaction with most basic Al–OH groups (3771 cm^{-1}). The corresponding analysis revealed that there is a strong tendency for Mo to form single-layer structures on the support surface. After sulfiding to produce the active MoS_2 phase, the IR spectra show that monolayer coverage is retained, but single-slab Mo sulfide structures may be more randomly distributed over the Al_2O_3 surface, which is in line with the X-ray absorption and electron microscopy results [54].

IR studies of the adsorption of NO and CO as probe molecules on HDS catalysts can be very informative (Fig. 9.5) [55–58]. The IR results show that NO adsorbs on edge sites of MoS_2 slabs (1786 and 1699 cm^{-1}). Moreover, from the NO band the adsorption on Mo could be easily distinguished from that on the promoter atoms (Co or Ni) [56]. When the promoter concentration increases, the number of Mo edge sites available for NO is reduced, indicating that Co is atomically dispersed at the MoS_2 nanocrystallite edges.

The sensitivity of the NO IR frequencies to the nature of the adsorption site is vital in discriminating between active and inactive surface species (Fig. 9.5) [57, 58]. There was no relationship found between HDS activity and the

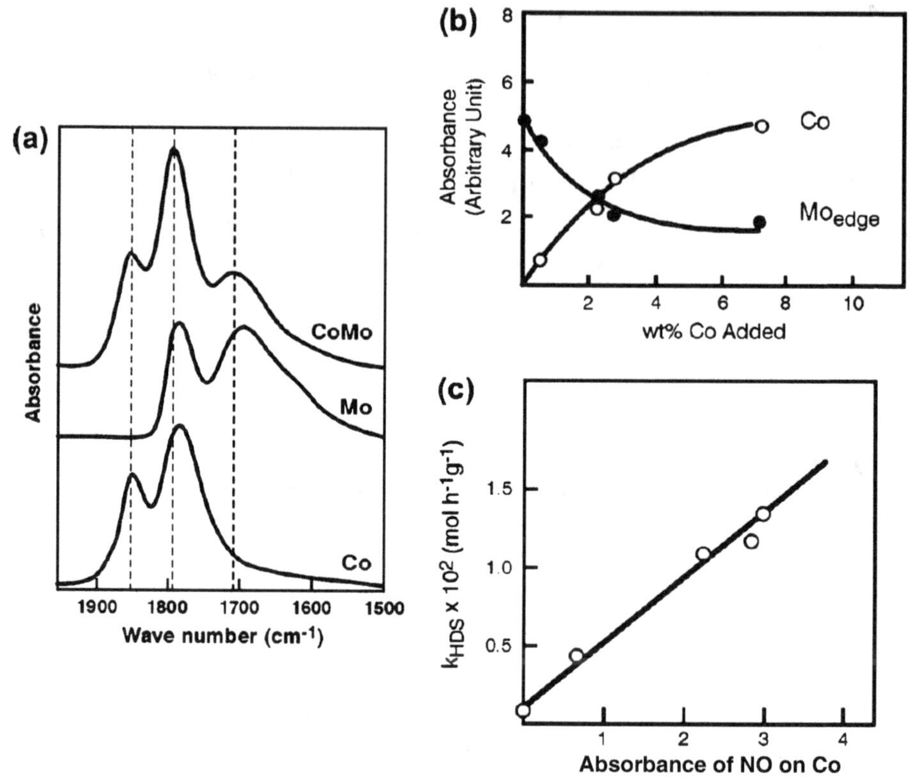

Figure 9.5. (a) IR spectra of NO adsorbed on sulfided 2% Co/Al_2O_3, 8% Mo/Al_2O_3, and 2% Co 8% Mo/Al_2O_3. (b) Absorbance of IR bands of NO adsorbed on Co (1850 cm^{-1}) and Mo (1690 cm^{-1}) plotted as a function of Co/Mo ratio at constant Mo loading. (c) Thiophene HDS activities versus absorbance of IR band of NO adsorbed on Co. Reprinted with permission from Topsøe, N.-Y., Tuxen, A., Hinnemann, B., Lauritsen, J. V., Knudsen, K. G., Besenbacher, F., Topsøe, H. (2011) *J. Catal.*, 279, 337. Copyright 2011 Elsevier.

total NO absorption; however, the fraction of NO adsorbed on the promoter site appears to correlate with the activity [57]. For example, for the sulfided Ni–Mo/Al_2O_3 catalyst, three bands due to adsorbed NO appear, at 1840, 1784, and 1685 cm^{-1}, whereas only the first band can be observed in the case of Ni/Al_2O_3. With an increase of the Ni content, the Ni-bound NO band intensity rises, accompanied by the reduction of the available Mo sites. The Ni–NO dependence on the Ni concentration is in line with thiophene HDS activity. Assignments of the band were verified recently by comparison with the DFT calculations for various possible adsorption complexes on promoted MoS_2 slabs [58].

9.9 SELECTIVE OXIDATION OF ALCOHOLS AND HYDROCARBONS

Selective oxidation of organic molecules on transition metal oxides has been a subject of intense scrutiny in the recent years. IR spectroscopy has been used extensively to characterize molecular structures formed under *in-situ* conditions. One should mention that the most promising approach in this application is the tandem of IR and Raman spectroscopies [59]; the latter is capable of directly probing oxide structures which do not exhibit intense or distinct IR spectral signatures or when the strong adsorption of the support interferes.

One of the most thoroughly studied catalysts for selective oxidation is supported vanadium oxide, V_2O_5. In the catalyst supported on titania and ceria, IR bands due to terminal V=O groups are observed at 1030 cm^{-1}, as well as the bridged V–O–V modes at ~930 cm^{-1} [59]. Besides fundamental modes, relatively weak and broad V=O overtone modes can be identified in the 1900–2100 cm^{-1} region. Upon exposure to methanol at 100°C, intensities of both V=O and V–O–V bands decrease; the former undergoes a redshift to ~1020 cm^{-1}, along with its overtone. The original band positions recover upon increase of the temperature to 300°C. Under reaction conditions, adsorbed methoxy species give rise to the bands in C–H stretch region (2930–2940 and 2815–2860 cm^{-1}). From the comparison with support materials, all bands were assigned to V–O–CH_3 groups. This assignment is verified by the analysis of the bands in the C–O spectral region. Raising the temperature leads to the decrease of the methoxy species band intensities and the progressive blue shift of their position. Changes in the V_2O_5 IR signals are consistent with the lengthening of the V=O bonds due to coordination or hydrogen bonding to adsorbed surface methoxy species.

In addition to the chemisorbed methanol, another set of IR bands observed upon adsorption on supported MoO_3 and WO_3 can be assigned to nondissociatively bound species. Absorption at 2959, 2852, and 2760 cm^{-1} was assigned to methanol on surface Lewis acid sites, which are abundant in these oxides [60]. Wachs et al. developed a method to elucidate densities of surface active sites for different metal oxides by quantification of IR spectra of different adsorbed methanol species [60] .

Multicomponent metal oxides are also widely used as oxidation catalysts. The complexity of their structure during the reaction leads to a continuous debate on the nature of the active sites. As an example, iron vanadate mixed oxide for methanol oxidation was investigated by Wachs et al. [61]. Upon methanol absorption on bulk V_2O_5 and Fe_2O_3, supported V_2O_5/Fe_2O_3 and $FeVO_4$, the IR bands due to CH_3–O–Me, and intact CH_3OH species can be detected. A comparison of band positions shows that in the case of mixed metal catalysts (V_2O_5/Fe_2O_3 and $FeVO_4$), V–OCH_3 and V–CH_3OH are predominantly formed, suggesting that the surface is enriched with VO_x species due to the high mobility of V^{5+} cations.

Another mixed metal oxide system, supported vanadia/niobia, was found to be active in the oxidation of propylene to acrolein, one of the top-ranked industrial chemical intermediates [62]. This reaction has been shown to follow the so-called Mars–van Krevelen mechanism, where the propylene methyl hydrogen is abstracted to form a symmetric allyl species that react with the bulk oxide oxygen. At room temperature, propylene is chemisorbed on the surface as σ- and π-complexes as can be seen from the bands at 1607 and 1100 cm^{-1}, respectively. With increase of the temperature, adsorbed propylene species are transformed into allyl species, which are the most abundant intermediates under reaction conditions. In the absence of oxygen, none of the IR bands due to allyl species are present, suggesting that gaseous oxygen is vital for the catalytic activity. It implies that the V^{5+} species have to be kept fully oxidized in order to catalyze the selective oxidation of propylene.

A number of studies have been devoted to the performance of vanadia catalysts in oxidation of aromatic molecules [63]. Besselmann et al. investigated adsorption and conversion of toluene on titania-supported V_2O_5 [63]. Since anatase is very absorbing in the range of V–O band vibrations, overtone bands were used as indicators for the transformations of the catalyst. Upon absorption of toluene, the bands at 2020 and 1972 cm^{-1} decrease due to the reduction of large V_2O_5 crystallites. If polymeric vanadia species are removed from the catalyst by washing with ammonia, one can detect an interaction of monomeric vanadyl species with toluene from the reduction of the band at 2048 cm^{-1}. Moreover, the surface hydroxyl groups form hydrogen bonds with adsorbed species, as inferred from the changes in the 3700–3600 cm^{-1} region. As to toluene, the aromatic system remains intact as can be seen from the ring vibrations at 1566, 1577, 1501, and 1475 cm^{-1}. Appearance of the bands due to CH_2 group at 2932 and 2877 cm^{-1} points out at the oxidation in the side chain of toluene toward aldehydic species upon absorption. Consecutive heating in inert or oxidative atmosphere leads to further oxidation of the adsorbate and the formation of the carboxylate bands (1540 and 1413 cm^{-1}). When heating is conducted in air, the vanadyl bands recover their intensity, indicating partial reoxidation of the catalyst. Differences in carboxylate band appearance between the untreated and washed catalyst were rationalized in terms of different absorption geometries on monomeric and polymeric/crystalline vanadia sites. These observations support the conclusion that the former is essential for the oxidative absorption of toluene, whereas further oxidation is promoted by the latter due to capability of fast oxygen insertion.

Catalytic oxidation of chlorinated benzenes, a minor yet very harmful component of waste incineration products on vanadia catalysts, has been studied by Lichtenberger and Amiridis [64]. FTIR spectra of *m*-dichrobenzene (m-DCB) adsorbed on the titania support and V_2O_5/TiO_2 catalyst at 250°C suggest the formation of surface phenolate species characterized by 1278 cm^{-1} (C–O stretching); 1252, 1158, and 1127 cm^{-1} (C–H in plane bending); and 1575 and 1468 cm^{-1} (C=C stretching vibration). No ring opening reaction can be detected. In the case of catalyst, additional bands can be detected, namely, surface maleates (bands at 1515, 1436, and 1162 cm^{-1}) and acetates (1444 and 1551 cm^{-1}). Upon exposure of the preadsorbed (m-DCB) on V_2O_5/TiO_2 to He/O_2 mixture, the surface species are oxidized further, giving rise to chlorinated acetates and/or acetyl halides, such as CH_3COCl^- or CH_2ClCOO^-, which can be detected from IR signals at 1616 and 1254 cm^{-1}. Based on the FTIR results and kinetic studies, the mechanism of the chlorinated benzenes on vanadia catalysts has been proposed [64], including (1) dissociative adsorption on the surface through nucleophilic substitution of chlorine atoms; (2) electrophilic substitution of partially chlorinated species leading to adsorbed benzoquinone; (3) further oxidation to maleates, acetates, and aldehydes; and finally, (4) formation of the gas-phase products (CO_2, H_2O, and HCl).

Oxidation of *o*- and *p*-chlorophenol on 3.6 wt% V_2O_5/TiO_2 catalyst appears to proceed through similar reaction steps [65]. Upon absorption, surface phenolates are formed through hydroxyl group coordination, which is a much faster process compared to nucleophilic substitution identified for chlorinated hydrocarbons [64]. At low temperatures, the rate of aromatic ring opening remains relatively low, leading to a high concentration of highly reactive phenolate species on the catalyst surface, making possible formation of the bulkier by-products through condensation, coupling, and chlorination/dechlorination reactions.

9.10 PROBING THE SOLID–LIQUID INTERFACE WITH ATR SPECTROSCOPY

When studying liquid-phase reactions with classical IR spectroscopy, one should always deal with high absorption of the solvent. For instance, less than 20 μm of water may completely obscure other bands in the 3600 and 1600 cm^{-1} regions, if the experiment is carried out in transmission; even careful background subtraction may cause distortions in the bands of interest. As discussed earlier, ATR methodology helps to overcome this limitation by probing a thin layer of medium in close vicinity to the internal reflection element (IRE). In what follows, several successful applications of ATR to problems in catalysis are discussed.

Mul et al. investigated photocatalytic activity of TiO_2–ZrO_2 in selective oxidation of cyclohexane to cyclohexanone, which is an important commercial product, used to obtain caprolactam for Nylon-6 production [66]. Nanocrystalline $Ti_{1-x}Zr_xO_2$ catalyst prepared by microemusltion method was spread on the ZnSe crystal. Upon cyclohexane adsorption on the catalysts, two new bands, 1157 and 1090 cm^{-1} compared to the liquid reagent, were registered, which correspond to the molecules absorbed on the catalysts surface. After UV irradiation, photocatalytic degradation of cyclohexane can be inferred from the decrease of the peaks at 1449, 1257, 1162, and 1128 cm^{-1}; formation of cyclohexanone leads to the bands at 1692 and 1713 cm^{-1}, which correspond to adsorbed and desorbed molecules, respectively. Products of further oxidation, that is, carbonates and carbonates, are also evident. Once the assignment of the band was complete, performance of the catalysts with different zirconia contents could be evaluated by following the evolution of the band intensities. Cyclohexanone production was found to decrease with increasing Zr content. This trend is explained by increasing hydrophilicity and acidity with Zr doping; adsorption of the nonpolar reagent on a hydrophilic surface is hindered. Furthermore, strong water adsorption hampers O_2 adsorption, favoring the electron–hole recombination process. The formed cyclohexanone is retained on the acidic surface, allowing for deeper oxidation and leading to catalyst deactivation.

In a recent report by the same group, accurate mechanistic information on cyclohexane oxidation over anatase was obtained by using $^{16}O_2$–$^{18}O_2$ isotope exchange [67]. When the catalyst was illuminated in 5% $^{18}O_2$/He gas-phase environment for 0, 15, 30, and 60 min, respectively, to prepare surfaces with variable amounts of Ti–^{18}OH active sites, almost no changes in the ATR spectra were registered. In the following oxidation in $^{18}O_2$, contributions of cyclohexanone with red-shifted carbonyl vibrations (1657 and 1642 cm^{-1} vs. 1690 and 1679 cm^{-1}) can be detected with a maximum ratio of about 1:1 between the nonlabeled and labeled products. When the catalyst was not exchanged prior to the reaction, mainly unlabeled products were obtained, suggesting that the oxygen in cyclohexanone originates from TiO_2-associated active sites and not from the dissolved oxygen. Based on these findings, the authors postulate that cyclohexyl hydroperoxide, observed spectroscopically, is a spectator species rather than a reaction intermediate as was commonly assumed, and the reaction proceeds via a Mars–van Krevelen mechanism.

Burgi and Baiker have explored the potential of modulation excitation ATR spectroscopy [68]. In this method, the chemical system is perturbed by periodically varying an external parameter such as temperature, pressure, or concentration of a reactant. If the species in the system are affected by this parameter, their properties will also change periodically at the same frequency as the perturbation, or harmonics thereof. The time-resolved response of the system is acquired by means of lock-in, or phase-sensitive detection, that is, extracting the signal at a given frequency filtering out direct-current (DC) components and noise. Typically, the concentration of the reagent in the flow cell mounted on the IRE is modulated with the period of several hundreds of seconds. To illustrate the strength of this approach, selective oxidation of 4-methoxy-6-methyl-2-pyrone to 5,6-dihydro-4-methoxy-6-methyl-2-pyrone was carried out on the Pd/TiO_2 catalyst. First, time-resolved ATR spectra were recorded and then the phase-sensitive correction scheme was applied to the spectral dataset. Subtracting the spectra of the reagent and product from the resulting spectra allows isolation of the spectra due to adsorbed species. Analysis of the phase resolved data also makes it possible to distinguish the bands belonging to different species in a crowded spectrum, since the transitions due to the same species vanish at the same phase settings.

9.11 ULTRA-FAST TIME-RESOLVED STUDIES

In order to follow the kinetics of a catalytic process, one can opt for a steady-state measurement or transient analysis. In a traditional approach, changing the reagent concentration will affect the population of the surface species and distribution of the reaction products. Correlation of these parameters allows determination of the kinetic interrelations governing the process. However, with the advances in modern technology, direct identification of short-lived intermediates has become feasible [69–71]. Very high time resolution can be achieved by using the "step-scan" technique, where single points in the interferogram are recorded after a triggering event, such as gas pulse or laser excitation. The spectra are obtained by Fourier transformation of the recomposed interferogram. Since thousands of points in the interferogram have to be taken, the process has to be highly reproducible.

Transient IR studies make it possible to pinpoint relevant surface intermediates, using pulse feeding of the reagents. For example, investigation of ethylene hydrogenation on an alumina-supported Pt catalyst has demonstrated the involvement of the CH_3–CH_2–Pt group [69]. With step-scan measurements with 25 μm resolution, the decay of the surface ethyl was found to correspond to the rise time of the ethane. A fraction of the product appears faster than the time resolution of the experiments, indicating the presence of more reactive ethyl species; heterogeneity of the surface accounts for this nonuniformity. In the case of carbon monoxide hydrogenation on a Ru catalyst, analysis of the band due to absorbed CO has shown different modes of adsorbate binding on the surface, with relative concentration varying in time. The linear CO site was shown to be catalytically relevant as a precursor for the products, methane and CO_2 [70].

Photocatalytic transformations can also be investigated with the step-scan methods, using pulsed radiation to initiate the reaction. Using ATR makes it possible to study reactions on solid–liquid interfaces. Mul et al. report on the photo-oxidation of cyclohexene of the vanadia supported on mesoporous material (TUD-1) [72]. Time-resolved measurements reveal the transient state in which the catalytic vanadyl site interacts with the reactive cyclohexenyl hydroperoxide intermediate. In another example, high time resolution (25 ns) assisted in identifying individual steps of the deNO_x reaction on an alumina-supported silver catalyst [71]. Femtosecond laser excitation served primarily to heat the sample rapidly and to trigger the process. Transformation of Ag-bound cyanate to isocyanate absorbed on alumina support underscores the importance of the nanoparticle–support interface.

9.12 IR MICROSPECTROSCOPY

Out of convenience and due to technical restrictions, the spot size in most IR instruments is maintained in the range of 0.1–1 cm, which is optimal for sizable powdered samples. When *in-situ* experiments are carried out in this manner, chemical information is obtained from the entire sample, therefore averaging it over whole the population of catalytic sites. It is, however, important to probe distinct areas on the catalytic surface, for example, crystalline faces, since they can exhibit varying catalytic activities and/or selectivities [73]. Correlating the structure of the surface in the individual regions of the active particles with their performance can help to advance our understanding of catalytic phenomena.

IR micro(spectro)scopy is the established method with widespread applications in life sciences [74]. A combination of an FTIR spectrometer and all-reflective focusing optics allows collecting high-quality spectra from areas as small as tens of microns. The first application of IR microscopy for catalysis has been reported by Schüth [75], where large silicalite-1 crystals were loaded with p-xylene and the spectra with polarized radiation were taken from the individual crystals. When the light polarization was aligned perpendicular to the crystal long axis, strong IR bands appeared; these transitions vanished at parallel orientation. Analysis of the vibration modes for p-xylene shows that the transition moment associated with these vibration modes changes along the long axis of the molecule. From these results it was inferred that the adsorbate molecules exclusively align parallel to the straight pores of the zeolite. Furthermore, diffusion characteristics can be evaluated from the spatially and time-resolved measurements of gas dynamics in microporous materials [76]. Diffusion boundaries associated with the surface of the crystals or intracrystalline features, such as intergrowth, can also be identified [77, 78].

Nowotny et al. report on *in-situ* investigation of the thermal decomposition of the template in single MFI crystals [79]. These organic compounds are added to zeolite crystallization solution to act as structure-directing agents in order to sustain the channel architecture of micro- and mesoporous materials. After preparation, the template has to be removed to vacate the pores for adsorbates and/or reactants, usually by high-temperature treatment. As-synthesized zeolites at ambient temperatures showed broad lattice bands as well as the vibrations identical to the IR spectra of tetrapropylammonium ions in solution. The presence of asymmetric CH deformation bands similar to those in the solid state indicated relatively low mobility of the template molecules. Upon heating, the ratio of the band intensities corresponding to CH_3 and CH_2 groups decreased, pointing out at the elongation of the hydrocarbon chains due to partial oligomerization. Mobility of the ions increases in the temperature range up to 613 K as evidenced by the CH deformation region. Above 633 K, the CH stretching vibrations intensified with remarkable differences between samples with varying Al concentrations. Further on, the template molecules decompose via Hoffman elimination, that is, removal of the alkyl chain in the form of alkene, and formation of tripropyl amine in Al-deficient zeolite samples or tripropyl ammonium cation in Al-enriched zeolite samples due to the interaction with acid sites. Further decomposition follows the same route leading to the formation of alkenes and ammonia.

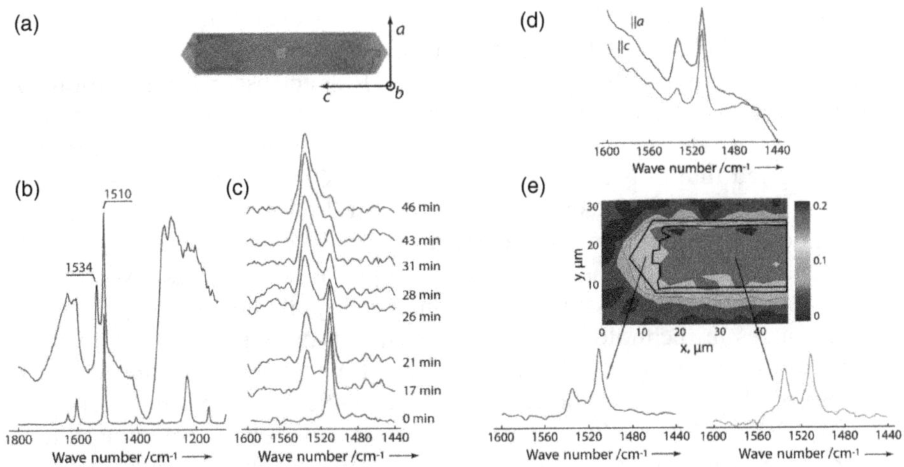

Figure 9.6. (a) Optical microphotograph of the H-ZSM-5 crystal after reaction with 4-fluorostyrene with the square probed area marked. (b) IR spectrum of an individual H-ZSM-5 crystal contacted with 4-fluorostyrene (top) and the IR spectrum of liquid 4-fluorostyrene (bottom). (c) Region of the IR spectra (1440–1600 cm^{-1}) taken *in situ* during the 4-fluorostyrene oligomerization reaction; the spectra are background corrected. (d) *In-situ* IR spectra taken with two different light polarizations. (e) Intensity of the IR band at 1534 cm^{-1} mapped over the crystal after reaction and the IR spectra taken form the edge and the body of the crystal, demonstrating differences in the intensity ratio of the bands. (See color insert.)

IR microscopy benefits greatly from replacing the thermal source radiation with synchrotron light. Despite the comparable total flux, the extremely high brilliance of the synchrotron radiation allows improvement of the spatial resolution of the method by an order of magnitude, nearing diffraction-limited values (3–10 µm, depending on the wavelength). The first successful demonstration of *in-situ* synchrotron IR microscopic measurements was aimed at the investigation of styrene oligomerization on an acidic zeolite crystal [80]. Formation of the dimeric and trimeric species was detected in a spatially resolved manner, and the concentration profile of the product was correlated with the orientation of porous systems in the crystal segments. Intrinsically polarized synchrotron light made possible the observation of the vibrational linear dichroism effects in the spectra of the oligomers due to the molecular alignment imposed by the zeolite channels (Fig. 9.6). Finally, the exact chemical nature of the transient species was identified by the comparison of the experimental band positions and DFT calculations [81].

9.13 CONCLUDING REMARKS

As the reader can infer from the case studies discussed earlier, IR spectroscopy is an exceptionally powerful method capable of delivering a wealth of relevant chemical information on the catalytic process. However, one should exercise caution when planning an IR experiment or interpreting the results. As early as in 1984, Peri, one of the pioneers of the IR spectroscopy application for catalysis, expressed his view that identification of truly active species is unlikely since their concentration is below the detection limit of the technique [82]. It is often argued that only spectator species, that is, the intermediates that are formed in parallel to the reaction of interest but do not participate directly in it, could be detected, leading to incorrect mechanistic conclusions.

However, the method has made enormous progress in the past two decades and the multitude of successful applications has dispelled the initial pessimism. First, advances in technology permit carrying out an experiment at relevant temperatures and pressures. Instrumentation for the *in-situ* or *operando* IR studies is readily available and relatively affordable. Verifying the interpretation of the IR results with kinetic experiments and modeling, including isotope exchange experiments, has become ubiquitous. IR spectroscopy has been also combined with other techniques, including Raman spectroscopy and X-ray absorption spectroscopy and diffraction (see Reference [83] and Chapter 15). Complementary information obtained in such a combined *in-situ* experiment provides a comprehensive view on the physicochemical transformations associated with catalytic reaction.

REFERENCES

[1] Eischens, R. P., Plisken, W. A., Francis, S. A. (1954) *J. Chem. Phys.*, 22, 1786.

[2] Eischens, R. P., Pliskin, W. A. (1958) The infrared spectra of adsorbed molecules, in Eley, D. D., Frankenburg, W. G., Komarewsky, V. I. (eds.), *Advances in Catalysis and Related Subjects*, Vol. 10, Academic Press, New York, p. 2.

[3] Ryczkowski, J. (2001) *Catal. Today*, 68, 263.

[4] Zecchina, A., Spoto, G., Bordiga, S. (2005) *Phys. Chem. Chem. Phys.*, 7, 1627.

[5] Stavitski, E., Weckhuysen, B. M. (2010) Vibrational spectroscopy and related in situ studies of catalytic reactions within molecular sieves, in Cejka, J., Corma, A., Zones, S. I. (eds.), *Zeolites and Catalysis: Synthesis, Reactions and Applications*, Wiley-VCH, Weinheim, Germany, p. 209.

[6] Lamberti, C., Zecchina, A., Groppo, E., Bordiga, S. (2010) *Chem. Soc. Rev.*, 39, 4951.

[7] Vimont, A., Thibault-Starzyk, F., Daturi, M. (2010) *Chem. Soc. Rev.*, 39, 4928.

[8] Andanson, J.-M., Baiker, A. (2010) *Chem. Soc. Rev.*, 39, 4571.

[9] Stavitski, E., Weckhuysen, B. M. (2010) *Chem. Soc. Rev.*, 39, 4615.

[10] Roedel, E., Urakawa, A., Kureti, S., Baiker, A. (2008) *Phys. Chem. Chem. Phys.*, 10, 6190.

[11] Lesage, T., Verrier, C., Bazin, P., Saussey, J., Daturi, M. (2003) *Phys. Chem. Chem. Phys.*, 5, 4435.

[12] Bürgi, T., Baiker, A. (2006) *Adv. Catal.*, 50, 227.

[13] Mul, G., Hamminga, G. M., Moulijn, J. A. (2004) *Vib. Spectrosc.*, 34, 109.

[14] Schneider, M. S., Grunwaldt, J. D., Burgi, T., Baiker, A. (2003) *Rev. Sci. Instrum.*, 74, 4121.

[15] Gilson, J., Fernandez, C., Thibault-Starzyk, F. (2009) *J. Mol. Catal. A: Chem.*, 305, 54.

[16] Hunger, M. (2005) *Microporous Mesoporous Mater.*, 82, 241.

[17] Paukshtis, E. A., Yurchenko, E. N. (1983) *Russ. Chem. Rev.*, 52, 242.

[18] Lercher, J. A., Gründling, C., Eder-Mirth, G. (1996) *Catal. Today*, 27, 353.

[19] Knozinger, H., Huber, S. (1998) *J. Chem. Soc. Faraday Trans.*, 94, 2047.

[20] Makarova, M. A., Dwyer, J. (1993) *J. Phys. Chem.*, 97, 6337.

[21] Holm, M. S., Svelle, S., Joensen, F., Beato, P., Christensen, C. H., Bordiga, S., Bjorgen, M. (2009) *Appl. Catal. A*, 356, 23.

[22] Parry, E. P. (1963) *J. Catal.*, 2, 371.

[23] Lavalley, J. C. (1996) *Catal. Today*, 27, 377.

[24] Maache, M., Janin, A., Lavalley, J. C., Joly, J. F., Benazzi, E. (1993) *Zeolites*, 13, 419.

[25] Datka, J., Gil, B., Kubacka, A. (1997) *Zeolites*, 18, 245.

[26] Buzzoni, R., Bordiga, S., Ricchiardi, G., Lamberti, C., Zecchina, A., Bellussi, G. (1996) *Langmuir*, 12, 930.

[27] Bevilacqua, M., Alejandre, A. G., Resini, C., Casagrande, M., Ramirez, J., Busca, G. (2002) *Phys. Chem. Chem. Phys.*, 4, 4575.

[28] Oliviero, L., Vimont, A., Lavalley, J. C., Sarria, F. R., Gaillard, M., Mauge, F. (2005) *Phys. Chem. Chem. Phys.*, 7, 1861.

[29] Nesterenko, N., Thibault-Starzyk, F., Montouillout, V., Yuschenko, V., Fernandez, C., Gilson, J., Fajula, F., Ivanova, I. (2004) *Microporous Mesoporous Mater.*, 71, 157.

[30] Armandi, M., Bonelli, B., Garrone, E., Ardizzi, M., Cavani, F., Dal Pozzo, L., Maselli, L., Mezzogori, R., Calestani, G. (2007) *Appl. Catal. B*, 70, 585.

[31] Datka, J., Sulikowski, B., Gil, B. (1996) *J. Phys. Chem.*, 100, 11242.

[32] Thibault-Starzyk, F., Stan, I., Abelló, S., Bonilla, A., Thomas, K., Fernandez, C., Gilson, J., Pérez-Ramírez, J. (2009) *J. Catal.*, 264, 11.

[33] Ordomsky, V. V., Murzin, V. Y., Monakhova, Y. V., Zubavichus, Y. V., Knyazeva, E. E., Nesterenko, N. S., Ivanova, I. I. (2007) *Microporous Mesoporous Mater.*, 105, 101.

[34] Martins, G. A. V., Berlier, G., Coluccia, S., Pastore, H. O., Superti, G. B., Gatti, G., Marchese, L. (2007) *J. Phys. Chem. C*, 111, 330.

[35] Korotkikh, O., Farrauto, R. (2000) *Catal. Today*, 62, 249.

[36] Song, C. S. (2002) *Catal. Today*, 77, 17.

[37] Ratnasamy, C., Wagner, J. P. (2009) *Catal. Rev. Sci. Eng.*, 51, 325.

[38] Jacobs, G., Graham, U. M., Chenu, E., Patterson, P. M., Dozier, A., Davis, B. H. (2005) *J. Catal.*, 229, 499.

[39] Azzam, K. G., Babich, I. V., Seshan, K., Lefferts, L. (2007) *J. Catal.*, 251, 153.

[40] Daly, H., Ni, J., Thompsett, D., Meunier, F. C. (2008) *J. Catal.*, 254, 238.

[41] Meunier, F. C., Reid, D., Goguet, A., Shekhtman, S., Hardacre, C., Burch, R., Deng, W., Flytzani-Stephanopoulos, M. (2007) *J. Catal.*, 247, 277.

[42] Meunier, F. C., Goguet, A., Hardacre, C., Burch, R., Thompsett, D. (2007) *J. Catal.*, 252, 18.

[43] Grunwaldt, J.-D., Maciejewski, M., Becker, O. S., Fabrizioli, P., Baiker, A. (1999) *J. Catal.*, 186, 458.

[44] Assmann, J., Narkhede, V., Khodeir, L., Löffler, E., Hinrichsen, O., Birkner, A., Over, H., Muhler, M. (2004) *J. Phys. Chem. B*, 108, 14634.

[45] Alayoglu, S., Nilekar, A. U., Mavrikakis, M., Eichhorn, B. (2008) *Nat. Mater.*, 7, 333.

[46] del Campo, A. E. S., Gayubo, A. G., Aguayo, A. T., Tarrio, A., Bilbao, J. (1998) *Ind. Eng. Chem. Res.*, 37, 2336.

[47] Palumbo, L., Bonino, F., Beato, P., Bjorgen, M., Zecchina, A., Bordiga, S. (2008) *J. Phys. Chem. C*, 112, 9710.

[48] Park, J. W., Seo, G. (2009) *Appl. Catal. A*, 356, 180.

[49] Roy, S., Baiker, A. (2009) *Chem. Rev.*, 109, 4054.

[50] Topsoe, N. Y. (2006) *Catal. Today*, 113, 58.

[51] Macleod, N., Lambert, R. M. (2003) *Appl. Catal. B*, 46, 483.

[52] Elizundia, U., López-Fonseca, R., Landa, I., Gutiérrez-Ortiz, M., González-Velasco, J. (2007) *Top. Catal.*, 42–43, 37.

[53] Su, Y., Amiridis, M. D. (2004) *Catal. Today*, 96, 31.

[54] Topsøe, N. Y., Topsøe, H. (1993) *J. Catal.*, 139, 631.

[55] Topsøe, N. Y., Topsøe, H. (1982) *J. Catal.*, 75, 354.

[56] Mauge, F., Duchet, J. C., Lavalley, J. C., Houssenbay, S., Payen, E., Grimblot, J., Kasztelan, S. (1991) *Catal. Today*, 10, 561.

[57] Topsøe, N.-Y., Topsøe, H. (1983) *J. Catal.*, 84, 386.

[58] Topsøe, N.-Y., Tuxen, A., Hinnemann, B., Lauritsen, J. V., Knudsen, K. G., Besenbacher, F., Topsøe, H. (2011) *J. Catal.*, 279, 337.

[59] Burcham, L., Deo, G., Gao, X., Wachs, I. (2000) *Top. Catal.*, 11–12, 85.

[60] Burcham, L. J., Briand, L. E., Wachs, I. E. (2001) *Langmuir*, 17, 6164.

[61] Routray, K., Zhou, W., Kiely, C. J., Wachs, I. E. (2010) *ACS Catal.*, 1, 54.

[62] Zhao, C., Wachs, I. E. (2006) *Catal. Today*, 118, 332.

[63] Besselmann, S., Löffler, E., Muhler, M. (2000) *J. Mol. Catal. A: Chem.*, 162, 401.

[64] Lichtenberger, J., Amiridis, M. D. (2004) *J. Catal.*, 223, 296.

[65] Hetrick, C. E., Lichtenberger, J., Amiridis, M. D. (2008) *Appl. Catal. B*, 77, 255.

[66] Hernández-Alonso, M. D., Almeida, A. R., Moulijn, J. A., Mul, G. (2009) *Catal. Today*, 143, 326.

[67] Almeida, A. R., Moulijn, J. A., Mul, G. (2011) *J. Phys. Chem. C*, 115, 1330.

[68] Bürgi, T., Baiker, A. (2002) *J. Phys. Chem. B*, 106, 10649.

[69] Wasylenko, W., Frei, H. (2005) *J. Phys. Chem. B*, 109, 16873.

[70] Wasylenko, W., Frei, H. (2007) *Phys. Chem. Chem. Phys.*, 9, 5497.

[71] Thibault-Starzyk, F., Seguin, E., Thomas, S., Daturi, M., Arnolds, H., King, D. A. (2009) *Science*, 324, 1048.

[72] Mul, G., Wasylenko, W., Hamdy, M. S., Frei, H. (2008) *Phys. Chem. Chem. Phys.*, 10, 3131.

[73] Weckhuysen, B. M. (2009) *Angew. Chem. Int. Ed.*, 48, 4910.

[74] Miller, L. M., Dumas, P. (2006) *Biochim. Biophys. Acta*, 1758, 846.

[75] Schüth, F. (1992) *J. Phys. Chem.*, 96, 7493.

[76] Lehmann, E., Chmelik, C., Scheidt, H., Vasenkov, S., Staudte, B., Kärger, J., Kremer, F., Zadrozna, G., Kornatowski, J. (2002) *J. Am. Chem. Soc.*, 124, 8690.

[77] Chmelik, C., Varmla, A., Heinke, L., Shah, D. B., Kärger, J., Kremer, F., Wilczok, U., Schmidt, W. (2007) *Chem. Mater.*, 19, 6012.

[78] Chmelik, C., Kärger, J., Wiebcke, M., Caro, J., van Baten, J. M., Krishna, R. (2009) *Microporous Mesoporous Mater.*, 117, 22.

[79] Nowotny, M., Lercher, J. A., Kessler, H. (1991) *Zeolites*, 11, 454.

[80] Stavitski, E., Kox, M. H. F., Swart, I., de Groot, F. M. F., Weckhuysen, B. M. (2008) *Angew. Chem. Int. Ed.*, 47, 3543.

[81] Stavitski, E., Pidko, E. A., Kox, M. H. F., Hensen, E. J. M., van Santen, R. A., Weckhuysen, B. M. (2010) *Chem. Eur. J.*, 16, 9340.

[82] Peri, J. B. (1984) Infrared spectroscopy in catalysis research, in Anderson, J. R., Boudart, M. (eds.), *Catalysis, Science and Technology*, Vol. 5, Springer-Verlag, West Berlin, Germany, p. 171.

[83] Newton, M. A., van Beek, W. (2010) *Chem. Soc. Rev.*, 39, 4845.

10

STRUCTURAL CHARACTERIZATION OF CATALYSTS BY *OPERANDO* RAMAN SPECTROSCOPY

GERHARD MESTL AND MIGUEL A. BAÑARES

> The frequency differences determined from the spectra, the width and character of the lines appearing in them, and the intensity and state of polarization of the scattered radiations enable us to obtain an insight into the ultimate structure of the scattering substance.
>
> —C.V. Raman, Nobel award lecture, 1930

10.1 INTRODUCTION

Catalysis is the key technology in chemical industries for the conversion of raw material streams into intermediates or chemical end products. It is the method of choice for environmental remediation of toxic emissions and finds increasing application in energy generation among other applications. The importance of catalysis in chemical industries is the motivation for the determination of relationships between catalytic activity and catalyst structure, being one of the keystones for further technological improvements. Spectroscopic techniques for the characterization of catalysts in the working state are powerful in this respect because they provide fundamental information about catalyst structures, including their surfaces.

In-situ *Characterization of Heterogeneous Catalysts*, First Edition.
Edited by José A. Rodriguez, Jonathan C. Hanson, and Peter J. Chupas.

Catalysis is a complex multidisciplinary science that cannot be fully understood without applied spectroscopy. Spectroscopy is thus the enabling tool for knowledge-based design of highly efficient and stable catalysts. A recent thematic issue was published in *Chemical Society Reviews* on the advances of *in-situ* characterization of catalysts [1]. The reaction environment shapes the state of a working catalyst leading to an information gap between its structures as determined before or after operation (*ex situ*), in controlled atmosphere (*in situ*), or under real working conditions. The need for characterization of catalysts during operation has been emphasized and demonstrated by many authors [2–15]. Real-time spectroscopy during operation is the choice to bridge the gap [12, 16–19] and is named *operando* methodology, first proposed in the literature in 2002 [11, 20]. The term *operando* is Latin for "working" or "operating" and is used to emphasize the simultaneous evaluation of both catalyst structure and catalytic performance, that is, reaction kinetics that match those obtained in an ideal reactor [3, 9, 11, 12, 20, 21].

Among the spectroscopic techniques that can provide information about the catalytic active sites during reaction, vibrational spectroscopies (Raman and infrared [IR]) provide most detailed molecular information in all phases (gas, liquid, solid, and their mixtures) and over a wide range of temperatures and pressures [12–14, 19, 21–25]. Many reviews summarize applications, developments, and opportunities for Raman spectroscopy in catalysis under *in-situ* and *operando* conditions [6, 14, 16, 21, 22, 26–34] among others, which were summarized in Reference 12, and the more recent review was made by Wachs and Roberts [25]. This chapter essentially deals with the most recent progress in the *operando* characterization of catalysts at work.

10.2 EQUIPMENT FOR *OPERANDO* RAMAN SPECTROSCOPY

In-situ and *operando* studies should be distinguished, as *in-situ* Raman cells are not necessarily designed as catalytic reactors. *In-situ* cells thus might be limited by diffusional constrains, insufficient gas flow patterns, and significant flow, temperature, and pressure gradients, which render them inadequate to deliver catalytic data. The key concept of the operando methodology is to simultaneously obtain reliable structure and catalytic data, making it possible to assess structure–performance relationships. Hence, an *operando* spectroscopic cell has to be a catalytic reactor meeting the necessary design specifications, such as (1) absence of mass or energy transfer limitations and (2) absence of bypassing or back-mixing, for example, characterized by dead volume or preferential paths for gas flow [12]. Mass transfer effects have to be eliminated to determine the reaction kinetics and activation energies [35]. Already in 1990, Hill and coworkers reported a detailed design of a Raman cell to study iron molybdate catalysts during methanol oxidation to formaldehyde in a cell designed to minimize void volume and temperature gradients; thus, activity data were representative [36].

10.2.1 Solid–Gas Interaction

Through the years, a number of cells have been developed for advanced *in-situ* characterization of catalysts under controlled conditions (pressure, temperature, reactants) and even during reaction. As far as we know, these started in 1977, when Grasselli et al. [37, 38] reported a heated Raman cell to monitor transformations in molybdate catalysts under reaction treatment. Designing cells became much simpler with the introduction of Raman microscopes and fiber optic probes, which work in backscattering (180°) [12]. Bañares and Mestl summarized the cells that have been used to monitor catalysts during reaction [12]. Some highlights are briefly mentioned here. Wachs's group modified the rotating sample design and used it to investigate supported oxides during selective oxidation of alkanes [39–42] and of methanol [39, 43]. Through the years, leading groups have reported Raman spectra assessing fundamental molecular information about the formation of surface carbonaceous species during reaction [4, 44] or kinetic studies during NO_x removal [45–47].

Bañares's group has compared the performance of conventional reactors with *operando* reactors to confirm conversion–selectivity values [48, 49] as well as to determine activation energies [50]. Meunier has recently reviewed these aspects on several *operando* and *in-situ* cells [51]. The Raman reactor designed by Mestl and coworkers [52, 53] includes a fiber optic Raman probe and a double-catalyst bed configuration allowing the investigation of the catalyst at higher contact times. This concept was expanded to a so-called multi-*operando* setup in which eight *operando* reactors are used simultaneously with sequential Raman–gas chromatography/mass spectrometry (GC)/mass spectrometry (MS) analyses of each reactor [54]. In an independent approach, Zhang reports the use of another multiple *operando* setup analyzing simultaneously several catalyst beds [55].

10.2.2 Gas Phase

Whereas most Raman investigations in catalysis evaluate the nature of the catalyst and surface intermediates, it is also possible to monitor reactions in gas or liquid phase. The gas phase is several orders of magnitude less dense than solids or liquids; hence, the Raman signal is accordingly weaker, but it is possible to monitor gas-phase Raman signals in a wide pressure range even at a few millibars [56]. Raman spectroscopy has been used to investigate the combustion of methane in the 4- to 16-bar range at 500–1000°C [57]. Woo and Hill already constructed such autoclave reactors in 1982 [58, 59], and in 1985 they reported the characterization of a catalyst during operation in autoclave reactors [60]. Ever since, many developments have been achieved; these are already reported elsewhere [12].

10.2.3 Supercritical Condition

The absence of strong signals by CO_2 and H_2O in Raman spectra allows investigation of catalysis under supercritical conditions. Grünwaldt et al. reviewed cell designs for spectroscopic experiments under supercritical conditions [61]; a remarkable feature is that these cells are typically designed with minimum void volume to deliver reliable kinetics.

10.3 EFFECT OF WATER ON CATALYST RECONSTRUCTION

Water is ubiquituous: it is present in the environment, it is fed in many reactions, and it forms during reaction. Thus, the presence and role of water on the reaction and on the catalyst structure cannot be neglected. This is of particular relevance for supported metal oxides, which are widely used in the industry for chemical and environmental processes [19, 62–64]. Lewandowska et al. reported on the role of water vapor, partial pressure, and temperature in the early stages of the hydration process of dehydrated, low-loading vanadia–titania catalysts using Raman spectroscopy, density functional theory (DFT), and molecular dynamics calculations [65] (Fig. 10.1). They demonstrated that water coordinates to supported vanadia hydrolyzing the V–O–Ti bridging oxygen bonds, resulting in supported $OV(OH)O_2$ monomer with two V–O–Ti bonds, a V=O, a V–OH bond, and high affinity for water. This structure is consistent with the presence of Brønsted acidity along with redox properties. The OVO_3 pyramids typically described in literature are exclusively stable under absolutely dehydrated conditions. Calculated harmonic vibrations predict the experimentally observed redshift of the V=O mode upon progressive coordination with water [65], thus explaining the vanadyl Raman mode redshift already reported by Jehng et al. in 1996 [66].

10.4 PEROXIDE SPECIES

Li and Oyama already reported the presence of peroxide species during the oxidation of ethanol by ozone on manganese oxide catalysts [67]. They used transient Raman experiments at low temperatures (–196 to –73°C) [67, 68]. Tian and Ren reported on the use of surface Raman spectroscopy in the study of electrochemical interfaces [69]. Surface-enhanced Raman spectroscopy (SERS) is the technique of choice for *in-situ* investigation of oxygen species during electrochemical reaction. Li et al. applied SERS to transition metal electrodes, monitoring the C=O and Pt–C Raman bands during electrooxidation on Au/Pt core/shell film electrodes [70]. Au/Pt core/shell nanoparticle film electrodes afford more than two orders of magnitude of SERS enhancement than electrochemically roughened Pt electrodes.

Bell and coworkers monitored peroxide oxygen species on a gold electrode [71]. Raman bands at 815–830 cm^{-1} evidence the presence of surface-bound OOH species at potentials where molecular oxygen is released from the electrode [71]; this study suggests that Au–OOH species are the precursors to

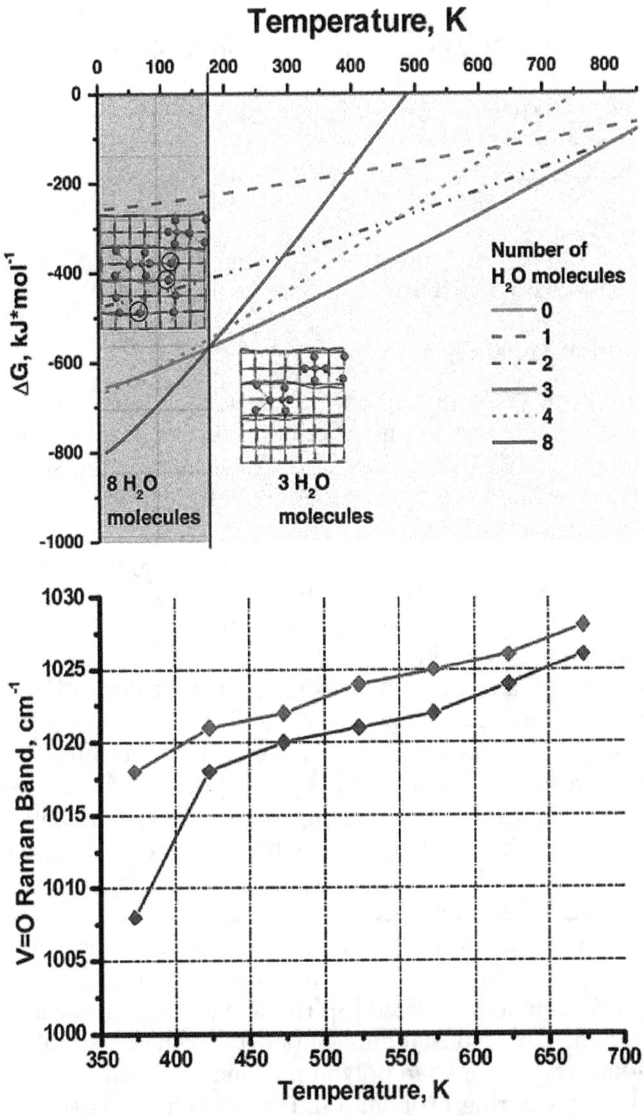

Figure 10.1. Top: Free energy ΔG of the supported vanadium oxide species on the titania surface as a function of temperature for a water partial pressure equivalent to the ambient air (1500 Pa). At the above-mentioned conditions, the structure with eight water molecules is the most stable until $T = 180$ K, followed by the structure containing three water molecules. Bottom: Raman V=O mode versus temperature for titania-supported vanadia (1.2 V atoms/nm^2) in dry (green trace) and humid (blue trace) air. Reprinted with permission from Lewandowska, A. E., Calatayud, M., Tielens, F., Bañares, M. A. (2011) Dynamics of hydration in vanadia-titania catalysts at low loading: a theoretical and experimental study, *J. Phys. Chem. C*, 115, 24133–24142. Copyright (2011) American Chemical Society. (See color insert.)

molecular oxygen, supporting conclusions from theoretical studies about oxygen evolution (Fig. 10.2) [72], thus being probably the first experimental evidence.

Recently, Raman spectroscopy has been used to monitor the concentration of H_2O_2 in selective oxidation reactions under supercritical conditions [73]. Figure 10.3 illustrates the Raman spectra recorded during this experiment.

10.5 CATALYSIS BY OXIDES

10.5.1 Supported Metal Oxides

Raman spectroscopy of supported oxides has been extensively reviewed elsewhere [12, 14, 21]. Here, we comment only on recent developments. For supported oxides, only a small fraction of the surface vanadia sites is reduced during the alkane oxidative dehydrogenation (ODH) reaction. Raman spectra recorded during ethane ODH do not show appreciable reduction under typical reaction conditions [11, 44]. More reducible oxide systems (e.g., alumina-supported chromia) exhibited partial reduction during reaction [3, 74–77]. The hydrocarbon-to-O_2 ratio determines the oxidation state of a catalyst, but the presence of additional reactants may further affect this property.

The time-of-flight values for alkane ODH strongly depend on the specific oxide support (titania > ceria > zirconia > niobia > alumina > silica) and do not correlate with the variations in the strength of the terminal V–O bond as measured by the respective Raman shifts [78, 79]. Actually, alkali metal doping of supported vanadia results in weakening of the terminal vanadyl bond, while reactivity and reducibility decrease, as confirmed experimentally by Raman–temperature-programmed reaction (TPR) experiments [3, 80] and by Raman–DFT modeling [81, 82]. Raman spectra of alumina-supported vanadia catalysts show that the population ratio of polymeric-to-isolated surface vanadia species decreases during reduction, whereas the total activity and selectivity in propane ODH essentially remain unaffected [3]. This suggested that the active sites for propane or ethane ODH on alumina-supported vanadia should be isolated surface vanadia sites, rather than polymeric vanadia species. The constancy of time-of-flight values during propane ODH with vanadia coverage on titania or zirconia [83] underlines the relevance of the bridging V–O–M support bond in the kinetically significant reaction step.

The higher reactivity of alkenes compared to alkanes leads to different states of the catalyst during reaction. The selective oxidation of propene to acrolein on supported vanadia catalysts leads to a more extensive reduction of vanadium sites than during alkane ODH. Zhao and Wachs investigated V_2O_5/Nb_2O_5 catalysts during propene oxidation to acrolein and detected a new Raman band at 978 cm^{-1}, which may be due to surface V^{4+} species [84]. This is due to the greater reducing power of alkenes relative to alkanes, leading to a dependence of reaction rates on O_2 partial pressure.

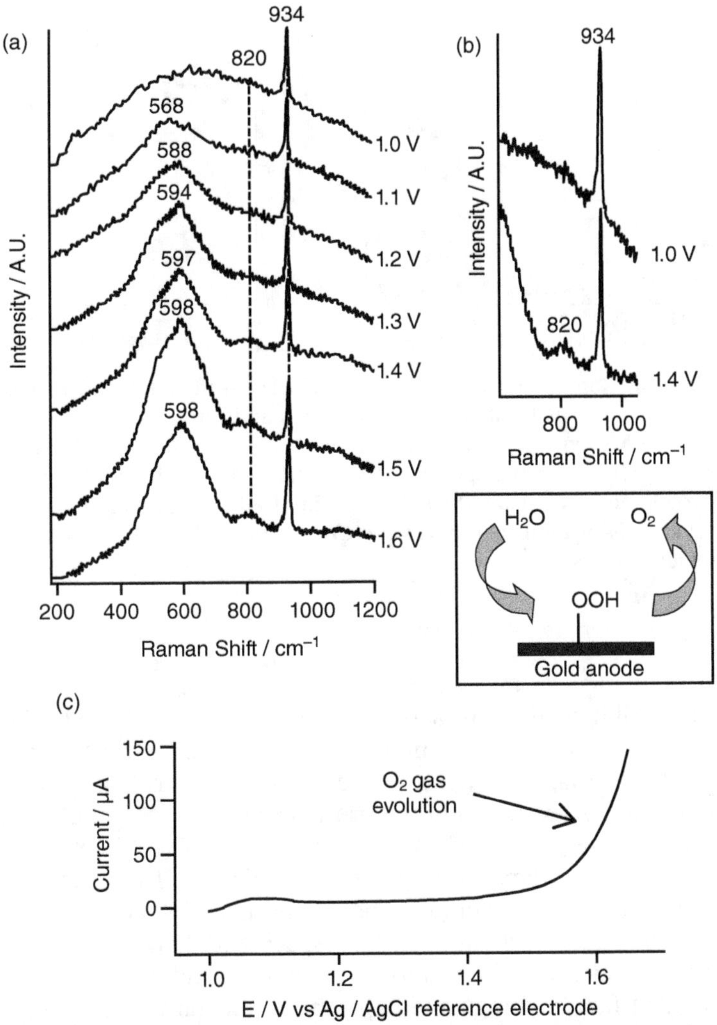

Figure 10.2. (a) Surface-enhanced Raman (SER) spectra of a gold electrode in 1 M $HClO_4$ during a linear sweep voltammetry scan from 1.0 to 1.65 V (2 mV/s). Each spectrum was acquired over 4 s at the beginning of the electrode potential indicated on the right of each spectrum. (b) Zoom-in of the spectra taken at 1.0 and 1.4 V. The insert shows the representation of the surface-bound OOH. (c) Current–potential curve measured during the linear sweep voltammetry. Reprinted with permission from Yeo, B. S., Klaus, S. L., Ross, P. N., Mathies, R. A., Bell, A. T. (2010) Identification of hydroperoxy species as reaction intermediates in the electrochemical evolution of oxygen on gold, *ChemPhysChem*, 11, 1854–1857. Copyright 2010 John Wiley & Sons.

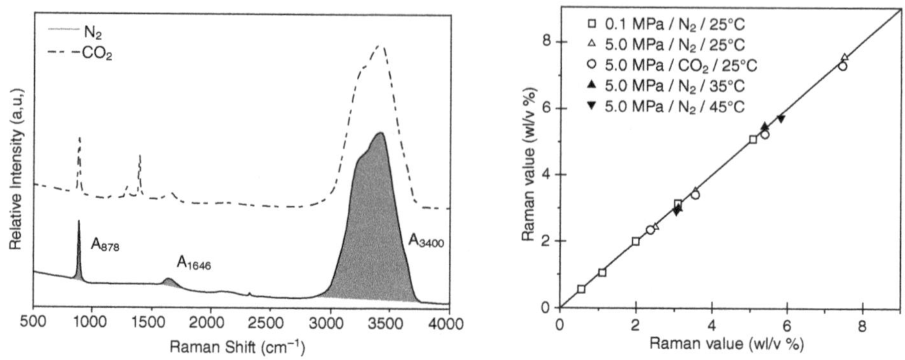

Figure 10.3. Left: Raman spectra of 2.5 wt/vol% H_2O_2 in water and N_2/CO_2. Right: Parity plot of data versus fit at different pressures and temperatures. Reprinted with permission from Moreno, T., Morán López, M. A., Huerta Illera, I., Piqueras, C. M., Sanz Arranz, A., García Serna, J., Cocero, M. J. (2011) Quantitative Raman determination of hydrogen peroxide using the solvent as internal standard: online application in the direct synthesis of hydrogen peroxide, *Chem. Eng. J.*, 166, 1061–1065. Copyright (2011) Elsevier.

Alcohols exhibit strong interaction with supported oxides. Bronkema and Bell used Raman spectroscopy combined with extended X-ray absorption fine structure (EXAFS) and X-ray absorption near-edge structure (XANES) to investigate methanol oxidation in the presence and absence of O_2 [85]. Formaldehyde decomposes to H_2 and CO in the absence of O_2, in line with observations of methanol reactivity on supported chromia [86]. The structural change of supported molybdena generates a new band near 760 cm^{-1} being assigned to partially reduced molybdenum oxide species [83]. The states of surface molybdenum species on silica during methanol oxidation depend on the molybdenum surface coverage and exposure to reactants.

Reducing conditions may rearrange supported oxides into segregated bulk domains. That was demonstrated using TPR–Raman for silica-supported vanadia close to its dispersion limit loading coverage [34]. During oxidation catalysis on supported oxides, a redox cycle is commonly observed; this imposes a reduction–reoxidation cycle on the catalyst, which leads to different average oxidation states, depending on the catalyst and on the specific reaction. This is not only the case for single supported oxides such as dispersed vanadia that transforms into segregated crystalline V_2O_5 [34]. But when more than one component is supported, and these possess structural reactivity, they may blend into segregated bulk mixed oxides out of their corresponding dispersed phases. This is the case in the propane ammoxidation reaction on alumina-supported antimony and vanadium oxides. The propane ammoxidation feed, which also contains ammonia (propane, ammonia, and air), results in a more reduced catalyst than that existing during propane ODH reaction (propane and air) [20]. This reduction to V^{4+} during ammoxidation is the driving force leading to catalyst rearrangement, that is, the formation of a rutile-type $SbVO_4$ phase [20, 87] from dispersed vanadium and antimony oxide species [20, 87]. This catalysis-induced restructuring is consistent with the structural reactivity of the rutile-type $SbVO_4$ phase [88] and appears related to the redox cycle of the catalyst [15, 87] and its remarkable redox reactivity [89]. Similar transformations occur for alumina-supported vanadium and phosphorous oxides during propane ammoxidation [48, 90].

This section commented on supported oxides, which may rearrange into segregated bulk oxides on their corresponding supports. The next two sections will deal with bulk oxides and supported-bulk oxides, and then comment about the interplay between molecularly dispersed oxides, bulk oxides, and supported-bulk oxides.

10.5.2 Bulk Mixed Metal Oxides for Oxidation

Bulk mixed oxides are among the earliest reported systems for *in-situ* and *operando* Raman studies [36, 91–93]. Early works already demonstrate the relevance of calcination treatments on the stability of the various mixed metal oxide catalysts and the use of oxygen isotopic exchange [6, 53, 76, 94–98], providing valuable information on the reactivity of lattice oxygen in bulk oxide catalysts. Mestl and coworkers used *operando* resonance Raman–GC/MS to understand Mo–V–W mixed oxide catalysts during the oxidation of methanol, acrolein, and propene [6, 14, 52, 53, 99] and demonstrated the presence of MoO_3- and Mo_5O_{14}-type oxides. A recent review by Kim et al. summarizes the uses of resonance Raman spectroscopy under *operando*, *in-situ*, and ambient conditions [100].

Mixed Ni–Nb oxide catalysts have recently proved to be efficient for the conversion of ethane into acetonitrile. Its reactivity appears related to some interaction between these two oxides [101]. An *operando* Raman–GC study demonstrates that the bulk Ni–Nb oxide is an efficient catalyst when the Ni–Nb oxide exhibits an incipient interaction as evidenced by Raman bands at 858, 775, and 970 cm^{-1} for a catalyst with a 90:10 atomic Ni:Nb population [102]. Heracleous and Lemonidou already described such bands [103]. As shown in an *operando* Raman–GC study, the increase of these Raman bands during reaction is directly related to acetonitrile formation during reaction [102]. The same study reveals that the system is no longer efficient for ethane ammoxidation when a more extensive Ni–Nb interaction leads to a well-defined $NiNb_2O_6$ mixed oxide phase.

Bulk volume oxygen species are almost exclusively monitored in the experiments described earlier. Only little, if any, information can be drawn about the catalytically active surface oxygen species [87, 104–108]. Thus, the question remains on the exact molecular nature of the active site which structure might be significantly different from the bulk [109]. Hence, the challenge remains to characterize such active surface species on bulk oxide catalysts. Recently, Zhao et al. report the presence of surface MoO_x and VO_x species on Mo–Nb and V–Nb mixed metal oxides by Raman spectroscopy [110] being consistent with the fact that the turn-over frequency of the reaction on dispersed molybdenum oxides is much higher than on molecularly dispersed SbO_x species [111]. This observation suggests that such a surface layer different from the bulk is directly related to catalytic properties. Following this suggestion, *so-called bulk oxide catalysts would have two different decisive tasks for reactivity: (1) they act as sometimes-rather-complicated supports to (2) generate, stabilize, and disperse the catalytically active surface species.*

10.5.3 Supported-Bulk Metal Oxides

If bulk oxide catalysts act as supports for dispersed phases, it is critical to characterize such surface species despite the presence of an overwhelming majority of bulk sites in the volume of the particles often possessing low surface-to-volume ratios. Moreover, the surface reactivity that might develop under catalytic reaction might be influenced by the bulk support properties as in case of classical supported catalysts. While the similarity between supported and bulk oxides is clear, supported oxides are, however, easier to characterize than surface sites on bulk oxides as bulk oxide catalysts often exhibit strong Raman bands that overlap overwhelmingly with those of the active surface layer. This is different for supported oxides since typical supports either do not exhibit Raman bands (γ-alumina), weak (silica) or insignificant Raman bands above ca. 700 cm^{-1} (α-alumina, titania, zirconia, ceria, zinc oxide, etc.). This is very convenient since the most informative Raman bands of supported oxides are located in the 800–1100 cm^{-1} range. The best approach to understand surface sites on bulk oxides is to increase their surface-to-volume ratios, and therefore the Raman signals of surface sites versus those of the bulk [16, 109, 112] by reducing their particle sizes. Handling these nanoscale materials is not easy as they too pose a health risk. For safe handling of nanoscaled catalysts and for keeping their catalytic properties, it is necessary to immobilize them on standard microscale supports, which also prevents their sintering [20, 48, 49, 87, 109, 113].

López-Medina et al. reported on the role of total coverage of Mo + V + Nb oxides on alumina and the role of tellurium promotion [113]. Such system is particularly efficient for propane selective oxidation [114–116]. *Operando* Raman–GC shows that the mixed phase forms during reaction. Since the supported phase consists of very small crystallites, these do not generate X-ray diffraction. Raman spectroscopy, however, is sensitive to changes occurring in the nanometer scale. Thus, the *operando* Raman–GC study delivers the structure–activity relationships during reaction [113]. Raman spectra confirm the formation of nanocrystalline MoO_3-type structures during reaction at 375°C (Fig. 10.4), which limits the formation of acrylic acid. This work also addresses the role of tellurium as dopant for Mo–V–Nb oxide catalysts on

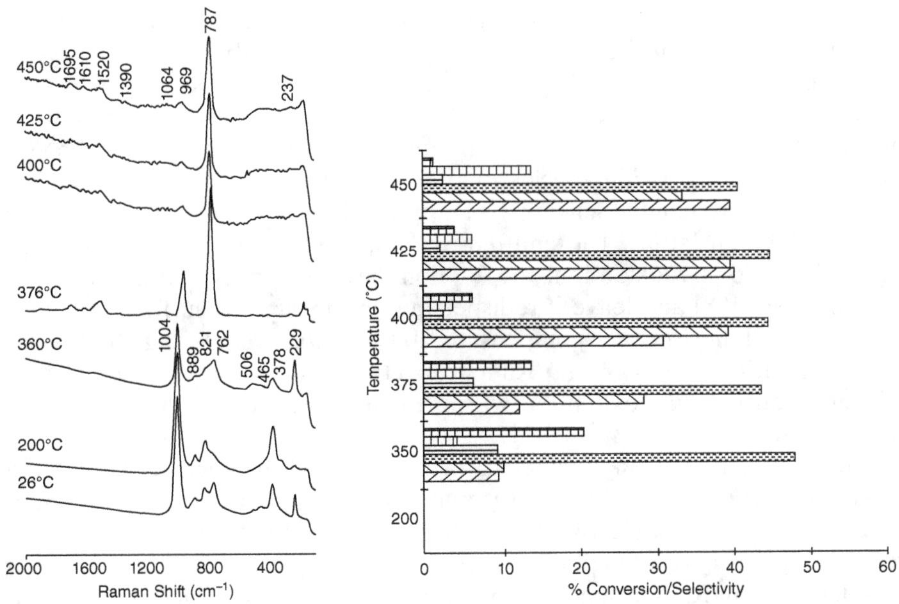

Figure 10.4. *Operando* Raman–GC spectra during the selective oxidation of propane on $12Mo_5V_4Nb_1$. Left: Raman spectra during reaction at the temperature indicated. Right: Simultaneous activity/selectivity data obtained during Raman spectra acquisition. $C_3H_8/O_2/H_2O/He$ = 12.5/20.4/15.9/51.2; 4800/h; 0.2 g of catalyst. Reprinted with permission from López-Medina, R., Fierro, J. L. G., Guerrero-Pérez, M. O., Bañares, M. A. (2011) Structural changes occurring at the surface of alumina-supported nanoscaled Mo-V-Nb-(Te)-O catalytic system during the selective oxidation of propane to acrylic acid, *Appl. Catal. A*, 406, 34–42. Copyright (2011) Elsevier.

alumina. Tellurium prevents the formation of MoO_3 during reaction, promoting a distorted rutile structure, providing in this case 25% acrylic acid yield at 400°C [113].

The rutile-type $SbVO_4$ phase is closely related to propane ammoxidation reactivity [108, 117, 118]. An *operando* Raman–GC study with support-stabilized, nanoscaled $SbVO_4$ delivers a link between phase formation and propane ammoxidation performance [20]. This study also revealed the interplay between segregated SbO_x and molecularly dispersed VO_x with the rutile-type $SbVO_4$ phase during propane ammoxidation and reoxidation cycles [15, 20, 87] $SbVO_4$ was absent in the fresh catalyst. With increasing reaction temperature, both propane ammoxidation and dispersed $SbVO_4$ phase formation occurs [15, 20, 87]. Most importantly, the propane ammoxidation performance of a monolayer of $SbVO_4$ on alumina was equivalent to that of bulk $SbVO_4$ [11], and the catalytic activity increases as the $SbVO_4$ Raman bands grow stronger [15]. This is consistent with DFT calculations that show that ammonia activates at vanadium sites surrounded by antimony sites via bridging oxygen [119]. Thus, this rutile $SbVO_4$ phase is intrinsically connected with ammonia activation. Moreover, there is a close relationship between the transformations of dispersed vanadia and segregated antimony oxide in the presence of $SbVO_4$ [87]. Dispersed V^{4+} oxide species migrate into the lattice of $SbVO_4$ during ammoxidation reaction (Scheme 10.1), promoting the emission of Sb^{5+} ions from the $SbVO_4$ lattice. The emitted Sb^{5+} ions enter into segregated Sb_2O_4 crystallites. Upon reoxidation, the V^{5+} ions are no longer stabilized in the $SbVO_4$ lattice [88] and leave it redispersing on the support surface. The emigrating vanadium ions are compensated by Sb^{5+} ions reentering the $SbVO_4$ [87]. In addition, the *operando* Raman spectra show the presence of surface alkoxides during ammoxidation, which are not detected in the absence of surface vanadium oxide species [120].

Similar trends most likely appear for other mixed oxide catalysts too, like vanadium phosphorus oxide (VPO) materials whose performance is related to a disordered overlayer [41, 121–123]. Raman spectroscopy uncovered the reaction chemistry of VPO phases [79, 106, 123–129]. Nanocrystalline δ-$VOPO_4$ was detected by Raman spectroscopy during butane oxidation [130, 131]. An amorphous V–P–O layer was synthesized by impregnating VO_x on a PO_x/TiO_2 substrate; this amorphous V–P–O layer selectively converted n-butane to maleic anhydride [132]. Thus, it is suggested that catalytically active species of unsupported $(VO)_2P_2O_7$, having V^{+4} volume sites, consist of surface redox V^{+5} and acid P^{+5} species in the amorphous surface layer. *Operando* Raman studies of supported VPO phases also provided confirming insight on the nature of the catalytically active site in this kind of catalyst [131].

The complex interplay of vanadium species located in the bulk and at the surface of VPO phases was studied during propane ammoxidation by Mikolajska et al. [90]. This work shows that acetonitrile forms when dispersed surface V^{5+} species are present, while it becomes selective to acetonitrile when V^{4+} species dominate in the system. This study suggests that V^{3+} is probably involved in redox processes during propane ammoxidation. *In-situ* studies show that the balance between these vanadium species populations is affected by the activation treatment.

10.5.4 How Different Are Supported and Bulk Oxide Catalysts?

In addition to the formation of bulk crystalline phases from supported oxides, structural transformations of surface oxides during reactions may lead to solid-state reactions that involve the support. In recent work, Martínez-Huerta et al. reported on the transformation of surface vanadium oxide species on ceria into $CeVO_4$ during ethane ODH catalysis [50] and, at the same time, the catalyst deactivated at temperatures above 500°C. A Raman investigation, however, showed that the formation of $CeVO_4$ was not related to catalyst deactivation as its formation was observed already at about 460°C. The Arrhenius plots

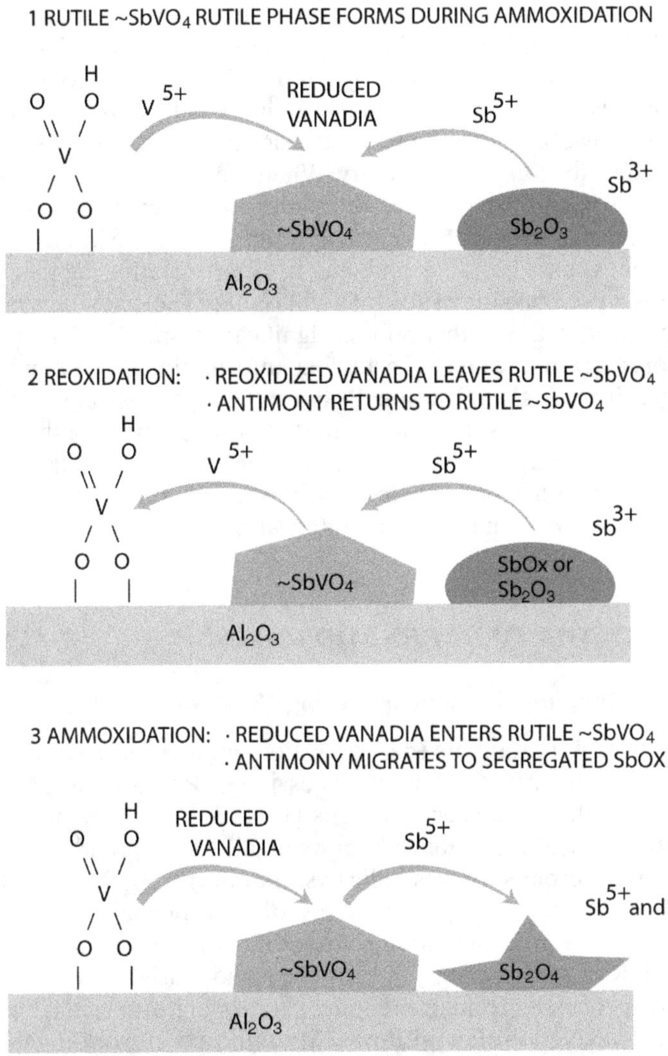

Scheme 10.1. Illustration of structural transformation of alumina-supported antimony and vanadium oxide during ammoxidation. (1) Ammoxidation on the fresh catalyst blends both components into rutile $SbVO_4$; (2) if the used system is reoxidized, reduced vanadium is not stable in the rutile lattice and leaves it, being replaced by Sb^{5+} ions; (3) during ammoxidation, vanadium ions reduce and locate in the rutile lattice, replacing Sb^{5+} ions.

determined for the reaction in the Raman cell show that the apparent activation energy did not significantly change as the catalyst altered from ceria-supported surface vanadia to ceria-supported $CeVO_4$. It was concluded that the V–O–Ce bonds present in both the fresh and aged catalysts were part of the active sites. However, despite their intrinsic chemical reactivity similarity, we must also consider long-term activity. When we move from laboratory to industrial process, runtimes become dramatically longer. Catalysts tend to loose some active phase, which eventually leads to deactivation. In a bulk catalyst, the nonexposed volume phase is not participating in the reaction directly, but it is a pool of active sites that will replace those lost during operation. In a supported catalyst, this pool is significantly smaller. This has to be taken into consideration when transferring information from fundamental investigations to industrial application. Both catalyst systems may exhibit comparable intrinsic reactivities, but global lifetimes might be grossly different. Many reactions are conducted on supported oxides, others on bulk oxides; the performance of supported-bulk oxides will in principle be similar, sometimes even better, than those of their bulk counterparts.

10.6 SPACE-RESOLVED *OPERANDO* RAMAN

10.6.1 Raman Mapping and Imaging during Catalysis

Catalysts might exhibit gradients (concentration, temperature, structural, oxidation state, etc.) on different length scales [133]. Raman mapping during reaction tries to understand such gradients [134, 135]. Early Raman imaging studies already revealed important phenomena, like the spreading of molybdenum oxide on different supports (silica vs. alumina) [136]. The capability to do high-throughput spectroscopic screening of multiple samples by Raman mapping is one of the main advantages for efficient optimization of chemical processes [137]. Raman mapping will further provide insight on reaction profiles in catalytic processes. Raman mapping profiles are also critical to understand catalyst deactivation. For instance, Moulijn and coworkers show that Raman spectroscopy is an effective tool to measure carbon deposit profiles within the pellets of spent industrial hydroprocessing catalysts [138]. The results suggest that activity is significantly reduced by exposure to heavy molecules forming coke. Baiker's group also designed a flow cell for a combined diffuse reflectance infrared Fourier transform spectroscopy (DRIFTS)–Raman imaging study during nitrogen storage–reduction (NSR) on Pt–Ba/alumina catalysts [137, 139]. The system reproduces NSR gas feed cycles between oxidative and reductive conditions. NO is oxidized to NO_2 on Pt and forms barium nitrate. Subsequently, fuel-rich conditions reduce it N_2. This combined DRIFTS–Raman study reveals chemical profiles of bulk NO_x storage along the catalyst bed, which help in understanding the mechanism of this reaction along the cell.

10.6.2 *Operando* Raman Spectroscopy in Microreactors

Novel microreactors and membrane microreactors are increasingly studied for they circumvent mass and heat transfer limitations [140–146]. Raman profiles in microreactors deliver valuable information, like assessing the degree of blending of two merging streams. Fletcher et al. used Raman to image the mixing of reactants at the junction of feed-in channels inside microreactors [147] during the synthesis of ethyl acetate from ethanol and acetic acid. Relative Raman signal intensities can be calibrated to relative concentrations using chemometrics to assess reactant space distribution and conversion along the channel (Fig. 10.5). Chemometric analyses of Raman spectra describe

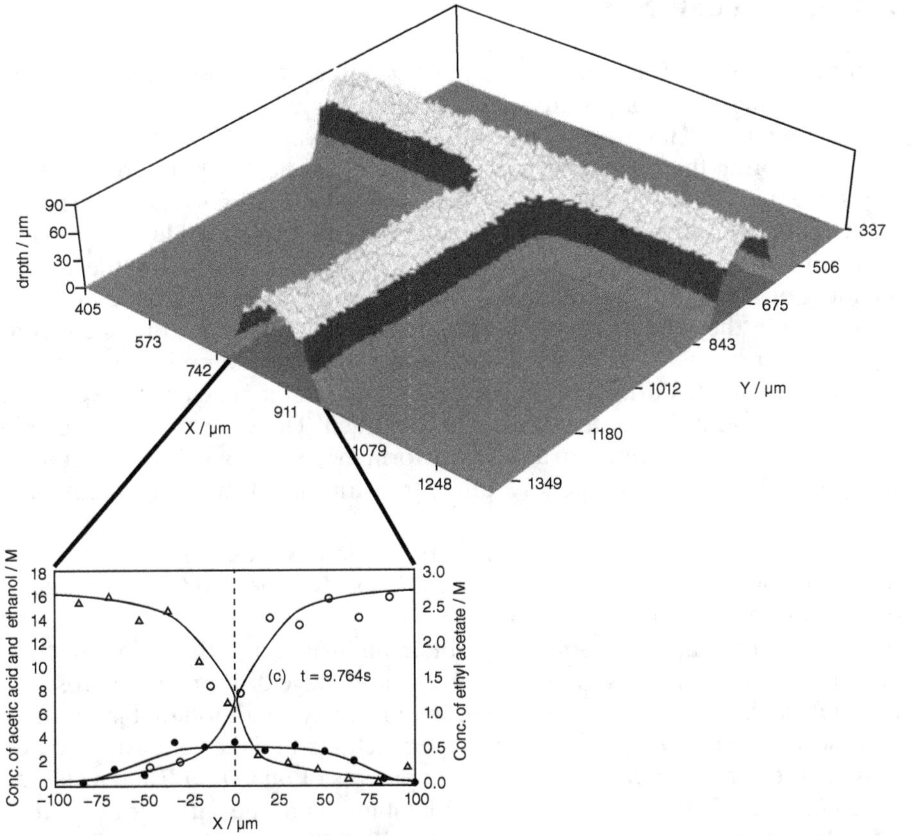

Figure 10.5. Top: Channel profile at the T-junction; ethanol enters from top-right channel, and acetic acid from right-end channel. Both streams blend and react into ethyl acetate along the channel that flows toward down-left. Bottom: Concentration profiles for (triangle) acetic acid, (open circle) ethanol and (solid circle) ethyl acetate across the downstream channel at a given *Y* and at the time indicated in the plot. Reprinted with permission from Paul, D. I., Fletcher, S. J., Haswell, X. Z. (2003) Monitoring of chemical reactions within microreactors using an inverted Raman microscopic spectrometer, *Electrophoresis*, 24, 3239–3245. Copyright (2003) John Wiley & Sons.

consumption of ethanol and acetic acid and their reaction. Raman images show that feed streams joining at the "T" junction combine into insufficiently mixed, coflowing laminar streams. Cao et al. used Raman profiles in a microreactor to monitor the evolution of atomic oxygen species in silver (640 cm^{-1} for dissolved oxygen in subsurface layers and 810 cm^{-1} for chemisorbed $O\gamma$) during methanol oxidation [148]. Urakawa et al. report the use of Raman profiles for optimizing microreactors during cyclohexene hydrogenation [149]. As commented earlier, chemometrics analysis converts Raman signal profiles into conversion–selectivity values along microreactor channels.

10.7 CONCLUSIONS

Heterogeneous catalysis is a complex phenomenon at the cross section between molecular chemistry, solid-state science, surface chemistry, and of chemical and physical sciences. Heterogeneous catalysis can never be fully comprehended by investigating the catalyst structures and the reactions separated. Catalysis per se requires the presence of its adequate reaction environment, that is, a reactor in operation, converting raw materials into products. Hence, chemical reaction and process engineering team up with spectroscopy for the understanding of heterogeneous catalysis. The multitude of spectroscopic methods developed in the last 50 years belongs to the basic toolbox for the scientist to unravel the processes during catalysis. Among these, most important are those that can be applied during the catalytic act, that is, in the reactor at the catalyst during operation: This is the *operando* approach. The described partnership of sciences is key to infer structure–performance relationships for catalytic processes, like molecular reactions, phase transitions, diffusional gradients, and the like.

During his Nobel award lecture in 1930, Sir C.V. Raman said that "The frequency differences determined from the spectra, the width and character of the lines appearing in them, and the intensity and state of polarization of the scattered radiations enable us to obtain an insight into the ultimate structure of the scattering substance." One great advantage of Raman spectroscopy is that it can be applied to samples under almost any condition that is sensitive to molecular vibrations of the catalyst, the reactants, and the acting adsorbates. This chapter has focused on recent developments of *operando* Raman studies, which provide a deeper and broader understanding of the catalytic act and its interplay between catalyst and reaction media in time and in space. Reactive surface species can be characterized, as well as the structure of the working catalyst (during reaction); thus, their mutual interactions can be understood. Such fundamental information has to be fed back for catalyst improvement, which constantly demands industry and the society.

The *operando* approach permits, provided cells perform like catalytic reactors [109, 150], the simultaneous collection of kinetic and characterization data during reaction. It is timely to remember another citation; the Spanish philosopher José Ortega y Gasset said, "the man is he and his circumstance"; we may say, "the catalyst is it and its circumstance." The recent progress in *operando* methodology allows us to understand how the catalyst depends on its circumstance, be it reaction condition (temperature, pressure, reactants) or be it within the reactive volume, that is, the shape of the catalyst or reactor channel. The *operando* methodology investigates the effect of the *circumstance* of the catalytic site versus time, position, and environment. To know the circumstance of the catalyst is to understand the catalyst, enabling the researcher to shape and design it for better performance.

> How much somebody understands depends on how much he has understood.
>
> —M. Richter

ACKNOWLEDGMENTS

G.M. acknowledges the support by Süd-Chemie AG. The financial support from the Spanish Ministry of Economy and Competitiveness/Ministry of Science and Innovation (CTQ2011-25517, CTQ2011-13343-E, and PIM2010-EEU-00138) is gratefully acknowledged.

REFERENCES

[1] Weckhuysen, B. M. (2010) Preface: recent advances in the in-situ characterization of heterogeneous catalysts, *Chem. Soc. Rev.*, 39, 4557–4559.

[2] Clausen, B. S., Steffensen, G., Fabius, B., Villadsen, J., Feidenhans, L. R., Topsøe, H. (1991) *In situ* cell for combined XRD and on-line catalysis tests: studies of Cu-based water gas shift and methanol catalysts, *J. Catal.*, 132, 524–535.

[3] García-Cortéz, G., Bañares, M. A. (2002) A Raman spectroscopy study of alumina-supported vanadium oxide catalyst during propane oxidative dehydrogenation with online activity measurement, *J. Catal.*, 209, 197–201.

[4] Kuba, S., Knözinger, H. (2002) Time-resolved *in situ* Raman spectroscopy of working catalysts: sulfated and tungstated zirconia, *J. Raman Spectrosc.*, 33, 325–332.

[5] Mestl, G., Linsmeier, C., Gottschall, R., Dieterle, M., Find, J., Herein, D., Jäger, J., Uchida, Y., Schlögl, R. (2000) Molybdenum oxide based partial oxidation catalyst: 1. Thermally induced oxygen deficiency, elemental and structural heterogeneity and the relation to catalytic performance, *J. Mol. Catal. A*, 162, 455–484.

[6] Mestl, G. J. (2002) *In situ* Raman spectroscopy for the characterization of MoVW mixed oxide catalysts, *J. Raman Spectrosc.*, 33, 333–347.

[7] Weaver, M. J. (2002) Surface-enhanced Raman spectroscopy as a versatile *in situ* probe of chemisorption in catalytic electrochemical and gaseous environments, *J. Raman Spectrosc.*, 33, 309–317.

[8] Weber, W. H. (2000) Raman scattering in materials science, *Springer Ser. Mater. Sci.*, 42, 233.

[9] Weckhuysen, B. M. (2002) Snapshots of a working catalyst: possibilities and limitations of *in situ* spectroscopy in the field of heterogeneous catalysis, *Chem. Commun.*, 97–110. (Feature Article).

[10] Weckhuysen, B. M. (2003) Determining the active site in a catalytic process: *operando* spectroscopy is more than a buzzword, *Phys. Chem. Chem. Phys.*, 5, 4351–4360.

[11] Bañares, M. A., Guerrero-Pérez, M. O., García-Cortéz, G., Fierro, J. L. G. (2002) Raman spectroscopy during catalytic operations with on-line activity measurement (*operando* spectroscopy): a method for understanding the active centres of cations supported on porous materials, *J. Mater. Chem.*, 11, 3337–3342.

[12] Bañares, M. A., Mestl, G. (2009) Structural characterization of operating catalysts by Raman spectroscopy, *Adv. Catal.*, 52, 43–128.

[13] Bañares, M. A., Wachs, I. E. (2010) Raman spectroscopy of catalysts, in Meyers, R. A. (ed.), *Encyclopedia of Analytical Chemistry*, Supplementary Vol. S1-S3, John Wiley & Sons, Hoboken, NJ, pp. 1895–1924.

[14] Mestl, G. (2000) *In situ* Raman spectroscopy—a valuable tool to understand operating catalysts, *J. Mol. Catal. A: Chem.*, 158, 45–65.

[15] Guerrero-Pérez, M. O., Bañares, M. A. (2004) Operando Raman-GC studies of alumina-supported Sb-V-O catalysts and role of the preparation method, *Catal. Today*, 96, 265–272.

[16] Bañares, M. A. (2005) *Operando* methodology: combination of *in situ* spectroscopy and simultaneous activity measurements under catalytic reaction conditions, *Catal. Today*, 100, 71–77.

[17] Thomas, J. M. (1980) Recent trends in surface science and their impact on catalyst charactersation, in Thomas, J. M., Lambert, R. M. (eds.), *Characterization of Catalysts*, John Wiley & Sons, New York, pp. 1–11.

[18] Thomas, J. M. (1999) Design, synthesis, and *in situ* characterization of new solid catalysts, *Angew. Chem. Int. Ed.*, 38, 3588–3628.

[19] Wachs, I. E. (2005) Recent conceptual advances in the catalysis science of mixed metal oxide catalytic materials, *Catal. Today*, 100, 79–94.

[20] Guerrero-Pérez, M. O., Bañares, M. A. (2002) Operando Raman study of alumina-supported Sb–V–O catalyst during propane ammoxidation to acrylonitrile with on-line activity measurement, *Chem. Commun.*, 1292–1293.

[21] Bañares, M. A., Wachs, I. E. (2002) molecular structures of supported metal oxide catalysts under different environments, *J. Raman Spectrosc.*, 33, 359–380.

[22] Bañares, M. A. (2004) Raman spectroscopy, in Weckhuysen, B. M. (ed.), *In-Situ Spectroscopy of Catalysts*, American Scientific Publishers, Stevenson Ranch, CA, pp. 47–92.

[23] Wachs, I. E. (1996) Raman and IR studies of surface metal oxide species on oxide supports: supported metal oxide catalysts, *Catal. Today*, 27, 437–455.

[24] Wachs, I. E. (2001) Raman spectroscopy of catalysts, in Lewis, I. R., Edwards, H. G. M. (eds.), *Handbook of Raman Spectroscopy*, CRC Press, Boca Raton, FL, pp. 799–833.

[25] Wachs, I. E., Roberts, C. A. (2010) Monitoring surface metal oxide catalytic active sites with Raman spectroscopy, *Chem. Soc. Rev.*, 39, 5002–5017.

[26] Cooney, R. P., Curthoys, G., Nguyen, T. T. (1975) Laser Raman spectroscopy and its application to the study of adsorbed species, *Adv. Catal.*, 24, 293–342.

[27] Delgass, W. N., Haller, G. L., Kellerman, R., Lunsford, J. H. (1979) *Spectroscopy in Heterogeneous Catalysis*, Academic Press, New York, p. 55.

[28] Knözinger, H. (1996) *In situ* Raman spectroscopy. A powerful tool for studies in selective catalytic oxidation, *Catal. Today*, 32, 71.

[29] Mestl, G., Knözinger, H. (1997) Vibrational spectroscopies, in Ertl, G., Knözinger, H., Weitkamp, J. (eds.), *Handbook of Heterogeneous Catalysis*, Vol. 2, Wiley-VCH, Weinheim, Germany, p. 539.

[30] Knözinger, H., Mestl, G. (1999) Laser Raman spectroscopy—a powerful tool for *in situ* studies of catalytic materials, *Top. Catal.*, 8, 45–55.

[31] Mestl, G., Srinivasan, T. K. K. (1998) Raman spectroscopy of monolayer-type catalysts: supported molybdenum oxides, *Catal. Rev. Sci. Eng.*, 40, 451–570.

[32] Stair, P. C. (2001) Advances in Raman spectroscopy methods for catalysis research, *Curr. Opin. Solid State Mater. Sci.*, 5, 365–369.

[33] Wachs, I. E. (2000) Historical perspective of Raman spectroscopy in catalysis, in Weber, W. H., Merlin, R. (eds.), *Raman Scattering in Materials Science*, Springer, Berlin, Germany, pp. 271–273.

[34] Bañares, M. A., Cardoso, J. H., Agulló-Rueda, F., Correa-Bueno, J. M., Fierro, J. L. G. (2000) Dynamic states of V-oxide species: reducibility and performance for methane oxidation on V_2O_5/SiO_2 catalysts as a function of coverage, *Catal. Lett.*, 64, 191–196.

[35] Burcham, L. J., Wachs, I. E. (1999) The origin of the support effect in supported metal oxide catalysts: *in situ* infrared and kinetic studies during methanol oxidation, *Catal. Today*, 49, 467–484.

[36] Wilson, J. A., Hill, C. G., Dumesic, J. A. (1990) Raman spectroscopy of iron molybdate catalyst systems-Part III-*in situ* studies of supported and bulk catalysts under reaction and redox conditions, *J. Mol. Catal. A*, 61, 333–352.

[37] Grasselli, J. G., Hazle, M. A. S., Wolfram, L. E. (1977) Application of Raman spectroscopy in the petroleum industry, in West, A. R. (ed.), *Molecular Spectroscopy*, Heyden and Son, London, pp. 200–224.

[38] Grasselli, J. G., Snavely, M. K., Bulkin, J. J. (1981) *Chemical Applications of Raman Spectroscopy*, Wiley Interscience, New York.

[39] Bañares, M. A., Hu, H., Wachs, I. E. (1995) Genesis and stability of silicomolybdic acid on silica-supported molybdenum oxide catalysis: *in situ* structure–selectivity study on selective oxidation reactions, *J. Catal.*, 155, 249–255.

[40] Bañares, M. A., Martínez-Huerta, M. V., Gao, X., Fierro, J. L. G., Wachs, I. E. (2000) Dynamic behavior of supported vanadia catalysts in the selective oxidation of ethane *in situ* Raman, UV–vis DRS and reactivity studies, *Catal. Today*, 61, 295–301.

[41] Guliants, V. V., Benziger, J. B., Sundaresan, S., Yao, N., Wachs, I. E. (1995) Evolution of the active surface of the vanadyl pyrophosphate catalysts, *Catal. Lett.*, 32, 379–386.

[42] Sun, Q., Jehng, J.-M., Hu, H., Herman, R. G., Wachs, I. E., Klier, K. (1997) *In situ* Raman spectroscopy during the partial oxidation of methane to formaldehyde over supported vanadium oxide catalysts, *J. Catal.*, 165, 91–101.

[43] Burcham, L. J., Deo, G., Gao, X., Wachs, I. E. (2000) *In situ* IR, Raman, and UV-vis DRS spectroscopy of supported vanadium oxide catalysts during methanol oxidation, *Top. Catal.*, 11/12, 85–100.

[44] Mul, G., Bañares, M. A., van der Linden, B., Khatib, S. J., Moulijn, J. A. (2003) MultiTRACK and *operando* Raman-GC study of oxidative dehydrogenation of propane over alumina-supported vanadium oxide catalysts, *Phys. Chem. Chem. Phys.*, 5, 4378–4383.

[45] Haller, G., Lunsford, J. H., Laane, J. (1996) Temperature dependence of the Raman spectrum of barium peroxide, *J. Phys. Chem.*, 100, 551–555.

[46] Mestl, G., Rosynek, M. P., Lunsford, J. H. (1997) Decomposition of nitric oxide over barium oxide supported on magnesium oxide. 3. *In situ* Raman characterization of the role of oxygen, *J. Phys. Chem. B*, 101, 9329–9334.

[47] Xie, S., Mestl, G., Rosynek, M. P., Lunsford, J. H. (1997) Decomposition of nitric oxide over barium oxide supported on magnesium oxide. 1. Catalytic results and *in situ* Raman spectroscopic evidence for a barium-nitro intermediate, *J. Am. Chem. Soc.*, 119, 10186–10191.

[48] Mikolajska, E., Rasmussen, S. B., Lewandowska, A. E., Bañares, M. A. (2011) *Operando* and *in situ* Raman studies of alumina-supported vanadium phosphate catalysts in propane ammoxidation reaction: activity, selectivity and active phase formation, *Phys. Chem. Chem. Phys.*, 14, 2128–2136.

[49] Bañares, M. A., Khatib, S. J. (2004) Structure-activity relationships in alumina-supported molybdena-vanadia catalysts for propane oxidative dehydrogenation, *Catal. Today*, 96, 251–257.

[50] Martínez-Huerta, M. V., Deo, G., Fierro, J. L. G., Bañares, M. A. (2008) Operando Raman-GC Study on the structure-activity relationships in V_5+/CeO_2 catalyst for ethane oxidative dehydrogenation: the formation of $CeVO_4$, *J. Phys. Chem. C*, 112, 11441–11447.

[51] Meunier, F. (2010) The design and testing of kinetically-appropriate operando spectroscopic cells for investigating heterogeneous catalytic reactions, *Chem. Soc. Rev.*, 39, 4602–4614.

[52] Ovsitser, O., Uchida, Y., Mestl, G., Weinberg, G., Blume, A., Jäger, J., Dieterle, M., Hibst, H., Schlögl, R. (2002) Molybdenum oxide based partial oxidation catalyst: part 3. Structural changes of a MoVW mixed oxide catalyst during activation and relation to catalytic performance in acrolein oxidation, *J. Mol. Catal. A*, 185, 291–303.

[53] Dieterle, M., Mestl, G., Jäger, J., Uchida, Y., Schlögl, R. (2001) Mixed molybdenum oxide based partial oxidation catalyst 2. Combined X-ray diffraction, electron microscopy and Raman investigation of the phase stability of $(MoVW)_5O_{14}$-type oxides, *J. Mol. Catal. A*, 174, 169–185.

[54] Bañares, M. A., Prieto, J., Goberna-Selma, C., Guerrero-Perez, M. O., Lewandowska, A. E., Garcia-Casado, M. (2007) *Operando* Raman Methodology for the determination of structure-activity relationships in supported oxide catalysts, *Prepr. Pap. Am. Chem. Soc. Div. Pet. Chem.*, 52, 62–66.

[55] Li, G., Hu, D., Xia, G., Zhang, Z. C. (2009) Methanol partial oxidation on MoO_3/SiO_2 catalysts: application of vibrational spectroscopic imaging techniques in a high throughput operando reactor, *Top. Catal.*, 52, 1381–1387.

[56] Tejeda, G., Maté, B., Fernandez-Sánchez, J. M., Montero, S. (1996) Temperature and density mapping of supersonic jet expansions using linear Raman spectroscopy, *Phys. Rev. Lett.*, 76, 34.

[57] Reinke, M., Mantzaras, J., Schaeren, R., Bombach, R., Inauen, A., Schenker, S. (2004) High-pressure catalytic combustion of methane over platinum: *in situ* experiments and detailed numerical predictions, *Combust. Flame*, 136, 217–240.

[58] Woo, S. I., Hill, C. G. (1982) Raman study of dicobalt octacarbonyl covalently attached to a phosphinated silica, *J. Mol. Catal.*, 15, 309–325.

[59] Woo, S. I., Hill, C. G. (1984) Vibrational spectroscopy studies of $Co_2(CO)_6(PPh_3)_2$/Al_2O_3 interacting with CO, H_2 and C_3H_6, *J. Mol. Catal.*, 24, 165–188.

[60] Woo, S. I., Hill, C. G. (1985) *In situ* Raman spectroscopy studies of the hydroformylation of propylene, *J. Mol. Catal.*, 29, 231–258.

[61] Grünwaldt, J. D., Wandeler, R., Baiker, A. (2003) Supercritical fluids in catalysis: opportunities of *in situ* spectroscopic studies and monitoring phase behavior, *Catal. Rev.*, 45, 1–96.

[62] Grzybowska-Swierkosz, B. (1997) Vanadia-titania catalysts for oxidation of o-xylene and other hydrocarbons, *Appl.Catal. A*, 157, 263–310.

[63] Forzatti, P., Tronconi, E., Elmi, A. S., Busca, G. (1997) Methanol oxidation over vanadia-based catalysts, *Appl. Catal. A*, 157, 387–408.

[64] Topsøe, N.-Y., Dumesic, J. A., Topsøe, H. (1995) Vanadia-titania catalysts for selective catalytic reduction of nitric-oxide by ammonia: I.I. Studies of active sites and formulation of catalytic cycles, *J. Catal.*, 151, 241–252.

[65] Lewandowska, A. E., Calatayud, M., Tielens, F., Bañares, M. A. (2011) Dynamics of hydration in vanadia-titania catalysts at low loading: a theoretical and experimental study, *J. Phys. Chem. C*, 115, 24133–24142.

[66] Jehng, J.-M., Deo, G., Weckhuysen, B. M., Wachs, I. E. (1996) Effect of water vapor on the molecular structures of supported vanadium oxide catalysts at elevated temperatures, *J. Mol. Catal. A: Chem.*, 110, 41–54.

[67] Li, W., Oyama, S. T. (1997) Ethanol oxidation using ozone over supported maganese oxide catalysts: An *in situ* laser Raman study, *Stud. Surf. Sci. Catal.*, 110, 873.

[68] Li, W., Gibbs, G. V., Oyama, S. T. (1998) Mechanism of ozone decomposition on a manganese oxide catalyst. 2. Steady-state and transient kinetic studies, *J. Am. Chem. Soc.*, 120, 9041.

[69] Tian, Z.-Q., Ren, B. (2000) Molecular-level investigation on electrochemical interfaces by Raman spectroscopy, *Chin. J. Chem.*, 18, 135–146.

[70] Li, J.-F., Yang, Z.-L., Ren, B., Liu, G.-K., Fang, P.-P., Jiang, Y.-X., Wu, D.-Y., Tian, Z.-Q. (2006) Surface-enhanced Raman spectroscopy using gold-core platinum-shell nanoparticle film electrodes: Toward a versatile vibrational strategy for electrochemical interfaces, *Langmuir*, 22, 10372.

[71] Yeo, B. S., Klaus, S. L., Ross, P. N., Mathies, R. A., Bell, A. T. (2010) Identification of hydroperoxy species as reaction intermediates in the electrochemical evolution of oxygen on gold, *ChemPhysChem*, 11, 1854–1857.

[72] Rossmeisl, J., Logadottir, A., Nørskov, J. K. (2005) Electrolysis of water on (oxidized) metal surfaces, *Chem. Phys.*, 319, 178–184.

[73] Moreno, T., Morán López, M. A., Huerta Illera, I., Piqueras, C. M., Sanz Arranz, A., García Serna, J., Cocero, M. J. (2011) Quantitative Raman determination of hydrogen peroxide using the solvent as internal standard: online application in the direct synthesis of hydrogen peroxide, *Chem. Eng. J.*, 166, 1061–1065.

[74] Puurunen, R., Weckhuysen, B. M. (2002) Spectroscopic study on the irreversible deactivation of chromia/alumina dehydrogenation catalysts, *J. Catal.*, 210, 418–430.

[75] Puurunen, R., Beheydt, B. G., Weckhyuysen, B. M. (2001) Monitoring chromia/alumina catalysts *in situ* during propane dehydrogenation by optical fiber UV-visible diffuse reflectance spectroscopy, *J. Catal.*, 204, 253–257.

[76] Wu, Z., Kim, H. S., Stair, P. C., Rugmini, S., Jackson, S. D. (2005) On the structure of vanadium oxide supported on aluminas: UV and visible Raman spectroscopy, UV-visible diffuse reflectance spectroscopy, and temperature-programmed reduction studies, *J. Phys. Chem. B*, 109, 2793–2800.

[77] Magg, N., Immaraporn, B., Giorgi, J. B., Schröder, T., Bäumer, M., Döbler, J., Wu, Z. L., Kondratenko, E., Cherian, M., Baerns, M., Stair, P. C., Sauer, J. (2004) Vibrational spectra of alumina- and silica-supported vanadia revisited: an experimental and theoretical model catalyst study, *J. Catal.*, 226, 88.

[78] Bañares, M. A. (1999) Supported metal oxide and other catalysts for ethane conversion: a review, *Catal. Today*, 51, 319–348.

[79] Wachs, I. E., Jehng, J.-M., Deo, G., Weckhuysen, B. M., Guliants, V. V., Benziger, J. B. (1996) *In situ* Raman spectroscopy studies of bulk and surface metal oxide phases during oxidation reactions, *Catal. Today*, 32, 47–55.

[80] Cortéz, G. G., Fierro, J. L. G., Bañares, M. A. (2003) Role of potassium on the structure and activity of alumina-supported vanadium oxide catalysts for propane oxidative dehydrogenation, *Catal. Today*, 78, 219–228.

[81] Si-Ahmed, H., Calatayud, M., Minot, C., Lozano Diz, E., Lewandowska, A. E., Bañares, M. A. (2007) Combining theoretical description with experimental *in situ* studies on the effect of potassium on the structure and reactivity of titania-supported vanadium oxide catalyst, *Catal. Today*, 126, 96–102.

[82] Lewandowska, A. E., Calatayud, M., LozanoDiz, E., Minot, C., Bañares, M. A. (2008) Combining theoretical description with experimental *in situ* studies on the effect of alkali additives on the structure and reactivity of vanadium oxide supported catalysts, *Catal. Today*, 139, 209–213.

[83] Christodoulakis, A., Machli, M., Lemonidou, A. A., Boghosian, S. (2004) Molecular structure and reactivity of vanadia-based catalysts for propane oxidative dehydrogenation studied by *in situ* Raman spectroscopy and catalytic activity measurements, *J. Catal.*, 222, 293–306.

[84] Zhao, C., Wachs, I. E. (2006) Selective oxidation of propylene to acrolein over supported V_2O_5/Nb_2O_5 catalysts: an *in situ* Raman, IR, TPSR and Kinetic Study, *Catal. Today*, 118, 332–343.

[85] Bronkema, J. L., Bell, A. T. (2007) Mechanistic studies of methanol oxidation to formaldehyde on isolated vanadate sites supported on MCM-48, *J. Phys. Chem. C*, 111, 420–430.

[86] Korhonen, S. T., Bañares, M. A., Fierro, J. L. G., Krause, A. O. I. (2007) Adsorption of methanol as a probe for surface characteristics of zirconia-, alumina-, and zirconia/alumina-supported chromia catalysts, *Catal. Today*, 126, 235–247.

[87] Guerrero-Pérez, M. O., Bañares, M. A. (2007) Operando Raman-GC Study of supported alumina Sb- and V-based catalysts: effect of Sb/V molar ratio and total Sb+V coverage in the structure of catalysts during propane ammoxidation, *J. Phys. Chem. C.*, 111, 1315–1322.

[88] Landa-Cánovas, A. R., García-García, F. J., Hansen, S. (2010) Structural flexibility in ~$SbVO_4$, *Catal. Today*, 158, 156–161.

[89] Guerrero-Pérez, M. O., Kim, T. J., Bañares, M. A., Wachs, I. E. (2008) Nature of catalytic active sites for Sb–V–O mixed metal oxides, *J. Phys. Chem. C*, 112, 16858–16863.

[90] Mikolajska, E., Rasmussen, S. B., Lewandowska, A. E., Bañares, M. A. (2012) *Operando* and *in-situ* Raman studies of alumina-supported vanadium phosphate catalysts in propane ammoxidation reaction: activity, selectivity and active phase formation, *Phys. Chem. Chem. Phys.*, 14, 2128–2136.

[91] Snyder, T. P., Hill, C. G., Jr. (1991) Stability of bismuth molybdate catalysts at elevated temperatures in air and under reaction conditions, *J. Catal.*, 132, 536–555.

[92] Hoefs, E. V., Monnier, J. R., Keulks, G. W. (1979) The investigation of the type of active oxygen for the oxidation of propylene over bismuth molybdate catalysts using infrared and Raman spectroscopy, *J. Catal.*, 57, 331–337.

[93] Kovats, W. D., Hill, C. G., Jr. (1986) Raman spectroscopy of bulk bismuth molybdate catalysts, *Appl. Spectrosc.*, 40, 1215–1223.

[94] Ueda, W., Moro-Oka, Y., Ikawa, T. (1982) ^{18}O tracer study of the active species of oxygen on Bi_2MoO_6 catalyst, *J. Chem. Soc. Faraday Trans. I*, 78, 495–500.

[95] Ozkan, U. S., Moctezuma, E., Driscoll, S. A. (1990) Transient response studies of C4 hydrocarbon oxidation over $MnMoO_4/MoO_3$ catalysts, *Appl. Catal.*, 58, 305.

[96] Stair, P. (2007) The application of UV Raman spectroscopy for the characterization of catalysts and catalytic reactions, *Adv. Catal.*, 51, 75–98.

[97] Tian, H., Wachs, I. E., Briand, L. E. (2005) Comparison of UV and visible Raman spectroscopy of bulk metal molybdate and metal vanadate catalysts, *J. Phys. Chem. B*, 109, 23491–23499.

[98] Ricchiardi, G., Damin, A., Bordiga, S., Lamberti, C., Spano, G., Rivetti, F., Zecchina, A. (2001) Vibrational structure of titanium silicate catalysts. A spectroscopic and theoretical study, *J. Am. Chem. Soc.*, 123, 11409–11419.

[99] Knobl, S., Zenkovets, G. A., Kryukova, G. N., Ovsitser, O., Niemeyer, D., Schlögl, R., Mestl, G. (2003) The synthesis and structure of a single-phase, nanocrystalline MoVW mixed-oxide catalyst of the Mo_5O_{14} type, *J. Catal.*, 215, 177–187.

[100] Kim, H., Kosuda, K. M., Van Duyne, R. P., Stair, P. (2010) Resonance Raman and surface- and tip-enhanced Raman spectroscopy methods to study solid catalysts and heterogeneous catalytic reactions, *Chem. Soc. Rev.*, 39, 4820–4844.

[101] Rojas, E., Guerrero-Pérez, M. O., Bañares, M. A. (2009) Direct ammoxidation of ethane: an approach to tackle the worldwide shortage of acetonitrile, *Catal. Commun.*, 10, 1555–1557.

[102] Guerrero-Pérez, M. O., Rojas, E., Gutiérrez-Alejandre, A., Ramírez, J., Sánchez-Minero, F., Fernández-Vargas, C., Bañares, M. A. (2011) *In situ* Raman studies during sulfidation, and operando Raman-CG during ammoxidation reaction using nickel-containing catalysts: a valuable tool to identify the transformations of catalytic species, *Phys. Chem. Chem. Phys.*, 13, 9260–9267.

[103] Heracleous, E., Lemonidou, A. A. (2006) Ni–Nb–O mixed oxides as highly active and selective catalysts for ethene production via ethane oxidative dehydrogenation. Part I: characterization and catalytic performance, *J. Catal.*, 237, 162–174.

[104] Gaigneaux, E. M., Dieterle, M., Ruiz, P., Mestl, G., Delmon, B. (1998)) Catalytic performances and stability of three Sb-Mo-O phases in the selective oxidation of isobutene to methacrolein, *J. Phys. Chem. B*, 102, 10542–10555.

[105] Conte, M., Budroni, G., Bartley, J. K., Taylor, S. H., Carley, A. F., Schmidt, A., Murphy, D. M., Girgsdies, F., Ressler, T., Schlögl, R., Hutchings, G. J. (2006) Chemically induced fast solid-state transition of w-$VOPO_4$ in vanadium phosphate catalysts, *Science*, 313, 1270–1273.

[106] Xue, Z.-Y., Schrader, G. L. (1999) *In situ* laser Raman spectroscopy studies of VPO catalyst transformations, *J. Phys. Chem. B*, 103, 9459–9467.

[107] Kiely, C. J., Hutchings, G. J. (2007) Adventures with vanadium phosphate catalysts: reflections on a long-standing collaboration with J.C. Volta, *Appl. Cat. A Gen.*, 325, 194–197.

[108] Centi, G., Perathoner, S. (1995) Modification of the surface reactivity of vanadium antimonate catalysts during catalytic propane ammoxidation, *Appl. Catal. A*, 124, 317–337.

[109] Bañares, M. A. (2011) *Operando* spectroscopy: the knowledge bridge to assessing structure-performance relationships in catalyst nanoparticlesrticles, *Adv. Mater.*, 23, 5293–5301.

[110] Zhao, Z., Gao, X., Wachs, I. E. (2003) Comparative study of bulk and supported V-Mo-Te-Nb-O mixed metal oxide catalysts for oxidative dehydrogenation of propane to propylene, *J. Phys. Chem. B*, 107, 6333–6342.

[111] Badlani, M., Wachs, I. E. (2001) Methanol: a "smart" chemical probe molecule, *Catal. Lett.*, 75, 137–149.

[112] Guerrero-Pérez, M. O., Bañares, M. A. (2006) From conventional *in situ* to operando studies in Raman spectroscopy, *Catal. Today*, 113, 48–57.

[113] López-Medina, R., Fierro, J. L. G., Guerrero-Pérez, M. O., Bañares, M. A. (2011) Structural changes occurring at the surface of alumina-supported nanoscaled Mo-V-Nb-(Te)-O catalytic system during the selective oxidation of propane to acrylic acid, *Appl. Catal. A.*, 406, 34–42.

[114] Shiju, N. R., Guliants, V. V. (2009) Recent developments in catalysis using nanostructured materials, *Appl. Catal. A*, 356, 1–17.

[115] Ueda, W., Sadakane, M., Ogihara, H. (2008) Recent developments in catalysis using nanostructured materials, *Catal. Today*, 132, 2–8.

[116] López Nieto, J. M. (2006) Selective oxidation of short-chain alkanes over hydrothermally prepared MoVTeNbO catalysts, *Top. Catal.*, 41, 3–15.

[117] Pantazidis, A., Burrows, A., Kiely, C. J., Mirodatos, C. (1998) Direct evidence of active surface reconstruction during oxidative dehydrogenation of propane over VMgO catalyst, *J. Catal.*, 177, 325–334.

[118] Zanthoff, H. W., Grünert, W., Buchholz, S., Heber, M., Stievano, L., Wagner, F. E., Wolf, G. U. (2000) Bulk and surface structure and composition of V-Sb mixed-oxide catalysts for the ammoxidation of propane, *J. Mol. Catal. A*, 162, 435–454.

[119] Rojas, E., Calatayud, M., Guerrero-Pérez, M. O., Bañares, M. A. (2010) Correlation between theoretical and experimental investigations of the ammonia adsorption process on the (110)-$VSbO_4$ surface, *Catal. Today*, 158, 178–185.

[120] Wickman, A., Andersson, A. (2011) Effect of Fe, Ga, Ti and Nb substitution in ≈$SbVO_4$ for propane ammoxidation, *Appl. Catal. A*, 391, 110–117.

[121] Hutchings, G. J., Desmartin-Chomel, A., Olier, R., Volta, J. C. (1994) Role of the product in the transformation of a catalyst to its active state, *Nature*, 368, 41–45.

[122] Cavani, F., Caldarelli, A., Luciani, S., Cortelli, C., Cruzzolin, F. (2012) Selective oxidation of o-xylene to phthalic anhydride: from conventional catalysts and technologies toward innovative approaches, *Catal. Spec. Period. Rep.*, 24, 204–222.

[123] Centi, G., Trifiro, F., Ebner, J. R., Franchetti, V. M. (1988) Mechanistic aspects of maleic anhydride synthesis from C_4 hydrocarbons over phosphorus vanadium oxide, *Chem. Rev.*, 88, 55–80.

[124] Moser, T. P., Schrader, G. L. (1985) Selective oxidation of n-butane to maleic anhydride by model VPO catalysts, *J. Catal.*, 92, 216–231.

[125] Gai, P. L., Kourtakis, K., Coulson, D. R., Sonnichsen, G. C. (1997) *J. Phys. Chem. B*, 101, 9916.

[126] Centi, G. (1993) Vanadyl pyrophosphate—a critical overview, *Catal. Today*, 16, 5–26.

[127] Abdelouahah, B., Olier, F., Guilhaume, R., Lefebvre, N., Volta, J. C. (1992) A study by *in situ* laser Raman spectroscopy of VPO catalysts for n-butane oxidation to maleic anhydride I. Preparation and characterization of pure reference phases, *J. Catal.*, 134, 151–167.

[128] Igarashi, H., Tsuji, K., Okuhara, T., Misono, M. (1993) Effects of consecutive oxidation on the production of maleic anhydride in butane oxidation over four kinds of well-characterized vanadyl pyrophosphates, *J. Phys. Chem.*, 97, 7065–7071.

[129] Wachs, I. E., Jehng, J.-M., Deo, G., Weckhuysen, B. M., Gualiants, V. V., Benziger, J. B., Sundaresan, S. (1997) Fundamental studies of butane oxidation over model-supported vanadium oxide catalysts: molecular structure-reactivity relationships, *J. Catal.*, 170, 75–88.

[130] Guliants, V. V., Holmes, S., Benziger, J. B., Heaney, P., Yates, D., Wachs, I. E. (2001) Selective oxidation of n-butane to maleic anhydride by model V-P-O catalysts, *J. Mol. Catal. A*, 172, 265–276.

[131] Cavani, F., Luciani, S., Esposti, E. D., Cortelli, C., Leanza, R. (2010) Surface dynamics of a vanadyl pyrophosphate catalyst for n-butane oxidation to maleic anhydride: an *in situ* Raman and reactivity study of the effect of the P/V atomic ratio, *Chem. Eur. J.*, 16, 1646–1655.

[132] Guliants, V. V., Benziger, J. B., Sundaresan, S., Wachs, I. E., Jehng, J.-M., Roberts, J. E. (1996) The effect of the phase composition of model VPO catalysts for partial oxidation of n-butane, *Catal. Today*, 28, 275–295.

[133] Calvino-Casilda, V., Bañares, M. A. (2012) Spectroscopic Raman imaging, *Catal.: Spec. Period. Rep. (R. Soc. Chem.)*, 24, 1–47.

[134] Gardiner, D. J., Littleton, C. J., Bowden, M. (1988) Automated mapping of high-temperature corrosion products on iron chromium alloy using Raman microscopy, *Appl. Spectrosc.*, 42, 15–19.

[135] Watanabe, M., Ogawa, T. (1988) Raman scattering and photoluminescence tomography, *Jpn. J. Appl. Phys.*, 27, 1066–1070.

[136] Leyrer, J., Mey, D., Knözinger, H. (1990) Spreading behavior of molybdenum trioxide on alumina and silica: a Raman microscopy study, *J. Catal.*, 124, 349–356.

[137] Urakawa, A., Baiker, A. (2009) Space-resolved profiling relevant in heterogeneous catalysis, *Top. Catal.*, 52, 1312–1322.

[138] Vogelaar, B. M., van Langeveld, A. D., Eijsbouts, S., Moulijn, J. A. (2007) Analysis of coke deposition profiles in commercial spent hydroprocessing catalysts using Raman spectroscopy, *Fuel*, 86, 1122–1129.

[139] Urakawa, A., Maeda, N., Baiker, A. (2008) Space- and time-resolved combined DRIFT and Raman spectroscopy: monitoring dynamic surface and bulk processes during NO_x storage reduction, *Angew. Chem. Int. Ed.*, 47, 9256–9259.

[140] Au, L. T. Y., Yeung, K. L. (2001) An investigation of the relationship between microstructure and permeation properties of ZSM-5 membranes, *J. Membr. Sci.*, 194, 33–55.

[141] Wan, Y. S. S., Chau, J. L. H., Yeung, K. L., Gavriilidis, A. (2001) Design and fabrication of zeolite-based microreactors and membrane microseparators, *Microporous Mesoporous Mater.*, 42, 157–175.

[142] Yeung, K. L., Yao, N. (2004) Scanning probe microscopy in catalysis, *J. Nanosci. Nanotechnol.*, 4, 647–690.

[143] Yeung, K. L., Zhang, X. F., Lau, W. N., Martin-Aranda, R. (2005) Experiments and modeling of membrane microreactors, *Catal. Today*, 110, 26–37.

[144] Kobayashi, J., Mori, Y., Kobayashi, S. (2005) Hydrogenation reactions using $scCO_2$ as a solvent in microchannel reactors, *Chem. Commun.*, 20, 2567–2568.

[145] Tiggelaar, R. M., Benito-Lopez, F., Hermes, D. C., Rathgen, H., Egberink, F. G., Mugele, D. N., Reinhoudt, A., Van Den Berg, W., Verboom, H., Gardeniers, R. J. M. (2007) Fabrication, mechanical testing and application of high-pressure glass microreactor chips, *Chem. Eng. J.*, 131, 163–170.

[146] Trachsel, F., Hutter, C., von Rohr, P. R. (2008) Transparent silicon/glass microreactor for high-pressure and high-temperature reactions, *Chem. Eng. J.*, 135, S309–S316.

[147] Fletcher, I., Haswell, S. J., Zhang, X. (2003) Monitoring of chemical reactions within microreactors using an inverted Raman microscopic spectrometer, *Electrophoresis*, 24, 3239–3245.

[148] Cao, E., Firth, S., McMillan, P. F., Gavriilidis, A. (2007) Application of microfabricated reactors for operando Raman studies of catalytic oxidation of methanol to formaldehyde on silver, *Catal. Today*, 126, 119–126.

[149] Urakawa, A., Trachsel, F., von Rohr, P. R., Baiker, A. (2008) On-chip Raman analysis of heterogeneous catalytic reaction in supercritical CO_2: phase behaviour monitoring and activity profiling, *Analyst*, 133, 1352.

[150] Khatib, S. J., Guil-Lopez, R., Peña, M. A., Fierro, J. L. G., Bañares, M. A. (2006) Alumina-supported V–Mo–O mixed oxide catalysts, the formation of phases involving aluminum: $AlVMoO_7$, *Catal. Today*, 118, 353–359.

11

IN-SITU ELECTRON PARAMAGNETIC RESONANCE OF POWDER CATALYSTS

Angelika Brückner

11.1 INTRODUCTION

Among the multitude of spectroscopic techniques used for *in-situ* characterization of solid catalysts, electron paramagnetic resonance (EPR) spectroscopy still is one that is less frequently applied. One reason originates probably from its detection principle, which requires paramagnetic species containing unpaired electrons to be present in the systems under study [1–3]. Yet for those catalysts EPR is a very valuable tool, since it provides information on the local environment of paramagnetic species, their dispersion, changes in their valence states and interactions with each other as well as with reactants contained in the surrounding gaseous or liquid phase, and, when applied under catalytic working conditions, it can reflect reaction-dependent changes of all these properties very sensitively. There are two main classes of species with relevance for catalysis that are accessible by *in-situ* EPR.

11.1.1 Paramagnetic Transition Metal Ions

These species belong to the systems by far most frequently studied in heterogeneous catalysis by *in-situ* EPR as they are of paramount importance for

In-situ *Characterization of Heterogeneous Catalysts*, First Edition.
Edited by José A. Rodriguez, Jonathan C. Hanson, and Peter J. Chupas.

many catalytic reduction/oxidation (redox) processes. Several reviews and book chapters are dedicated to this topic [4–11]. Many of these studies have been performed under conditions as close as possible to those of the catalytic reaction, that is, at elevated temperatures and under flowing gases. Some of those examples are discussed in more detail in Section 11.3.1. Yet not all paramagnetic transition metal ions (TMIs) can be analyzed under such conditions. Some of them have short relaxation times and require low registration temperatures of $T \leq 77$ K. This is true, for example, for Ni^{2+}, Co^{2+}, Ti^{3+}, Fe^{2+}, and V^{3+} species. Catalysts containing such ions have therefore been analyzed after thermal treatment in the presence of reactants and quenching to low spectra recording temperatures—an approach which can also provide valuable information on their redox and catalytic behavior, as shown for example by Dyrek et al. for supported Ni catalysts in butene dimerization [12] and by Pietrzyk and Sojka for Co–BEA zeolites used as catalysts for the abatement of NO from exhaust gases [13].

11.1.2 Radicals and Radical Ions

Apart from a few early *in-situ* EPR studies of noncatalytic carbonization and coal pyrolysis reactions in which carbon-based radicals of unknown structure have been detected, it was Lunsford et al. who confirmed for the first time the formation of short-lived CH_3 radical intermediates during oxidative coupling of methane over doped MgO catalysts and, thus, could propose a sound mechanism for this reaction [14]. Due to the short lifetime of these species, they could be detected only by a special matrix isolation technique at temperatures as low as 12 K. More recently, the first clear evidence for the formation of ethyl radicals on a heterogeneous Ziegler–Natta model catalyst containing $TiCl_4$ and an $Al(CH_3)$ activator was provided by *in-situ* EPR, though at 40 K [15].

Another opportunity for visualizing short-lived radical intermediates is using a spin trap [16]. These are usually organic N-oxides which readily capture the short-lived radicals and form paramagnetic adducts that are stable enough to be detected by *in-situ* EPR. This technique has been preferentially applied for homogeneous liquid-phase reactions in organic chemistry, biochemistry, and electrochemistry. However, in a recent review, opportunities for its application on heterogeneous TiO_2-based photocatalysts are summarized [17]. Apart from organic radicals formed by conversion of organic reactants, there is a variety of inorganic radicals such as $O^{\bullet -}$ and $O_2^{\bullet -}$ species which may be formed as intermediates in oxidation catalysis, for example, by electron transfer between the catalyst surface and adsorbed oxygen. Moreover, small molecules such as NO (being paramagnetic per se) and CO which, upon adsorption, can be converted to paramagnetic $CO^{\bullet -}$ and/or change the spin state of the adsorption site, are frequently used as probe molecules to analyze the local structure of special surface sites on solid catalysts. Such approaches have been recently reviewed [18]. An example of using NO and CO as probe molecules to elucidate the valence state of active Co sites in zeolites is discussed in Section 11.3.1, while the beneficial use of the spin-trapping technique during *in-situ* EPR studies of a working fuel cell will be presented in Section 11.3.2.

Besides these two most important classes of species accessible by *in-situ* EPR in solid catalysts, a few catalytically relevant materials may show conduction electron spin resonance (CESR). This is the case, for example, for conducting polymers such as polyanilines (PANIs) or certain carbon-based supports. However, *in-situ* EPR studies of such systems are very rare. Thus, a supported PtO_2/PANI catalyst has been monitored during low-temperature oxidation of CO [19]. With the onset of catalytic activity, a steep increase of the CESR signal was detected due to the transfer of electrons from CO oxidation to the conduction band of the catalyst. Surprisingly, in a working fuel cell, the CESR signal of the Pt/carbon electrode was found to decrease—an effect which was attributed to the formation of water in the system [20].

This introduction, in which the main application fields of *in-situ* EPR spectroscopy have just been mentioned, is followed by a short section on very basic theoretical and instrumental aspects. In due course, some application examples will be presented in more detail to illustrate the benefits of EPR for the study of solid catalysts. Since not all aspects of the variety of *in-situ* EPR applications in heterogeneous catalysis can be tackled in detail, some typical applications such as the study of TMIs in oxide catalysts during gas-phase reactions (Section 11.3.1) and a rather new application, namely monitoring of a working fuel cell (Section 11.3.2), have been selected. Finally, an attempt is made to judge the potential and limitations of the method together with possible future needs.

11.2 BASIC PRINCIPLE AND INSTRUMENTATION OF EPR

The theory of EPR spectroscopy is rather complex and cannot be treated exhaustively in this chapter, which focuses only on some basic features to facilitate understanding of the application examples. For more detailed presentations, the reader is referred to other textbooks [1–3]. Basic aspects particularly related to EPR of TMIs are also discussed in several reviews [4–6, 8–11, 21].

The simplest case of a paramagnetic species contains a molecule, atom, or ion with one single electron, and thus obeys a total electron spin of $S = 1/2$. This is the case for radicals and radical anions but also for a number of TMIs, among them V^{4+} (d^1), Mo^{5+} (d^1), Cu^{2+} (d^9), and low-spin Co^{2+} (d^7) in octahedral symmetry. When such samples are brought into a homogeneous external magnetic field B_0, the electron spin vectors S can take only two different orientations, more or less parallel or antiparallel to the field direction z, which corresponds to states of different energy and is called the Zeeman interaction. The energy values, a function of the magnetic field, can be obtained by solving the Schrödinger equation $H\Psi = E\Psi$, in which H is the so-called spin Hamiltonian. When the environment of the paramagnetic site is isotropic, H takes the form in Equation 11.1, in which S and L are the spin and angular momentum vectors, B is the magnetic field vector, β is the Bohr magneton, $g_e = 2.0023$ is the free electron g factor, λ is the spin-orbit coupling constant, and g is the g factor of the single electron bound in the orbital, which can deviate considerably from g_e, depending on λ:

$$H = \beta\vec{B}\left(\vec{L} + g_e\vec{S}\right) + \lambda\vec{L}\cdot\vec{S} = \beta\vec{B}\cdot g\cdot\vec{S}. \tag{11.1}$$

By irradiating the sample with microwaves, transitions of the electron spin between the two energy levels can be induced when the energy of the radiation matches exactly the energy difference of the two levels. In continuous-wave (cw) EPR—the technique most frequently used for *in-situ* studies of solid catalysts—radiation of fixed frequency is used while the magnetic field is swept until the resonance condition (Eq. 11.1) is fulfilled. Then, a single isotropic line would be recorded at the respective field value B_0.

When the nucleus of the atom to which the electron is attached has a nonzero nuclear spin I, dipolar coupling between the electron spin and the nuclear spin can split the electron energy levels into $(2I + 1)$ so-called hyperfine structure (hfs) levels. This is the case, for example, for vanadium ($I = 7/2$) (Fig. 11.1), copper ($I = 3/2$), and cobalt ions ($I = 7/2$). For a freely tumbling VO^{2+} vanadyl ion in solution, eight hfs sublines of equal distance and intensity

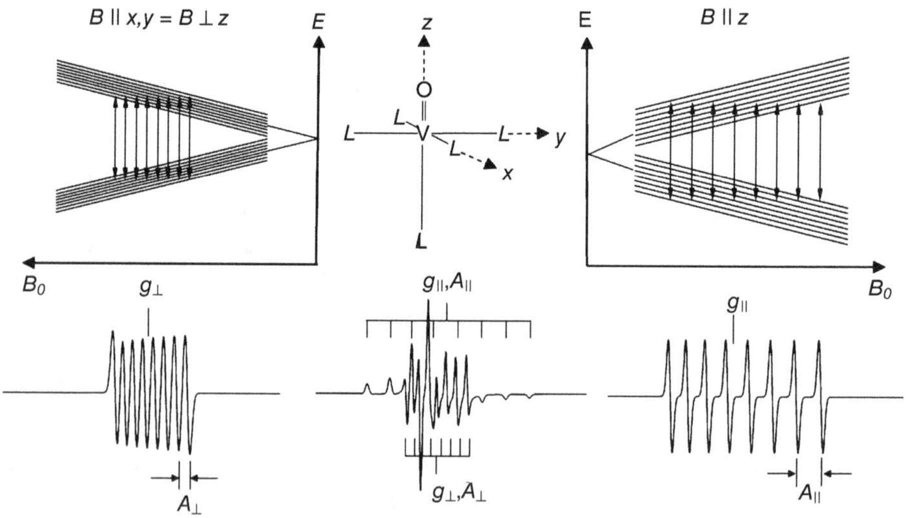

Figure 11.1. Energy level splitting and resulting EPR (sub)spectra for an axially symmetric VO^{2+} species. The schemes and subspectra on the left and right side correspond to the orientation of B_0 along (right) and perpendicular (left) to the z-axis of a distorted $O{=}V^{4+}O_5$ species (middle), observable only in single crystals while the sumspectrum (bottom middle) reflects the superposition seen in powder spectra.

would be observed. The distance between the eight hfs lines is characterized by the hyperfine coupling constant A, while the position of the signal is reflected by the g factor. As a matter of fact, the local environment of a vanadyl ion is not isotropic but axially symmetric with a short V=O bond (taken as the z-axis) and four bonds of equal length in the basal plane (Fig. 11.1). Therefore, the g and A parameters are anisotropic as well and depend on the angle between the z-axis and the direction of the external magnetic field. The entities in the spin Hamiltonian comprising Zeeman and hyperfine interactions of the single electron are then reflected by the first two terms in Equation 11.2, in which $\hat{S}$ and $\hat{I}$ are the spin matrices and g and A are no longer scalars but tensors, that is, 3 × 3 matrices which reduce to their diagonal form with g_{xx}, g_{yy}, and g_{zz}, and A_{xx}, A_{yy}, and A_{zz} as principal values while all off-diagonal elements vanish:

$$H = \beta \vec{B} \mathbf{g} \hat{S} + \hat{S} \mathbf{A} \hat{I}. \tag{11.2}$$

For an axial center such as VO^{2+}, $g_{xx} = g_{yy} = g_{\perp}$, $g_{zz} = g_{\parallel}$, $A_{xx} = A_{yy} = A_{\perp}$, and $A_{zz} = A_{\parallel}$. In the powder spectrum of a solid catalyst containing VO^{2+} single sites, in which anisotropy of g and A is not averaged out by free tumbling as in solution, two groups of eight hfs lines superimpose, giving rise to a very characteristic powder spectrum (Fig. 11.1, bottom middle). From this, the parameters $g_{\perp}$, $g_{\parallel}$, $A_{\perp}$, and $A_{\parallel}$ can be derived by spectra simulation and may be used as probes for reaction-dependent coordination changes [22].

Many catalytically relevant TMI such as, for example, Fe^{3+} (d^5), Mn^{2+} (d^5), and Cr^{3+} (d^3) contain more than one unpaired electron and, thus, $S > 1/2$. In this case, the mutual interaction between these electrons called fine structure (fs) must be taken into account and an additional fs term, $\hat{S}\mathbf{D}\hat{S}$, must be added to the spin Hamiltonian in Equation 11.2. This case, which gives rise to more than one allowed spin transition, is discussed in more detail for Fe^{3+} and Cr^{3+} elsewhere [8, 9].

The powder spectrum in Figure 11.1 (bottom middle) with the well-resolved hfs lines is characteristic of isolated VO^{2+} single sites which do not magnetically interact with each other. As the concentration of such sites increases and mutual magnetic interactions become dominating, those splittings are no longer resolved and the EPR spectra of such coupled paramagnetic species collapse into single lines, which are weak and broad when spin–spin exchange is weak, but narrow and intense when this exchange is strong. Alternatively, EPR signals can vanish completely due to spin pairing when the electron spins in a sample are antiferromagnetically coupled. Such phenomena have been observed in many bulk vanadyl compounds, and procedures for evaluating the line shape of such exchange-narrowed EPR signals by the method of moments and for deriving exchange energies from their temperature dependence have been used to elucidate the role of electronic couplings for catalytic reactions [8, 9, 11].

The temperature dependence of EPR signals is also a crucial issue for *in-situ* EPR studies. The signal intensity, which is proportional to the population difference of the two energy levels N_1 and N_0 involved in the spin flip (Boltzmann law, k = Boltzmann constant, Eq. 11.3) and to the magnetic susceptibility χ_m (Curie law, C = Curie constant, Eq. 11.4), decreases with rising measurement temperature T:

$$I \sim \frac{N_1}{N_0} \approx 1 - \frac{g\beta B_0}{kT}, \tag{11.3}$$

$$I \sim \chi_m = \frac{C}{T}, \tag{11.4}$$

and this can lead to low signal-to-noise ratios for *in-situ* measurements at elevated temperature. Therefore, in many EPR studies of catalysts, samples have been pretreated under reaction-like conditions but then cooled to room temperature or below for recording the EPR spectra [4–7]. Nevertheless, such experiments, even though not performed at true reaction temperature, can provide very valuable information on catalytic mechanisms as well. One such example is presented in Section 11.3.1.

As mentioned earlier, conventional cw-EPR measurements are performed by irradiating the sample with microwaves of a fixed frequency while sweeping the magnetic field. Virtually all *in-situ* EPR studies under working conditions at elevated temperature, the majority of which deals with heterogeneous catalytic gas-phase reactions, have been performed in so-called X-band with a microwave frequency of ≈9.5 GHz. Various setups for this purpose have been recently reviewed [8,9], comprising both a special home-built high-temperature/high-pressure cavity which serves at the same time as a catalytic reactor as well as home-built heatable flow reactors implemented in commercial EPR cavities. An example for the latter case, which allows along with the EPR spectra simultaneous recording of ultraviolet–visible (UV-Vis) and Raman spectra as well as analysis of the product stream by gas chromatography (GC), is shown in Figure 11.2A. For such kind of experiments, the label *operando* has been introduced a few years ago. In contrast to *in-situ* EPR, which has been used for a long time in a wider sense, this implies that inevitably together with the EPR spectra, catalytic activity must be measured to ensure that the catalyst is really in its working state. The reactor in Figure 11.2A can be heated at normal pressure up to about 830 K by a preheated stream of nitrogen. The special benefits of this setup, which was the first *operando* three-in-one spectroscopy, have been demonstrated for monitoring the oxidative dehydrogenation of propane over vanadium-containing catalysts [23, 24]. Equipped with a thick-walled (1.5 mm) quartz capillary (2.5 mm inner diameter) and a pressure controller/reducer on the inlet and outlet, respectively, the same setup has been used to monitor the dimerization of butenes up to a total pressure of 20 bar under flow conditions [25].

The first setup with a working fuel cell directly implemented in the X-band cavity of the EPR spectrometer has been presented by the group of Roduner (Fig. 11.2B) [19]. The reactor consists of two Teflon® (DuPont, Wilmington,

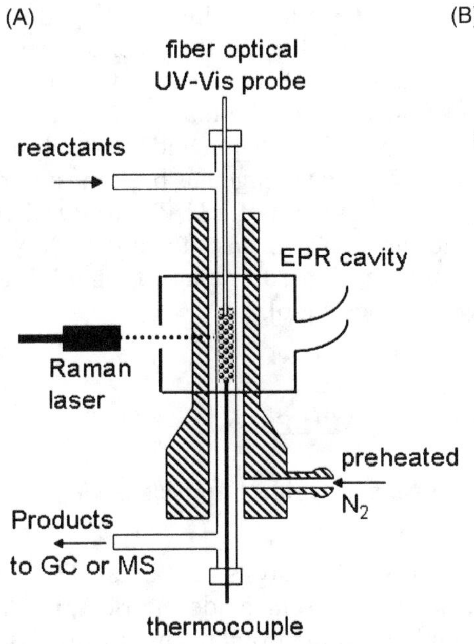

Figure 11.2. Schemes of setups for monitoring (A) heterogeneous catalytic gas phase reactions by simultaneous *operando* EPR/UV-Vis/Raman spectroscopy [23] and (B) working fuel cells by *in-situ* EPR spectroscopy [20]. MS, mass spectrometry; GC, gas chromatography.

DE) half cylinders, each of which contains an indention in which a Pt mesh (serving as spacer to allow gas distribution and electrical contact) and a gas diffusion electrode consisting of carbon-supported Pt particles is placed. Each half cylinder is equipped with two channels for gas inlet and outlet, which also contain silver wires for connecting the electrodes via a potentiostat. The latter are separated by a proton-conducting polymer electrolyte membrane (PEM). An application example of this setup is described in more detail in Section 11.3.2.

Commercial X-band EPR probe heads used for *in-situ* studies are rectangular cavities in which a standing microwave is adjusted by proper tuning [1]. This implies that the inner dimension of the cavity must fit to the wavelength which is 31 mm in X-band. The openings on top and bottom of such a cavity are 11 mm in diameter, which is large enough to accommodate a mantle Dewar including a 3 mm reactor tube inside (Fig. 11.2A). In certain cases, when the X-band EPR spectra are complex and contain several superimposed signals, for example, from hfs interaction, it can be very helpful to perform the measurements at a higher microwave frequency of ≈35 GHz (and consequently at higher magnetic field) in the so-called Q-band. This provides higher spectral resolution, since the splitting of the g tensor components (first term in Eq.

11.2) increases with rising magnetic field while the hfs splitting does not depend on the field (second term in Eq. 11.2) [10]. Yet the inner dimension of a Q-band cavity is much smaller than that of an X-band cavity, which imposes a technical challenge since the outer diameter of the mantle Dewar housing the flow microreactor cannot be larger than about 5 mm and the catalyst mass is restricted to about 2 mg. Very recently this challenge has been mastered [26, 27]. An *in-situ* EPR study in both X- and Q-band of a $H_4PVMo_{11}O_{40}$ heteropolyacid in the presence of methanol up to 523 K has shown that the superimposed spectra of Mo^{5+} and V^{4+} can be evaluated in a more reliable way by combining measurements at both frequencies [26].

11.3 SELECTED APPLICATION EXAMPLES

11.3.1 Monitoring Transition Metal Ions in Oxide Catalysts during Gas-Phase Reactions

The majority of *in-situ* EPR studies of solid catalysts containing TMIs deals with monitoring of vanadium species in different oxide matrices, mainly in selective oxidation reactions of aliphatic and aromatic hydrocarbons, and several reviews report on such investigations [7–11]. Thus, two industrial VO_x/Sb-TiO_2 catalysts containing 4–6 wt% vanadium and 9.5% antimony have been analyzed by *operando* EPR in the oxyhydrative scission of 1-butene and *n*-butane to acetic acid with online GC analysis of the effluent gas using the setup shown in Figure 11.2A [28]. First of all, as indicated by the higher total V^{4+} EPR intensity (compare Fig. 11.3A and Fig. 11.3B), a higher amount of initially pentavalent vanadium sites is reduced under reaction conditions with 1-butene as a reactant than with *n*-butane. In the presence of Sb, this reduction went down to EPR-silent V^{3+}. This has been concluded from the fact that the V^{4+} intensity increased by reoxidation of V^{3+} after exposing the catalyst to ambient air at the end of the *operando* EPR experiment. In the Sb-free VO_x/TiO_2 catalyst, partial reduction of V^{5+} stopped at the valence state +4, indicated by a continuously increasing V^{4+} EPR signal under reaction conditions, which remained constant after exposure to ambient air at the end (Fig. 11.3C). Interestingly, a strong decrease of 1-butene conversions was observed with time on stream over VO_x/Sb-TiO_2, but not over Sb-free VO_x/TiO_2. This has been related to the accumulation of hydrocarbon deposits on the surface of the former catalyst (detected by Fourier transform infrared [FTIR] spectroscopy) which might have been a consequence of the lower oxidation potential resulting from deep reduction to catalytically inactive V^{3+}. These inactive V^{3+} sites are supposed to dilute the active V^{4+} species which is evident from the partially resolved hfs in VO_x/Sb-TiO_2 after 5 h on stream. The Sb-free catalyst shows only an isotropic signal with rising amplitude and decreasing line width after 5 h (Fig. 11.3C), which arises from neighboring V^{4+} species connected by magnetic exchange interactions within clusters. Most probably, it is these clusters

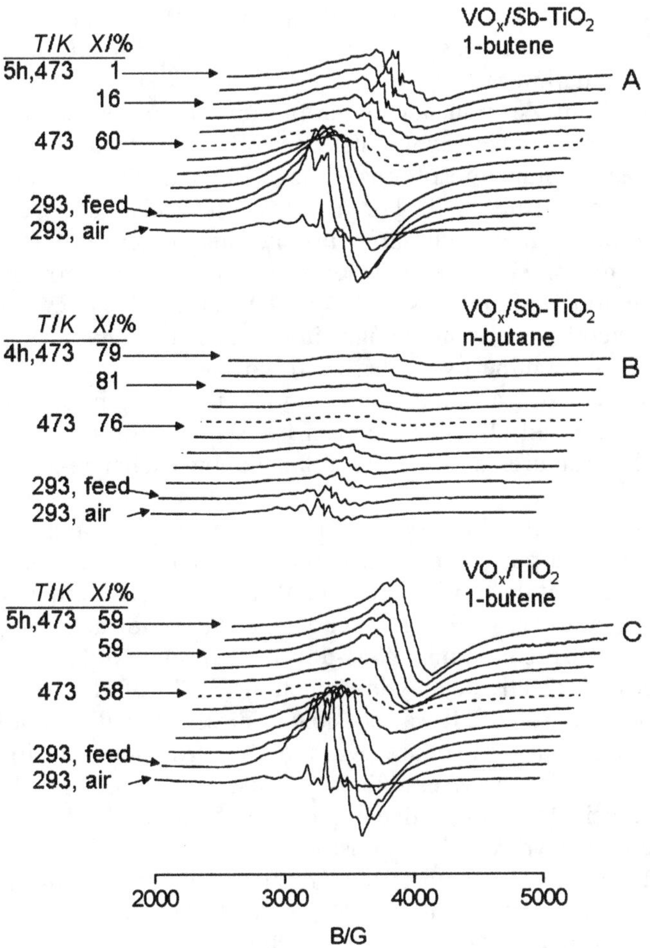

Figure 11.3. *Operando* EPR spectra of a 4.2 wt% V, 9.5 wt% Sb/TiO_2 catalyst in a flow of (A) 3% 1-butene, 8% O_2/N_2 and (B) 2.2% *n*-butane, 10% O_2/N_2, and (C) respective spectra of a 6.1 wt% V/TiO_2 catalyst in the same feed mixture as denoted in A. Temperature and hydrocarbon conversions are depicted on the left side of the spectra [28].

which account for an oxidizing power high enough to prevent accumulation of hydrocarbon deposits [28].

When *n*-butane was used instead of 1-butene as feed component, no such time-dependent deactivation was observed and also the overall conversion values were by about 30% higher over both types of catalysts (Fig. 11.3B). Remarkably, the total V^{4+} intensity remained significantly lower throughout the whole duration of the experiment with *n*-butane, suggesting a higher equilibrium V valence state probably very close to +5 in the working state, which might account for the higher activity in this case.

Frequently it has been claimed that useful EPR information can only be obtained from isolated paramagnetic sites but not from those coupled by effective magnetic interactions like in Figure 11.3B, since the latter average out splittings from Zeeman and hyperfine interaction which bear valuable structural information. However, it has been shown particularly for a series of different bulk vanadium phosphorus oxide (VPO) catalysts containing V^{4+} that this is not generally true [8, 11]. For such cases, a special data evaluation procedure based on the method of moments and the temperature dependence of the EPR line intensity has been developed, which allows to correlate reaction-dependent temporal alterations of the spin-exchange efficiency with catalytic activity, reflected by transient changes in the EPR line shape [8, 11]. Thus, it was found for crystalline $(VO)_2P_2O_7$ and various other bulk VPO catalysts used in selective hydrocarbon oxidation and ammoxidation that high catalytic performance required effective electron transport pathways throughout the catalyst bulk structure, which were reflected by exchange-narrowed EPR single lines.

Another important field of *in-situ* EPR studies in solid catalysts comprises zeolites and related porous materials containing TMIs such as Cu and Fe as active sites. These catalysts have been mainly applied in abatement of nitrogen oxides from exhaust gases. Thus, Kucherov et al. performed comprehensive investigations of Cu-containing ZSM-5 in the presence of gases relevant for feed mixtures of selective catalytic reduction (SCR) of NO_x. This work has been recently reviewed [8]. It was possible to discern two different Cu^{2+} single sites in square-planar and square-pyramidal environment, the former being more active due to better accessibility by reactants. This accessibility was also found to be crucial for being active in the case of Cu^{2+} in ZSM-5, ZSM-11, and ZSM-12 used for the direct decomposition of NO by Kustova et al. [29]. Both groups used the Bruker (Rheinstetten, Germany) high-temperature cavity ER 4114 HT equipped with a homemade flow cell. ZSM-11 and ZSM-12 were almost twice as active as ZSM-5 and it was claimed that this is due to the easier accessible Cu^{2+} sites in the straight channels of these matrices which do not exist in ZSM-5.

Besides Cu-containing zeolites, also those with Fe ions in cationic pore positions have proved to be very effective catalysts for SCR of nitrogen oxides, whereby different reducing agents such as NH_3, CO, and hydrocarbons have been tested [30, 31]. Even zeolites with very low Fe contents were highly active, for example, a 0.2% Fe–ZSM-5 [30, 32]. A long debate about the nature of the active Fe site has been ongoing, cycling mainly about the question whether the active Fe species is a single site, an Fe–O–Fe dimer, or a small oligonuclear cluster. Among the variety of characterization techniques used to answer this question, a combination of *in-situ* EPR and *in-situ* UV-Vis spectroscopy was most valuable since both techniques can distinguish single sites in different coordination environments as well as agglomerates of different size [32]. Here we will only consider the EPR spectra obtained from a 0.2% Fe–ZSM-5 during SCR of NO at 623 K (Fig. 11.4). The oxidized catalyst in air shows three types

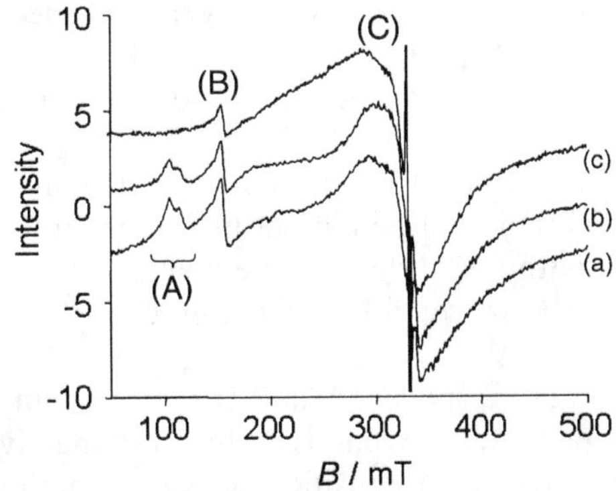

Figure 11.4. *In-situ* EPR spectra at 623 K of a 0.2% Fe/ZSM-5 catalyst (a) in air flow; (b) after switching from air to a flow of 0.1% NO, 0.1% NH_3, and 2% O_2/He; and (c) after switching from air to a flow of 0.1% NO, 0.1% isobutane, and 2% O_2/He [32]. Assignment of signals at effective *g* values of (A) 6.0 and 5.6, (B) 4.3, and (C) 2.0 is given in the text.

of signals at effective *g* values of (A) 6.0 and 5.6, (B) 4.3, and (C) 2.0. The theory behind their assignment is explained in more detail elsewhere [8, 9, 33]. Signals A and B arise from isolated Fe^{3+} ions in octahedral (A) and tetrahedral (B) coordination, while signal C can originate either from highly symmetric Fe^{3+} single sites, the coordination number of which cannot be discerned, or from oxidic Fe_xO_y clusters of different nuclearity. Clusters have been widely excluded for the sample shown in Figure 11.4, based on the results of UV-Vis spectra [32]. Thus, signal C in Figure 11.4 is supposed to arise mainly from Fe^{3+} single sites.

Upon switching from air flow to the SCR feed containing a mixture of 0.1% NO, 0.1% NH_3, and 2% O_2 in helium, a slight intensity decrease due to permanent reduction of Fe^{3+} to EPR-silent Fe^{2+} is observed for signals A and B but not for signal C, illustrating clearly that Fe sites located in different pore positions of the ZSM-5 matrix show different redox stability. Most sensitive against irreversible reduction are those sites giving rise to signal A (distorted octahedral coordination), followed by signal B (distorted tetrahedral coordination), while species C is most stable in the valence state +3. Wichterlova et al. have observed three different single sites with the same reducibility trend as reflected in Figure 11.4 in an *in-situ* UV-Vis study of Co–ZSM-5, based on which they assigned three distinct pore positions to these metal sites, namely a position in the straight channels (α), in the intersection between the straight and the sinusoidal channels (β), and in a boat-shaped site in the sinusoidal channels (γ) [34]. The same approach was later applied to Fe–ZSM-5 as well [35]. Presuming that the same relation between site distortion, reducibility, and pore position is true for the sample in Figure 11.4, the Fe sites reflected by

EPR signals A, B, and C have been tentatively assigned to the previously mentioned α-, β-, and γ-sites [36].

When isobutane is used instead of NH_3 as reducing agent, this permanent reduction is much more pronounced (Fig. 11.4). Interestingly, the catalytic activity under these conditions was markedly lower than with NH_3 [32]. Considering also *in-situ* FTIR results, which point to the preferential reaction of NO with Fe^{3+} but not with Fe^{2+}, it has been concluded that only those Fe sites, which are stable against irreversible reduction, catalyze the SCR of NO at 623 K.

Remarkably, the same Fe species A and B, which seem to be not relevant for the SCR of NO, since they are readily reduced to inactive Fe^{2+}, were found to be the active sites in the decomposition of N_2O to highly active $O^{\bullet-}$, which can even oxidize CO to CO_2 at room temperature. This has been shown on monitoring Fe–ZSM-5 during alternating exposure to N_2O and CO flow [36].

In a very elegant *in-situ* EPR study, in which they used CO as a probe molecule besides NO and propene reactants as well as dedicated adsorption/reaction experiments at elevated temperatures followed by spectra recording at 77 K, Pietrzyk and Sojka have shown for the first time that the active Co sites in Co–BEA and Co–ZSM-5 catalysts are passing through a Co^{2+}/Co^0 redox cycle during SCR of NO by propene [13]. After treatment in vacuum at 773 K, their zeolites contained essentially high-spin Co^{2+} ions (d^7, $S = 3/2$) in extraframework positions which, due to short relaxation times, did not show any EPR signal at 77 K. However, when CO was adsorbed at 77 K, a multiline spectrum was observed (Fig. 11.5A). By spectra simulation, three superimposed subsignals arising from species with one single electron ($S = 1/2$) and hyperfine coupling to a nucleus with $I = 7/2$ have been identified (Table 11.1), which were assigned to low-spin Co^{2+}–CO complexes in different pore positions, based on comparison with literature data of Co carbonyl complexes and the well-known fact that CO as a strong ligand forces the central metal ion to obey the low-spin state.

When NO was adsorbed instead of CO, a cobalt EPR signal with $S = 1/2$ and $I = 7/2$ was observed, too, yet with very different g and A parameters (Fig. 11.5B, Table 11.1). Since NO as a rather weak ligand does not favor the low-spin state of cobalt, this signal was assigned to a high-spin $[Co(NO)_2]^{2+}$ dinitrosyl complex formed by ligand-to-metal electron transfer, which leaves Co in a formal zerovalent state [13]. When this sample was evacuated at 573 K followed by adsorption of CO at 77 K, the EPR signal changed again (Fig. 11.5C), being now characterized by g and A values compatible with those of organometallic Co^0 complexes (Table 11.1). This is clear evidence that NO is able to reduce Co^{2+} to Co^0. Note that the EPR spectra in Figure 11.5A,C both arise from Co carbonyl complexes with $S = 1/2$, yet with Co in different valence states, illustrating impressively the suitability of *in-situ* EPR spectroscopy for discriminating between different Co valence states. In comparison to low-spin Co^{2+} (d^7, Fig. 11.5A), the signals of Co^0 (d^9, Fig. 11.5B,C) show lower g_{iso} and higher A_{iso} values.

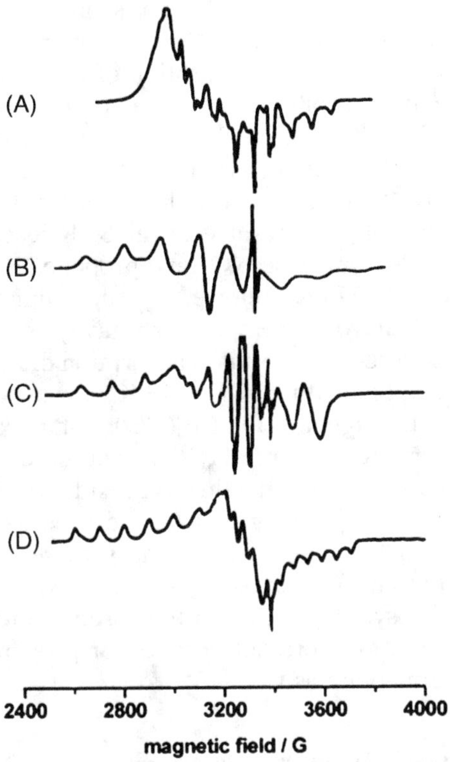

Figure 11.5. *In-situ* EPR spectra of a Co–BEA zeolite (A) after adsorption of CO at 77 K (signal arising from a low-spin Co^{2+}–CO complex, $S = 1/2$); (B) after adsorption of NO at 77 K (signal arising from a $Co^{2+}(NO)_2$ complex, d^9, $S = 1/2$); (C) after adsorption of CO at 77 K on the $Co^{2+}(NO)_2$ complex that had been heated to 573 K before (signal arising from a $Co^0(CO)_n$ complex); and (D) after adsorption of propene at 77 K on the $Co^{2+}(NO)_2$ complex that had been heated to 573 K before (signal arising from a Co^0–propene complex) [13].

TABLE 11.1. Spin Hamiltonian Parameters of Different Co Complexes ($S = 1/2$ in All Cases) in Co-BEA Derived by Spectra Simulation [10]

Species	g_{xx}	g_{yy}	g_{zz}	A_{xx} (mT)	A_{yy} (mT)	A_{zz} (mT)
$Co^{2+}(CO)$, d^7, low spin	2.34	2.179	2.016	4.4	3.4	7.6
	2.259	2.149	1.986	3.5	1.8	5.8
	2.239	2.155	2.066	4.5	3.7	7.6
$[Co(NO)_2]^{2+}$, d^9	2.081	2.195	2.086	18.2	10.2	3.1
$Co^0(CO)_n$, d^9	2.105	2.199	1.973	4.8	12.5	3.5

When propene was adsorbed on the initial sample containing exclusively EPR-silent high-spin Co^{2+} after thermal pretreatment in vacuum, no EPR signal was observed [13]. However, adsorption of propene on a sample prereduced by adsorption of NO followed by evacuation at 573 K gave rise to the spectrum plotted in Figure 11.5D, which has been assigned to a Co^0(propene) complex. Note that the g_{iso} and A_{iso} values are comparable to those of the $Co^0(CO)_n$ and $Co^0(NO^+)_2$ complexes (Table 11.1). Finally, the authors have confirmed the relevance of these results for the SCR reaction by treating the catalysts with propene/NO, propene/NO_2, and propene/NO/O_2 mixtures in the temperature range of 293–773 K, followed by quenching to 77 K for spectra acquisition. Treatment at room temperature gave rise to the signal of a $[Co(NO^+)_2]^{2+}$ complex observed, too, upon adsorption of NO only (Fig. 11.5A). This line disappeared upon heating to $T \leq 473$ K. However, at 573 K when the SCR reaction started, the signal of the $Co^0(CO)_n$ complex known from separate adsorption experiments in Figure 11.5C was observed, which disappeared again at higher temperature, probably due to reoxidation of Co^0 to Co^{2+} (confirmed by subsequent adsorption of NO on the used catalyst restoring the signal of the$[Co(NO^+)_2]^{2+}$ complex). In summary, these experiments provide conclusive evidence that the SCR of NO by propene over Co zeolites proceeds via formation of dinitrosyl intermediates that promote reduction of Co^{2+} to Co^0, which in turn serve as coordination sites for propene and CO being a possible reaction intermediate as well [13].

11.3.2 Radical-Induced Membrane Damage in Fuel Cells

PEM fuel cells are gaining more and more attention as clean future energy sources since they do not emit CO_2 in contrast to fossil fuels, when hydrogen is used as fuel [37]. The membrane is the core of such a device and separates the cathode and anode compartments. With hydrogen as fuel, the following processes occur:

$$H_2 \rightarrow 2H^+ + 2e^- \text{ (anode)}, \tag{11.5}$$

$$\frac{1}{2}O_2 + 2e^- + 2H^+ \rightarrow H_2O \text{ (cathode)}, \tag{11.6}$$

where the protons penetrate through the membrane while the electrons create an electrical current when the circuit between the two electrodes is closed. Instead of H_2, other fuels such as methanol or ethanol are also used. Catalysts are usually noble metal particles such as Pt which are deposited on both sides of the membrane. A major problem is the limited stability of the membrane which has to withstand rather harsh operation conditions due to the attack of *in-situ* formed highly reactive $^{\bullet}OH$ radicals. For rational development of such membranes, detailed knowledge on the formation and reactions of such aggressive intermediates would be desirable. However, the frequently used conventional electrochemical methods such as cyclovoltammetry or current–

voltage analysis can neither identify such radicals nor easily differentiate between the various reaction steps. This has been for the first time achieved by Roduner and his group, who were able to monitor the reactions proceeding in a working fuel cell by *in-situ* EPR using the tailor-made setup depicted in Figure 11.2B [20].

They used Pt nanoparticles as a catalyst and two different membrane materials, Nafion® 115 (DuPont), which is a copolymer of perfluorosulfonic acid and polytetrafluoroethylene, and a fluorine-free cross-linked membrane consisting of polyetherketone and polyethersulfone [38]. Since the concentration of the formed radicals was too low and their lifetime was too short to be detected directly, the spin traps α-(4-pyridyl-1-oxide)-*N-tert*-butylnitrone (POBN) and 5,5-dimethyl-1-pyrroline-1-oxide (DMPO) were used to trap the radicals as stable paramagnetic spin adducts. To analyze the anode and cathode processes separately, aqueous solutions of the spin trap were introduced either into the anode or the cathode compartment.

With POBN at the anode, an EPR signal reflecting the super-hyperfine structure (shfs) interaction of the single electron with the nuclear spins of one N atom ($I = 1$) and two equivalent H atoms ($I = 1/2$) was observed after 5 min of operation (Fig. 11.6A), which was independent of the membrane material. Interestingly, coupling to one H and one D atom was observed upon replacing D_2O for H_2O as solvent for the spin trap, while this was not observed when H_2 as fuel was replaced by D_2. This confirmed that electrons from the fuel (Eq. 11.5) reduce the spin trap on the anode surface which then abstracts a proton from the solvent (Fig. 11.6A). The authors have also deposited a solution of POBN in a water/ethanol mixture on the anode to check for the reaction of ethanol as a potential fuel of the PEM. In this case, the EPR spectrum shows the presence of two radicals, the one observed already in Figure 11.6A (left) and another one showing shfs of the single electron with the nuclear spin of N but only with one H atom. From this observation it has been concluded that ethanol is oxidized on the anode, forming a C-centered radical that is trapped by POBN (Fig. 11.6B). In contrast to a POBN$^•$-H or POBN$^•$-D adduct, a POBN$^•$-R adduct does not give rise to additional shfs from the C of the R group.

To test the stability of the membrane, the same experiments have also been performed with deposition of the aqueous POBN solution onto the cathode [20]. In this case, the typical EPR spectrum of a surface-immobilized POBN$^•$-R radical with shfs from N and one H nucleus was found, the intensity of which was much larger when an F-free polymer membrane was used, while it was negligible with a Nafion membrane. It has been supposed that this radical was formed as a consequence of trapping degradation products of the membrane originating from an attack of $^•$OH radicals formed by oxygen reduction on aromatic rings of the polymer (Eq. 11.6). The fact that these radical signals showed much broader and partly anisotropic lines in comparison to the respective POBN$^•$-R radicals generated photochemically in an ethanol/water solution of POBN was taken as an indication that they are immobilized on the membrane surface after having been formed by membrane degradation.

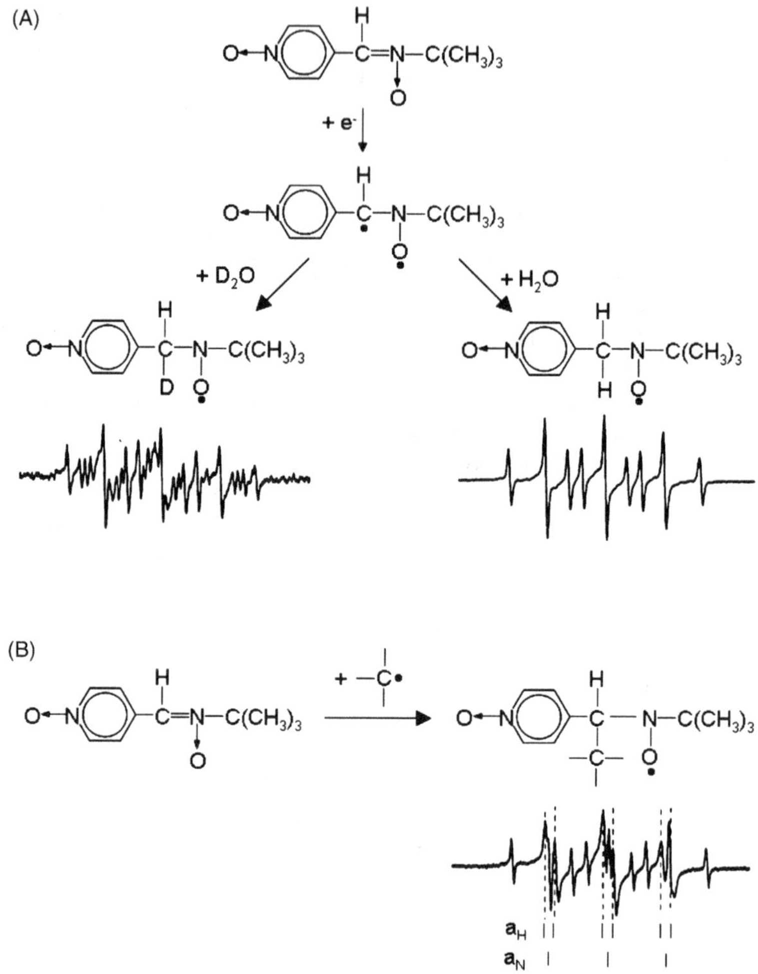

Figure 11.6. (A) Reduction of the spin trap POBN at the anode surface, followed by abstraction of H^+ or D^+ from H_2O or D_2O used as solvent for POBN and (B) trapping of a C-centered radical formed by oxidative degradation of ethanol at the anode surface [38].

To further explore the possibility of membrane damage by *in-situ*-formed $^{\bullet}OH$ radicals, Roduner et al. studied the reaction of different polyaromatic and Nafion membranes in the presence of Fenton's reagens (well known to produce highly reactive $^{\bullet}OH$ and $^{\bullet}OOH$ radicals) at room temperature and in the presence of H_2O_2 up to 80°C [39]. Only for the fluorine-free polyaromatic membranes, immobilized radicals on the surface deriving from membrane damage could be detected, while this was not the case for the Nafion membrane. During *in-situ* EPR measurements with this membrane in the working fuel cell (i.e., under closed-circuit conditions), the characteristic spectrum of the DMPO-H adduct but not the one of DMPO-OH was detected on both the anode and the cathode side when DMPO was used as a spin trap, yet the radical concentration was about twice as high on the anode side. Its formation is supposed to occur by reaction of reduced DMPO with water (Fig. 11.6). Interestingly, the number of the resulting DMPO-H species was found to be only 1 ppm of that of the produced Faraday electrons, which indicated that the fuel cell was properly working and the extent of side reaction leading to radical formation and potential membrane damage was negligible, at least within the period of time studied (several hours).

Recently, very much the same *in-situ* EPR fuel cell setup as that introduced by Roduner et al. was used to study the influence of Ce(III) on the stability of a Nafion 117 membrane [40]. In this case, 10% of the protons in the membrane had been neutralized by Ce(III) and the behavior of this membrane was compared to the Ce-free analogue. DMPO was used as a spin trap to identify radicals on both the anode and the cathode during operation of the fuel cell. For the Ce-free reference membrane, results similar to those of Roduner et al. were obtained: On the anode, only the H-Adduct of the spin trap was detected, while on the cathode, trapped $^{\bullet}$OH and carbon-centered $^{\bullet}$R radicals (originating from membrane degradation) were found.

The introduction of Ce(III) into the membrane led, first of all, to a lower abundance of all radicals on both sides of the membrane [40]. Furthermore, trapped $^{\bullet}$R radicals reflecting membrane damage as well as $^{\bullet}$OH radicals could not be detected anymore. It has been argued that Ce(III) quenches the highly reactive (and therefore destructive) $^{\bullet}$OH radicals (Eq. 11.7), which in turn hinders membrane damage via $^{\bullet}$R radical formation. Different from the Ce-free Nafion membrane, no trapped $^{\bullet}$OH but less reactive $^{\bullet}$OOH radicals were detected on the cathode of the Nafion/Ce(III) membrane, the formation of which was attributed to the reaction of Ce(VI) with H_2O_2 (Eq. 11.8). Thus, the operating Ce(III)/Ce(IV) redox couple is regarded to be essential for rapid quenching of the highly destructive $^{\bullet}$OH radicals and, thus, stabilization of the membrane:

$$^{\bullet}\mathrm{OH} + \mathrm{Ce(III)} + \mathrm{H}^+ \rightarrow \mathrm{H_2O} + \mathrm{Ce(IV)}, \tag{11.7}$$

$$\mathrm{Ce(IV)} + \mathrm{H_2O_2} \rightarrow \mathrm{Ce(III)} + {}^{\bullet}\mathrm{OOH} + \mathrm{H}^+. \tag{11.8}$$

In summary, these studies have shown that *in-situ* EPR investigations are possible even under the harsh conditions of a working fuel cell, in which proper tuning of the cavity must be realized in the presence of an electrical current and certain amounts of water, and that the spin-trapping technique can beneficially be applied to visualize the different extent of membrane damage depending on the membrane material.

11.4 CONCLUSIONS AND OUTLOOK

In this chapter, the versatility of *in-situ* and *operando* EPR studies of heterogeneous catalytic systems for deriving structure–reactivity relationships and details of reaction mechanisms has been exemplified. Despite its restriction to systems containing unpaired electrons, this technique bears valuable potential for the wealth of oxide catalysts containing TMIs, since almost any reaction catalyzed by such materials implies changes of the valence states. Many of such ions are accessible by EPR, when they contain single electrons. If such reaction-dependent redox cycles involve TMIs in their highest oxidation states without single electrons, simultaneous coupling of EPR with UV-Vis spectroscopy will turn out to be very beneficial, since such highly oxidized TMIs are sensitively detected via their charge-transfer transitions in the UV-Vis spectra—an aspect that is not considered in this chapter but in several previous papers [2, 5, 23].

Besides paramagnetic TMI, *in-situ* EPR is the method of choice when radical intermediates are involved in the reactions under study. Unfortunately, such radical intermediates are frequently very reactive and, thus, too short-lived to be accessible by *in-situ* EPR at elevated temperatures. Cleverly designed spin-trapping experiments can help in this case to identify the radical intermediates, yet care must be taken to select a suitable scavenger molecule, the spin adduct of which allows to deduce the nature of the mother radical from the g factor and/or the hyperfine coupling pattern. This has been illustrated in Section 11.3.2. Moreover, with the fuel cell example in that section, it has been demonstrated that EPR is pretty tolerant against the variety of *in-situ* setups implemented in the cavity as long as the samples do not absorb microwaves too much and proper tuning of the cavity can be ensured. Apart from the setups depicted in Figure 11.2 and cited in Section 11.2 [8, 9, 25], in which energy is transferred to the sample by heating, commercial cavities with grids in the front side have been used to study powder samples under light irradiation. There is a wide variety of photocatalytic *in-situ* EPR studies, the majority of which is focused on monitoring TiO_2 while shining UV and/or visible light on it either in the absence or in the presence of reactants [17, 41]. Taking account of the growing importance of sustainable hydrogen production (e.g., by photocatalytic water splitting) and the need to find new ways for converting CO_2 to valuable products (e.g., by artificial photosynthesis), it is expected that such studies will gain even more attention in the near future. Very recently, the first simultaneous *in-situ* EPR/Raman study of a homogeneous photocatalytic water reduction system comprising an iridium photosensitizer and an iron carbonyl water reduction catalyst has been published [42]. This new variant of *in-situ* EPR is now being adapted for monitoring heterogeneous photocatalytic water splitting and CO_2 converting systems with solid catalysts.

This chapter is restricted to the application of EPR spectroscopy in the continuous-wave operation mode, since only with this technique are *in-situ* studies at elevated temperatures possible, which are required in the vast majority of heterogeneous catalytic processes. Although sophisticated time-resolved pulse EPR techniques usually provide a much higher time resolution down to the microsecond range and a more detailed insight into the interactions of single paramagnetic species with their environment (neighboring sites in the solid and/or adsorbed molecules), they can only be applied at very low T (usually well below liquid nitrogen temperature) and on systems that contain essentially single paramagnetic sites with negligible mutual magnetic interactions. Both requests are essential to ensure the detection of a spin echo which is the basis for pulse EPR techniques such as echo-detected EPR (ED-EPR) and electron spin-echo envelope modulation (ESEEM). However, despite this limitation, pulse EPR techniques are highly valuable for investigating the structure or/and interaction of potentially active sites in model catalysts (e.g., transition metal species in zeolites [43, 44] or well-defined oxide phases such as heteropolyacids [45]) with various adsorbates that could serve as reactants in catalytic processes.

Finally, it should be mentioned that EPR imaging studies, when applied *in situ* during heterogeneous catalytic reactions, could open a new perspective to gain spatially resolved information on the participation of, for example, different regions of a catalyst pellet in a certain reaction. This has already been realized with other methods such as nuclear magnetic resonance (NMR), X-ray absorption and diffraction, optical spectroscopy, and electron microscopy [46]. Although there are numerous examples of EPR imaging, particularly in biology and medicine but also in materials science [47, 48], its introduction into *in-situ* studies of solid catalysts is still a dream of the future, though it does not seem to be impossible since there have been respective approaches already years ago. In fact, this technique has already been used to study the distribution of Mo^{5+} in a supported $H_3Mo_{12}O_{40}/SiO_2$ catalyst [49].

In summary, it has been shown in this chapter that continuous-wave *in-situ* EPR spectroscopy is a valuable technique to study the behavior of paramagnetic sites in working solid catalysts under a wide variety of conditions such as elevated temperature, flowing gases or liquids, and under irradiation with light. Future challenges remain in adapting the experimental setups more closely to real industrial reaction conditions, for example, to elevated pressures and in introducing spatial resolution.

REFERENCES

[1] Weil, J. A., Bolton, J. R., Wertz, J. E. (1994) *Electron Paramagnetic Resonance, Elementary Theory and Practical Applications*, John Wiley & Sons Inc., New York.

[2] Pilbrow, J. R. (1990) *Transition Ion Electron Paramagnetic Resonance*, Clarendon Press, Oxford, UK.

[3] Gerson, F., Huber, W. (2003) *Electron Spin Resonance Spectroscopy of Organic Radicals*, Wiley-VCH, Weinheim, Germany.

[4] Louis, C., Lepetit, C., Che, M. (1994) EPR characterization of oxide supported transition metal ions: relevance to catalysis, *Molecular Engineering*, 4, 3–38.

[5] Dyrek, K., Che, M. (1997) EPR as a tool to investigate the transition metal chemistry on oxide surfaces, *Chem. Rev.*, 97, 305–331.

[6] Sojka, Z. (1995) Molecular aspects of catalytic reactivity. Application of EPR spectroscopy to studies of the mechanism of heterogeneous catalytic reactions, *Catal. Rev. Sci. Eng.*, 37, 461–512.

[7] Labanowska, M. (2001) EPR monitoring of redox processes in transition metal oxide catalysts, *ChemPhysChem*, 2, 712–731.

[8] Brückner, A. (2004) Electron paramagnetic resonance, in Weckhuysen, B. M. (ed.), *In-situ Spectroscopy of Catalysts*, American Scientific Publishers, Stevenson Ranch, CA, pp. 219–252.

[9] Brückner, A. (2009) Electron paramagnetic resonance: a powerful tool for monitoring working catalysts, *Adv. Catal.*, 52, 265–308.

[10] Brückner, A. (2010) *In situ* electron paramagnetic resonance: a unique tool for analyzing structure-reactivity relationships in heterogeneous catalysis, *Chem. Soc. Rev.*, 39, 4673–4684.

[11] Brückner, A. (2003) Looking on heterogeneous catalytic systems from different perspectives: multitechnique approaches as a new challenge for *in situ* studies, *Catal. Rev. Sci. Eng.*, 45, 97–150.

[12] Dyrek, K., Kiessling, D., Labanowska, M., Wendt, G., Widziszewska, J. (1993) Interaction of butenes with NiO-Al_2O_3/SiO_2 catalyst studied by electron paramagnetic resonance, *Colloids Surf. A*, 72, 183–190.

[13] Pietrzyk, P., Sojka, Z. (2007) Co^{2+}/Co^0 redox couple revealed by EPR spectroscopy triggers preferential coordination of reactants during SCR of NO_x with propene over cobalt-exchanged zeolites, *Chem. Commun.*, 1930–1932.

[14] Matir, W., Lunsford, J. H. (1981) The formation of gas-phase π-allyl radicals from propylene over bismuth oxide and γ-bismuth molybdate catalysts, *J. Am. Chem. Soc.*, 103, 3728–3732.

[15] Risse, T., Schmidt, J., Hamann, H., Freund, H.-J. (2002) Direct observation of radicals in the activation of Ziegler–Natta catalysts, *Angew. Chem. Int. Ed.*, 41, 1517–1520.

[16] Rehorek, D. (1991) Spin trapping of inorganic radicals, *Chem. Soc. Rev.*, 20, 341–353.

[17] Wang, Z., Ma, W., Chen, C., Ji, H., Zhao, J. (2011) Probing paramagnetic species in titania-based heterogeneous photocatalysis by electron spin resonance (ESR) spectroscopy—a mini review, *Chem. Eng. J.*, 170, 353–362.

[18] Chiesa, M., Giamello, E., Che, M. (2010) EPR characterization and reactivity of surface-localized inorganic radicals and radical ions, *Chem. Rev.*, 110, 1320–1347.

[19] Klasovsky, F., Hohmeyer, J., Brückner, A., Bonifer, M., Arras, J., Steffan, M., Lucas, M., Radnik, J., Roth, C., Claus, P. (2008) Catalytic and mechanistic investigation of polyaniline supported PtO2 nanoparticles: a combined *in situ*/operando EPR, DRIFTS, and EXAFS study, *J. Phys. Chem C*, 112, 19555–19559.

[20] Pachenko, A., Dilger, H., Möller, E., Sixt, T., Roduner, E. (2004) *In situ* EPR investigation of polymer electrolyte membrane degradation in fuel cell applications, *J. Power Sour.*, 127, 325–330.

[21] Vedrine, J. C. (1984) Electron spin resonance, in Delannay F. (ed.), *Characterization of Heterogeneous Catalysts*, Marcel Dekker, Inc., New York, p. 161.

[22] Janke, C., Schneider, M., Bentrup, U., Radnik, J., Martin, A., Scholz, G., Brückner, A. (2011) Impact of phosphorus and nitrogen on structure and catalytic performance of VZrPON oxynitrides in the ammoxidation of 3-picoline, *J. Catal.*, 277, 196–207.

[23] Brückner, A. (2005) Killing three birds with one stone, *Chem. Commun.*, 1761–1763.

[24] Brückner, A., Scholz, G., Heidemann, D., Schneider, M., Herein, D., Bentrup, U., Kant, M. (2007) Structural evolution of $H_4PVMo_{11}O_{40}xH_2O$ during calcination and isobutane oxidation: new insights into vanadium sites by a comprehensive *in situ* approach, *J. Catal.*, 245, 369–380.

[25] Brückner, A., Bentrup, U., Zanthoff, H., Maschmeyer, D. (2009) The role of different Ni sites in supported nickel catalysts for butene dimerization under industry-like conditions, *J. Catal.*, 266, 120–128.

[26] Stösser, R., Marx, U., Herrmann, W., Jabor, J., Brückner, A. (2010) *In situ* EPR study of chemical reactions in Q-band at higher temperatures—a challenge for elucidating structure-reactivity relationships in catalysis, *J. Am. Chem. Soc.*, 132, 9873–9880.

[27] Stösser, R., Herrmann, W., Marx, U., Brückner, A. (2011) Spin exchange in solutions of TEMPOL in *n*-octanol and 1-methyl-3-octylimidazolium hexafluorophosphate in the temperature range from 300 to 500 K, *J. Phys. Chem. A*, 115, 2939–2952.

[28] Brückner, A., Bentrup, U., Fait, M., Kubias, B. (2005) Structure-reactivity relationships in supported VO_x catalysts for the oxyhydrative scission (OHS) of 1-butene and n-butane to acetic acid: a comprehensive catalytic and *in situ* study, *Catal. Today*, 99, 123–129.

[29] Kustova, M. Y., Kustov, A. K., Christiansen, S. E., Leth, K. T., Rasmussen, S. B., Christensen, C. H. (2006) Cu-ZSM-5, Cu-ZSM-11, and Cu-ZSM-12 catalysts for direct NO decomposition, *Catal. Commun.*, 7, 705–708.

[30] Schwidder, M., Santhosh Kumar, M., Klementiev, K., Pohl, M. M., Brückner, A., Grünert, W. (2005) Selective reduction of NO with Fe-ZSM-5 catalysts of low Fe content: I. Relations between active site structure and catalytic performance, *J. Catal.*, 231, 314–330.

[31] Pérez-Ramírez, J., Groen, J., Brückner, A., Kumar, M. S., Bentrup, U., Debbagh, M. N., Villaescusa, L. A. (2005) Evolution of isomorphously substituted iron zeolites during activation: comparison of Fe-beta and Fe-ZSM-5, *J. Catal.*, 232, 318–334.

[32] Santhosh Kumar, M., Schwidder, M., Grünert, W., Bentrup, U., Brückner, A. (2006) Selective reduction of NO with Fe-ZSM-5 catalysts of low Fe content: part II. Assessing the function of different Fe sites by spectroscopic *in situ* studies, *J. Catal.*, 239, 173–186.

[33] Aasa, R. (1970) Powder line shapes in the electron paramagnetic resonance spectra of high-spin ferric complexes, *J. Chem. Phys.*, 52, 3919–3931.

[34] Dědeček, J., Kaucký, D., Wichterlová, B. (2000) Co^{2+} ion siting in pentasil-containing zeolites, part 3. Co^{2+} ion sites and their occupation in ZSM-5: a VIS diffuse reflectance spectroscopy study, *Microporous Mesoporous Mater.*, 35–36, 483–494.

[35] Čapek, L., Kreibich, V., Dědeček, J., Grygar, T., Wichterlová, B., Sobalík, Z., Martens, J. A., Brosius, R., Tokarová, V. (2005) Analysis of Fe species in zeolites

by UV-VIS-NIR, IR spectra and voltammetry. Effect of preparation, Fe loading and zeolite type, *Microporous Mesoporous Mater.*, 80, 279–289.

[36] Berrier, E., Ovsitser, O., Kondratenko, E. V., Schwidder, M., Grünert, W., Brückner, A. (2007) Temperature-dependent N_2O decomposition over Fe-ZSM-5: identification of sites with different activity, *J. Catal*, 249, 67–78.

[37] Wang, Y., Chen, K. S., Mishler, J., Cho, S. C., Adroher, X. C. (2011) A review of polymer electrolyte membrane fuel cells: technology, applications, and needs on fundamental research, *Appl. Energy*, 88, 981–1007.

[38] Pachenko, A., Dilger, H., Kerres, J., Hein, M., Ullrich, A., Kaz, T., Roduner, E. (2004) *In situ*-spin trap electron paramagnetic resonance study of fuel cell processes, *Phys. Chem. Chem. Phys.*, 6, 2891–2894.

[39] Vogel, B., Aleksandrova, E., Mitov, S., Krafft, M., Dreizler, A., Kerres, J., Hein, M., Roduner, E. (2008) Observation of fuel cell membrane degradation by ex situ and *in situ* electron paramagnetic resonance, *J. Electrochem. Soc.*, 155, B570–B574.

[40] Danilczuk, M., Schlick, S., Coms, F. D. (2009) Cerium (III) as a stabilizer of perfluorinated membranes used in fuel cells: *in situ* detection of early events in the ESR resonator, *Macromolecules*, 42, 9843–8949.

[41] Dimitrijevic, N. M., Vijayan, B. K., Poluektov, O. G., Rajh, T., Gray, K. A., He, H., Zapol, P. (2011) Role of water and carbonates in photocatalytic transformation of CO_2 to CH_4 on titania, *J. Am. Chem. Soc*, 133, 3964–3971.

[42] Hollmann, D., Gärtner, F., Ludwig, R., Barsch, E., Junge, H., Blug, M., Hoch, S., Beller, M., Brückner, A. (2011) New insights into the mechanism of photocatalytic water reduction by DFT-supported *In situ*-EPR/Raman-Spectroscopy, *Angew. Chem. Int. Ed.*, 50, 10246–10250.

[43] Baldansuren, A., Dilger, H., Eichel, R. A., van Bokhoven, J. A., Roduner, E. (2009) Interaction and reaction of ethylene and oxygen on six-atom silver clusters supported on LTA zeolite, *J. Phys. Chem. C*, 113, 19623–19632.

[44] Umamaheswari, V., Hartmann, M., Pöppl, A. (2005) Pulsed ENDOR Study of Cu(I)-NO adsorption complexes in Cu-L zeolite, *J. Phys. Chem. B*, 109, 10842–10848.

[45] Kaminker, I., Goldberg, H., Neumann, R., Goldfarb, D. (2020) High-field pulsed EPR spectroscopy for the speciation of the reduced $[PV_2Mo_{10}O_{40}]^{6-}$ polyoxometalate catalyst used in electron-transfer oxidations, *Chem. Eur. J.*, 16, 10014–10020.

[46] Weckhuysen, B. M. (2009) Chemical imaging of spatial heterogeneities in catalytic solids at different length and time scales, *Angew. Chem. Int. Ed.*, 48, 4910–4943.

[47] Yokoyama, H. (2009) *EPR Imaging and Its In Vivo Application*, Nova Science Pub. Inc., New York.

[48] Lucarini, M., Pedulli, G. F., Lazzari, D., Vitali, M., Andrews, S. M. (2002) EPR imaging determination of high molecular weight nitroxide radicals in the UV degradation of polycarbonate-poly(acrylonitrile-butadiene-styrene) polymers, *Macromol. Chem. Phys.*, 203, 2239–2244.

[49] Xiang, Z., Xu, Y. (1997) The status quo and prospect of ESR imaging applications to study on catalysts, *Appl. Magn. Reson.*, 12, 69–79.

12

APPLICATION OF AMBIENT-PRESSURE X-RAY PHOTOELECTRON SPECTROSCOPY FOR THE *IN-SITU* INVESTIGATION OF HETEROGENEOUS CATALYTIC REACTIONS

David E. Starr, Hendrik Bluhm, Zhi Liu,
Axel Knop-Gericke, and Michael Hävecker

12.1 INTRODUCTION

Surface chemistry plays an important role in many atmospheric and environmental chemical processes and in corrosion, tribology, and catalysis [1–3]. On the surface of a catalyst, molecules undergo a series of chemical transformations that must be identified and understood at a fundamental level [1–3]. Detailed molecular-level information about surface chemical processes can be garnered from surface-sensitive experimental techniques that use charged particles (electrons or ions) with short mean free paths in solid materials [4]. These charged particles also have short mean free paths through gases at elevated pressures, limiting many surface-sensitive experimental techniques to ultra-high-vacuum (UHV) conditions [1, 3]. Linking UHV studies of surface chemical processes to technical systems, in particular catalysts, that operate at higher pressures relies on the assumption that the most important effect of higher pressure is to increase the coverage of adsorbates on the surface.

In-situ *Characterization of Heterogeneous Catalysts*, First Edition.
Edited by José A. Rodriguez, Jonathan C. Hanson, and Peter J. Chupas.

However, the chemical composition or phase of a solid surface is intimately connected to the chemical composition and potential (i.e., pressure and temperature) of the gas phase surroundings. Since the difference in gas phase chemical potential between UHV conditions (e.g., 25°C and 10^{-10} mbar) and those relevant for catalysis (e.g., 500°C and 1000 mbar) can be as large as 75 kJ/mol (e.g., for an ideal gas), the gas phase chemical potential may provide a significant contribution to the surface free energy. This implies that the chemical phases formed on the surface of an operating catalyst could be fundamentally different than those investigated at UHV conditions and that investigations of surface chemical processes relevant to catalysis should be conducted as close as possible to technical operating conditions [3].

Since surface-sensitive experimental techniques provide molecular-level information about surfaces and adsorbates, they are extremely powerful tools for surface chemical analysis. X-ray photoelectron spectroscopy (XPS) is particularly useful for surface chemical analysis since it is chemically specific and quantitative [4]. In XPS, an X-ray photon impinges on a sample, a photoelectron is emitted, and its kinetic energy is measured. The kinetic energy of the photo-emitted electron is a sensitive measure of both the element from which the electron was emitted and the chemical surroundings of that element [4]. The desire to carry out XPS measurements at elevated pressure conditions has existed for decades and numerous research groups have contributed to the development of a modern form of XPS that can be used at gas pressures up to about 7 mbar [5–19]. This technique is currently known as ambient-pressure X-ray photoelectron spectroscopy (AP-XPS).

In the following sections, we will provide an overview of the historical development of AP-XPS addressing specific technical advances and their relevance. Following this, examples of the application of AP-XPS will be provided. These examples will demonstrate the versatility of this technique for the investigation of catalytically relevant systems ranging from structurally well-defined systems composed of adsorbates on single crystal surfaces and nanoparticles, the catalytic growth of nanotubes, to operating electrochemical cells and industrial heterogeneous catalysts.

12.2 HISTORICAL PERSPECTIVE AND TECHNICAL ADVANCES

12.2.1 Initial AP-XPS Instrument Designs

The general concept for AP-XPS was developed by Hans and Kai Siegbahn nearly 40 years ago [5–8] and remains the same today. The sample is enclosed in an elevated pressure volume and since both the X-ray source and electron energy analyzer must be kept at UHV conditions they are isolated from the elevated pressure volume. The X-ray source (either laboratory or synchrotron based) is separated by having the X-rays enter the elevated pressure volume through an X-ray transparent window (e. g. a thin silicon nitride or aluminum membrane), or by using multiple differential pumping stages. The electron energy analyzer is isolated from the elevated pressure volume by having the photo-emitted electrons leave the volume through a small aperture into a differentially pumped analyzer. The initial designs of Hans and Kai Siegbahn were based on a magnetic spectrometer. These instruments were predominantly used to study organic solvents and limited to a pressure of about 1 mbar. Later improvements to these instruments broadened the range of liquid solvents that could be studied with photoelectron spectroscopy [9, 10].

Soon after the initial developments of Siegbahn and Siegbahn, Joyner, Roberts, and Yates developed a photoelectron spectrometer that could be used for solid samples [11]. This instrument enclosed the sample in a gas cell, had a single differential pumping stage around the cell, and could also operate at pressures up to ~1 mbar. Approaches similar to those used by Joyner, Roberts, and Yates are still used today [12, 13]. Later, Ruppender et al. incorporated multiple differential pumping stages between the elevated pressure volume and the electron energy analyzer [14]. By doing so, higher pressures in the elevated pressure volume could be kept while maintaining UHV conditions in the analyzer. Despite this technical achievement, the highest pressure for these experiments remained limited to ~1 mbar. This is because the major obstacle to increasing the pressure for AP-XPS experiments is the attenuation of the photoelectrons from elastic and inelastic scattering by the gas phase. To overcome this limitation, the combination of an intense, focused X-ray source, which produces substantially more photoelectrons, and the ability to decrease their path length through the gas phase, was required. The use of synchrotron radiation and an electrostatic lens system incorporated into a multistage differentially pumped analyzer has greatly facilitated in overcoming this barrier [15–20].

12.2.2 Modern Synchrotron-Based AP-XPS

Modern synchrotron-based AP-XPS systems are able to operate at pressures higher than 7 mbar [15–20]. While this may appear as only a modest improvement over past designs, this improvement in pressure is quite significant since it allows the investigation of surfaces in equilibrium with water vapor at more technically relevant temperatures (the equilibrium vapor pressure of water at the triple point is 6.1 mbar [21]). Water is a common component in a broad range of environmental and technological systems and a reactant in many catalytic processes. This increase in operating pressure of AP-XPS is a result of two improvements: (1) the use of synchrotron light which provides a high flux of photons in a focused spot on the sample (i.e., a high brightness) and (2) an electrostatic lens system that refocuses the photo-emitted electrons on the small apertures between the differential pumping stages and increases the transmission of the photoelectrons through the differential pumping stages.

The attenuation of photoelectrons by gas phase scattering follows a Beer's law dependence (i.e., $I/I_{vac} = \exp(-z\sigma(E)p/kT)$), where z is the path length of

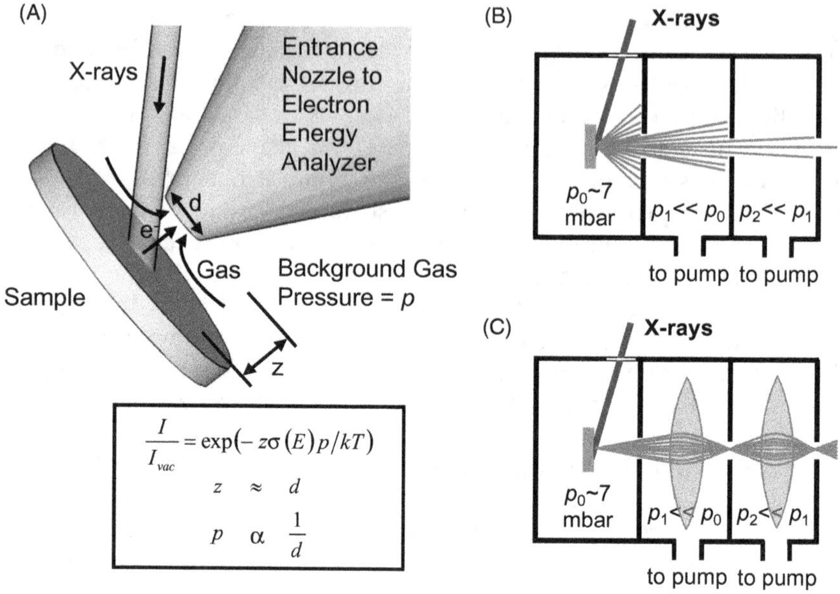

Figure 12.1. (A) Definition of parameters used in the text and geometry of an AP-XPS experiment showing the sample, incident X-rays and entrance nozzle to the differentially pumped electrostatic lens system. (B) Inserting multiple differential pumping stages between the sample and the final aperture of the differential pumping system decreases the collection plane half-angle and therefore the transmission of the photoemitted electrons through the differential pumping stages. (C) Electrostatic lenses inserted into the differential pumping stages refocuses the photoelectrons onto the apertures between the stages and increases the collection efficiency of electrons.

the photoelectrons through the gas, $\sigma(E)$ is the energy-dependent electron scattering cross section of the gas, p is the gas pressure, and I_{vac} is the photoelectron intensity under UHV conditions which is proportional to the X-ray flux (Fig. 12.1A) [15]. The distance z is limited to values approximately equal to the diameter of the aperture d to ensure that the gas pressure at the sample is not reduced, since gas is pumped through the entrance aperture to the electron energy analyzer. A general rule is that at distance of one aperture diameter ($z = d$), the pressure at the sample is 0.95 that of the background gas [15]. This results in the maximum operating pressure scaling inversely with the aperture diameter, or as $1/d$. However, simply reducing the aperture size will not increase the effective operating pressure since the photoemission signal will decrease as $\sim 1/d^2$ unless there is a simultaneous increase in photon flux density (photons per second per square centimeter) [15]. This implies that a high brightness source, such as can be achieved at a synchrotron, is desirable. By combining aperture sizes with diameters, d, of 1 mm or less with the high flux and ability to focus the X-ray beam to spot sizes on the sample less than 1 mm of a synchrotron, operating pressures of 7 mbar or higher have been achieved in modern AP-XPS experiments [15–20].

The other major improvement of modern AP-XPS designs over previous designs is the incorporation of an electrostatic transfer lens in the differential pumping stages between the elevated pressure sample volume and the electron energy analyzer [15]. The insertion of multiple differential pumping stages between the sample volume and the electron energy analyzer increases the distance between the final aperture of the differential pumping system and the sample. While this allows for higher pressures in the sample volume while maintaining UHV conditions in the electron energy analyzer, it also reduces the solid angle of the photo-emitted electrons subtended by the final aperture (Fig. 12.1B). The collection efficiency of photoelectrons is determined by the solid angle, Ω, collected by the analyzer. This depends on the collection plane half angle, a, via: $\Omega = 4\pi \sin^2(a/2)$ [22]. For modern AP-XPS differential pumping designs, inserting the differential pumping stages decreases the half-angle, a, from a few degrees (e.g., from 5 to 8°) to a fraction of a degree in the absence of an electrostatic lens system. This would lead to a reduction in the collection efficiency by more than three orders of magnitude [15]. The electrostatic lens system refocuses photo-emitted electrons onto the apertures between the differential pumping stages, increasing the transmission of the photoelectrons through the differential pumping stages (Fig. 12.1C). The overall affect of this is to increase the acceptance half-angle of the lens system. By using electrostatic lenses in the differential pumping stages, some of the photoelectrons that were lost in previous designs are recovered and results in orders of magnitude greater photoelectron collection efficiencies.

In addition to providing a more focused X-ray spot on the sample, adaptation of AP-XPS for use at synchrotron sources has other advantages. These include the tunability of the incident X-ray energy which broadens the range of spectroscopic techniques that can be used (e.g., X-ray absorption spectroscopy) and allows depth-dependent chemical analysis of samples (i.e., depth profiling) [23–27]. In addition, the increased photon intensity of a synchrotron leads to reduced data acquisition times compared to lab-based X-ray sources, often by multiple orders of magnitude. These advantages, as well as those listed earlier, have allowed modern AP-XPS to have a significant impact on linking surface chemical analysis conducted with traditional UHV surface-science techniques to catalytic systems at near operating conditions in the decade since its development.

12.3 USING AP-XPS TO STUDY MOLECULAR ADSORPTION ON SINGLE CRYSTAL SUBSTRATES

12.3.1 General Considerations

Due to the well-defined structure and periodicity of single crystals, experiments on single crystal surfaces under UHV conditions have a long history of providing detailed molecular level information about adsorbate–surface interactions [1–3]. For AP-XPS experiments on single crystal surfaces, this level of detail is retained, but the elevated pressure conditions extend the thermodynamic phase space that can be explored with photoelectron spectroscopy and may open new kinetic pathways for reactions not found in UHV conditions. Adsorption studies on single crystal surfaces provide information on, for example, site-specific (e.g., a top, bridge, or hollow) adsorption and occupation, as well as the formation of new chemical phases on surfaces upon exposure to gases. In addition, by changing surface orientation and/or miss-cut angle to form vicinal surfaces with varying step densities, the influence of face specificity (e.g., (100) versus (111)) or the role of defects on adsorption and reactions can be systematically addressed. AP-XPS experiments on single crystal surfaces provide a direct connection between the vast knowledge gained from research on single crystals under UHV condition and how these systems evolve as the pressure increases. This helps to bridge the decades old pressure gap between UHV studies and catalytically relevant conditions.

12.3.2 Surface Crystallographic Dependence for Water Adsorption on Cu Surfaces

A recent study of H_2O adsorption on Cu surfaces using AP-XPS highlighted the dependence of surface reactivity on the crystallographic orientation of the surface [28, 29]. The adsorption and reaction of water on Cu catalysts is important in a variety of industrially relevant reactions, including the water–gas shift reaction and the synthesis of methanol [30, 31]. The requirement of OH groups to stabilize molecular water adsorption on Cu surfaces has been directly illustrated by comparing AP-XPS results for the adsorption of water on the Cu(110) surface to those on Cu(111) [32]. For a relative humidity (RH) up to 32% (1.3 mbar, –5°C, $\mathrm{RH} = 100 \times P_{\mathrm{H2O}}/P^{O}_{\mathrm{H2O}}$, where P^{O}_{H2O} is the saturation vapor pressure of water at temperature T), the Cu(111) surface remains free of both molecular H_2O and OH. In contrast to this, AP-XPS experiments for the H_2O/Cu(110) system at similar conditions revealed the presence of pure OH (RH < ~0.01%) and mixed H_2O/OH phases (RH > ~0.01%). The presence of water molecules at such low RH was attributed to H-bonding between the OH groups and water molecules. This is a direct consequence of the higher H_2O dissociation barrier of ~0.3 eV on the (111) surface compared with the (110), surface which allows for the dissociation of H_2O on Cu(110) to form OH and act as anchoring sites for molecular H_2O adsorption.

12.3.3 CO-Induced Restructuring of Stepped Pt Surfaces

Tao et al. recently investigated the restructuring of stepped Pt surfaces, specifically (557) and (332), at CO pressures up to 0.7 mbar [33]. Using AP-XPS, they determined that at 0.7 mbar the CO coverage is approximately 1.0, nearly twice the amount of CO adsorbed on the Pt(557) surface at 7×10^{-9} mbar. Along with the increase in CO coverage, O 1s and Pt 4f spectra showed a substantial increase in intensity at high binding energies (BEs), as indicated by the arrows in Figure 12.2, left panel [33]. In general, O 1s and C 1s BEs shift to higher values as the coordination of CO to the surface decreases. For example, on Pt(111), the O 1s BE of CO bound to Pt bridge sites is 531.0 eV and is 532.7 eV for a-top bound CO [34]. Similarly, a lower coordinated Pt atom should lead to the observation of higher Pt 4f BEs. Therefore, the higher BEs observed by Tao et al. would be consistent with CO bound to low-coordinated Pt sites. The additional features observed in the Pt 4f and O 1s spectra were reversible as evidenced by the direct correlation between peak intensity and pressure changes between 7×10^{-9} and 0.7 mbar. The observed CO coverage cycled between ~1.0 and 0.5 between these same pressures. Complimentary scanning tunneling microscopy (STM) experiments indicated a dramatic surface restructuring at pressures above 0.1 mbar and the formation of triangular nanoclusters of approximately 2.2×2.1 nm in size [33] (see Fig. 12.2). The formation of these nanoclusters leads to an increase in the number of undercoordinated Pt atoms on the surfaces which act as new adsorption sites for CO at elevated pressures. Such dramatic restructuring of the Pt (557) surface was proposed to be driven by a relaxation of repulsive CO–CO interactions as the CO coverage increased to nearly 1.0 and confirmed by density functional theory (DFT) calculations.

These results highlight the dynamic nature of Pt at elevated pressure conditions. Further, the level of detail in these studies provides a molecular-level understanding of the mechanism responsible (i.e., CO–CO repulsion) for such dynamic behavior. By extending these findings to studies on structurally more complex model and technical catalysts, the dynamic nature of the Pt surface and its role in the reactivity of supported Pt catalysts may be understood.

12.4 USING AP-XPS TO STUDY ADSORPTION ONTO WELL-DEFINED NANOPARTICLES

12.4.1 General Considerations

Nanoparticles are vital components of modern catalysts. They offer several advantages over their bulk counterparts, in particular an increased surface area and thus higher reactivity at a lower total amount of material [1–3]. This also reduces the costs of the catalyst, which very often contains precious and rare metals. An additional advantage of nanocatalysts is the tunability of their properties through variation of their size, composition, and the choice of the support material (most often an oxide). This creates a wide parameter space that needs to be evaluated in order to optimize the catalytic performance.

AP-XPS is a versatile method to investigate the properties of nanoparticles under reaction conditions in the presence of a gas or gas mixture. As we shall discuss in the following sections, in particular the elemental composition and chemical nature (e.g., oxidation state) of the nanoparticles can be investigated. This information can further be separated into surface versus bulk composition

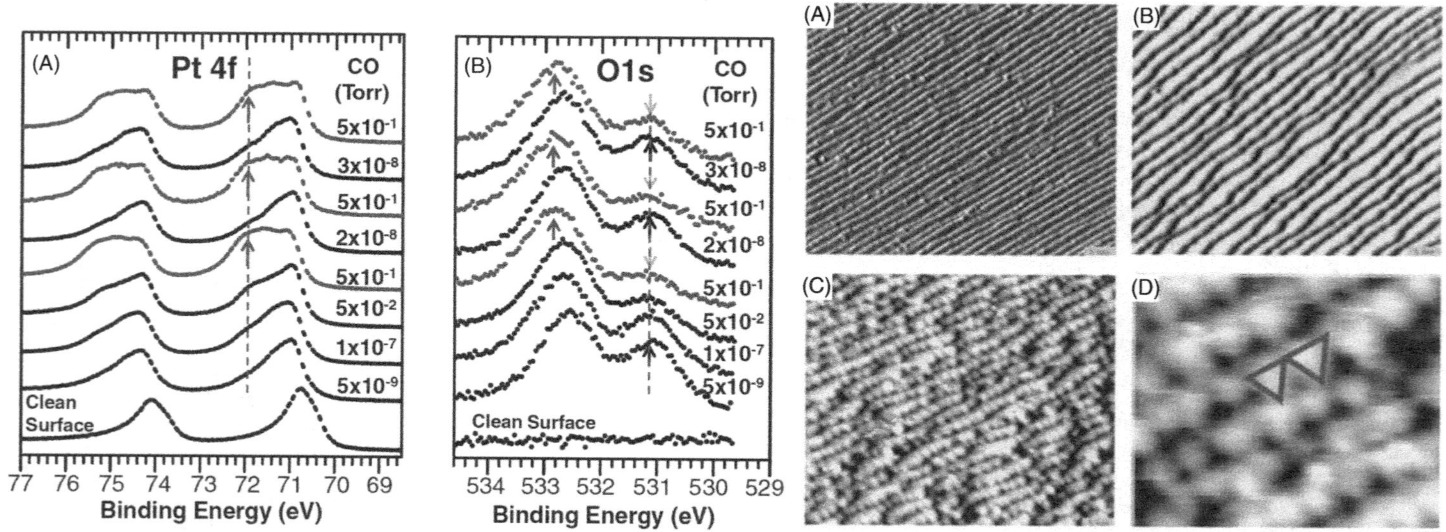

Figure 12.2. Left: Pt 4f and O 1s spectra for CO adsorption on Pt(557) at CO pressures indicated in the figure. The presence of high BE peaks indicated by the arrows in the Pt 4f and O 1s spectra, respectively, are likely due to the adsorption of CO at undercoordinated Pt atoms on the stepped surface. Right: STM images at UHV conditions (A), 7×10^{-8} mbar CO (B), and 1 mbar CO pressure (C and D) showing the formation of Pt nanoclusters at elevated pressure conditions (Reproduced with permission from Reference [33]).

if photoelectron spectra are measured as a function of electron kinetic energy, by making use of the kinetic energy dependence of the inelastic mean free path of electrons in matter. (Depth profiling by variation of the detection angle is not a suitable method in the case of nanoparticles due to their three-dimensional geometry.) Since at typical electron kinetic energies in an XPS experiment (100–1000 eV) the mean free path of the electrons in a solid is on the order of 5–20 Å, both the surface and interior of nanoparticles with sizes of a few nanometers are fully accessible [35].

To date, AP-XPS investigations have been performed on a variety of nanoparticle catalysts, including CO oxidation on Pt/CeO_x [36, 37], ethane dehydrogenation on Pt(Sn)/Mg(Al)O [38], CO oxidation on Au nanoparticles deposited on TiO_2 [39, 40], and the oxidation and reduction of Rh nanoparticles on SiO_x/Si substrates [41]. Several investigations have focused on bimetallic or trimetallic alloy nanoparticles, which are of special interest since they can adjust the chemical composition of their surface to the gas phase conditions (e.g., oxidizing vs. reducing conditions). In particular, experiments on Pd, Rh, and Pt bimetallic and trimetallic nanoparticles, supported on SiO_x/Si wafers, were carried out using AP-XPS [23–25].

12.4.2 Surface Chemistry of RhPd Nanooparticles in the Presence of NO, CO, and O_2

A study by Grass et al. investigated the surface chemistry of 15 ± 2 nm $Rh_{0.5}Pd_{0.5}$ nanoparticles in the presence of NO, CO, and O_2. The data obtained on the nanoparticles were compared with measurements on a bulk $Rh_{0.5}Pd_{0.5}$ polycrystal under the same conditions [25]. Although the RhPd crystal has a nominal composition of 50% Rh and 50% Pd, it was found that after a standard surface preparation by sputtering and flash annealing to 750°C, the surface was ~90% Pd and only ~10% Rh. However, a stoichiometric surface can be prepared by alternate Ar^+ sputtering and vacuum annealing at 200°C. The $Rh_{0.5}Pd_{0.5}$ nanoparticles with diameters of 15 ± 2 nm were prepared by means of colloidal chemistry methods and deposited on SiO_x/Si wafers [23, 24]. The nanoparticles show a Rh-rich surface composition (Rh 90%, Pd 10%) after preparation. Upon alternating exposures to either pure NO (0.1 mbar) or mixtures of 0.1 mbar NO + 0.1 mbar CO at 300°C, both samples show qualitatively the same behavior: under more reducing conditions (NO + CO) the surface is enriched in Pd, while under oxidizing conditions (pure NO) the amount of Rh is increased. Due to the initial larger enrichment in Rh in case of the nanoparticles the Rh concentration oscillates between 48% (reducing) and 67% (oxidizing) compared with 32% and 48% for the bulk crystal.

While the qualitative changes to the surface composition of bulk crystal and nanoparticles are similar upon changes to the gas phase environment, the kinetics of these changes may differ. In order to investigate this in more detail, Grass et al. prepared Rh-deficient surfaces by heating both samples in 0.1 mbar

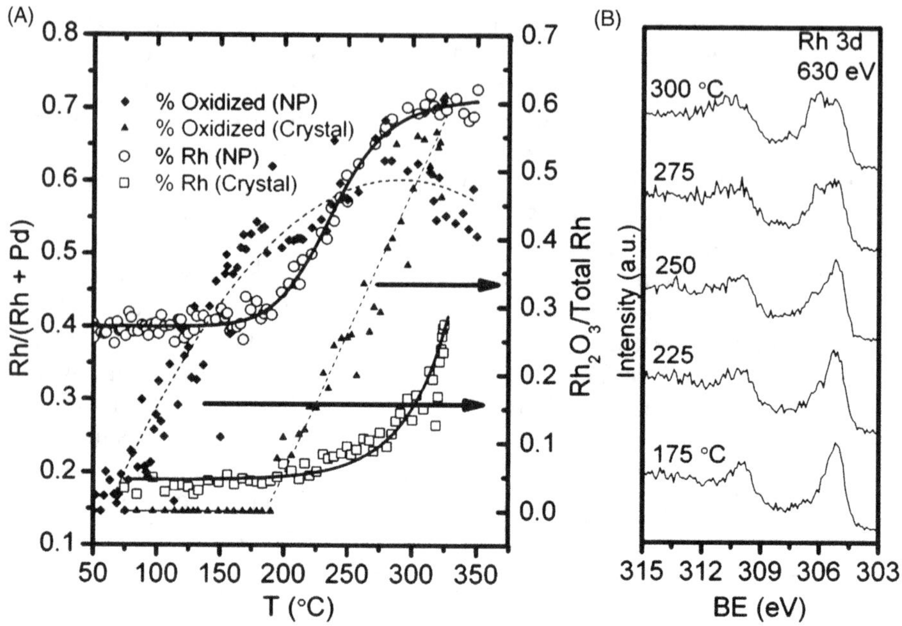

Figure 12.3. (A) Oxidation and Rh surface segregation is monitored for both nanoparticles (solid diamonds and open circles) and a bulk crystal (solid triangles and open squares) while heating in 0.1 mbar NO after previous CO exposure. The lines are guides to the eye. (B) Selected Rh 3d spectra for the $Rh_{0.5}Pd_{0.5}$ bulk crystal. Reprinted (adapted) with permission from Grass, M. E., Park, M., Aksoy, F., Zhang, Y., Kunz, M., Liu, Z., Mun, B. S. (2010) Effect of O2, CO, and NO on surface segregation in a Rh0.5Pd0.5 bulk crystal and comparison to Rh0.5Pd0.5 nanoparticles, *Langmuir*, 26, 16362–16367. Copyright (2010) American Chemical Society.

of pure CO at 300°C, resulting in Rh concentrations of about 20% and 40% for the bulk crystal and the nanoparticles, respectively. To study the kinetics of the oxidation and segregation of Rh, the samples were then heated in 0.1 mbar of NO from 25 to 325°C at a rate of 4°C/min. Figure 12.3 summarizes the results. The amount of oxidized Rh is determined from Rh 3d spectra (see Fig. 12.3B), by comparing the intensity of the metallic Rh (BE ~305 eV) and oxidized Rh (BE ~306 eV) peaks. The summary of the data in Figure 12.3A clearly shows that the onset of the oxidation reaction is much lower (~70°C) in the case of the nanoparticles, compared with ~200°C for the bulk crystal. Likewise, the surface segregation of Rh starts at 200°C and is completed at 350°C, while it is still ongoing at this temperature for the bulk crystal. The reduction kinetics can be monitored in a similar way (see Fig. 12.4). Following the reaction in 0.1 mbar NO at up to 350°C, the samples were exposed to 0.1 mbar CO, again starting at 25°C, and heated to 325°C at 4°C/min. Both the

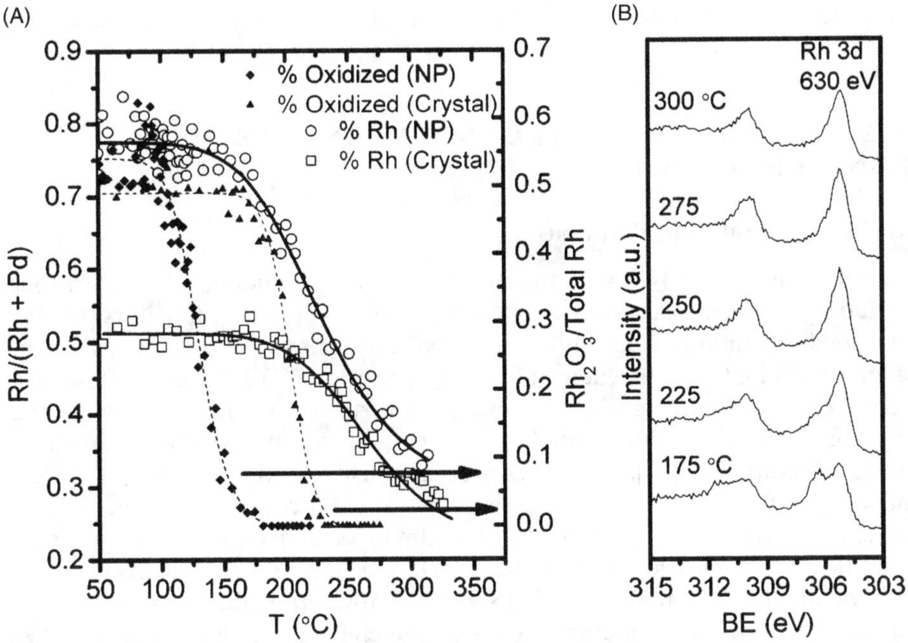

Figure 12.4. (A) Reduction and Pd surface segregation is monitored for both nanoparticles (solid diamonds and open circles) and a bulk crystal (solid triangles and open squares) while heating in 0.1 mbar CO after previous exposure to NO. The lines are guides to the eye. (B) Selected Rh 3d spectra for the $Rh_{0.5}Pd_{0.5}$ bulk crystal. Reprinted (adapted) with permission from Grass, M. E., Park, M., Aksoy, F., Zhang, Y., Kunz, M., Liu, Z., Mun, B. S. (2010) Effect of O2, CO, and NO on surface segregation in a Rh0.5Pd0.5 bulk crystal and comparison to Rh0.5Pd0.5 nanoparticles, *Langmuir*, 26, 16362–16367. Copyright (2010) American Chemical Society.

desegregation of Rh from the surface as well as the reduction of Rh are faster in the case of the nanoparticles compared to the bulk sample.

The increased surface segregation and oxidation reaction kinetics for nanoparticles, compared to the bulk sample, can be explained by a higher defect concentration at the surface and in the interior of the nanoparticles. Diffusion along grain boundaries as well as along surfaces is much faster than in the bulk, which proceeds via substitutional self-diffusion [42]. The faster and more complete oxidation of nanoparticles may also be explained by an increased concentration of surface defects, as well as by thermodynamic arguments [41, 43]. In summary, the previous discussion showed that nanoparticles display a distinctly different response to changes in the chemical potential of the gas phase. This offers new possibilities for more energy- and cost-efficient catalytic materials, but also poses a challenge to investigate and understand the fundamental properties of the materials at the nanoscale and under reaction conditions.

12.5 USING AP-XPS TO STUDY METAL-CATALYZED NANOTUBE GROWTH

12.5.1 General Considerations

Carbon nanotubes (CNTs) find many, diverse technological applications including interconnects for integrated circuits [44–46], supercapacitors [47, 48], and field-emission devices [49]. These applications require the vertical alignment of CNTs. Catalytic chemical vapor deposition (CVD) [50] is often used as a growth technique and produces dense, vertically aligned millimeter-long mats of forests of multi- and single-walled carbon nanotubes (SWNTs) on SiO_2 and Al_2O_3 supports [51]. The CVD process consists of three steps: first, a few nanometer-thick layer of a catalyst is evaporated onto a substrate; second, the layer is nanostructured by annealing the film in vacuum to form nanoparticles; and, third, the obtained nanoparticles are exposed to a hydrocarbon atmosphere at elevated temperature [51]. CNT formation depends on several parameters such as the nanoparticle size and chemical state of the catalyst as well as the strength of the interaction between the nanoparticle and the support [51]. The active state of standard catalysts, such as Fe and Ni, remains in question and has been proposed to be oxidic [52], carbidic [53], or metallic [54] under CVD conditions. AP-XPS is a promising approach to provide insight into the electronic structure and active state of the catalyst nanoparticles under CNT growth conditions.

12.5.2 Effects of Nanoparticle–Support Interactions on CNT Growth

The catalyst–support interaction of Fe nanoparticles on silica and alumina supports was investigated by XPS. Fe2$p_{3/2}$ spectra of Fe nanoparticles supported on alumina show the formation of Fe^{2+} and Fe^{3+} states in addition to the metallic state [55]. Since no reduction of the alumina support was detected, the authors concluded that there is a strong interaction between Fe and the oxygen atoms of the alumina support. Due to this strong interaction, the Fe nanoparticles do not agglomerate and therefore the particle size distribution is relatively sharp and particles do not exceed sizes greater than 10 nm on the alumina support. In contrast, Fe particles on a silica support do not have as strong an interaction with the support as on alumina. This weaker interaction results in a broader Fe nanoparticle size distribution on silica compared to alumina [55]. Due to its influence on the Fe nanoparticle morphology, the interaction strength between the Fe nanoparticles and the support has strong implications for the growth of the CNTs. Vertically aligned, carpet-like CNTs grow on alumina-supported Fe particles, whereas the mobility of the Fe particles on silica leads to the growth of spaghetti-like CNTs which are not aligned [55].

12.5.3 Using AP-XPS to Investigate the Growth of Carbon Nanotubes on Insulating and Conducting Substrates

The catalyst–gas phase interaction during the growth of CNTs was investigated by AP-XPS [56]. A 0.25 nm Fe layer was deposited on SiO_2. To remove potential carbon impurities the layer was oxidized in 1×10^{-3} mbar of oxygen and subsequently reduced in 0.8 mbar of ammonia at 500°C. Ethylene was used as a carbon source for CNT growth. *Ex-situ* Raman and scanning electron microscopy (SEM) investigations demonstrated a spaghetti-like growth of CNTs under the conditions used in this study [56]. The C1s core level spectra obtained in an atmosphere of 5×10^{-3} mbar of ethylene after 30 s of growth are characterized by two peaks at 284.65 and 285.4 eV. These BEs are consistent with partially sp^3-hybridized, disordered C which may also be hydrogenated [56]. Such defects are often observed for CNTs, graphite powder, and nanocarbon samples [57]. After 7 min, the C1s peak shifts to 284.45 eV, the typical BE of sp^2-bonded carbon present in CNTs. The $Fe2p_{3/2}$ spectra show indications of Fe^{2+} and Fe^{3+} which decrease with time during the CVD process. The authors concluded that these species are located at the Fe particle-SiO_2 support interface. Additional experiments with deliberately oxidized iron nanoparticles did not show growth of CNTs and the oxidized Fe did not reduce in the ethylene atmosphere at 500°C [56]. The C1s spectra are dominated by a peak at 285.3 eV with a shoulder at 284.7 eV, but the peak assigned to graphitic sp^2-bonded carbon at 284.45 eV is absent. *Ex-situ* investigations by SEM and Raman spectroscopy did not indicate the presence of any CNTs.

For applications of CNTs as interconnects in integrated circuits or field emission devices, the growth of CNTs on conductive substrates such as metals, metal nitrides, and metal silicides is required [58]. The growth of CNTs on conducting substrates is more difficult and less studied than on insulating substrates. Challenges associated with the growth of CNTs on conducting substrates arise from metal carbide formation due to the support reacting with the gas used for the CVD process and the catalyst alloying with the support. Ta is often used as a diffusion barrier for Cu on Si in microelectronics. As revealed in the literature, the comparison between Ta and oxides as catalyst supports for CNT growth is controversial. AP-XPS was used to investigate the interaction of the Ta support with the catalyst during CNTs' growth [58].

Figure 12.5 shows the $Ta4f_{7/2}$ and $Fe2p_{3/2}$ AP-XPS spectra of a 1.1 nm thick Fe catalyst film deposited on a 100 nm Ta film supported on a silicon wafer. The sample was exposed to air prior to transferring it into the AP-XPS setup. This results in the Fe being completely oxidized and the Ta being in a mixed metallic and oxide state at the surface. During heat treatment of the sample at 600°C in vacuum, changes in the chemical state were observed. The Ta4f7/2 peak shifted 0.3 eV to higher BE, indicating alloy formation between the Ta and Fe at the surface [59]. Part of the Ta remains oxidized. Unlike Fe on the Al_2O_3 or SiO_2 supports, the Fe on Ta is reduced during heat treatment in vacuum in the absence of a reducing gas [56]. The same heat treatment of a

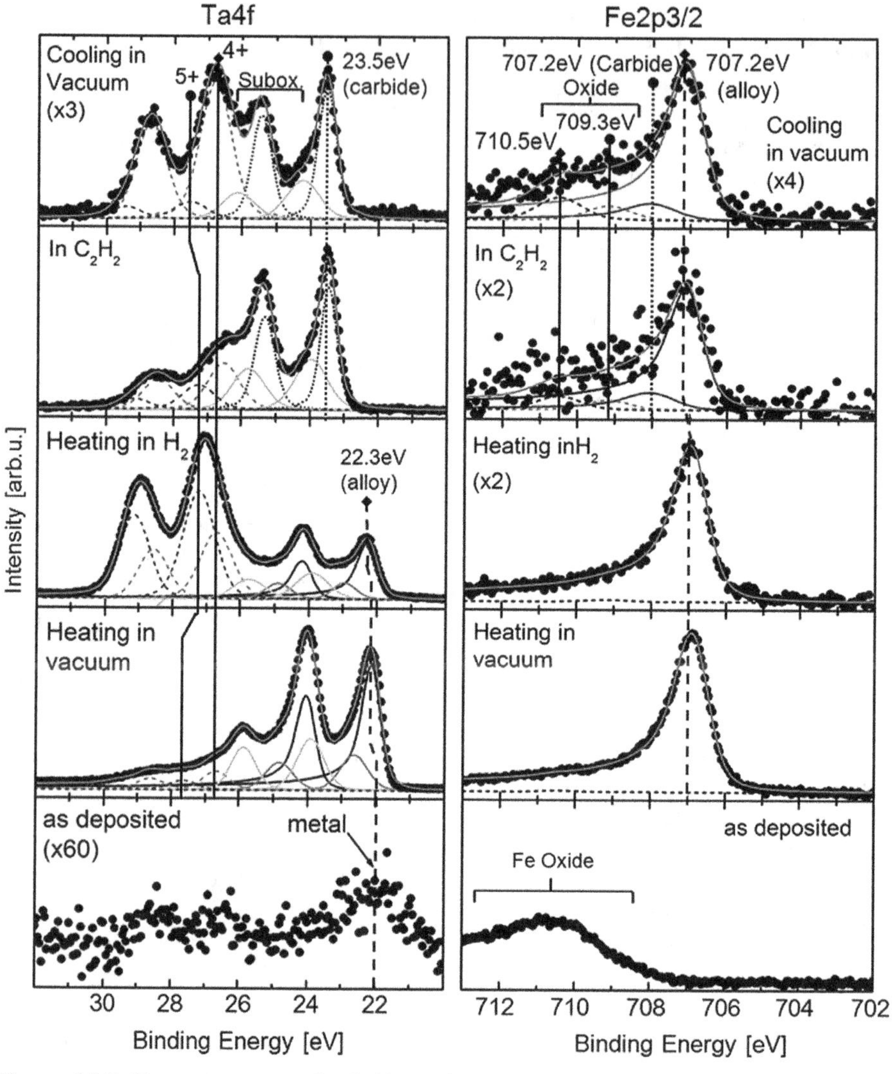

Figure 12.5. Process-step-resolved (from bottom to top) AP-XPS spectra of Ta4f ($h\nu$ = 330 eV) and Fe2p ($h\nu$ = 1030 eV) regions during CVD (2×10^{-2} mbar C_2H_4 at 650°C) of Fe-covered (1.1 nm) Ta (100 nm). (Fit lines are for Ta4f: Ta–Fe alloy, solid black; Ta carbide, dotted black; and oxides, dotted gray. Note that the numbers in brackets indicate multipliers for intensities of separate scans). Reprinted (adapted) with permission from Bayer, B. C., Hofmann, S., Castellarin-Cudia, C., Blume, R., Baehtz, C., Esconjauregui, S., Wirth, C. T., Oliver, R. A., Ducati, C., Knop-Gericke, A., Schlögl, R., Goldoni, A., Cepek, C., Robertson, J. (2011) Support-catalyst-gas interactions during carbon nanotube growth on metallic Ta films, *Journal of Physical Chemistry C*, 115, 4359–4369. Copyright 2011 American Chemical Society.

bare Ta reference sample reveals an almost complete reduction of Ta to the metallic state. Upon H_2 introduction, residual water/oxygen contamination in the H_2 line caused a partial reoxidation of the Ta [60] but not of the alloyed Fe [58]. These results show that small amounts of oxygen contamination are sufficient to oxidize the Ta. Also, these observations imply that not only does the presence of Fe enhance Ta oxidation but also the presence of Ta promotes Fe reduction. During exposure to an ethylene atmosphere at 2×10^{-2} mbar at 650°C, Ta carbide with some remaining Ta oxide is formed. The Fe exhibits a small shift to 707.2 eV, which indicates a stronger interaction between the now carbidic Ta surface and the Fe. In addition, a small Fe carbide component can be detected. The C1s spectra show the formation of a peak at 284.3 eV assigned to graphitic sp^2-bonded carbon (not shown). A bare Ta sample showed a similar evolution, but with a lack of the Ta–Fe alloy shift and a strongly decreased graphitic sp^2 C1s peak associated with no CNT growth. The oxide component in the reference bare Ta sample significantly decreased after the growth compared to the Ta–Fe alloy, implying again that the presence of the Fe promotes the oxidation of the Ta surface [58].

In summary, AP-XPS was used to investigate the growth of CNTs. The interaction of the catalysts and the support has a strong impact on the directed growth of nanotubes and the resulting CNT morphology. A strong catalyst–support interaction implies a forest-like growth of CNT. The chemical state of the catalyst and the support changes as a function of the process step and Ta supports the reduction of Fe.

12.6 AP-XPS APPLIED TO ELECTROCHEMISTRY

12.6.1 General Considerations

The need for clean, secure, and sustainable energy has created a surge in the research and development of electrochemical devices, such as batteries, fuel cells, and super capacitors. Many roadblocks to higher performing electrochemical devices exist, including challenges in engineering design and a limited scientific understanding of these devices. For example, the fuel cell was invented more than 100 years ago but it has not seen widespread use, despite several key advantages for current and future energy technologies. These include: high efficiency, cleanliness, high resistance to catalyst poisoning, and the possibility to use hydrocarbon fuels either through reforming or burning them directly. This is particularly true for solid oxide fuel cells (SOFCs).

Many significant advances have been made in the last several decades through the use of traditional techniques for electrochemical evaluation of electrode overpotentials, particularly voltammetric techniques and electrochemical impedance spectroscopy (EIS) employing reference electrodes. These techniques provide valuable information regarding global electrode overpotentials and resistances. Despite advances in electrochemical measurement and modeling, our understanding of many aspects of SOFC operation remains circumstantial. These aspects include: the rate-limiting step in SOFC operation, the cathode oxygen reduction reaction mechanisms, the physics governing electrode overpotential losses, and oxidation/reduction at and the length scales of the electrochemically active region of mixed ionic–electronic conducting (MIEC) electrodes. This is largely due to the inherently convoluted nature of electrochemical and chemical processes and the lack of suitable *in-situ* experimental techniques. As pointed out by Adler in 2004 [61], "New in-situ analytical techniques are needed, particularly which can be applied at ambient pressures, that can probe what is happening in an electrode as a function of temperature, P_{O2}, polarization, local position, and time."

As discussed before, core level photoemission spectroscopy (or XPS) is a powerful tool that gives qualitative and quantitative analysis of elements and their chemical states. For investigating SOFCs, XPS can also be used as a noninvasive tool for measuring local surface potentials. The electrons of a given core level from a negatively biased sample or area will have a higher kinetic energy than those from a positively biased or nonbiased sample or area. Such kinetic energy shifts of core-level photoelectron spectra provide a direct measure of the local surface potentials without the use of a reference electrode and the introduction of an extra interface. Thus, photoelectron kinetic energy shifts can be used to measure important properties, such as local overpotentials across exposed interfaces. The development of new AP-XPS end stations has made it possible to carry out such experiments not only *in situ* but also *in operando* with elemental and chemical specificity [16]. A device such as an electrochemical cell can be monitored under operation conditions.

12.6.2 AP-XPS for *In-Operando* Studies of SOCs

In 2008, the first experiment using AP-XPS as an *in-operando* tool to study solid oxide electrochemical cells (SOCs) was performed [62, 63]. By scanning a small X-ray spot across the surface or using the electron energy analyzer in an imaging mode, local elemental and chemical information across the sample surface can be obtained. In addition, local electrical surface potential changes can be determined from the changes in the kinetic energy of the core level photoelectron peak. The correlations between local chemical processes and local surface potentials associated with such changes under operating conditions are crucial for developing a detailed understanding of SOC device operation.

The first one-chamber SOC cell consisted of Au–ceria working electrodes (WEs) and Pt paste counter electrodes (CEs) deposited on a single-crystal yttria-stabilized zirconia (YSZ) electrolyte disk. Using such a planar cell design, all cell components are exposed and located on the same side of the electrolyte disk to enable AP-XPS to access the electrode–electrolyte interfaces. The cell was mounted and heated to 750°C in 1:1 H_2 and H_2O mixture at a total pressure of ~1 mbar. By changing the bias between the CE and WEs, the ceria electrode can either perform electrolysis (water splitting) or hydrogen oxidation (i.e., operate as a fuel cell). A gas environment was selected to ensure a stable gas composition during the whole experiment. These experiments successfully demonstrated that the mixed ionic–electronic conducting ceria electrode undergoes Ce^{3+}/Ce^{4+} oxidation–reduction changes with different applied biases [62, 63].

Later, these techniques were expanded to include a new version of ceria–YSZ–Pt SOC [64] and Ni–YSZ–Pt SOCs [65, 66]. A new end station at the Advanced Light Source (ALS) Beamline 9.3.2 [20] and a special sample holder were utilized as well [67]. In all the experiments, the WE (the ceria/Au or Ni) was grounded and the bias voltage was put on the Pt CE with a potentiostat. Two-probe linear sweep voltammetry (LSV) and EIS experiments could be conducted simultaneously while XPS data were taken.

Figure 12.6A,B shows a schematic of the planar ceria/YSZ/Pt cell geometry and simplified experimental setup [65]. Ceria WEs of different thickness (50, 100, and 250 nm) are sputtered on top of a gold current collector and extend onto the YSZ electrolyte toward the Pt counter electrode. Such a cell design mandates that oxygen ions move in the vertical direction through the ceria

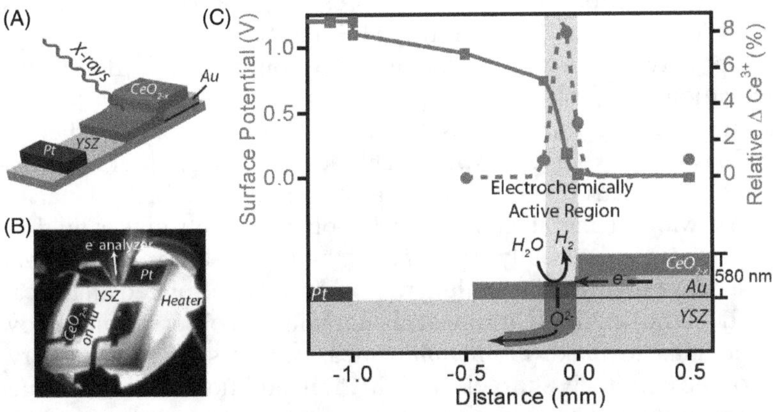

Figure 12.6. Electrochemical activity spreads over 150 μm of the ceria anode. (A), Solid oxide cells have a 200 nm Pt counter electrode, a 300 nm Au current collector on top of a 30 nm alumina film (black), and a 50, 100, or 250 nm ceria WE patterned onto a polycrystalline YSZ substrate. This geometry exposes all cell components to the X-ray beam. The drawing is not to scale. (B) During operation, the cell is heated close to the detector aperture in 1 mbar H_2 and H_2O. (C) A 250-nm-thick ceria anode converts H_2O to H_2 and O_2– in a 150 μm region at +1.2 V cell potential. *In-situ* AP-XPS reveals local surface potentials (red squares) and the relative shift of Ce oxidation state out of equilibrium (green circles) in this region. (Reproduced with permission from Reference [64]). (See color insert.)

and electrons (polarons) move in the lateral direction across the ceria. Therefore, the ionic and electronic potential changes can be separated and measured individually. Using this specially fabricated single-chamber SOC, the authors have demonstrated that the active electrochemical region on ceria extends 150 μm away from the current collector and that significant shifts from the equilibrium surface concentrations of Ce^{3+}/Ce^{4+} are needed to drive the electrooxidation of H_2 and the electrolysis of H_2O. The correlations between local potential losses and local chemical state changes were obtained directly from working SOC devices. A detailed description of this work can be found in Reference 65.

The study of a Ni–YSZ–Pt SOC [66], was the first use of the new end station at ALS Beamline 9.3.2 [20]. For this work, the spectrometer was optimized to perform one-dimensional (1-D) spatially resolved AP-XPS. In this study, the individual overpotentials (such as between Ni and YSZ, and between YSZ and Pt) in SOC devices were measured directly. From the measurement of such a model cell, the origins of changes in the individual overpotentials with applied bias can be interpreted using different electrocatalytic activities of Ni and Pt for the H_2O splitting and H_2 oxidation reactions. This study found without ambiguity that H_2O splitting is faster than H_2 oxidation on Ni, while on Pt H_2 oxidation proceeds more rapidly than H_2O splitting.

Currently several groups are utilizing AP-XPS end stations at the ALS as noncontact tools to probe processes at the electrode and electrode–electrolyte interface "as a function of temperature, pressure, polarization, local position, and time." A joint effort between the ALS and Sandia National Laboratory to develop a two-chamber model SOFC is ongoing (H. Bluhm, private communication).

In addition to the *in-operando* SOC experiments mentioned earlier, the use of AP-XPS is being expanded to *in-situ* characterization of other electrochemical devices and electrode and electrolyte materials as well. A planar lithium ion battery with Li metal anode, block polymer electrolyte, and $LiFePO_4$ cathode has been successfully tested *in situ* and the cell was charged and discharged successfully. Metallic lithium was observed to be plated on the cathode during and after charging (Hudson et al., unpublished observations). However, it is challenging to make *in-operando* measurements on such battery cells because of the extremely short electron MFP. In addition, attempts to measure local surface potential and chemical state across an electrolyte were hindered by beam-induced damage to the polymer electrolyte. To the authors' knowledge, at least two hard X-ray AP-XPS end stations are being developed to increase the probing depth for such studies.

The electrochemical research community using AP-XPS is rapidly growing and new AP-XPS instruments and techniques are being developed for electrochemical research at several synchrotron facilities. More detailed information between local chemical processes and local surface potentials under real cell operating conditions are expected to be obtained in future. Such information should provide fundamental mechanistic information on electrochemical devices that is currently inaccessible by other experimental means. Studies in this area will no doubt facilitate significant advances in electrochemical material and devices.

12.7 AP-XPS INVESTIGATIONS OF OPERATING TECHNICAL CATALYSTS

12.7.1 General Considerations

Direct selective oxidation of propane to acrylic acid,

$$C_3H_8 + 2O_2 \rightarrow H_2C{=}CHCOOH + 2H_2O \quad \Delta_R H^\circ = -852\ \text{kJ/mol},$$

has recently attracted increasing attention (see Reference [68] and references therein). Technical realization of alkane oxidation processes is generally difficult due to insufficient selectivity. Limited selectivity arises from the desired product and intermediate reaction products usually having a higher reactivity (e.g., total oxidation to CO_2) than the alkane feed. To obtain a high degree of conversion while maintaining high selectivity requires highly complex catalysts such as quinternary MoVTeNb oxides [69]. Using these mixed oxide catalysts composed of an orthorhombic phase called M1, selectivities to acrylic acid around 80% at 60% conversion have been achieved [70]. Results seem to indicate that it is beneficial to suppress crystallographic phases other than M1. For example, the so-called M2 phase has been shown to be inactive for propane activation but has revealed activity in subsequent reaction steps [71–73].

Although detailed models of the bulk crystal structure of the M1 phase exist [74, 75], not much is known about the surface termination of the MoVTeNb oxide catalyst. In particular, there is a lack of characterization of the surface under selective propane oxidation conditions that correlate peculiarities of the active surface structure to the catalytic function. This is of utmost significance due to the observation that the elemental surface composition is distinctly different from the bulk composition and that the element abundance of fresh and used catalysts differs as well [76, 77]. *In-situ* depth profiling by variation of the photoelectron kinetic energy with synchrotron-based XPS revealed a depletion of Mo and an enrichment of Te at the outermost surface under propane oxidation conditions in the presence of the feed of $C_3H_8/O_2/H_2O_{(g)}$ [70, 78]. It is this formation of a thin active surface layer that necessitates the application of an experimental technique, such as synchrotron-based AP-XPS, that combines high surface sensitivity with the possibility to apply it under reaction conditions of high temperature and the presence of the gas feed. The well-defined and stable, highly crystalline bulk structure of phase-pure M1 MoVTeNb oxide provides an appropriate basis for studying such surface dynamic processes.

12.7.2 Surface Evolution of a Phase-Pure M1 MoVTeNb Oxide Catalyst under Selective Propane Oxidation Conditions

Pure-phase M1 MoVTeNb oxide powder catalysts have been obtained using a precipitation–purification procedure as described in References 78 and 79 and studied under propane oxidation conditions by AP-XPS at the ISISS facility (Innovative Station for In situ Spectroscopy) at Helmholtz Zentrum Berlin für Materialien und Energie (HZB/BESSY II). The acrylic acid abundance in the effluent gas has been determined by online proton transfer mass spectrometry (PTR-MS) to verify the catalytic functionality of the powder catalyst in the AP-XPS setup. Figure 12.7 displays the acrylic acid evolution while the catalyst was heated stepwise from room temperature to 240 and 420°C in a feed of $C_3H_8/O_2/He$ and $C_3H_8/O_2/H_2O_{(g)}$, respectively, at a total pressure of 0.25 mbar. It was found that there is an immediate strong increase in acrylic acid abundance when steam was added to the feed. This is in good qualitative agreement with studies at atmospheric pressure in fixed bed reactors where the concentration of steam has a pronounced impact on acrylic acid formation and thus verifies that the active state of the catalyst is probed during the AP-XPS experiment. After the addition of steam, the acrylic acid signal steadily increased with time on stream (TOS) and leveled off only after approximately 5 h. This stresses the necessity of long-term experiments while simultaneously verifying the catalyst performance to obtain meaningful information about the equilibrium surface state of bulk powder catalysts. The catalyst's performance to acrylic acid remained rather constant for the rest of the experiment (10 h).

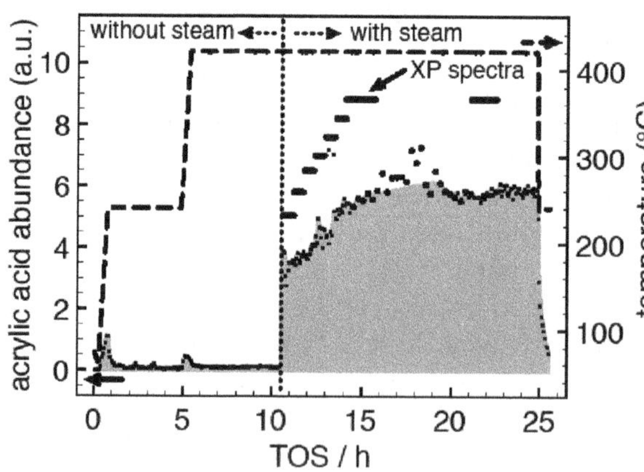

Figure 12.7. Evolution of the acrylic acid abundance (dots) with time on stream (TOS) during AP-XPS measurements when the temperature was increased stepwise from 60 to 240°C and finally to 420°C. The temperature program is shown as a dashed line. After approximately 10 h TOS in dry feed ($C_3H_8/O_2/He = 2/1/2.8$ sccm), steam has been added to the feed ($C_3H_8/O_2/H_2O_{(g)} = 2/1/2.8$ sccm) as indicated by the dashed vertical line and arrows. The horizontal bars define time spans when AP-XPS spectra have been obtained.

The evolution of the surface composition has been characterized by AP-XPS while the activity to acrylic acid increased. Time spans when sets of AP-XPS spectra have been recorded are indicated by horizontal bars in Figure 12.7. AP-XPS core level spectra of all elements have been obtained with a low kinetic energy of 150 eV, resulting in a high surface sensitivity due to a photoelectron inelastic mean free path of only approximately 0.6 nm. A depletion of Mo and, accordingly, enrichment in Te, V, and Nb was observed with increasing activity to acrylic acid. This resulted in a significant deviation from the M1 bulk composition (e.g., as determined by energy-dispersive X-ray spectroscopy [EDX] [70, 78]) on the outermost surface under steady-state propane oxidation conditions. This suggests that the active state of the catalysts is not part of the M1 bulk structure but is instead a thin surface layer on top of the crystalline MoVTeNb oxide bulk. This active surface layer is absent after preparation but formed *in situ* under reaction conditions. Surprisingly, the Te/V element abundance ratio depicted in Figure 12.8 remains constant at 1.4 during the reconstruction of the surface. EDX showed a Te/V ratio in the bulk of 0.38. This suggests that active ensembles with Te and V in close proximity are created on the surface.

The spectral shape of $Mo3d_{5/2}$, $Te3d_{5/2}$, and $Nb3d_{5/2}$ core levels did not reveal any significant modifications to these elements during the course of the reaction, suggesting a single as-prepared valence of Mo^{6+}, Te^{4+}, and Nb^{5+} deduced from the measured BEs of 232.5 ± 0.1 eV, 576.4 ± 0.1 eV, and 206.8 ± 0.1 eV, respectively. In contrast, the $V2p_{3/2}$ core level peak broadened to the high BE side with TOS. A consistent deconvolution of the $V2p_{3/2}$ XP spectra suggests the presence of two components with a BE difference of 1.05 eV, resulting in a BE of about 516.1 ± 0.1 eV and 517.2 ± 0.1 eV that can readily be assigned

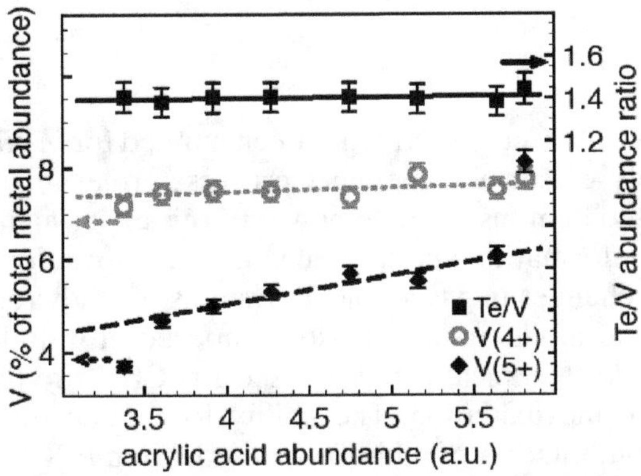

Figure 12.8. Relationship between the acrylic acid abundance and the surface element content of V^{4+} (grey empty dots), V^{5+} (black filled diamonds) species, and the Te/V element abundance ratio (black filled squares), respectively, while the catalytic performance approaches steady state. Lines are included as guides to the eye, error bars as indicated.

to V^{4+} and V^{5+} species, respectively [78]. The abundance of V^{4+} and V^{5+} species versus the acrylic acid abundance after steam has been added to the feed while the catalyst approaches a stationary state (compare with Fig. 12.7) is plotted in Figure 12.8 to study the impact of the vanadium oxidation state on the catalytic performance. Obviously, a V^{5+} abundance increase goes along with an enhanced formation of acrylic acid while the amount of V^{4+} species remained rather constant.

The experimental observations revealed by AP-XPS can be summarized as:

- The concentration of steam in the feed has the most pronounced impact on acrylic acid formation as shown earlier and is in agreement with fixed bed reactor studies.
- Steam in the feed causes a depletion of molybdenum (Mo^{6+}) on the catalyst surface and an increased abundance of V^{5+} species. The correlation between the formation of V^{5+} oxo species and the acrylic acid productivity provides direct experimental evidence for V^{5+} sites being involved in the activation of propane based on chemical reasoning [80]. The rather constant Te/V surface element ratio suggests a reconstruction of the surface under propane oxidation conditions with catalytically active ensembles with V^{5+} and Te^{4+} oxo sites in close vicinity, as suggested previously [70, 80].

In conclusion, AP-XPS provided direct detailed experimental insight into the reduction/oxidation (redox) chemistry of a complex oxidation catalyst such as MoVTeNb oxide. Thus, AP-XPS is a probe that can substantially assist other methods to obtain a detailed understanding not only of model systems but also materials that are close to a technical catalyst.

12.8 CONCLUSION

Modern synchrotron-based AP-XPS has been applied for *in-situ* investigations of a broad variety of surface chemical processes relevant to catalysis. The examples presented demonstrate the power of this experimental technique to provide detailed information about the chemical composition of surfaces and their response to changes in gas ambient. Systems of study range from single-crystal surfaces, and model nanocatalysts composed of well-defined nanoparticle distributions, to technical catalysts used for CNT production, fuel cells, and selective propane oxidation. The ability to conduct these experiments stems from the combination of a highly focused, intense X-ray spot, as found at a synchrotron, and the incorporation of electrostatic lenses into the differential pumping stages of the electron energy analyzer. This combination is the culmination of decades of work by numerous research groups to increase the operating pressure of XPS. The developments in this field during the past few years demonstrate the versatility of this technique for *in-situ* and *in-operando* investigations of the chemical composition of surfaces and ensure that synchrotron-based AP-XPS will find new exciting experimental applications and systems to investigate for years to come.

ACKNOWLEDGMENTS

This work was supported in part by the Office of Science, U.S. Department of Energy under Contract No. DE-AC02-05CH11231. The Center for Functional Nanomaterials, Brookhaven National Laboratory, is supported by the U.S. Department of Energy, Office of Basic Energy Sciences, under Contract No. DE-AC02-98CH10886. The Advanced Light Source is supported by the Director, Office of Science, Office of Basic Energy Sciences, of the US Department of Energy under Contract No. DE-AC02-05CH11231. A. Knop-Gericke thanks S. Hofmann, B.C. Bayer, and R. Blume for their contributions to the project. This work was partially funded through the European Union (Technotubes Project, www.technotubes.eu). The staff of the Helmholtz-Zentrum Berlin is acknowledged for their continual support of the ambient pressure electron spectroscopy activities of the FHI at BESSY II. M. Hävecker thanks A. Trunschke and R. Schlögl for stimulating discussions. The preparation of the MoVTeNb oxide catalyst by Y.V. Kolen'ko is appreciated. This work was partially supported by the German Research Foundation (Deutsche Forschungsgemeinschaft, DFG) through the cooperated research center "Structure, dynamics, and reactivity of transition metal oxide aggregates" (Sonderforschungsbereich 546, http://www.chemie.hu-berlin.de/sfb546).

REFERENCES

[1] Somorjai, G. A. (1994) *Introduction to Surface Chemistry and Catalysis*, John Wiley and Sons, Inc., New York.

[2] Adamson, A. W. (1990) *Physical Chemistry of Surfaces*, John Wiley and Sons, Inc., New York.

[3] Masel, R. I. (1996) *Principles of Adsorption and Reaction on Solid Surfaces*, John Wiley and Sons, Inc., New York.

[4] Briggs, D., Seah, M. P. (eds.) (1990) *Practical Surface Analysis, 2nd Ed. Volume 1 Auger and X-ray Photoelectron Spectroscopy*, John Wiley and Sons, Inc., New York.

[5] Siegbahn, H., Siegbahn, K. (1973) ESCA applied to liquids, *Journal of Electron Spectroscopy and Related Phenomena*, 2, 319–325.

[6] Siegbahn, H., Asplund, L., Kelfve, P., Hamrin, K., Karlsson, L., Siegbahn, K. (1974) ESCA applied to liquids. II. Valence and core electron spectra of formamide, *Journal of Electron Spectroscopy and Related Phenomena*, 5, 1059–1079.

[7] Siegbahn, H., Asplund, L., Kelfve, P., Siegbahn, K. (1975) ESCA applied to liquids. III. ESCA phase shifts in pure and mixed organic solvents, *Journal of Electron Spectroscopy and Related Phenomena*, 7, 411–419.

[8] Fellner-Feldegg, H., Siegbahn, H., Asplund, L., Kelfve, P., Siegbahn, K. (1975) ESCA applied to liquids. IV. A wire system for ESCA measurements on liquids, *Journal of Electron Spectroscopy and Related Phenomena*, 7, 421–428.

[9] Siegbahn, H., Svensson, S., Lundholm, M. (1981) A new method for ESCA studies of liquid-phase samples, *Journal of Electron Spectroscopy and Related Phenomena*, 24, 205–213.

[10] Siegbahn, H. (1985) Electron spectroscopy for chemical analysis of liquids and solutions, *The Journal of Physical Chemistry*, 89, 897–909.

[11] Joyner, R. W., Roberts, M. W., Yates, K. (1979) A "high-pressure" electron spectrometer for surface studies, *Surface Science*, 87, 501–509.

[12] Pantförder, J., Pöllmann, S., Zhu, J. F., Borgmann, D., Denecke, R., Steinrück, H.-P. (2005) New setup for in-situ x-ray photoelectron spectroscopy from ultrahigh vacuum to 1 mbar, *The Review of Scientific Instruments*, 76, 014102.

[13] Bukhtiyarov, V. I., Kaichev, V. V., Prosvirin, I. P. (2005) X-ray photoelectron spectroscopy as a tool for in-situ study of the mechanisms of heterogeneous catalytic reactions, *Topics in Catalysis*, 32, 3–15.

[14] Ruppender, H. J., Grunze, M., Kong, C. W., Wilmers, M. (1990) In-situ X-ray photoelectron spectroscopy of surfaces at pressures up to 1 mbar, *Surface and Interface Analysis*, 15, 245–253.

[15] Ogletree, D. F., Bluhm, H., Lebedev, G., Fadley, C. S., Hussain, Z., Salmeron, M. (2002) A differentially pumped electrostatic lens system for photoemission studies in the millibar range, *The Review of Scientific Instruments*, 73, 3872–3877.

[16] Ogletree, D. F., Bluhm, H., Hebenstreit, E. B., Salmeron, M. (2009) Photoelectron spectroscopy under ambient and temperature conditions, *Nuclear Instruments & Methods in Physics Research. Section A, Accelerators, Spectrometers, Detectors and Associated Equipment*, 601, 151–160.

[17] Bluhm, H. (2010) Photoelectron spectroscopy of surfaces under humid conditions, *Journal of Electron Spectroscopy and Related Phenomena*, 177, 71–84.

[18] Bluhm, H., Hävecker, M., Knop-Gericke, A., Kleimenov, E., Schlögl, R., Teschner, D., Bukhtiyarov, V. I., Ogletree, D. F., Salmeron, M. (2004) Methanol oxidation on a copper catalyst investigated using in-situ X-ray photoelectron spectroscopy, *Journal of Physical Chemistry B*, 108, 14340–14347.

[19] Salmeron, M., Schlögl, R. (2008) Ambient pressure photoelectron spectroscopy: a new tool for surface science and nanotechnology, *Surface Science Reports*, 63, 169–199.

[20] Grass, M. E., Karlsson, P. G., Aksoy, F., Lundqvist, M., Wannberg, B., Mun, B. S., Hussain, Z., Liu, Z. (2010) New ambient pressure photoemission endstation at Advanced Light Source beamline 9.3.2, *The Review of Scientific Instruments*, 81, 053106.

[21] West, R. C., ed. (1979) *CRC Handbook of Chemistry and Physics*, 60th ed., CRC Press Inc., Boca Raton, FL.

[22] Kleimenov, E. (2005) High pressure X-ray photoelectron spectroscopy applied to vanadium phosphorus oxide catalysts under reaction conditions. PhD Thesis, Technical University, Berlin, Germany.

[23] Tao, F., Grass, M. E., Zhang, Y., Butcher, D. R., Renzas, J. R., Liu, Z., Chung, J. Y., Mun, B. S., Salmeron, M., Somorjai, G. A. (2008) Reaction-driven restructuring of Rh-Pd and Pt-Pd core-shell nanoparticles, *Science*, 322, 932–934.

[24] Tao, F., Grass, M. E., Zhang, Y., Butcher, D. R., Aksoy, F., Aloni, S., Altoe, V., Alayoglu, S., Renzas, J. R., Tsung, C.-K., Zhu, Z., Liu, Z., Salmeron, M., Somorjai, G. A. (2010) Evolution of structure and chemistry of bimetallic nanoparticle catalysts under reaction conditions, *Journal of the American Chemical Society*, 132, 8697–8703.

[25] Grass, M. E., Park, M., Aksoy, F., Zhang, Y., Kunz, M., Liu, Z., Mun, B. S. (2010) Effect of O_2, CO, and NO on surface segregation in a $Rh_{0.5}Pd_{0.5}$ bulk crystal and comparison to $Rh_{0.5}Pd_{0.5}$ nanoparticles, *Langmuir*, 26, 16362–16367.

[26] Hävecker, M., Mayer, R. W., Knop-Gericke, A., Bluhm, H., Kleimenov, E., Liskowski, A., Su, D., Follath, R., Requejo, F. G., Ogletree, D. F., Salmeron, M., Lopez-Sanchez, J. A., Bartley, J. K., Hutchings, G. J., Schlögl, R. (2003) In-situ investigation of the nature of the active surface of a vanadyl pyrophosphate catalyst during n-butane oxidation to maleic anhydride, *Journal of Physical Chemistry B*, 107, 4587–4596.

[27] Hävecker, M., Cavalleri, M., Herbert, R., Follath, R., Knop-Gericke, A., Hess, C., Hermann, K., Schlögl, R. (2009) Methodology for the structural characterization of V_xO_y species supported on silica under reaction conditions by means of in situ O K-edge X-ray absorption spectroscopy, *Physica Status Solidi (B)*, 246, 1459–1469.

[28] Andersson, K., Ketteler, G., Bluhm, H., Yamamoto, S., Ogasawara, H., Pettersson, L. G. M., Salmeron, M., Nilsson, A. (2008) Autocatalytic water dissociation on Cu(110) at near ambient conditions, *Journal of the American Chemical Society*, 130, 2793–2797.

[29] Andersson, K., Ketteler, G., Bluhm, H., Yamamoto, S., Ogasawara, H., Pettersson, L. G. M., Salmeron, M., Nilsson, A. (2007) Bridging the pressure gap in water and hydroxyl chemistry on metal surfaces: the Cu(110) case, *Journal of Physical Chemistry C*, 111, 14493–14499.

[30] Schumacher, N., Boisen, A., Dahl, S., Gokhale, A. A., Kandoi, S., Grabow, L. C., Dumesic, J. A., Mavrikakis, M., Chorkendorff, I. (2005) Trends in low-temperature water-gas shift reactivity on transition metals, *Journal of Catalysis*, 229, 265–275.

[31] Yoshihara, J., Campbell, C. T. (1996) Methanol synthesis and reverse ater-gas shift kinetics over Cu(110) model catalysts: structural sensitivity, *Journal of Catalysis*, 161, 776–782.

[32] Yamamoto, S., Andersson, K., Bluhm, H., Ketteler, G., Starr, D. E., Schiros, T., Ogasawara, H., Pettersson, L. G. M., Salmeron, M., Nilsson, A. (2007) Hydroxyl-induced wetting of metals by water at near-ambient conditions, *Journal of Physical Chemistry C*, 111, 7848–7850.

[33] Tao, F., Dag, S., Wang, L.-W., Liu, Z., Butcher, D. R., Bluhm, H., Salmeron, M., Somorjai, G. A. (2010) Break-up of stepped platinum catalyst surfaces by high CO coverage, *Science*, 327, 850–853.

[34] Björnehom, O., Nilsson, A., Tillborg, H., Bennich, P., Sandell, A., Hernnäs, B., Puglia, C., Mårtensson, N. (1994) Overlayer structure from adsorbate and substrate core-level binding-energy shifts—CO, CCH3 and O on Pt(111), *Surface Science*, 315, L983–L989.

[35] Seah, M. P., Dench, W. A. (1979) Quantitative electron spectroscopy of surfaces: a standard data base for electron inelastic mean free paths in solids, *Surface and Interface Analysis*, 1, 2–11.

[36] Pozdnyakova, O., Teschner, D., Wootsch, A., Kröhnert, J., Steinhauer, B., Sauer, H., Toth, L., Jentoft, F. C., Knop-Gericke, A., Paál, Z., Schlögl, R. (2006) Preferential CO oxidation in hydrogen (PROX) on ceria-supported catalysts, part I: oxidation state and surface species on Pt/CeO2 under reaction conditions, *Journal of Catalysis*, 237, 1–16.

[37] Teschner, D., Wootsch, A., Pozdnyakova-Tellinger, O., Kröhnert, J., Vass, E. M., Hävecker, M., Zafeiratos, S., Schnörch, P., Jentoft, P. C., Knop-Gericke, A., Schlögl, R. (2007) Partial pressure dependent in situ spectroscopic study on the preferential CO oxidation in hydrogen (PROX) over Pt/ceria catalysts, *Journal of Catalysis*, 249, 318–327.

[38] Virnovskaia, A., Jørgensen, S., Hafizovic, J., Prytz, Ø., Kleimenov, E., Hävecker, M., Bluhm, H., Knop-Gericke, A., Schlögl, R., Olsbye, U. (2007) In situ XPS investigation of Pt(Sn)/Mg(Al)O catalysts during ethane dehydrogenation experiments, *Surface Science*, 601, 30–43.

[39] Willneff, E. A., Braun, S., Rosenthal, D., Bluhm, H., Hävecker, M., Kleimenov, E., Knop-Gericke, A., Schlögl, R., Schroeder, S. L. M. (2006) Dynamic electronic structure of a Au/TiO_2 catalyst under reaction conditions, *Journal of the American Chemical Society*, 128, 12052–12053.

[40] Herranz, T., Deng, X., Cabot, A., Alivisatos, P., Liu, Z., Soler-Illia, G., Salmeron, M. (2009) Reactivity of Au nanoparticles supported over SiO_2 and TiO_2 studied by ambient pressure photoelectron spectroscopy, *Catalysis Today*, 143, 158–166.

[41] Grass, M. E., Zhang, Y., Butcher, D. R., Park, J. Y., Li, Y., Bluhm, H., Bratlie, K. M., Zhang, T., Somorjai, G. A. (2008) A reactive oxide overlayer on rhodium nanoparticles during CO Oxidation and its size dependence studied by in situ ambient-pressure X-ray photoelectron spectroscopy, *Angewandte Chemie (International Ed. In English)*, 47, 8893–8896.

[42] Porter, D., Easterling, K. (1992) *Phase Transformations in Metals and Alloys*, 2nd ed., CRC Press, Boca Raton, FL.

[43] Campbell, C. T., White, J. M. (1978) Adsorption, desorption, and reactions of CO and O_2 on Rh, *Journal of Catalysis*, 54, 289–302.

[44] Robertson, J., Zhong, G., Hofmann, S., Bayer, B. C., Esconjauregui, C. S., Telg, H., Thomsen, C. (2009) Use of carbon nanotubes for VLSI interconnects, *Diamond and Related Materials*, 18, 957–962.

[45] Yamazaki, Y., Katagiri, M., Sakuma, N., Suzuki, M., Sato, S., Wada, M., Matsunaga, N., Sakai, T., Awano, Y. (2010) Synthesis of a closely packed carbon nanotube forest by a multi-step growth method using plasma-based chemical vapor deposition, *Applied Physics Express*, 3, 55002–55004.

[46] Dijon, J., Fournier, A., Szkutnik, P. D., Okuno, H., Jayet, C., Fayolle, F. (2010) Carbon nanotubes for interconnects in future integrated circuits: the challenge of the density, *Diamond and Related Materials*, 19, 382–388.

[47] Simon, P., Gogotsi, Y. (2008) Materials for electrochemical capcitors, *Nature Materials*, 7, 845–854.

[48] Futaba, D. N., Hata, K., Yamada, T., Hiraoka, T., Hayamizu, Y., Kakudate, Y., Tanaike, O., Hatori, H., Yumura, M., Lijma, S. (2006) Shape-engineerable and highly densely packed single-walled carbon nanotubes and their applications as super-capacitor electrodes, *Nature Materials*, 5, 987–994.

[49] Teo, K. B. K., Minoux, E., Hudanski, L., Peauger, F., Schnell, J.-P., Gangloff, L., Legagneaux, P., Dieumegardt, D., Amaratunga, G. A. J., Milne, W. I. (2005) Microwave devices—carbon nanotubes as cold cathodes, *Nature*, 437, 968.

[50] Hata, K., Futaba, D. N., Mizuno, K., Namai, T., Yumura, M., Iijima, S. (2004) Water-assisted highly efficient synthesis of impurity-free single-walled carbon nanotubes, *Science*, 306, 1362–1364.

[51] Cantoro, M., Hofmann, S., Pisana, S., Scardaci, V., Parvez, A., Ducati, A., Ferrari, A. C., Blackburn, A. M., Wang, K.-Y., Robertson, J. (2006) Catalytic chemical vapor deposition of single-wall carbon nanotubes at low temperatures, *Nano Letters*, 6, 1107–1112.

[52] de los Arcos, T., Garnier, M. G., Seo, J. W., Oelhafen, P., Thommen, V., Mathys, D. (2004) The influence of catalyst chemical state and morphology on carbon nanotube growth, *Journal of Physical Chemistry B*, 108, 7728–7734.

[53] Nishimura, K., Okazaki, N., Pan, L. J., Nakayama, Y. (2004) In situ study of iron catalysts for carbon nanotube growth using X-ray diffraction analysis, *Japanese Journal of Applied Physics Part 2—Letters and Express Letters*, 43, L471–L474.

[54] Hofmann, S., Sharma, R., Ducati, C., Du, G., Mattevi, C., Cepek, C., Cantoro, M., Pisana, S., Parvez, A., Cervantes-Sodi, F., Ferrari, A. C., Dunin-Borkowski, R., Lizzit, S., Petaccia, L., Goldoni, A., Robertson, J. (2007) In situ observation of catalyst dynamics during surface-bound carbon nanotube nucleation, *Nano Letters*, 7, 602–608.

[55] Mattevi, C., Wirth, C. T., Hofmann, S., Blume, R., Cantoro, M., Ducati, C., Cepek, C., Knop-Gericke, A., Milne, S., Castellarin-Cudia, C., Dolafi, S., Goldoni, A., Schlögl, R., Robertson, J. (2008) In-situ X-ray photoelectron spectroscopy study of catalyst-support interactions and growth of carbon nanotube forests, *Journal of Physical Chemistry C*, 112, 12207–12213.

[56] Hofmann, S., Blume, R., Wirth, C. T., Cantoro, M., Sharma, R., Ducati, C., Hävecker, M., Zafeiratos, S., Schnörch, P., Oestereich, A., Teschner, D., Albrecht, M., Knop-Gericke, A., Schlögl, R., Robertson, J. (2009) State of transition metal catalysts during carbon nanotube growth, *Journal of Physical Chemistry C*, 113, 1648–1656.

[57] Larciprete, R., Lizzit, S., Botti, S., Cepek, C., Goldoni, A. (2002) Structural reorganization of carbon nanoparticles into single-wall nanotubes, *Physical Review B*, 66, 1214021–1214024.

[58] Bayer, B. C., Hofmann, S., Castellarin-Cudia, C., Blume, R., Baehtz, C., Esconjauregui, S., Wirth, C. T., Oliver, R. A., Ducati, C., Knop-Gericke, A., Schlögl, R., Goldoni, A., Cepek, C., Robertson, J. (2011) Support-catalyst-gas interactions during carbon nanotube growth on metallic Ta films, *Journal of Physical Chemistry C*, 115, 4359–4369.

[59] Yang, D. H., Jiang, H., Ott, R., Minor, K., Grant, J., Varga, L., Barnard, J. A., Doyle, W. D. (1999) XPS study of polycrystalline and epitaxial FeTN films deposited by d.c. reactive magnetron sputtering, *Surface and Interface Analysis*, 27, 259–272.

[60] Khanuja, M., Sharma, H., Mehta, B. R., Shivaprasad, S. M. (2009) XPS depth-profile of the suboxide distribution at the native oxide/Ta interface, *Journal of Electron Spectroscopy and Related Phenomena*, 169, 41–45.

[61] Adler, S. B. (2004) Factors governing oxygen reduction in solid oxide fuel cell cathodes, *Chemical Reviews*, 104, 4791–4843.

[62] DeCaluwe, S. C., Jackson, G. S., Farrow, R. L., McDaniel, A. H., El Gabaly, F., McCarty, K. F., Nie, S., Linne, M. A., Bluhm, H., Newberg, J. T., Liu, Z., Hussain, Z. (2009) In situ XPS for evaluating ceria oxidation states in SOFC anodes, *ECS Transactions*, 16, 253.

[63] DeCaluwe, S. C., Grass, M. E., Zhang, C., El Gabaly, F., Bluhm, H., Liu, Z., Jackson, G. S., McDanial, A. H., McCarty, K. F., Farrow, R. L., Linne, M. A., Hussain, Z., Eichhorn, B. W. (2010) In situ characterization of ceria oxidation states in high-temperature electrochemical cells with ambient pressure XPS, *Journal of Physical Chemistry C*, 114, 19853–19861.

[64] Zhang, C. J., Grass, M. E., McDaniel, A. H., DeCaluwe, S. C., El Gabaly, F., Liu, Z., McCarty, K. F., Farrow, R. L., Linne, M. A., Hussain, Z., Jackson, G. S., Bluhm, H., Eichhorn, B. W. (2010) Measuring fundamental properties in operating solid oxide electrochemical cells by using in situ X-ray photoelectron spectroscopy, *Nature Materials*, 9, 944–949.

[65] McDaniel, A. H., El Gabaly, F., Akhadov, E., Farrow, R. L., McCarty, K. F., Linne, M. A., Decaluwe, S. C., Zhang, C., Eichhorn, B., Jackson, G. S., Liu, Z., Grass, M., Hussain, Z., Bluhm, H. (2009) In-situ investigation of SOFC patterned electrodes using ambient-pressure X-ray photoelectron spectroscopy, *ECS Transactions*, 25, 335.

[66] El Gabaly, F., Grass, M., McDaniel, A. H., Farrow, R. L., Linne, M. A., Hussain, Z., Bluhm, H., Liu, Z., McCarty, K. F. (2010) Measuring individual overpotentials in an operating solid-oxide electrochemical cell, *Physical Chemistry Chemical Physics*, 12, 12138–12145.

[67] Whaley, J. A., McDaniel, A. H., El Gabaly, F., Farrow, R. L., Grass, M. E., Hussain, Z., Liu, Z., Linne, M. A., Bluhm, H., McCarty, K. F. (2010) Note: fixture for characterizing electrochemical devices in-operando in traditional vacuum systems, *The Review of Scientific Instruments*, 81, 086104.

[68] Trunschke, A. (2011) Propane-selective oxidation to acrylic acid, in Hess C., Schlögl R. (eds.), *Nanostructured Catalysts: Selective Oxidations*, RSC Nanoscience & Nanotechnology, Vol. 19, Royal Society Chemistry, Cambridge, UK, pp. 56–95.

[69] Ushikubo, T., Nakamura, H., Koyasu, Y., Wajiki, S. (1991) United States patent 5049692.

[70] Sanfiz, A. C., Hansen, T. W., Teschner, D., Schnörch, P., Girgsdies, F., Trunschke, A., Schlögl, R., Looi, M. H., Hamid, S. B. A. (2010) Dynamics of the MoVTeNb oxide M1 phase in propane oxidation, *Journal of Physical Chemistry C*, 114, 1912–1921.

[71] Ushikubo, T., Oshira, K., Kayo, A., Hatano, M. (1997) Ammoxidation of propane over Mo-V-Nb-Te mixed oxide catalysts, in Li C., Xin Q. (eds.), *Studies in Surface Science and Catalysis*, Vol. 112, Elsevier, Amsterdam, The Netherlands, p. 473.

[72] Baca, M., Pigamo, A., Dubois, J. L., Millet, J.-M. M. (2002) Propane oxidation on MoVTeNbO mixed oxide catalysts: study of the phase composition of active and selective catalysts, *Topics in Catalysis*, 23, 39–46.

[73] Botella, P., Lopez-Nieto, J. M. L., Solsona, B. (2002) Preparation, characterization and catalytic behaviour of a new TeVMoO crystalline phase, *Catalysis Letters*, 78, 383–387.

[74] DeSanto, P., Buttrey, D. J., Grasselli, R. K., Lugmair, C. G., Volpe, A. F., Toby, B. H., Vogt, T. (2004) Structural aspects of the M1 and M2 phases in MoVNbTeO propane ammomidation catalysts, *Zeitschrift fur Kristallographie*, 219, 152–165.

[75] Murayama, H., Vitry, D., Ueda, W., Fuchs, G., Anne, M., Dubois, J. L. (2007) Structure characterization of orthorhombic phase in MoVTeNbO catalyst by powder X-ray diffraction and XANES, *Applied Catalysis A: General*, 318, 137–142.

[76] Teschner, D., Vass, E. M., Schlögl, R. (2009) Photoelectron spectroscopy of catalytic oxide materials, in Jackson S. D., Hargreaves J. S. J. (eds.), *Metal Oxide Catalysis*, Wiley-VCH Verlag GmbH & Co. KGaA, Weinheim, Germany, pp. 243–298.

[77] Guliants, V. V., Bhandara, R., Hughett, A. R., Bhatt, S., Schuler, B. D., Brongersma, H. H., Knoester, A., Gaffney, A. M., Han, S. (2006) Probe molecule chemisorption low energy ion scattering study of surface active sites present in the orthorhombic Mo-V-(Te-Nb)-O catalysts for propane (amm)oxidation, *Journal of Physical Chemistry B*, 110, 6129–6140.

[78] Hävecker, M., Wrabetz, S., Kröhnert, J., Csepei, L., Naumann d'Alnoncourt, R., Kolen'ko, Y. V., Girgsdies, F., Schlögl, R., Trunschke, A. (2011) Surface chemistry of phase-pure M1 MoVTeNb oxide during operation in selective oxidation of propane to acrylic acid, *Journal of Catalysis*, 285, 48–60.

[79] Deniau, B., Bergeret, G., Jouguet, B., Dubois, J. L., Millet, J. M. M. (2008) Preparation of single M1 phase MoVTe(Sb) NbO catalyst: study of the effect of M2 phase dissolution on the structure and catalytic properties, *Topics in Catalysis*, 50, 33–42.

[80] Grasselli, R. K., Buttrey, D. J., DeSanto, P., Burrington, J. D., Lugmair, C. G., Volpe, A. F., Jr., Weigand, T. (2004) Active centers in Mo-V-Nb-Te-O-x (amm)oxidation catalysts, *Catalysis Today*, 91–92, 251–258.

13

COMBINED X-RAY DIFFRACTION AND ABSORPTION SPECTROSCOPY IN CATALYSIS RESEARCH: TECHNIQUES AND APPLICATIONS

Anatoly I. Frenkel and Jonathan C. Hanson

13.1 INTRODUCTION

The investigation of the structure, dynamics, and electronic properties of catalysts in real time and under reaction conditions is essential for elucidating reaction pathways and kinetics. Among synchrotron-based techniques, X-ray absorption fine-structure spectroscopy (XAFS) and X-ray diffraction (XRD) are perhaps the ones most commonly used for *in-situ* and *in-operando* research [1–10]. Time-resolved XRD allows one to get information concerning the following properties: phase identification, fractional composition, long-range crystalline properties and molecular structure of catalysts under reaction conditions, as well as kinetics of crystallization of nanoparticles and bulk solids [1] (see Chapter 4). Time-resolved XAFS can be used to follow electronic and local structural changes associated with catalytic reactions with characteristic times ranging from minutes to tens of milliseconds. Most applications of time-resolved XAS use either energy-dispersive [2–6] (see Chapter 3) or energy-scanning modes [7–10](see Chapter 1).

The pioneering work of Couves et al. [11], Sankar et al. [12], and Clausen et al. [9] on combining XRD and XAFS in one single experiment with enough

In-situ *Characterization of Heterogeneous Catalysts*, First Edition.
Edited by José A. Rodriguez, Jonathan C. Hanson, and Peter J. Chupas.

time resolution to follow the kinetics of structural changes of solids under reaction conditions opened the door to the employment of a powerful characterization tool for unraveling complex structural transformations of catalysts. Further developments in the technique increased the resolution of both the XRD and XAFS measurements [13]. In this chapter, we introduce the instrumentation used at the Daresbury [11], DORIS [9], European Synchrotron Radiation Facility (ESRF) [13] and National Synchrotron Light Source (NSLS) [14] for such measurements and demonstrate its applicability for *in-situ* studies of catalytic reactions. Examples discussed in the succeeding sections will emphasize the complementary qualities of each technique by pointing out what information revealed by one technique is not accessible by the other. These complementary qualities will be exploited to advance our understanding of the *actual* structure, which combines local and average information of the catalytic systems.

13.2 INSTRUMENTATION

All current implementations of combined X-ray diffraction and X-ray absorption fine structure actually measure XRD and XAFS consecutively, not simultaneously, because XRD is normally done with fixed energy and XAFS must be done while scanning the energy. The beamline optics must have a way of varying the energy for the XAFS measurements and keeping a fixed energy for the XRD data acquisition.

The first combined XRD and XAFS setup was at the Daresbury Synchrotron [11]. In this setup, the diffraction data were collected with an Inel curved linear detector (Artenay, France) when the optics was set up to provide a single wavelength of X-rays illuminating the sample. The XAFS was measured in the dispersive extended X-ray absorption fine structure (EXAFS) mode (see Chapter 3) using a range of wavelengths around the absorption edge which are focused at the sample and redispersed on photodiode array.

The setup at the DORIS synchrotron in Hamburg provided energy-scanning EXAFS in transmission mode with ion chambers located before and after the sample, and diffraction with a position-sensitive detector (PSD) on a 2θ arm. The authors of that work also presented a description of new information obtained by combining the techniques [9]. That study used an example of *in-situ* reduction of a CuO/ZnO/alumina catalyst. The appearance of the Cu^0 diffraction was slightly delayed from a Cu^0 feature in the XAFS [9]. This demonstrates that the amorphous Cu^0 appears before the crystalline phase.

The setup at Swiss–Norwegian beamlines (SNBL), ESRF (Fig. 13.1) has three detectors for diffraction. The first detector provides high-resolution patterns by using a set of detectors with analyzer crystals before them. This high resolution is critical when small distortions in the lattice are expected during the catalysis. For limited theta ranges (see later), a scan can be completed in 5 min. The second detector is an image plate detector that allows for collecting

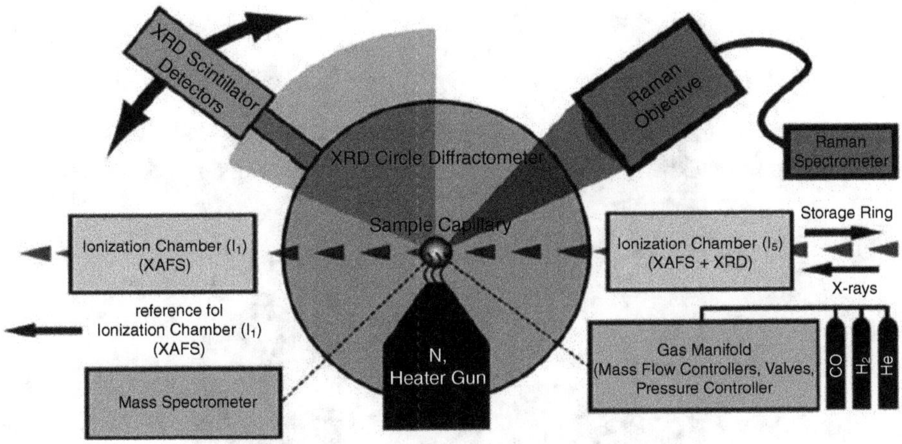

Figure 13.1. In the setup at SNBL at ESRF the XRD is collected in high-resolution mode with multiple detectors. The XAFS is collected in continuous scanning mode with ion chambers. Reprinted with permission from de Smit, E., et al. (2010) *J. Am. Chem. Soc.*, 132, 14928–14941. Copyright (2010) American Chemical Society.

a full pattern with better statistics. The image plate detector has an 80- or 100-s delay while the image plate is being scanned. The third detector is a small two-dimensional (2-D) detector with fast readout. This detector allows rapid measurements (<200 ms) of part of the diffraction pattern. The beamline has separate monochromators for XRD and XAS and the switch between the XRD and XAFS monochromators can be carried out in 10 s, which allows XRD and XAFS measurement at different energies without the delay of driving the monochromator.

The recently built XRD/XAFS setup at the NSLS beamline X18A [14] has a channel-cut monochromator (CCM) with the regular and quick EXAFS scanning modes. Focusing to a spot size of 0.5×0.5 mm^2 is achieved by a rhodium-coated toroidal mirror.

The X-ray detectors that were used in the XRD measurements at NSLS X18A include a curved linear detector (Inel), a silicon linear detector (SiLD), and a PerkinElmer (PE; Waltham, MA) amorphous silicon 2-D detector. The SiLD is a microstrip array of silicon diodes (800 elements, 0.125×4 mm^2 in area per element), designed and made at Brookhaven National Laboratory (BNL) [15]. It is mounted on the 2θ arm and can be used in stationary mode. It has an approximately 10° acceptance angle, the readout time (low limit) is 0.03 s, and the typical exposure time in the experiments described later is 30 s. A 2-D area detector (PE) is the one most commonly used. It has 2048×2048 pixels and a 200×200 μm^2 pixel size. The detector records patterns with an area of 40×40 cm^2. The detector is mounted on the 2θ arm, the typical 2θ range is 5–48°, and the typical exposure in the experiments (described later)

Figure 13.2. (a) The Inel detector mounted on the 2θ arm of goniometer. Alternatively, the 2-D amorphous silicon detector or the silicon linear detector can be mounted on this arm. The transmission and fluorescence XAFS detectors are shown. The flow cell, heater and PIPS fluorescence detector, and blower heater are also shown. (b) Detail of the flow cell, heater, and fluorescence XAFS detector.

was 30 s. The XAFS data are often collected in the fluorescence mode with a passivated implanted planar silicon (PIPS) detector (Fig. 13.2b). Alternatively, ion chambers can be used for transmission XAS.

All the previously mentioned experiments could be carried out with a plug flow cell [9, 13, 16]. The flow control is provided and the reaction products were measured with a mass spectrometer or gas chromatograph. Various hot air and resistive heaters were used to control the temperature.

13.3 EXAMPLES

In the following subsections, examples of combined XRD/XAFS that demonstrate the usefulness of the nearly simultaneous measurement are presented and discussed.

13.3.1 Structural Transformations of $CuFe_2O_4$ under CO Flow at Different Temperatures

This set of experiments shows how XAFS can see an amorphous metallic phase during the reaction $3CuFe_2O_4 + 2CO \rightarrow 2CO_2 + 2Fe_3O_4 + 3Cu$ before the diffraction of Cu appears. The copper spinel $CuFe_2O_4$ is potentially useful for the water–gas shift (WGS) reaction [17, 18]. The reduction of this copper spinel with hydrogen and carbon monoxide has been previously studied by separate XRD and XAFS measurements. These experiments show that the copper cations migrate out of the structure under reducing conditions as the temperature is raised and, under oxidizing conditions, copper readily reincorporates back into the spinel structure. Its structural transformation upon reduction cannot be fully understood by XRD technique alone, due to the likelihood of forming metastable, short living, strongly dispersed, and/or disordered phases during the formation of pure Cu. Only by combining XRD and XAFS in a single experiment, where the sample is subjected to the same temperature/gas treatment conditions, can one detect both the short-range and the long-range structure of this system and ensure that the two results describe the same chemical and physical states of the sample.

The following paragraphs present a study of $CuFe_2O_4$ during the reduction with CO. The XRD data were measured with the SiLD detector mounted on the 2θ arm of the diffractometer. The sample was loaded in a 1/8 in. Kapton tube (Concentric Microtubing, Cornelius, NC) in a flow cell. The XAFS was measured with a PIPS detector (the sample was too thick to allow for measurement of transmission or reference signal). The temperature was controlled with an air blower that was placed below the sample (see Fig. 13.2b). The sample was "activated" by heating at 200°C in a 10% O_2/He flow. Then, the system was reduced at different temperatures while taking XRD and XAFS measurements (one after the other) on the same sample under identical conditions. The Cu-edge XAFS, Fe-edge XAFS, and the diffraction measurements are presented along with a comparison of the results from different techniques.

Figure 13.3a shows that the energy space X-ray absorption near-edge structure (XANES) data of Cu edge in CO flow at various temperatures. It shows that with increasing temperature, the white line (at ca. 8995 eV) shifts to lower

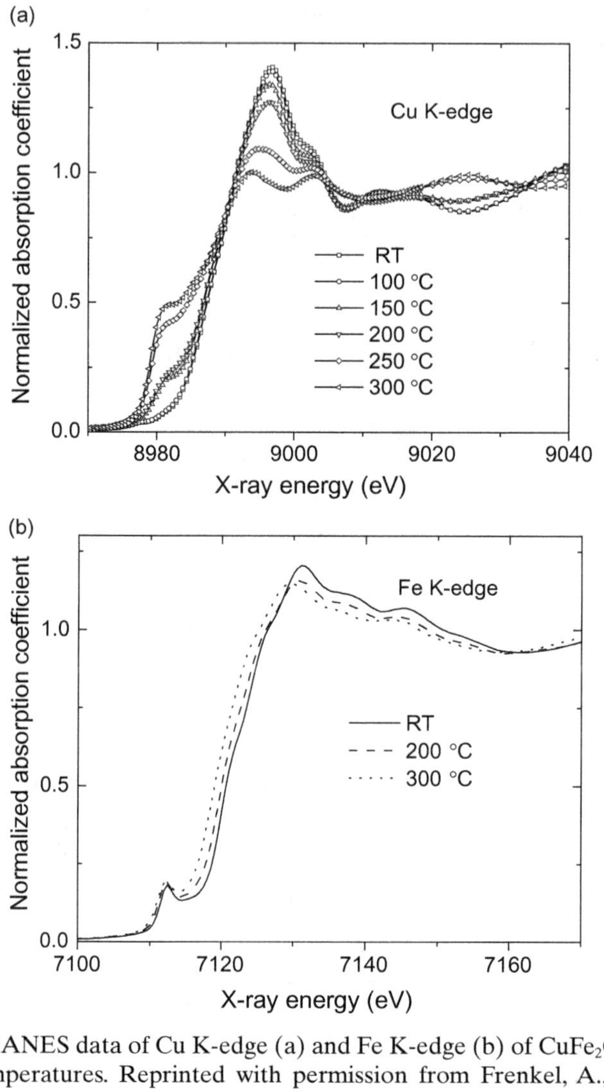

Figure 13.3. XANES data of Cu K-edge (a) and Fe K-edge (b) of $CuFe_2O_4$ in CO flow at various temperatures. Reprinted with permission from Frenkel, A., et al. (2011) *J. Phys. Chem. C*, 115, 17884–17890. Copyright (2011) American Chemical Society.

energy and decreases in intensity, indicating that the reaction gradually changes copper to lower oxidation states than the initial Cu(II). A characteristic shoulder (at ca. 8990 eV) indicating the formation of metallic copper emerges in the Cu-edge data at 150°C. The XANES features at 300°C are already close to those of a pure metallic copper. During the entire temperature series, the data feature several isosbestic points, indicating a one-step transformation.

The Fe K-edge XANES data (Fig. 13.5b) reveal a gradual shift to lower energies and a reduction in the peak intensity with temperature, consistent

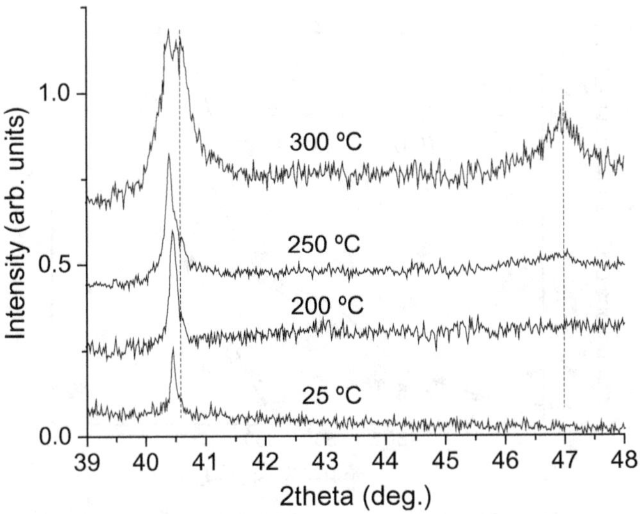

Figure 13.4. XRD data at different temperatures under CO reducing conditions. The Cu^0 phase appears near 40.5 and 47°. The peak near 40.5° overlaps with a spinel peak, but the peak near 47° is only from (200) Cu^0. The normalized integrated areas of this peak at different temperatures are presented in Table 13.1. Reprinted with permission from Frenkel, A., et al. (2011) *J. Phys. Chem. C*, 115, 17884–17890. Copyright (2011) American Chemical Society.

with the gradual reduction of iron as temperature increases. The average oxidation state of iron in $CuFe_2O_4$ is 3, but the average oxidation state in Fe_3O_4 is 8/3. The rate of Fe reduction parallels the rate of Cu reduction which is indeed confirmed by the quantitative data analysis of the Cu and Fe XANES (see later).

Figure 13.4 shows the XRD data in the region where Cu peaks appear during reduction of the copper spinel. In this plot, there is only one spinel peak and two metallic copper peaks. There are slight changes in the diffraction pattern when the Cu^{2+} cations leave the oxide and the pure magnetite phase is formed. There is also a small shift on the peak position due to a cell dimension change when copper segregates from the oxide. This shift can, therefore, be related to the relative amounts of cuprospinel and magnetite. The amount of the metallic copper phase can be estimated from the (200) peak height ratio as discussed in the next section.

Quantitative analyses of the XAFS measurements were done as follows. The volume mixing fractions of reduced and unreduced Cu and Fe were obtained from linear combination fits (LCFs) of the Cu K-edge and Fe K-edge XANES data, respectively (Fig. 13.5 and Table 13.1). The references for the LCFs were based on the room temperature (RT; most oxidized) data and 300°C (most reduced) data to the XANES signals. The fractions of the Cu^0

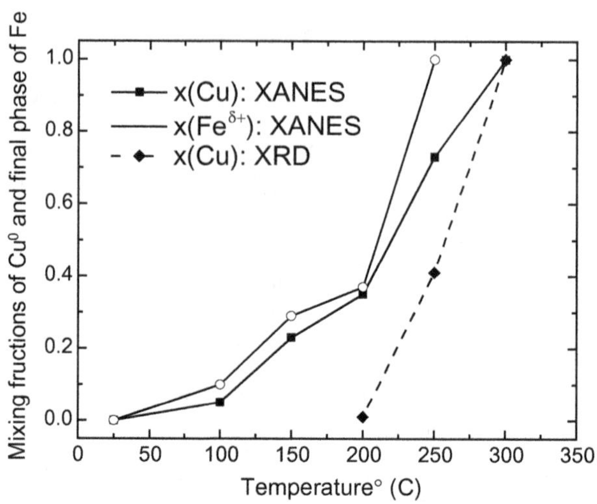

Figure 13.5. Comparison of Cu^0 and $Fe^{\delta+}$ fractions determined by XRD and XAS at different temperatures. Reprinted with permission from Frenkel, A., et al. (2011) *J. Phys. Chem. C*, 115, 17884–17890. Copyright (2011) American Chemical Society.

TABLE 13.1. Temperature-Dependent Volume Mixing Fractions of the Final States of Cu (Metal) and Fe (Consistent with Magnetite) during the *In-situ* XAFS Experiments

Temperature (°C)	x(Cu) XAFS	x($Fe^{\delta+}$) XAFS	x(Cu) XRD
25	0	0	0
100	0.05	0.1	
150	0.23	0.29	
200	0.35	0.37	0.01
250	0.73	1	0.41
300	1	1	1

XRD-derived fraction of crystalline reduced metal Cu is given for comparison.

species in the XRD data were determined from the integrated peak intensity of the Cu 200 reflection [19]. The relative amounts of metallic copper phase obtained by XRD are given in Table 13.1.

EXAFS analysis provided unique information about the rates of reduction of Fe and Cu in the complex system. The fact that the temperature dependences for transforming both of these species are in good agreement with each other suggests a specific mechanism in which Fe occupies sites vacated by Cu.

It should also be pointed out that both EXAFS and XRD analyses results give no hint of the presence of any copper species other than the initial spinel and the final metallic Cu^0. The comparison between the EXAFS and XRD results leads to the conclusion that there is an underestimation of the relative amount of Cu^0 by XRD at the intermediate temperature range. This is consistent with Cu^0 appearing as an amorphous, and/or low-dimensional phase prior to the appearance of the crystalline phase.

13.3.2 *In-situ* Studies of Structural Transformations of a Cu/Ceria Inverse Catalyst

The previous example demonstrates that important conclusions about the phase composition of a mixed system during reduction can be made by cointerpreting the results of nearly simultaneous XRD and XAFS *in-situ* measurements. In that case, we ruled out a reaction intermediate and concluded in favor of a one-step process. The following example demonstrates how the XRD and XAFS measurements, at first glance, appear to contradict each other in the results they report.

Traditionally, nanoparticles of metals on metal oxides (e.g., copper on ceria) have been used as WGS catalysts [20]. Recently, an inverse copper–ceria catalyst where the nanoparticle oxide is deposited on a second metal oxide has been synthesized [21]. This system is a new WGS catalyst that is attractive in mobile applications due to its high activity at low temperatures and stability during reduction/oxidation (redox) cycles [22]. This inverse system has also proven to be a highly efficient catalyst for the preferential CO oxidation reaction, since the presence of bulk-like copper structures quenches the hydrogen oxidation reaction, therefore providing a wider range of operating temperatures than a conventional Cu/CeO_2 system [21]. In this experiment, we investigated structural transformations in the $CuO/Cu_xCe_{1-x}O_2$ ("inverse Cu/ceria") catalyst during CO reduction and subsequent reoxidation with O_2. The as-prepared sample is a mixture of a CuO phase as a support with a $Cu_xCe_{1-x}O_2$ mixture on top. An overall Cu : Ce ratio of 6 : 4 was determined by X-ray fluorescence [21, 22]. The inverse nature of the catalyst is illustrated from the diffraction patterns that show broad peaks due to the nanosize of CeO_2 crystals and well-defined peaks for the CuO phase. The morphology has been also independently confirmed by high-resolution transmission electron microscopy (HR-TEM) [21]. Upon reduction, the signal of Cu^0 produced very sharp peaks, indicating the bulk nature of the metallic phase (Fig. 13.6a). The XANES of the starting mixture illustrates that the Cu is found in both components. Figure 13.8b shows the observed XANES and the linear combination of the XANES for pure CuO and pure $Cu_{0.1}Ce_{0.9}O_2$ (it also shows the individual components contributing to the linear combination).

Previous studies of the system by separate XAFS and XRD experiments showed a strong interaction between copper and ceria crystallographic phases due to the high mobility of oxygen interchange between them, but the existing data were inconclusive as to the presence, and the nature, of intermediate transitions. The combined XRD/XAFS measurements can quantify the composition of Cu in the two phases and characterize their structures (including

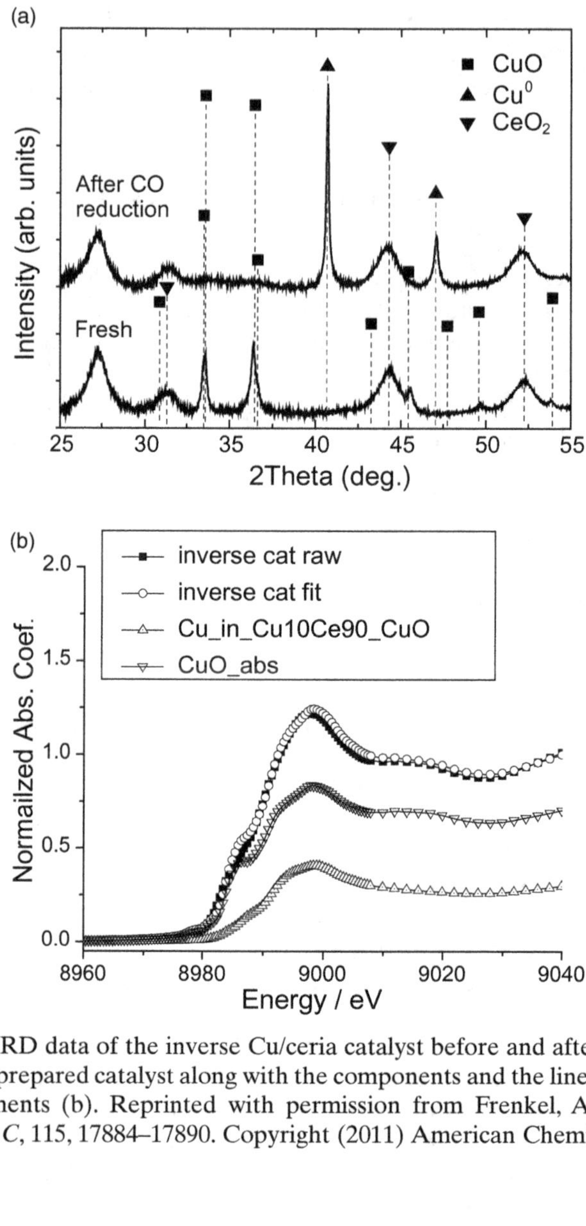

Figure 13.6. XRD data of the inverse Cu/ceria catalyst before and after reduction (a) XANES of as-prepared catalyst along with the components and the linear combination of the components (b). Reprinted with permission from Frenkel, A., et al. (2011) *J. Phys. Chem. C*, 115, 17884–17890. Copyright (2011) American Chemical Society.

the local structure in the disordered and/or low-dimensional phases) and oxidation states.

XRD patterns (measured with 30 s exposure time) were collected at 20 keV during reduction and oxidation measurements. The reduction experiments were carried out by flowing 5% CO/He mixture at the flow rate of 10 cc/min while increasing the temperature from RT to 200°C and then holding the temperature at 200°C. For the oxidation measurements, the gas was switched to a 20% O_2/He mixture with 10 cc/min flow rate, and the temperature was held at 200°C, then increased and held at 250 and 300°C. XAFS scans were collected before and after the CO reduction at 25 and 200°C. XAFS scans were also collected during the reoxidation experiment (at the end of the 30-min exposure to O_2) and at 300°C (after the extended exposure to O_2). Gases leaving the reaction chamber were analyzed with a quadrupole mass spectrometer (QMS). The evolved gases were consistent with the structural changes observed by XRD and XAFS.

Two-dimensional powder XRD patterns were integrated using the FIT2D code [23]. Lattice parameters and phase fractions were determined through Rietveld refinement [24] by using the General Structure Analysis System (GSAS) software [25, 26]. The X-ray refinement was complicated by the fact that the model had to include the substituted Cu^{2+} near the Ce^{4+} sites. The fractional Cu^{2+} and Ce^{4+} fractional occupancies were refined while being restrained to a total of 1.0. The Cu^{2+} was shifted approximately 1 Å from the Ce^{4+} site based on the difference electron density map. The reliability of the fractional content was limited by the fact that the two atoms are close to one another and the data were only collected to a q of 6 Å^{-1}. The mole fraction of Cu in ceria was determined to be 0.10(1), which is lower than was expected (the number in parentheses is the standard deviation from the refinement). During the CO reduction, the fractional content of the Cu in ceria went from 0.1 to 0.04. (Refinements from a separate measurement of the same material during reduction in CO with data to a q of 12 Å^{-1} showed an expected range of 0.24–0.02.)

The results of the time-resolved XRD measurements are shown in Figure 13.7 and Table 13.2. From Figure 13.7a, it can be seen that CuO in the fresh inverse catalyst was directly reduced to Cu metal in 5% CO/He below 200°C. No intermediate phase (Cu_2O) was observed.

The observed and calculated XANES data at 25°C before reduction and at 200°C after reduction are shown in Figure 13.7b. The fits were obtained as a linear combination of Cu^0, Cu^{2+} in CeO_2, Cu^{2+} in CuO, and Cu^+ in Cu_2O where the fractional combination was determined from fitting the XANES spectrum; the values are shown in Table 13.2. The XANES fit overestimated and the XRD fit underestimated the Cu^{2+} in CeO_2.

Figure 13.8a shows the XRD fit of the mole fractions of the Cu, CuO, and Cu_2O during, before, and after the gas was switched from 5% CO in He to 20% O_2 in He, and during step-wise heating to 300°C. The reoxidation resulted in the mixture of CuO and Cu_2O at 200°C in 20% O_2/He, and Cu_2O was found to be fully oxidized at 300°C. The results demonstrate that the Cu_2O intermediate structure is stable in the reoxidation step.

The diffraction measurements were stopped at the end of the 200°C step and at the end of the 300°C step to collect EXAFS scans. The raw data and LCFs are shown in Figure 13.10b and the mole fractions are listed in Table 13.2.

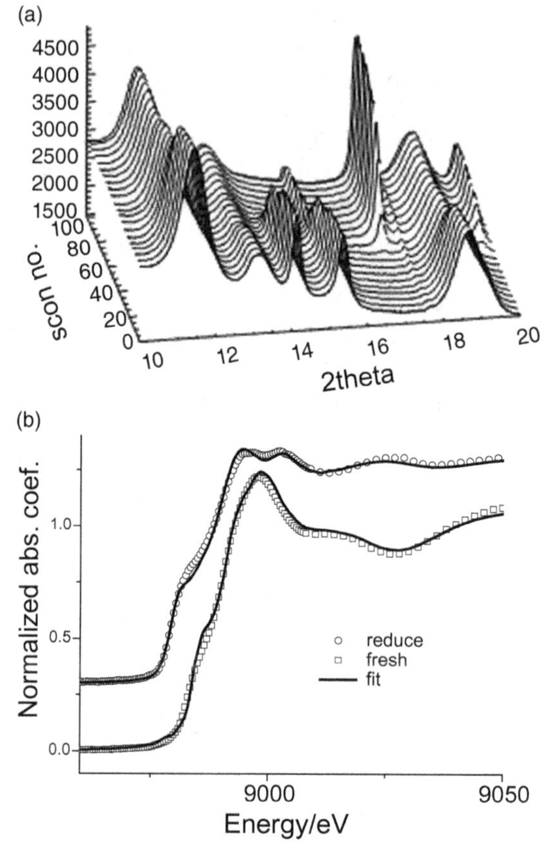

Figure 13.7. (a) Series of powder patterns collected with the detector at 2θ = 40°. Exposures were 30 s, but they were grouped in averages of 5 for clarity. The total time for the figure was 53 min [5]. (b) Fit XANES spectra at 25 and 200 while flowing 5% CO in He. Reprinted with permission from Frenkel, A., et al. (2011) *J. Phys. Chem. C*, 115, 17884–17890. Copyright (2011) American Chemical Society.

TABLE 13.2. Mixing Fractions of Cu^0, Cu^+, and Cu^{2+} in the Inverse Cu/Ceria Catalysts Obtained during CO Reduction and O_2 Oxidation by XAFS and XRD Techniques in a Single Experiment

	Cu^0		Cu^+ in Cu_2O		Cu^{2+} in CuO		Cu^{2+} in CeO_2	
Sample	XAFS	XRD	XAFS	XRD	XAFS	XRD	XAFS	XRD
Fresh	0	0	0	0	0.73	0.90	0.27	0.1
Reduced at 200°C	0.7	0.96	0.17	0	0.13	0	0	0.04
Reoxidized at 200°C	0.23	0	0.30	0.31	0.47	0.67	0	0.02
Reoxidized at 300°C	0	0	0		0.70	0.95	0.30	0.05

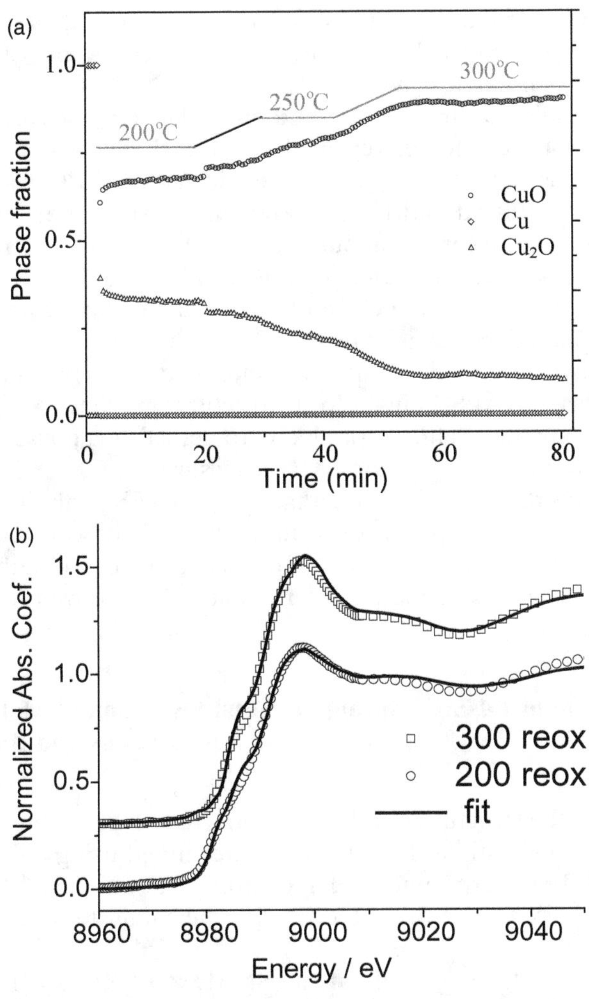

Figure 13.8. (a) Molar phase fraction of copper species during reoxidation; (b) phase fractions from profile refinement, K-edge XANES data, and LCFs at 200 and 300 while flowing 20% O_2 in He. Reprinted with permission from Frenkel, A., et al. (2011) *J. Phys. Chem. C*, 115, 17884–17890. Copyright (2011) American Chemical Society.

The fresh sample was found to be dominated by Cu^{2+}, and the reoxidized sample at 300°C was found to be identical with the fresh sample, in the XANES region (Fig. 13.8b). LCFs of XANES data provided the mixing fractions of Cu^0, Cu^+, and Cu^{2+} in the reduced and reoxidized samples. For the linear combination analysis, the XANES data measured in Cu metal foil, CuO, and Cu_2O powder data were used as standards for the Cu^0, Cu^+, and Cu^{2+} oxidation states, respectively. In addition, a standard for the doped phase $Cu:CeO_2$ was also used [20]. To account for the temperature difference between the Cu/ceria data and the standards (collected at RT), a minor smoothing was applied to the XANES data of the Cu and CuO standards before the LCF was carried out.

Numerical results for the XANES and XRD data analyses are summarized in Table 13.2. Both techniques revealed that the chemical state of Cu in the starting sample, as well as in the sample reoxidized at 300°C, was Cu^{2+}. In the sample that was reduced and then subsequently oxidized at 200°C, the two techniques reported different amounts of Cu^0, Cu^+, and Cu^{2+}. In both cases, XRD underestimated the amounts of *minority* phases in the sample (Cu^+ and Cu^{2+} in the reduced sample and Cu^0 and Cu+ in the oxidized sample). One can conclude that the reduced sensitivity of XRD to these contributions was likely to be caused by the enhanced disorder and/or low dimensionality of these undetected phases. XAFS, being a local structural method, is sensitive to the local structure only, and is thus more likely to detect such phases (Table 13.1). However, without the simultaneous XRD measurement, XAFS results would not indicate their disordered nature, that is, the XAFS analysis results would have been incomplete or, worse, even misleading. Only when combining the two measurements together in one self-consistent analysis methodology can the new information about the disordered nature of the transient species be obtained.

13.3.3 Structure of Pd/ZrO_2 during the Total Oxidation of Methane Using Combined *In-situ* XRD, X-ray Absorption Spectroscopy and Raman Spectroscopy

In this example, the structure of Pd/PdO supported on ZrO_2 during the catalysis of the total combustion of methane is determined using combined XRD/XAFS [27, 28]. These experiments demonstrate that one needs the combination of XRD and XAFS to fully characterize the catalyst under varying reaction conditions.

The measurements were made at the SNBL at ESRF (see Fig. 13.1). The transmission XANES was measured from 24.3 to 24.5 keV in approximately 200 s. The XRD measurements were made with the high-resolution mode (multidetector with analyzers) at 24.79 keV (0.5 Å). The reactions took place in a plug flow reactor with gas flow control, blower heater temperature control, and product analysis with a mass spectrometer analyzer.

A gas mixture of 1% CH_4–4% O_2 in He was passed through the catalyst at series of temperatures and the XRD and XAFS were collected. The methane conversion varies greatly with temperature (Fig. 13.9). There is a large drop in conversion in step 4 which is the highest temperature tested. The conversion increases again during the cooling step 6.

The XRD patterns at the various temperatures are plotted in Figure 13.10. Even though the Pd concentration is 10%, diffraction lines for the PdO are not observed. This indicates that the PdO was highly dispersed. The XRD of Pd metal is observed during the highest temperature step 4 and during cooling

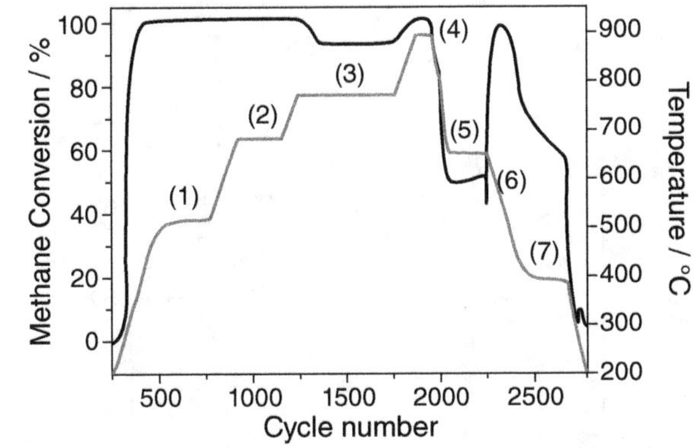

Figure 13.9. Methane conversion during temperature-controlled reaction of 1% CH_4–4% O_2 in He. The temperature program is also shown and the steps are marked for alignment of Figure 13.12 and Figure 13.13. Reproduced by permission of The Royal Society of Chemistry from Grunwaldt, J. D., et al. (2007) *Chem. Comm.*, 4635–4637.

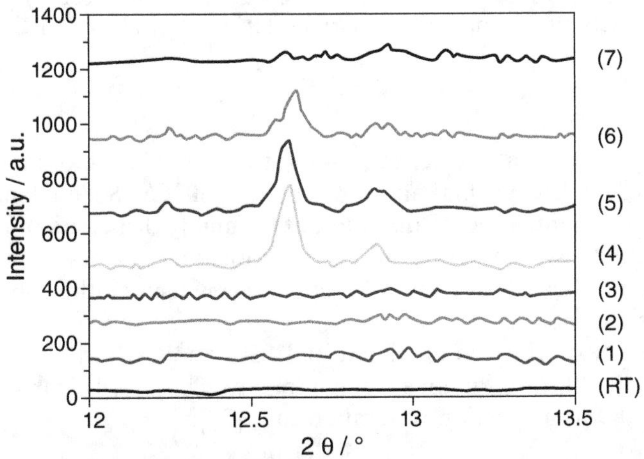

Figure 13.10. XRD patterns during methane combustion over 10% Pd/ZrO_2 in 1% CH_4/4% O2/He (RT) after reduction in 5%H_2/He; (1) 150°C, (2) 500°C, (3) 670°C, (4) 770°C, (5) after 890°C down to 770°C, (6) 630°C, and (7) 400°C. Reproduced by permission of The Royal Society of Chemistry from Grunwaldt, J. D., et al. (2007) *Chem. Comm.*, 4635–4637.

down until step 7 (500°C). This points to a decrease in activity when the Pd is reduced. The XANES (Fig. 13.11) data qualitatively give the oxidation at all the steps. The spectra indicate that the Pd is oxidized until the spectra at 735°C, when the white line starts to decrease and the reduced spectra is fully obtained by 770°C.

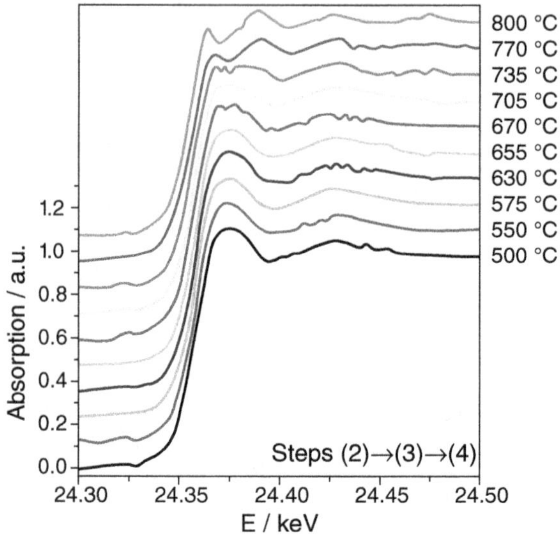

Figure 13.11. XANES spectra at different temperatures are shown during methane combustion over 10% Pd/ZrO_2 in 1% CH4/4% O2/He. Reproduced by permission of The Royal Society of Chemistry from Grunwaldt, J. D., et al. (2007) *Chem. Comm.*, 4635–4637.

This study shows that the active phase for total oxidation of methane is the PdO. The lack of XRD signal and the presence of XAFS signal, measured in the same experiment, indicate that the PdO is highly dispersed or amorphous. This experiment also shows how this information which is vital for understanding mechanism of catalysis cannot be obtained by using either one technique or the other, but both.

Subsequent experiments at SNBL at ESRF took advantage of the additional capability of simultaneous Raman spectra. This more clearly showed the amount of PdO as a function of temperature [27].

13.3.4 Carbide Catalyst Phases in Fischer–Tropsch Synthesis [29]

The Fischer–Tropsch synthesis (FTS) converts CO and H_2 to short chain alkanes. A (qausi-)simultaneous XRD/XAFS study was done at the SNBL at ESRF (see Fig. 13.3). Fe K-edge (7.112 keV) XAFS data were collected in transmission mode. EXAFS and XANES spectra were recorded using the monochromator in constant speed mode, resulting in an acquisition time of typically 5 min for one XAFS spectrum. XRD data were acquired using a Si(111) CCM tuned to a wavelength of 0.50000 Å. Diffracted (scattered) X-rays were detected by a circle diffractometer equipped with six scintillation detectors, each placed after a Si analyzer crystal. The intrinsic resolution (full width at half maximum, FWHM) of the diffractometer was approximately

TABLE 13.3. Physicochemical and Catalytic Properties of Catalyst Materials before and after 6 h FTS at 10 bar and 250°C

	XAFS (mol%)	XRD (vol%)	Raman	FTS Performance
Before FTS	χ-Fe_5C_2 (76%) Fe_xC (24%)	χ-Fe_5C_2 (90%) ε-Carbides (10%)	Some graphite C	
After FTS	χ-Fe_5C_2 (74%) Fe_xC (12%) Fe_3O_4 (14%)	χ-Fe_5C_2 (57%) ε-Carbides (5%) Fe_3O_4 (38%)	No increase in graphite C	High conversion and selectivity

0.01°. The diffractometer was calibrated before the experiments using a Si reference. The acquisition time of a 9.0–27.0° 2θ diffraction pattern was typically 15 min. The station is now equipped with two monochromators: a high-resolution powder diffraction (HRPD)-dedicated CCM and a second double-crystal monochromator (DCM) for XAFS. Both monochromators can be easily and rapidly (within 10 s) interchanged. A pair of alternating XRD and XAFS measurements could be collected in 20 min on the same sample.

In this study, the catalyst precursor is Fe_2O_3, which is pretreated with CO at 250°C (high μ_c) or 1% CO in H_2 at 350°C (low μ_c) to first produce other iron oxides and then different iron carbides (see Table 13.3). To simplify the description, only the results from the high μ_c precursor will be discussed.

The EXAFS fitting analysis of the iron carbides was facilitated by preparation of reference carbides. The XRD indicated that the precursor was primarily χ-Fe_5C_2 with less crystallinity than in reference carbides. The EXAFS fitting analysis of the pretreated sample showed a significant difference in the coordination numbers from the χ-Fe_5C_2 reference sample. In addition, the LCFs of the XANES suggested a significant contribution from an amorphous iron carbide phase (Fe_xC in the XAFS column of Table 13.3). The authors argue that the amorphous carbide is related to the ε-carbides.

After pretreatment the sample was cooled to 250°C FTS conditions (CO/H_2 5;5mL/min and pressure of 10 bar) The sample pretreated under high μ_c conditions started out with 24% amorphous iron carbide, which reduced to 12% and contributed to the formation of Fe_3O_4. The relative amount of χ-Fe_5C_2 also decreased. Figure 13.12 shows the evolution of the Fe_3O_4 under FTS conditions.

Figure 13.13 shows the evolution of the Fe_3O_4 in the phase-corrected k_1-weighted Fourier-transformed Fe K-edge data during FTS. It shows the evolution of the Fe_3O_4. The catalysts prepared under "high" μ_c conditions contains mainly χ-Fe_5C_2 and is more active and selective (based on product analysis from the *in-situ* sample), but is susceptible to oxidation.

This example illustrates a quantitative use of nearly simultaneous XRD and XAFS because they performed least squares fitting of models to the data.

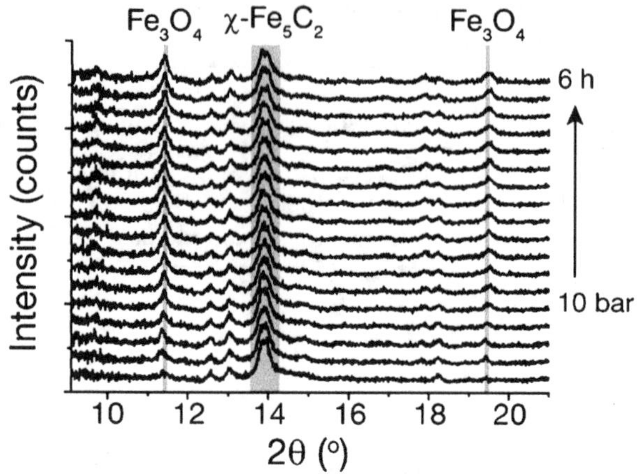

Figure 13.12. XRD patterns of catalysts prepared from Fe_3O_4 in 1 bar of CO at 250°C (high μ_c). The series of patterns was taken over 6 h, while the FTS conditions (CO/H_2 5;5mL/min and pressure of 10 bar) were maintained. Reprinted with permission from de Smit, E., et al. (2010) *J. Am. Chem. Soc.*, 132, 14928–14941. Copyright (2010) American Chemical Society.

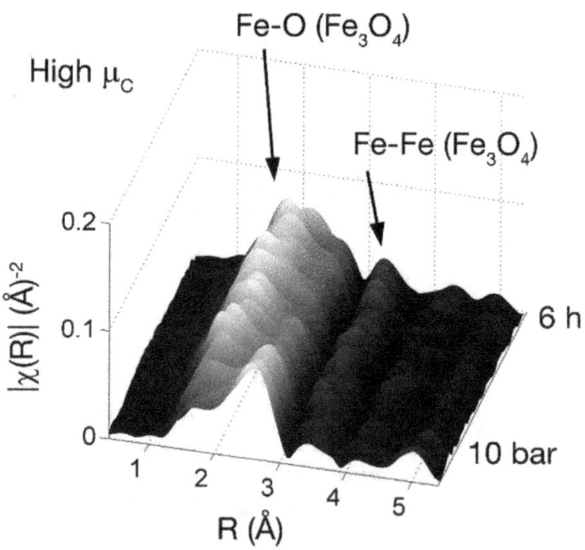

Figure 13.13. Phase-corrected, k_1-weighted Fourier-transformed Fe K-edge EXAFS while the FTS conditions (CO/H_2 5;5mL/min and pressure of 10 bar) were maintained. Reprinted with permission from de Smit, E., et al. (2010) *J. Am. Chem. Soc.*, 132, 14928–14941. Copyright (2010) American Chemical Society.

Table 13.3 is an adaptation of table 5 for the high μ_c pretreatment from Reference 27. The larger contribution of the amorphous phase as determined by the XAFS is a result of the fact that XAFS is based on the local order in the catalyst. This is similar to the result discussed earlier for the Cu spinel results.

The authors of the paper conclude: "The catalysts were characterized under FTS conditions and it was found that amorphous phases play a significant role in Fe-based FTS catalysts. Therefore the XAFS/XRD proved to be a valuable combination of techniques, allowing the characterization of amorphous and crystalline phases at realistic conditions" [29].

13.4 SUMMARY AND OUTLOOK

13.4.1 Is Combination of XRD and XAFS Justified?

Why combine XAFS and XRD in the same experiment? Can the experiment be reproduced at the two different beamlines, optimized for XAFS and for XRD? These are important questions that we would like to address with an emphasis on synchrotron-based catalysis studies. Due to the sample heterogeneities, possible effects of radiation damage, and irreversibility of chemical states before and after the reaction, it is not always possible to repeat exactly the same process at different beamlines. In some cases, however, the reactor cell design and the nature of the process do allow reproducing the experiment in virtually identical conditions on the two beamlines, but the combination may still be preferred, if not for any other reason than being a high-throughput data collection method. In some cases, however, the all-in-one instrument is less than ideal for high-precision structural analyses by, for example, Rietveld refinement of the XRD data, or multiple-scattering analysis of the EXAFS data, requiring relatively long integration times, and special instruments on dedicated XRD and XAFS beamlines. The examples discussed in this chapter show that the combination of XAFS and XRD in a single experiment is best suited for kinetics studies where the goal is twofold: to detect and identify reaction intermediates. Simultaneous XRD and EXAFS experiments capture the same state of the sample from two different perspectives, and thus give a more complete understanding than each of them individually.

13.4.2 Summary

The technique of (quasi-)simultaneous measurements of XRD and XAFS data in catalysts under *in-situ* conditions has been developed and continuously improved over the last 20 years. Current implementations allow for quantitative measurements with XRD and XAFS cycle times from 1 to 30 min. This capability has been critical for understanding many important catalytic mechanisms. The main reason for the success of this method is that such combination of techniques, in real time and in real operating conditions, allows for measurements of both short- and long-range order from the same system at nearly the same time.

Experiments which combine two techniques can suffer from the compromises made to enable the combination. For instance, the XAS is measured at the most highly absorbing wavelengths and consequently, the diffraction patterns are difficult to correct for absorption. An obvious solution to this problem is moving away from the absorption edge when the diffraction is measured. However, this makes the time gap between XRD and XAS measurements larger because it takes time to scan from one wavelength to another. (This problem is overcome at SNBL at ESRF where there are two monochromators and they can be switched in less than 10 s.)

13.4.3 Outlook

In the future, if there is sufficient flux and fast detectors, a diffraction pattern can be collected at each XAS point and the diffraction data can be corrected for changing wavelength. Figure 13.14 shows the raw time-resolved diffraction data initially at constant energy followed by an energy scan which was a measurement during a part of a single XAS scan of $CuFe_2O_4$ during CO reduction. Figure 13.14 is part of the data during the first scan of multiple scans during the reduction of $CuFe_2O_4$ during reduction in 5% CO in He. Figure 13.15 shows the evolution of the Cu^0 phase in diffraction patterns extracted from this at identical wavelengths but at different times during the reduction process.

Additional important insights can be obtained using the diffraction anomalous fine structure (DAFS) technique. The new experimental setup at the X18A beamline at the NSLS offers an exciting new possibility to study heterogeneous catalysis by *in-situ* DAFS [30]. This method, combining the long-range order sensitivity of XRD and short-range order specificity of XAFS within the same technique, has been used in the past for discriminating between the nanocrystalline and bulk amorphous phases of the same atomic species [31]. In DAFS, a specific phase of a system is selected by the Bragg diffraction

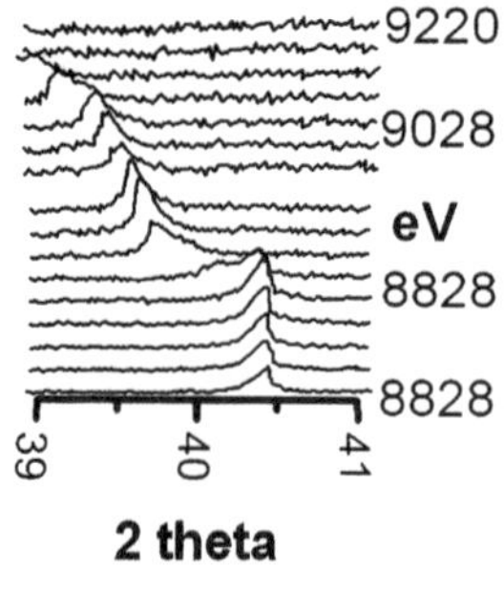

Figure 13.14. The raw time-resolved data at constant energy and then during the changes during part of an EXAFS scan.

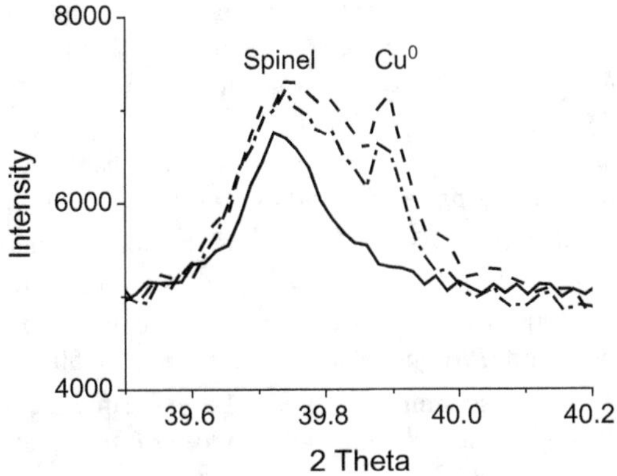

Figure 13.15. Powder diffraction patterns extracted from continuous simultaneous measurement of XRD and XAFS during reduction of $CuFe_2O_4$.

peak, as in a conventional XRD experiment, and the local structure of that phase is interrogated by spectroscopy, as in a conventional XAFS experiment. Measurement of the local structure of ordered phases is possible by DAFS because in that experiment, Bragg peak intensity is measured as a function of X-ray energy in the resonant scattering region [31]. This hybrid technique allows one in principle to perform *in-situ*, real-time chemical speciation and kinetics studies in the system containing reduced and oxidized states of the same catalyst.

ACKNOWLEDGMENTS

A.I.F. acknowledges the support of this work by the U.S. Department of Energy (DOE) Grant No. DE-FG02-03ER15476. J.C.H. acknowledges the support of this work under contract with DOE contract DE-AC02-98CH10886. The use of the NSLS beamlines was supported by U.S. DOE Contract No. DE-AC02-98CH10886. Beamlines X18A and X18B at the NSLS are supported in part by the Synchrotron Catalysis Consortium, U.S. DOE Grant No DE-FG02-05ER15688.

REFERENCES

[1] Norby, P., et al. (1998) Hydrothermal synthesis of the microporous aluminophosphate CoAPO-5; in situ time-resolved synchrotron X-ray powder diffraction studies, *Catalysis Today*, 39, 301–309.

[2] Phizackerley, R. P., et al. (1983) An energy-dispersive spectrometer for the rapid measurement of X-ray absorption-spectra using synchrotron radiation, *Journal of Applied Crystallography*, 16, 220–232.

[3] Tolentino, H., et al. (1990) Aberration-free and harmonic-free optics for time-resolved X-ray absorption-spectroscopy using synchrotron radiation, *Nuclear Instruments & Methods in Physics Research. Section A, Accelerators, Spectrometers, Detectors and Associated Equipment*, 289, 307–316.

[4] Allen, P. G., et al. (1993) A 4-point crystal bender for dispersive-X-ray absorption-spectroscopy, *Journal of Applied Crystallography*, 26, 172–179.

[5] Aquilanti, G., et al. (2005) EXAFS study of the local structure of InAs up to 80 GPa, *Journal of Physics. Condensed Matter*, 17, 1811–1824.

[6] Mathon, O., et al. (2004) Dynamics of the magnetic and structural alpha-epsilon phase transition in iron, *Physical Review Letters*, 93, 255503.

[7] Frahm, R. (1988) Quick scanning EXAFS—1st experiments, *Nuclear Instruments & Methods in Physics Research. Section A, Accelerators, Spectrometers, Detectors and Associated Equipment*, 270, 578–581.

[8] Richwin, M., et al. (2001) Piezo-QEXAFS: advances in time-resolved X-ray absorption spectroscopy, *Journal of Synchrotron Radiation*, 8, 354–356.

[9] Clausen, B. S., et al. (1993) A combined QEXAFS XRD method for online, in-situ studies of catalysts—examples of dynamic measurements of Cu-based methanol catalysts, *Catalysis Letters*, 20, 23–36.

[10] Lee, J. M., et al. (1998) Design and feasibility of quick EXAFS scans for a "broom-stick" double-crystal monochromator at PLS beamline, *Journal of Synchrotron Radiation*, 5, 524–526.

[11] Couves, J. W., et al. (1991) Tracing the conversion of aurichalcite to a copper catalyst by combined X-ray absorption and diffraction, *Nature*, 354, 465–468.

[12] Sankar, G., et al. (1999) In situ combined X-ray absorption spectroscopic and X-ray diffractometric studies of solid catalysts, *Topics in Catalysis*, 8, 1–21.

[13] van Beek, W., et al. (2011) SNBL, a dedicated beamline for combined in situ X-ray diffraction, X-ray absorption and Raman scattering experiments, *Phase Transitions: A Multinational Journal*, 84, 726–732.

[14] Ehrlich, S. N., et al. (2011) Combined XRD and XAS, *Nuclear Instruments & Methods in Physics Research. Section A, Accelerators, Spectrometers, Detectors and Associated Equipment*, 649, 213–215.

[15] Siddons, D. P., et al. (2007) A Guinier camera for SR powder diffraction: high resolution and high throughput, *AIP Conference Proceedings*, 879, 1767–1770.

[16] Chupas, P. J., et al. (2008) A versatile sample-environment cell for non-ambient X-ray scattering experiments, *Journal of Applied Crystallography*, 41, 822–824.

[17] Estrella, M., et al. (2009) In situ characterization of CuFe2O4 and Cu/Fe3O4 water-gas shift catalysts, *Journal of Physical Chemistry C*, 113, 14411–14417.

[18] Frenkel, A., et al. (2011) Combining X-ray absorption and X-ray diffraction techniques for in situ studies of chemical transformations in heterogeneous catalysis: advantages and limitations, *Journal of Physical Chemistry C*, 115, 17884–17890.

[19] Toby, B. H. (2005) CMPR—a powder diffraction toolkit, *Journal of Applied Crystallography*, 38, 1040–1041.

[20] Wang, X. Q., et al. (2006) In situ studies of the active sites for the water gas shift reaction over Cu-CeO2 catalysts: complex interaction between metallic copper and oxygen vacancies of ceria, *Journal of Physical Chemistry B*, 110, 428–434.

[21] Hornes, A., et al. (2010) Inverse CeO2/CuO catalyst as an alternative to classical direct configurations for preferential oxidation of CO in hydrogen-rich stream, *Journal of the American Chemical Society*, 132, 34.

[22] Barrio, L., et al. (2010) Unraveling the active site in copper-ceria systems for the water-gas shift reaction: in situ characterization of an inverse powder CeO2-x/CuO-Cu catalyst, *Journal of Physical Chemistry C*, 114, 3580–3587.

[23] Hammersley, A. P., et al. (1994) Calibration and correction of spatial distortions in 2D detector systems, *Nuclear Instruments & Methods in Physics Research. Section A, Accelerators, Spectrometers, Detectors and Associated Equipment*, 346, 312–321.

[24] Rietveld, H. M. (1969) A profile refinement method for nuclear and magnetic structures, *Journal of Applied Crystallography*, 2, 65.

[25] Toby, B. H. (2001) EXPGUI, a graphical user interface for GSAS, *Journal of Applied Crystallography*, 34, 210–213.

[26] Larson, A. C., et al. (1995) *GSAS General Structure Analysis System*, Los Alamos National Laboratory, Los Alamos, NM.

[27] Grunwaldt, J.-D., et al. (2009) Insight into the structure of Pd/ZrO2 during the total oxidation of methane using combined in situ XRD, X-ray absorption and Raman spectroscopy, *Journal of Physics: Conference Series*, 190, 012160.

[28] Grunwaldt, J. D., et al. (2007) Insight into the structure of supported palladium catalysts during the total oxidation of methane, *Chemical Communications (Cambridge, England)*, 44, 4635–4637.

[29] de Smit, E., et al. (2010) Stability and reactivity of epsilon-chi-theta iron carbide catalyst phases in Fischer-Tropsch synthesis: controlling mu(c), *Journal of the American Chemical Society*, 132, 14928–14941.

[30] Stragier, H., et al. (1992) Diffraction anomalous fine-structure—a new X-ray structural technique, *Physical Review Letters*, 69, 3064–3067.

[31] Frenkel, A. I., et al. (2002) Direct separation of short range order in intermixed nanocrystalline and amorphous phases, *Physical Review Letters*, 89, 285503–285506.

14

COMBINING INFRARED SPECTROSCOPY WITH X-RAY TECHNIQUES FOR INTERROGATING HETEROGENEOUS CATALYSTS

MARK A. NEWTON AND MARCOS FERNÁNDEZ-GARCÍA

14.1 INTRODUCTION

For many years both infrared (IR) spectroscopy [1], along with a whole battery of techniques based upon X-rays, have been used to study heterogeneous catalysts [2]. IR is primarily used as a sensitive probe of surface molecular functionality (see Chapter 8 and Chapter 9 in this book). X-ray techniques, based upon X-ray absorption or scattering, provide different information, being direct probes of structure and chemical state of elements within the catalyst. Historically, for the most part, the IR and X-ray worlds have existed separately, first in different laboratory-based instrumentation and then, and especially in the case of X-rays, within progressively advanced X-ray facilities such as synchrotrons. That such a combination of probes is intrinsically valuable and powerful in the study of heterogeneous catalysts is easily evidenced: numerous publications [3] continue to appear that make specific use of IR and X-ray techniques (most commonly extended X-ray absorption fine structure [EXAFS]) in combined efforts to understand a variety of aspects of the genesis

In-situ *Characterization of Heterogeneous Catalysts*, First Edition.
Edited by José A. Rodriguez, Jonathan C. Hanson, and Peter J. Chupas.

and behavior of catalytic materials. In these studies the IR and X-ray components remain remote from each other and the notion of "combined" means only that several different experiments have been made using different pieces of equipment, cells, people, and such like in order to stitch together a multi-technique view of the problem to hand.

For some considerable time, however, it has been realized that studying the behavior of many different types of material using a range of *simultaneously applied* and complementary probes [4–13] brings with it many potential benefits. The level of internal consistency guaranteed by a well-designed experiment of this nature should remove the possibility of erroneous correlations and deductions being arrived at through the comparison of a number of individual, and often different, measurements. Further, the overall duty cycle of the investigative process should be massively improved. Such single, holistic, experiments should, therefore, provide at least the sum of information provided by a linear sum of several separate ones and much more efficiently. It is only in the last few years, however, that a substantial drive toward synthesizing new experimental methodologies dedicated to this end, and within the overall umbrella of *in-situ* and *operando* [7, 8, 10, 12] study of heterogeneous catalysts, has occurred.

Within this overall trend, the hard X-rays that synchrotrons provide have played a highly significant role, bringing with them direct access to the details of materials structure and structural change over a very wide range of length and timescales. Importantly, most of the common hard X-ray probes, by which we mean X-ray absorption spectroscopy (XAFS), X-ray scattering, and X-ray diffraction (XRD), also lend themselves to relevant *in-situ/operando* study given the inherent ability of hard X-rays to penetrate matter. This means that they can be applied almost irrespective of sample environment. In the case of heterogeneous catalysis, this is highly important. It allows a great deal of flexibility in how an experiment can be parameterized such that the processes desired to be studied can be properly respected. In many cases, X-ray techniques allow one to enter into arenas where techniques based on the detection of electrons simply cannot go. As such, they are able to get closer to the arguably unattainable, but nonetheless stimulating, ideal that underpins the *operando* philosophy.

What hard X-ray techniques cannot generally do, however, is to specifically address the surface molecular speciation that is central to any catalytic process: this is the (potentially quantitative) forte of vibrational spectroscopies such as IR and Raman. Therefore, in order to gain a more holistic, and one would hope more insightful, window into the details of how catalysts work, applying of X-ray techniques and vibrational spectroscopies in direct combination would seem to be a very worthwhile thing to do [11].

In addition to works by Newton et al. and Bentrup [11, 12], where both IR and Raman combinations with X-rays are considered, what can be done specifically using couples between Raman spectroscopy and XRD is dealt with by van Beek et al. in Chapter 15 of this book. Here we will investigate what has been achieved, and what may yet still be done, in terms of synchronously applying IR spectroscopy alongside the battery of techniques that modern synchrotron sources provide us with for unraveling the inner workings of heterogeneous catalysis.

14.2 BRAIDING INFRARED SPECTROSCOPIES WITH X-RAY TECHNIQUES: AN OVERVIEW

IR spectroscopy can be applied to the study of catalysts in a number of distinct ways, these being: transmission (TR); diffuse reflectance (DRIFTS), attenuated total reflection (ATR); and reflection–absorption (RAIRS or IRRAS). Each of these techniques samples the system under study in a different way and is subject to different physical constraints as a result of this. While we will not analyze these fundamentals in depth—there is ample literature that already does this [14, 15]—in considering constructing an experiment that combines one of the IR methods synchronously with a complementary X-ray technique, there are certain factors worth noting.

Prime among these are how can one obtain a satisfactory spatial (or indeed volumetric) convergence of the two different parts of the electromagnetic spectrum while retaining the optimal geometries for each probe. It has become evident from a number of spatiotemporal vibrational spectroscopic [16–18] and X-ray-based studies [19–22] of working catalysts that sampling two spatially separated volumes, even within the same catalyst bed, has the potential to be extremely misleading.

A first point worth noting is that it is evident that, for the most part, materials used in mid-IR as windows are not at all compatible for use with X-rays in the most commonly used X-ray energy range (ca. 5–30 keV) [11]. This conclusion is reached on two bases: first, their composition leads to unacceptably high levels of either X-ray absorption or scattering; and second, even if that is not the case (for instance diamond, sapphire), that they are often highly crystalline materials that will, of themselves, diffract X-rays very efficiently. Such parasitic X-ray scattering may lead to any number of unwanted complications to the X-ray part of the experimental couple.

At very hard X-ray energies approximately >30–40 keV, wherein a number of catalytically relevant K-edges become accessible for X-ray absorption studies, and techniques like pair distribution function (PDF) analysis of total X-ray scattering data (see later) become increasingly viable, these restrictions are progressively ameliorated. Nonetheless, for the most part, the most likely means to achieve the desired couple will be the modification of the available IR experimental setup such that it may be made transmissive to X-rays that do follow the same optical path as the IR light.

The next issue, therefore, is how to ensure convergence of sampling volumes between the X-ray and IR lights. This in turn leads to the question of the sampling depth of the individual IR methods. For ATR, which uniquely among the IR variants relies upon the generation and reflection of an evanescent wave emanating from the surface of the ATR crystal, one can simply state that the likely sampling depth of the measurement is intrinsically restricted and will be on the order of a micrometer at best [23–25]. For DRIFTS and TR, the actual sampling depth that may be achieved in the midrange IR is highly wavelength and sample dependent: the exact chemical nature of the sample, its texture (particle sizes), and density of packing will all affect the sampling depth achieved in any case. Moreover, if any of these parameters change significantly during the experiment, then the effective sampling depth can hypothetically change a lot. A pertinent example would be connected to a phase transformation (for instance, from a less absorbing oxidic phase to a strongly absorbing metallic phase of one of the elements of interest).

Though midrange IR sampling depths for DRIFTS of up to 3 mm can be found in the literature [14], such length scales are observed in materials that are not generally representative of heterogeneous catalysts. Such length scales are ubiquitously associated with samples that are highly diluted in the IR transparent KBr. Sample dilution of this type, while common in the broader world of DRIFTS spectroscopy, is something that any *in-situ* catalysis experiment is unlikely to support on X-ray absorption and/or chemical reactivity grounds [11]. Intuitively, therefore, it seems reasonable that such sampling depths are going to be the exception rather than the rule when considering typical catalyst materials, and that this needs to be kept in mind when designing a combined experiment.

Though quantitative measurements to establish the actual sampling depth of (particularly) a DRIFTS experiment can be made, they appear to be not so common in the catalysis literature. In the current context, though, it is very instructive to consider the results of one of the very rare examples [26] where this has been done. Mondelli et al. [26] elegantly used variations of DRIFTS bed thickness (A) over an underlying boron nitride disk, and a cross-correlation of DRIFTS and quantitative mass spectrometry (MS) data (B), to definitively establish that the DRIFTS samples (at best) less than 200 μm into the surface of a catalyst bed comprised of CeO_2 or $CeZrO_4$. The results of these studies are shown in Figure 14.1.

As such, either these sorts of measurements should be carried out on a case-by-case basis or one should always seek to combine the X-rays with the DRIFTS such that the X-ray beam samples within a few hundred micrometers of the surface of the catalyst bed. This is easily achievable on most modern X-ray beamlines and seems to be the most reasonable way to attempt to ensure that the X-rays and the DRIFTS are as convergent as possible in their sampling of the catalyst bed.

That one is able to achieve this is also important for a second reason: it is only at the very top of the sample bed that, a priori, the catalyst sees the reacting gases as applied. Depending upon the activity of the sample, the more one probes below this region the less one knows about exactly what sort of a gaseous environment the sample is being experienced by the sample.

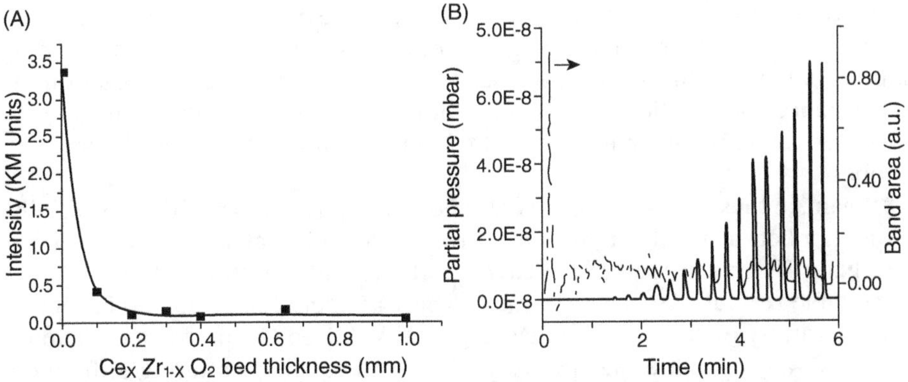

Figure 14.1. The two approaches used by Mondelli et al. [26] to establish the sampling depth of a DRIFTS experiment when using catalysts based upon a $CeZrO_4$ support material. (A) Variation of sample bed thickness over an underlying BN disk. (B) Correlation of the appearance of IR visible bands due to CO adsorption by the catalyst—right-hand ordinal axis—with the breakthrough of gas phase (pulsed CO) observed using a downstream mass spectrometer. Both methods conclude that in the midrange IR the DRIFTS samples ≤200 μm into the surface of the sample. Reproduced from Mondelli, C., Dal Santo, V., Trovarelli, A., Boaro, M., Fusi, A., Psaro, R., Recchia, A. (2006) An *operando* DRIFTS-MS study on model $Ce_{0.5}Zr_{0.5}O_2$ redox catalyst: a critical evaluation of DRIFTS and MS data on CO abatement reaction, *Catal. Today*, 113, 81–86, with permission from Elsevier.

Another aspect of both IR spectroscopy and X-ray methods that is worth considering is the timescales upon which they may be used. Catalysts and catalytic processes are intrinsically dynamic on a variety of timescales. Catalysts may evolve over very long periods of time during operation, yet catalytic events are intrinsically molecular and therefore ultimately controlled by extremely fast chemical dynamics (nanoseconds or less). In between these two extremes lie the kinetic timescales (ca. from microseconds to minutes) which are of considerable mechanistic interest in practical application. All of these timescales can be addressed by IR and X-ray techniques.

Ultrafast chemical dynamics on the timescale of microseconds to a few tens of nanoseconds has been achievable using step-scan IR techniques for many years [27]; the application of such techniques within the milieu of heterogeneous catalysis has been, however, rather rare [28, 29]. The same can be said for ultrafast X-ray scattering, absorption, or emission measurements [30]. The capacity for such X-ray experiments has been hugely augmented by the appearance of third-generation synchrotron sources but this has not yet filtered through in any significant way to the study of fundamental aspects of heterogeneous catalysis.

Measurements using step-scan and or pump–probe techniques using X-rays do, however, require an absolute reproducibility, or very facile repetition, of a given chemical event (in response to a rapid stimulus or "pump") to be successful. In many catalytic systems, modes of experimentation, or indeed aspects of chemical reactivity (e.g., spontaneously oscillatory and/or chaotic kinetic behavior), this is not the case. As such, time-resolved investigation of, for instance, the kinetic evolution of a catalytic system in such cases must take place on the basis of what can be obtained in a single shot from the applied probes.

Again, in these sorts of regimes, IR and numerous X-ray probes—especially when applied at third-generation sources—are well matched. Modern time-resolving IR spectrometers are capable of standard spectral repetition rates at suitable resolutions (by and large 2–4 cm^{-1}) of between 20 and 40 Hz. Similarly, a variety of X-ray techniques (e.g., EXAFS [see Chapter 1 and Chapter 3], small-angle scattering [SAXS] [31], wide-angle scattering/diffraction [WAXS] [32], and total X-ray scattering [33]) have also been demonstrated to be capable of single-shot acquisition on these (25–50 ms) timescales. In both IR and X-ray cases, what one might actually get from a single shot, in terms of useful data, using such rapid acquisition will be highly system and experiment dependent; nonetheless, this capacity exists and has been exploited.

14.3 CURRENTLY DEMONSTRATED COUPLES OF IR AND X-RAY METHODS AND THEIR UTILIZATION

Although as early as the mid-1990s successful attempts were made to directly combine IR spectroscopic measurements with both transmission infrared (TRIR)/EXAFS [34–36] and ATR/SAXS/WAXS [37], these were not experiments designed for the study of heterogeneous catalysts: it is only in 2004 that we see such an experiment appear [38]. This experiment is based upon the coupling of DRIFTS to TR-EXAFS and, until very recently, remained the only experiment of its type. While in the intervening years some interesting experiments that have used ATR and EXAFS [39, 40], or fused IR/ultraviolet–visible spectroscopy (UV-Vis) and Raman probes with X-rays [12, 41], have appeared, they have not placed the two probes in the same place and at the same time.

Within the specific remit of this review it is only in the last several years that Innocenzi et al. [42], Frenkel et al. [43], and Bando et al. [44] have added to this first DRIFTS/EXAFS experiment: Innocenzi et al. elegantly combining grazing incidence SAXS with TRIR [42]; Frenkel et al. evolving their own version [43] of the original DRIFTS/EXAFS experiment due to Newton et al. [38]; and Bando et al. coupling TRIR with TR-EXAFS [44]. As such, this section aims to delineate how these experiments have been used for the study of heterogeneous catalysts.

14.3.1 Coupling DRIFTS with Transmission-Based X-ray Measurements

Figure 14.2 shows the three currently existing designs that have been adopted to combine TR X-ray techniques (primarily EXAFS) with DRIFTS. This first

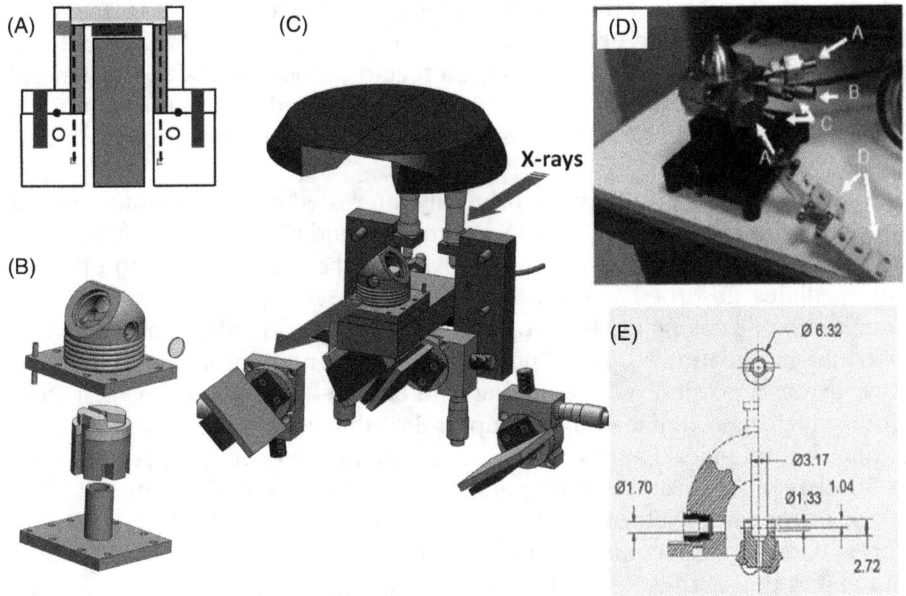

Figure 14.2. Schematic examples of the three designs currently demonstrated for coupling DRIFTS with TR X-ray techniques. (A) The ultra-low-volume cell implemented by Newton and coworkers [38]; (B) the approach based upon the modification of a commercial Spectra-Tech DRIFTS cell developed and implemented at the ESRF [48]; (C) the overall optical arrangement within which (A) and (B) are implemented; (D) the modified Harrick system developed by Marinkovic et al. (see Reference [43] for label details); and (E) a cross section of the Harrick cell system detailing a number of dimensions and the mounting system for the X-ray transmissive windows. Reactors (A) and (D) couple DRIFTS with TR-EXAFS. In addition to this, Reactor (B) permits DRIFTS to be coupled to HXRD (see text and References [49], [59], and [63]). Elements reproduced/adapted from Newton, M. A., Jyoti, B., Dent, A. J., Fiddy, S. G., Evans, J. (2004) Synchronous, time resolved, diffuse reflectance FT-IR, energy dispersive EXAFS (EDE) and mass spectrometric investigation of Rh catalysts during NO reduction by CO, *Chem. Comm. (Camb.)*, 2382–2383; Newton, M. A., Dent, A. J., Fiddy, S. G., Jyoti, B., Evans, J. (2007) Combining diffuse reflectance infrared spectroscopy (DRIFTS), dispersive EXAFS, and mass spectrometry with high time resolution: potential, limitations and application to the study of NO interaction with supported Rh catalysts, *Catal. Today*, 126, 64–72; and Marinkovic, N. S., Wang, Q., Frenkel, A. I. (2011) In situ diffuse reflectance IR spectroscopy and X-ray absorption spectroscopy for fast catalytic processes, *J. Synchrotron Radiat.*, 18, 447–455, by permission of (respectively) The Royal Society of Chemistry, Elsevier, and the International Union of Crystallography.

experiment (Fig. 14.2A) that coupled IR measurements with TR (dispersive) EXAFS [38, 45] was largely inspired by the IR-based work of McDougall and coworkers [46, 47]. The flat-top DRIFTS reactor, which sits inside the general optical arrangement shown in Figure 14.2C, constructed for this first combined X-ray/IR catalysis experiment, is essentially a (suitably adapted) copy of their novel DRIFTS system.

McDougall et al. [46, 47] were primarily interested in demonstrating how a concentration modulation DRIFTS approach could be used to great effect to tease out the molecular nature of reactive, and therefore difficult to observe, intermediates adsorbed upon working catalysts. As such their system was premised on achieving the lowest possible internal dead volume and best gas-switching/modulation response, possible. This design therefore was extremely attractive as a possibility for coupling with energy-dispersive EXAFS: a technique which, as is delineated in Chapter 3 of this book, is very well adapted to interrogating local catalyst structure on very short (<<1 s) timescales.

Subsequent to this first design, another variation was implemented at ID24 at the European Synchrotron Radiation Facility (ESRF) [48] (Fig. 14.3B,C). This is based upon a commercial (Spectra-Tech, Stamford, CT) DRIFTS cell and makes the sacrifice of increased dead volume (of the order of 2.5–3 cm^3) for a higher potential upper operating temperature (up to 773 K has been demonstrated) and a plug-flow configuration, that is, the applied feedstock is

Figure 14.3. A summary of results obtained from a time-resolved and synchronous DRIFTS/dispersive EXAFS/MS investigation of the interaction of NO with reduced (ca. 11 Å diameter) Rh nanoparticles supported upon γ-Al_2O_3 [45, 52–54]. (A) Selectivity toward production of N_2O and net NO consumption (per Rh atom present in the sample). (B) Representative DRIFTS spectra obtained after 60 s of exposure to 5% NO/He at two different temperatures. Each spectrum is acquired in approximately 64 ms. (C) and (D) show k^3-weighted Rh K-edge EXAFS obtained on the same timescales prior to NO exposure (C) and after 60 s exposure to 5% NO/He, again as a function of temperature. (E) Correlation of the intensity of the IR band observed at 1745 cm^{-1} with the formation of Rh^{III} within 60 s of exposure to NO and as a function of sample temperature. (F) Correlation of selectivity toward N_2 production (MS, filled circles, right ordinal axis) with the intensity of the 1745 cm^{-1} IR band and the level of Rh oxidation observed at a reaction temperature of 573K. This experiment utilized the sample cell configuration shown in Figure 14.2A. From Newton, M. A., Dent, A. J., Fiddy, S. G., Jyoti, B., Evans, J. (2007) Identification of the surface species responsible for N2O formation from the chemisorption of NO on Rh/alumina, *Phys. Chem. Chem. Phys.*, 9, 246–249. Reproduced by permission of the PCCP Owner Societies, and Newton, M. A., Dent, A. J., Fiddy, S. G., Jyoti, B., Evans, J. (2007) Combining diffuse reflectance infrared spectroscopy (DRIFTS), dispersive EXAFS, and mass spectrometry with high time resolution: potential, limitations and application to the study of NO interaction with supported Rh catalysts, *Catal. Today*, 126, 64–72, reproduced with permission of Elsevier.

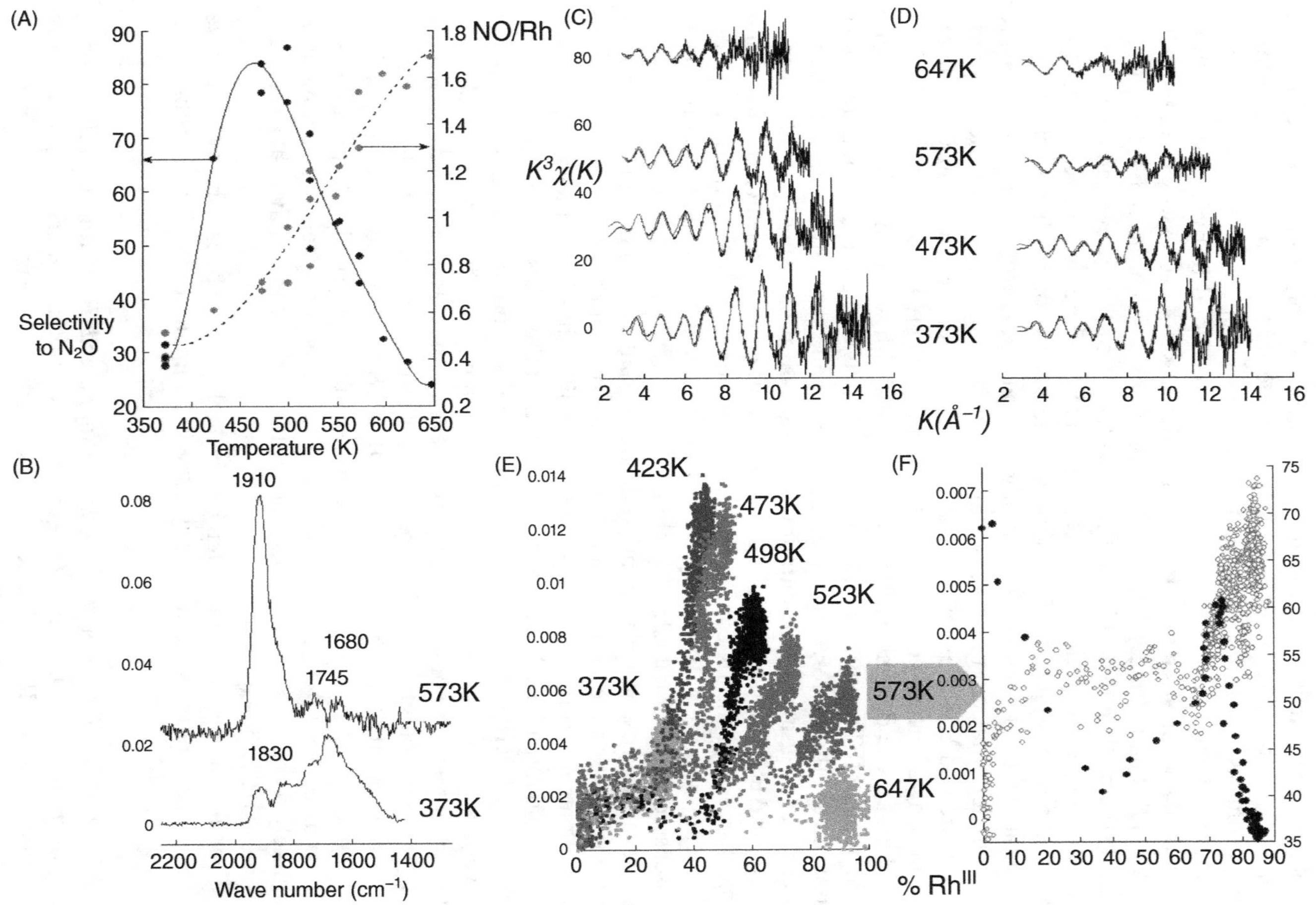
(A)
Selectivity to N_2O
NO/Rh
Temperature (K)
(B)
1910
1680
1745
1830
573K
373K
Wave number (cm^{-1})
(C)
$K^3\chi(K)$
(D)
647K
573K
473K
373K
$K(Å^{-1})$
(E)
423K
473K
498K
523K
373K
573K
647K
(F)
% Rh^{III}

obliged to pass through the sample rather than simply over it (as was the case for the original and modified McDougall design). In catalysis, the latter is important as it permits a contact time to be specified.

Very recently these two designs have been added to by a third, this time based on modification of a standard Harrick DRIFTS accessory (Pleasantville, NY) [43] (Fig. 14.3D,E). This retains a plug-flow configuration but has a larger dead volume than that of its predecessors (ca. 14 cm^3). However, and in its favor, a better upper operating pressure and upper temperature are achievable [43].

These three systems could be seen together as comprising a very complementary suite of solutions depending on what exactly one wishes to measure and under what circumstances. Further, at least in the case of the currently implemented ESRF experiment, this design leads to the possibility (see later) of making a further couple of IR to hard XRD [11, 48, 49].

All of these designs exploit the simplicity with which the TR X-ray probe and the DRIFTS may be combined in a mutually orthogonal manner that does not disrupt the preferred geometry of each probe. Aside from some peculiarities regarding sample presentation and normalization issues specific to dispersive EXAFS (Chapter 3, [50]), the only potential pitfall to watch out for is that the X-rays and the IR components of the experiment have a reasonably convergent sampling of the catalyst bed, axially speaking , as had been previously discussed in Section 14.2.

A further reason for wishing to do this is that in each design the temperature of the sample is measured using a thermocouple inserted into the top millimeter or so of the bed. Whether one likes it or not, thermal gradients (axial and radial) are likely to exist within all of these designs (not to say pretty much any DRIFTS accessory). By measuring as close to the thermocouple in a vertical sense, one is seeking to minimize any potential errors in deduction that might arise as of axial gradients in temperature. Radial thermal gradients are more difficult to exclude, especially for larger diameter sample beds. Here we note only that in combined X-ray/DRIFTS measurements, sample beds of ≤5 mm diameter are used. These are comparable to or smaller than those used in commercial DRIFTS systems. As such, the combined measurement is unlikely to be in any way worse off in this respect than any other (conventional) DRIFTS experiment.

Taking each system in order of chronological application, we now give some cases studies as to how they have been applied thus far to problems in catalysis.

14.3.1.1 The Dynamic Behavior of Supported Rh Nanoparticles during Exposure to Nitric Oxide (NO) As has been noted, the first couple of X-ray techniques with IR applied to heterogeneous catalysts appeared in a 2004 study [38] that compared and contrasted the behavior of 5 wt% Rh catalysts derived from Cl-containing and nonchlorinated precursors as they were heated under a feedstock containing CO and NO. At that time, however, the dispersive EXAFS component of the experiment was somewhat hamstrung by the charge-coupled device (CCD) camera used for X-ray detection, one that was intrinsically limited by a 300-ms readout time. As such, this initial demonstration was limited to relatively slow measurements (every ca. 6 s) that did not fully make use of the temporal resolving capacity of either the IR or the dispersive EXAFS.

Nonetheless, from this first measurement it was easy to see how this synchronous combination of approaches could be extremely powerful. Subsequent to this, the limitation due to detector readout at ID24 of the ESRF was removed through the implementation of the fast readout low noise detector (FReLoN) camera [51], which reduced readout limitations (to ca. 1.6 milliseconds per frame) on the X-ray side of the equation. As a result, the real potential of the single-shot time-resolving IR/dispersive EXAFS couple could be better explored [45, 47–60]; both usable DRIFTS (4 cm^{-1} resolution) and dispersive EXAFS at, for instance, the Rh K-edge could now be applied with synchronized spectral repetition rates of close to 20 Hz to investigate the behavior of 5 and 2 wt% Rh/Al_2O_3 catalysts [45, 51–55].

Figure 14.3 shows some typical results pertaining to the interaction of gas-phase NO with supported Rh particles of average particle diameter of approximately 11 Å [45, 52–54], and brings together results from the three simultaneously applied probes—Rh K-edge dispersive EXAFS, DRIFTS, and online MS—as a function of the sample temperature. Each EXAFS spectrum (shown in Fig. 14.3C,D in k^3-weighted form) is a single acquisition of 64 ms duration, as are the two representative DRIFTS spectra shown in Figure 14.3B.

Figure 14.3A summarizes the net uptake of NO and the selectivity toward N_2O formation observed during a 60 s exposure of reduced Rh nanoparticles to NO as a function of temperature. The temporal variation of IR intensity observed at 1745 cm^{-1} (Rh(NO^-) and/or Rh$(NO)_2$) shown in Figure 14.3E for seven separate temperatures clearly establishes an overall link with the selectivity-versus-temperature curve (black line, Fig. 14.3A) revealed by the MS. Uniquely, using this experiment, this may be precisely correlated with the degree to which the Rh nanoparticles are oxidizing (from the Rh K-edge EXAFS or X-ray absorption near-edge structure [XANES]). For NO exposure at 573 K, the temporal dependence of the IR feature at 1745 cm^{-1} is further correlated (Fig. 14.3F) to the measured selectivity to N_2 production (black circles) observed during the NO exposure.

To summarize, this experiment provides a unique viewpoint, showing how the conversion of NO, to either N_2 or N_2O, is mediated as a function of the competition between molecular and dissociative adsorption of the NO; the latter leads to the oxidation of the Rh and is increasingly dominant above 473 K. It can be shown that this also leads to the conclusion that there are two routes to N_2O formation. At intermediate temperatures, the Rh(NO^-) species is implicated as being the source of the major (and unwanted) formation of N_2O observed. At progressively higher temperatures, high levels of oxidation (> ca. 70%) of the Rh phases, another route that leads to N_2O formation, appears; N_2O production is now associated with either disproportionation of

$Rh(NO^+)$ species (1910 cm^{-1}) or the interaction of this species with gas phase or weakly absorbed molecular NO species. At 573 K, for instance, either of these possibilities lead to the very transient formation of an $Rh(NO)_2$ species which readily collapses to yield N_2O and advances the oxidation of the Rh phase slowly toward 100%.

14.3.1.2 EXAFS/IR Interrogation of Pd-Based Catalysts in Redox Cycling Situations That a high-speed and parallel acquisition of both IR and XAFS data can be used very effectively to tease out and explain some of the complexity of a fundamental gas–solid interaction is well demonstrated by the preceding example. In terms of more practical catalysis, however, a 5% loading of metals such as Rh, Pt, and Pd cannot be regarded as representative in many areas of catalysis wherein these metals find their most abundant practical usage, even though in the case of the previous example the average particle size—as independently verified using transmission electron microscopy (TEM) [61]—may well be.

The next example of the application of this method, this time using the system developed at the ESRF, is discussed [48]. This approach tries to push the experiment further in terms of the level of dilution of the metal (in this case, <2 wt% Pd) that is studied, the achievable upper operating temperature, and attempts to better meet the requirements of "plug flow" operation.

Figure 14.4 summarizes the result of a combined DRIFTS/dispersive EXAFS investigation of how Pd nanoparticles within a 1 wt%Pd/10 wt% $CeZrO_4/Al_2O_3$ system behave during CO/NO cycling at 673 K [59]. Figure 14.4A shows the normalized absorption data (300 ms per spectrum) in three cases. The inset shows the corresponding Fourier transform of the k^3 weighted EXAFS. The top spectrum (Figure 14.4A) refers to the as prepared catalysts that the XANES and EXAFS identify as nanoparticulate PdO, and therefore ca. 100% Pd^{II}. The middle and bottom spectra show how the Pd appears in XAFS under 5%CO/He (middle) and then 5%NO/He (bottom) environment at 673 K.

In contrast to the previous example, the Pd K-edge XANES and EXAFS data (Fig. 14.4A) tell us that we are not dealing with large-scale oxidation and reduction of the Pd phase as a function of its environment. Instead, the overall EXAFS envelope remains "metallic" in nature irrespective of the reactive environment; in this case, the Pd nanoparticles are responding in more subtle ways to the switching feedstock. Moreover, these subtle responses can be dynamically and directly correlated to a number of surface-bound IR active species, such as molecular CO and NO adsorbed on the surface of the Pd nanoparticles, and adsorbed NCO and carbonates. These can be directly correlated to, for instance, the production of CO_2 in the gas phase (Fig. 14.4B).

The outcomes of this and subsequent experiments [49, 56–60] using the same general approach has been to reveal several new aspects of the structural reactive chemistry of supported Pd nanoparticles; by and large, such details were not known of before the application of the synchronous approach to

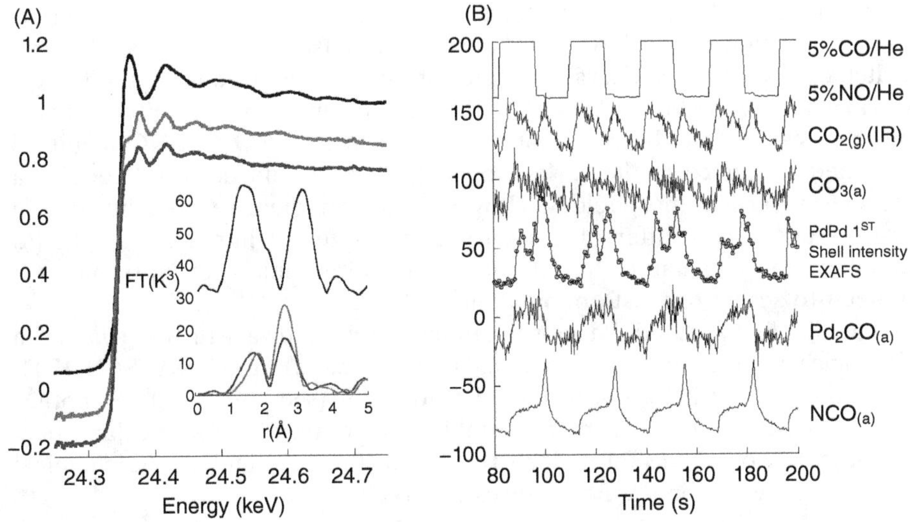

Figure 14.4. Summary of a synchronous dispersive EXAFS/DRIFTS/MS experiment concerning the behavior of small Pd nanoparticles supported (at 1 wt%) upon 10 wt% $CeZrO_4/Al_2O_3$ catalyst at 673K [59] and using the approach to coupling DRIFTS with transmission EXAFS shown in Figure 14.2B. (A) Normalized Pd K edge dispersive EXAFS at ca. 3 Hz spectral repetition rate from: (top spectrum) the as prepared sample at RT under flowing He; (middle) under flowing 5%CO/He at 673 K; (bottom) under flowing 5%NO/He at 673 K. The insert shows the corresponding Fourier transforms of the k^3 weighted spectra. (B) A synopsis of some of the raw results (from DRIFTS and EXAFS) obtained during the redox (CO/NO) cycling experiment for a number of observables as indicated. Reproduced from *Nature Materials*, 2007, 6, 528–532, by permission from the Nature Publishing group.

combining time-resolved IR with dispersive EXAFS. The apparently oscillating behavior of the first-shell Pd–Pd coordination derived from EXAFS (Fig. 14.4B), that is dependent upon both the feedstock experienced, was initially [56] interpreted as evidence for size and/or shape change driven by the different types of adsorbate present under each feed; even more significant variations, prior to the eventual observation of Pd^{II}, were observed when the oxidizing feed was changed to a 5% ($5O_2$:1NO)/He mix rather than simply 5%NO in He [57].

Following further development and experimentation (see later) it now seems that the notion of particle size change during the gas-switching events and as a result of gas adsorption/reaction seems unlikely. Particle reshaping as a function of the adsorbates present is likely to occur, especially within the notion of the effects that differing adsorbates may induce in the nanoparticle [49, 56, 59]; morphological change results in a corresponding change in the overall level of static disorder (frequency and type of surface sites). Alternatively, the changes in EXAFS and XANES intensities observed at the gas-switching events have general character of that expected from the theoretical work of Yevick and Frenkel [61]. This theoretical approach is premised entirely on the local surface disorder that adsorbates such as CO and NO might be expected to induce in the surface of a Pd particle and does not require a change in particle morphology. However, the complete reversibility of the process under two different gases (e.g., two different interactions with Pd metal, of covalent nature for CO and dominantly ionic with NO) makes it difficult to accept this last option alone.

That the first-shell Pd–Pd coordination was found to vary further *within* the CO cycle was at this time not explained upon the basis of the EXAFS/DRIFTS combination, and it took a further development (see later [49]) of another DRIFTS/X-ray couple to correctly establish the source of this particular aspect of the behavior of the Pd nanoparticles and how it relates to the chemical speciation present on the Pd particles observed by IR.

14.3.2 DRIFTS Coupled to High-Energy X-ray Diffraction (HXRD)

As a result of the adoption of the modified Spectra-Tech cell (Fig. 14.2B,C) for combining DRIFTS with TR–EXAFS, a further couple of DRIFTS to a rather different X-ray probe becomes possible. The adoption of 5 mm circular X-ray exit windows in this design permits a (albeit small) solid angle to be collected. By recognizing that as the energy of the X-rays applied increases $Q(\text{Å}^{-1})/2\theta$ space is progressively compressed into smaller and smaller solid angles, a new experiment becomes possible; that of coupling DRIFTS to high-energy X-ray diffraction (HXRD). Practically, the only other modification that needs be made to the DRIFT/EXAFS setup is to remove an appropriate section of the ellipsoid top mirrors (as shown in Fig. 14.2C); this can be done with no adverse effects on the IR component of the experiment.

It needs to be realized that to make such a couple viable with an exit window that subtends only ±10° the X-rays must be of a very high energy (> ca. 75 keV). This fact, however, plays directly to one of the intrinsic strengths of a high-energy (≥6 GeV) X-ray source such as the ESRF (France), PETRA III (Positron-Electron Tandem Ring Accelerator III, Germany), the APS (Advanced Photon Source, USA, 7 GeV), or Spring 8 (8 GeV). Such third-generation sources, coupled to state-of-the-art insertion devices, can easily access such X-ray energies while retaining a high photon flux. When combined with flat panel X-ray detectors that have recently been implemented for hard X-ray studies at synchrotrons [33, 62], a time-resolving couple between DRIFTS and HXRD results. This has recently been demonstrated for the study of working catalysts at ID15B of the ESRF and leads to a number of further and intriguing possibilities [49, 63].

What this experimental couple allows us to achieve is shown in Figure 14.5, this time for a 2 wt% Pd/Al_2O_3 catalyst (average particle size ca. 3 nm [49,

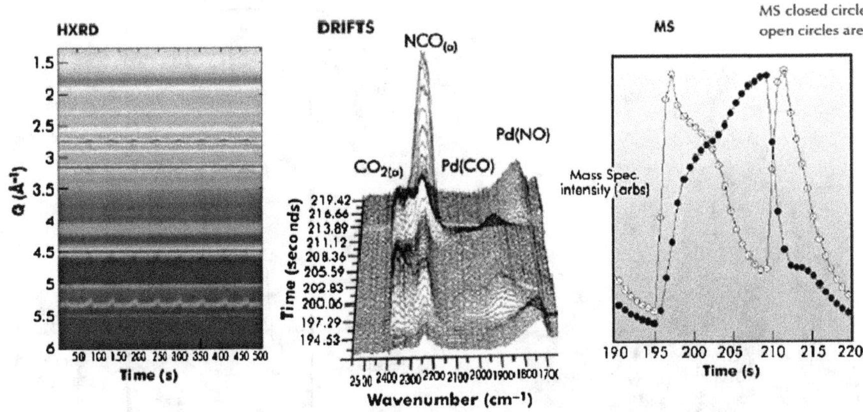

Figure 14.5. Results obtained through synchronously coupling HXRD, DRIFTS and MS using the reactor/optics design shown in Figure 14.2B and C [49]. The sample is a 2 wt% Pd/Al_2O_3 catalyst maintained at 673 K and subject to the same (CO/NO) redox cycling as for the 1 wt% Pd sample shown in figure 4. HXRD shows changes in the position of Bragg reflections due to the Pd nanoparticles (ca. 3 nm diameter), during redox (CO/NO) cycling. These "sawtooth" changes in the position of the Pd reflections occur within the period of exposure to 5% CO/He. DRIFTS shows the speciation concomitantly observed during a single NO–CO–NO cycle; MS shows the evolution of the gas phase during a single NO–CO–NO switch. The filled circles track $CO_{(g)}$, the open circles show the production of $CO_{2(g)}$. Reprinted with permission from Newton, M. A., Di Michiel, M., Kubacka, A., Fernández-García, M. (2010) Combining time resolved hard X-ray diffraction (XRD) and diffuse reflectance infrared spectroscopy (DRIFTS) to illuminate CO dissociation and transient carbon storage by supported Pd nanoparticles during CO/NO cycling, *J. Am. Chem. Soc.*, 132, 4540–4541. Copyright (2010) American Chemical Society. (See color insert.)

59]). How this relates to the equivalent dispersive EXAFS measurement and to the occupancy of linear and bridged CO sites during the CO part of the CO/NO cycling experiment is depicted in Figure 14.6.

The application of the HXRD in tandem with the DRIFTS and MS immediately identifies what lies behind the observed changes in EXAFS and XANES intensities observed during the central part of the CO step in the cycling experiment, as shown in Figure 14.4B, Figure 14.5, and Figure 14.6. The sawtooth variation in the d-spacing of Bragg reflections from the Pd nanoparticles (most clearly seem in this system for the [311] peak between 5 and 5.5 $Å^{-1}$) shows that what we observe most easily in the EXAFS, that is, changes in intensities (Fig. 14.3 and Fig. 14.6), are actually correlated to a lattice expansion (on the order of ca. ≤1%) in the supported nanoparticles. Together with the MS data, the source of this change is pinned down to the dissociation, largely unexpected for Pd, of a portion of the adsorbed CO species. This in

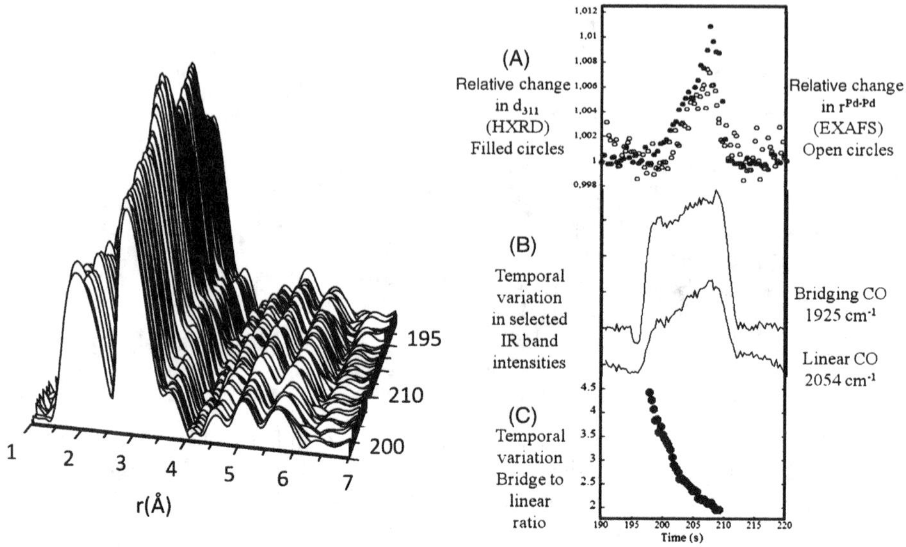

Figure 14.6. Left: Fourier transform representation of k^3-weighted dispersive EXAFS (300 ms per spectrum) obtained from a 2 wt% Pd/Al_2O_3 catalyst during CO/NO cycling at 673 K showing the temporal variation in the intensity of the first-shell Pd–Pd coordination number (vis-à-vis Figure 14.4B, right panel). Right: (A) comparison of the relative change in the d-spacing of the Pd (311) Bragg reflection observed in HXRD, with the relative variation in first-shell Pd–Pd bond distance derived from analysis of the dispersive EXAFS; (B) temporal variation in bridging and linear molecular CO species adsorbed upon the Pd nanoparticle during CO exposure; (C) the temporal variation of the "bridge-to-linear" ratio for CO adsorbed upon the Pd nanoparticles during the reducing (5%CO/He) part of a single NO–CO–NO cycle. Reprinted with permission from Newton, M. A., Di Michiel, M., Kubacka, A., Fernández-García, M. (2010) Combining time resolved hard X-ray diffraction (XRD) and diffuse reflectance infrared spectroscopy (DRIFTS) to illuminate CO dissociation and transient carbon storage by supported Pd nanoparticles during CO/NO cycling, *J. Am. Chem. Soc.*, 132, 4540–4541. Copyright (2010) American Chemical Society.

turn leads to transient storage of atomic carbon within the Pd and a concomitant dimensional expansion of the nanoparticles.

This small lattice expansion is much more easily detected by the HXRD than the EXAFS, the former having approximately one order of magnitude more sensitivity [49] in this respect. However, as is also shown in Figure 14.6B (top panel), this information does lie within the dispersive EXAFS data and (not shown) also within the continuum resonances in the Pd K-edge XANES structure [64]. However, the HXRD unambiguously establishes that the transient dip in the intensity for the first-shell Pd–Pd interaction (Fig. 14.4B and Fig. 14.6A) results from an increased level of static disorder in the Pd that accompanies the storage of atomic carbon within the Pd [64].

14.3.2.1 Studying Oxygen Storage and Release in Pd/CeZrO$_4$/Al$_2$O$_3$ Catalysts A principal strength of the EXAFS approach is in its elemental specificity. However, in multielemental systems, this generally means that only one component of a working system may be addressed at any given time (there are one or two exceptions to this general statement where the intrinsic nature of the dispersive EXAFS experiment permits simultaneous investigation of more than one elemental edge; see, for instance, References [3(j)] and [55]). XRD is not, per se, elementally specific, but if the phases present together in the catalyst have sufficient order to produce detectable Bragg diffraction peaks, then XRD can in principle be used to study more than one phase at a time.

A great many catalyst systems use several components working together to achieve the desired end point. A case in point is the use of promoter oxide phases alongside noble metals such as Pd, Pt, and Rh within the formulation of three-way exhaust catalysts. Figure 14.7 shows how the combination of DRIFTS with HXRD can be used in a case such as this to gain a more global insight into how the synergy between the nanosized Pd and$CeZrO_4$ (supported upon Al_2O_3) works during model deNOx (CO/NO cycling) catalysis [63].

Figure 14.7A shows HXRD data as a color map over the duration of a reduction/oxidation (redox) cycling experiment of the type shown in Figure 14.5 for a Pd/Al_2O_3. The catalysts in this case are comprised of 4 wt% Pd/33 wt% $CeZrO_4$/Al_2O_3 (4Pd33ZCA). The structural–reactive behavior observed in this system is clearly very different from that observed in the simpler Pd/Al_2O_3. Structurally, the XRD allows us to see that the dynamism of this system has shifted from the Pd nanoparticles to the $CeZrO_4$ phase as both are visible in the HXRD.

In essence, what now dominates the XRD in this case are the oxygen storage and release functions of the $CeZrO_4$. The formation and consumption of Ce^{3+} provides the physical basis for the changing lattice parameter of this phase, which we can follow and quantify as it occurs using this approach: it can be shown that the promoter oxide oscillates between states that we can approximate to $Ce_2Zr_2O_{7.5}$ and $Ce_2Zr_2O_7$ (i.e., a 50 wt% reduction of the Ce) in the presence of the Pd nanoparticles; in the absence of Pd, levels of stoichiometric change in the $CeZrO_x$ phase are much reduced and kinetically the processes are rather different in character.

Further, it can be easily shown that one of the effects of the oxygen transfer is to curtail the formation of the PdC_x phase observed in the PdAl_2O_3 case [63]. If we compare the simultaneously obtained DRIFTS shown in Figure 14.7B with that shown in Figure 14.5 for the Pd/Al_2O_3 case, it is also evident that the presence of the $CeZrO_4$ phases grossly changes a number of aspects of the surface functionality that can be found during this redox cycling. At least six IR observables are significantly altered compared to the Pd/Al_2O_3 case. Figure 14.7C,D shows two examples of how the contact of the $CeZrO_4$ phase with the Pd, and its oxygen storage and release function, affect the speciation of molecular adsorbates such as CO (7(C)) and NCO (7(D)) species.

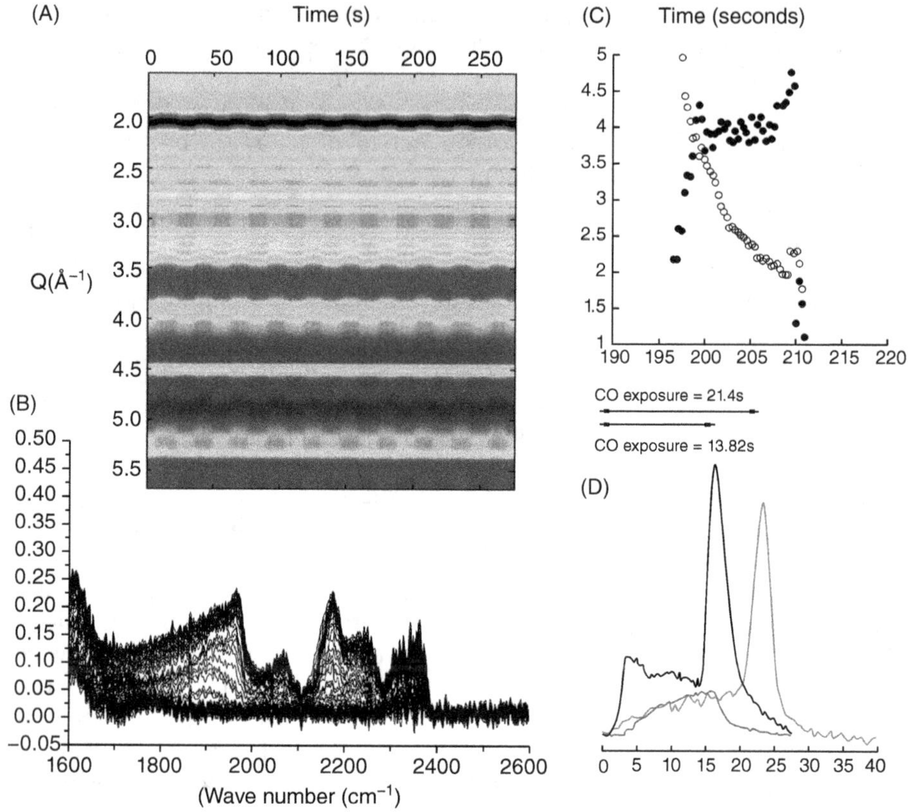

Figure 14.7. (A) HXRD obtained during redox (CO/NO) cycling of 4Pd33ZCA catalysts sample at 673 K [63]; (B) examples of the synchronously obtained DRIFTS spectra; (C) a comparison of the bridge-to-linear ratio for molecularly adsorbed CO observed in the 4Pd33ZCA case (filled circles) compared to that seen in the equivalent 2 wt% $PdAl_2O_3$ case (see also Fig. 14.6); and (D) comparison of the temporal production of adsorbed NCO species (ca. 2240 cm^{-1}) during a single NO–CO–NO cycle at 673 K for three cases: 2 wt% Pd/Al_2O_3, short (13.86 s) CO exposure (black); 4Pd33ZCA, short (13.86 s) CO exposure (blue); and 4Pd33ZCA, long (21.4 s) CO exposure (red). Reproduced from Newton, M. A., Di Michiel, M., Kubacka, A., Iglesias-Juez, A., Fernández-García, M. (2012) Observing oxygen storage and release at work during cycling redox conditions: synergies between noble metal and oxide promoter, *Angew. Chem. Int. Ed.*, 51, 2363–2367, with permission of John Wiley and Sons. (See color insert.)

The variation of the bridge-to-linear ratio of molecular CO adsorbed upon the Pd nanoparticles within a 13.86-s exposure to 5 wt% CO/He is observed to be the inverse of that observed in the Pd/Al_2O_3 case. The reason for this lies in the oxygen transfer from the $CeZrO_x$ phase which we are observing concurrently using the HXRD. As long as oxygen transfer persists, atomic carbon arising from the dissociation of molecular CO is efficiently shepherded back into the desired catalytic route (to form CO_2) before a PdC_x phase can form.

Figure 14.7D shows that the production of adsorbed NCO species is also greatly affected by the oxygen release properties of the system. For a 13.86-s cycle time the production of this species, which is so abundant at the CO-to-NO switch in the Pd/Al_2O_3 case (Fig. 14.5), is greatly diminished. However, if the gas-switching period is increased to 21.4 s over the 4Pd33ZCA system, it reappears with a magnitude essentially equivalent to the case where the $CeZrO_x$ is not present. Though not shown here this behavior can be precisely correlated with the oxygen storage capacity of the $CeZrO_x$ being exhausted as a result of the longer period of operation under reducing conditions. As a result, and from this specific perspective, the system reverts to the behavior observed in the absence of the promoter oxide and NCO is once again produced in copious quantities at the CO-to-NO switch. This behavior can be correlated to the reappearance of carbon storage by the Pd nanoparticles and constitutes a very strong case for the presence of a route to NCO in such systems that is based upon CO dissociation and insertion molecular NO [59]; this to compete alongside the conventional route that involves addition of atomic N to molecular CO.

14.3.3 DRIFTS/XAFS Couples at Lower X-ray Energies: Pt L_3- and Ni K-Edges

Returning to the coupling of X-ray absorption spectroscopy with DRIFTS, for some considerable time this couple, as implemented at the ESRF, was only applied to Rh and Pd chemistry. Such relatively high-energy X-ray edges played well to many of the intrinsic strengths of both the ESRF and ID24 (see Chapter 3) while avoiding some of the old ID24's limitations; further, they make the coupling of EXAFS to IR relatively easy. At these X-ray energies (23–25 keV), sample bed diameters of up to 5 mm can generally be used for the TR X-ray experiment, and such target areas present no potential problems for the IR component of the experiment.

At softer X-ray edges, or indeed for increasingly absorbing/scattering support materials such as CeO_2 or $CeZrO_4$, the usable TR X-ray path length through the sample, and therefore the target surface available for the DRIFTS, dwindles rapidly. Nonetheless, both at the ESRF and at National Synchrotron Light Source (NSLS) at Brookhaven there have been recent demonstrations that this synchronous methodology can be successfully used at much lower X-ray energies: Becker and coworkers [65] studying the dynamic behavior of Pt/Al_2O_3 catalysts (Pt L_3-edge = 11.564 keV) during methane oxidation; and Marinkovic et al. succeeding [43] in making combined measurements in their DRIFTS/EXAFS system (Fig. 14.2D) at both Pt L_3- and Ni K-edges (Ni K-edge = 8.34 keV). These examples also nicely show some of the relative merits and limitations of the quick-scanning and dispersive (as respectively implemented at Brookhaven and on the preupgrade ID24 at the ESRF) EXAFS approaches (see also, for instance, Chapter 1 and Reference [47]) within this general milieu.

Figure 14.9 shows simultaneously obtained Pt L_3-edge XANES and DRIFTS data during periodic (420 s of alternating feeds comprising 0.1% CH4/He and 0.1% CH4/1.5%O_2/He) operation of a 4 wt% Pt/Al_2O_3 catalyst for low temperature CH_4 oxidation catalysis. Individual XANES spectra were collected using 100 ms of X-ray acquisition, corresponding to approximately 4 Hz spectral repetition, with the IR data being collected at a similar rate [65].

For beamline (ID24) specific reasons outlined in Chapter 3, though the dispersive EXAFS may be collected extremely rapidly and with good quality in the XANES, EXAFS from samples of this sort at the Pt L_3-edge could not be obtained. However, in this case the experimental capacity is sufficient to obtain a great deal of information, and a temporally robust X-ray/IR dataset, that may be used for subsequent in-depth kinetic analysis. In this system, the IR allows similar datasets to be obtained that principally highlight the formation/removal of surface species such as linear and bridged carbonyls and carbonates. In a pertinent exemplar, however, of some of the intrinsic limitations of the DRIFTS measurement, IR bands (be they gas-phase or from surface-bound species) from the low concentrations of CH_4 employed are not detectable. The same was found to be true for, for instance, surface-bound OH groups. The high level of noise observable in the above approximately 3000 cm^{-1}—attributable to a rapidly decreasing detector response function and the use of an undersampling filter that removes any contribution to the spectra from light >4000 cm^{-1}—precludes such observations.

In spite of these limitations, some of which will be alleviated as a result of the current ID24 upgrade, and with the addition of a quantitative analysis of the online MS, a very holistic and detailed structural reactive view of this system and its behavior results.

Figure 14.9 compiles some results from a scanning EXAFS/DRIFTS experiment recently implemented at X18A at NSLS, Brookhaven, USA. These measurements use the modified Harrick cell shown in Figure 14.2D,E. While these are very much demonstration measurements, a comparison of the Pt L_3-edge data that may be obtained in conjunction with the DRIFTS using this system, compared to that obtained using the equivalent dispersive EXAFS experiment using this system, is very instructive.

In contrast to the XANES-limited dispersive EXAFS experiment, EXAFS can be obtained for 1 wt% Pt/Al_2O_3 catalyst (bottom panels) and for a Ni/Al_2O_3 catalyst of 4 wt% Ni loading. The top panels of Figure 14.9 show DRIFTS and Pt L_3-edge XANES obtained during temperature-programmed desorption of CO that had been previously adsorbed upon a 5 wt% Pt/Al_2O_3 catalyst.

In this demonstration, however, the XAFS data were accumulated using 600 s of acquisition time, over three orders of magnitude slower than the data shown in Figure 14.8. As such, though the quick-scanning EXAFS approach

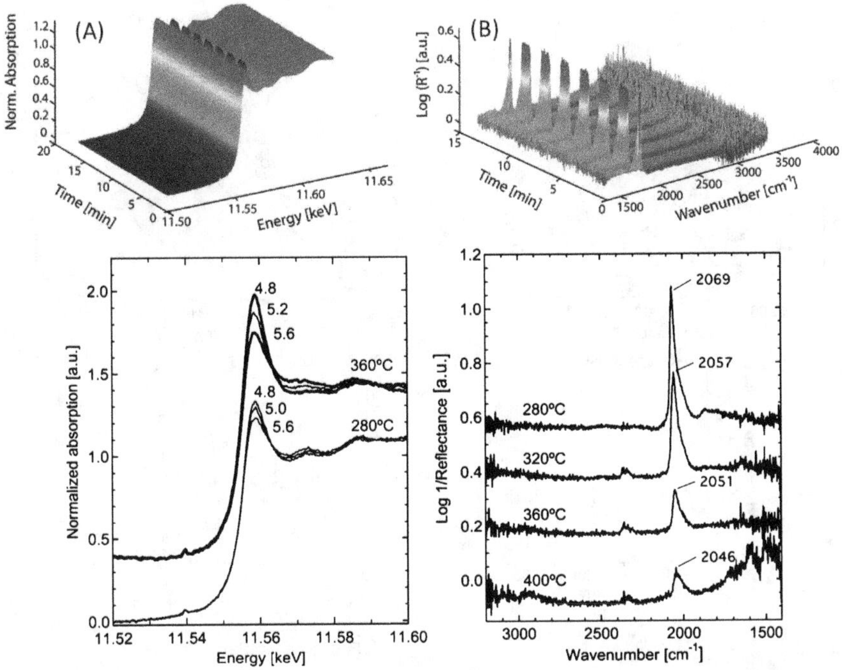

Figure 14.8. Synchronous time-resolved dispersive XANES (A) and DRIFTS (B) during CH_4 oxidation over 4 wt% Pt/Al_2O_3 [65]. The upper panels show a global view of the "periodic" operation of the experiment. The lower panels show more explicitly the overall data quality and changes/obtained observed by both spectroscopies *in situ* and for two temperatures of operation as indicated. Dispersive EXAFS and DRIFTS were collected *in situ* at approximately 4 Hz. Reprinted with permission from Becker, E., Carlsson, P.-A., Kylhammar, L., Newton, M. A., Skoglundh, M. (2011) In situ spectroscopic investigation of low-temperature oxidation of methane over alumina-supported platinum during periodic operation, *J. Phys. Chem. C*, 115, 944–951. Copyright (2011) American Chemical Society. (See color insert.)

clearly currently permits EXAFS data in such a system to be obtained, EXAFS on timescales of approximately 1 s or below still remains to be demonstrated (for different reasons) using either (scanning or dispersive) EXAFS approaches.

Further, the intrinsic dead volume of the Harrick system as currently implemented (ca. 14 cm^3 [43] vs. ~2.5–3 cm^3 [48] for the ESRF, Spectra-Tech-based model) limits this arrangement in terms of the speed of the kinetic processes it may be ultimately used to follow. Nonetheless, that the Brookhaven DRIFTS/EXAFS system is demonstrated to work well for EXAFS at both Pt L_3- and Ni K-edges (for 4 wt% Ni/Al_2O_3) EXAFS (as indicated in Fig. 14.9) is currently unique.

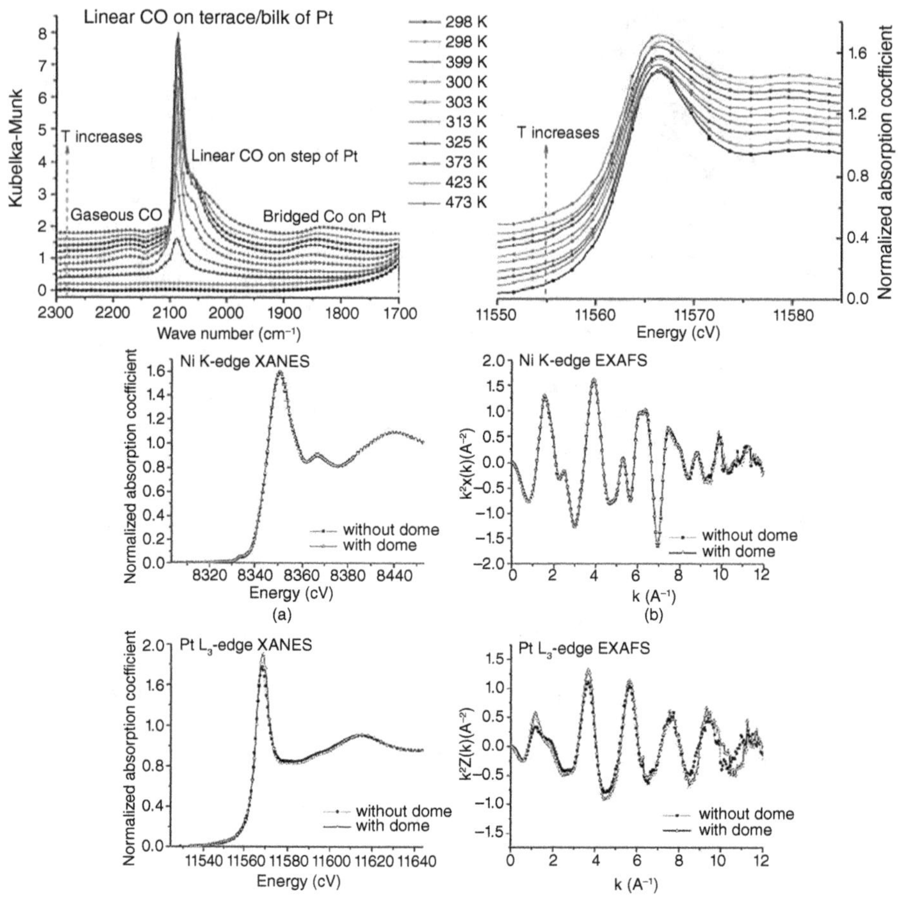

Figure 14.9. Results derived from the first demonstration of the DRIFTS/EXAFS system shown in Figure 14.2D and E due to Marinkovic and coworkers [43]. The top two graphs show DRIFTS (left) and Pt L_3-edge XANES (right) obtained during temperature-programmed desorption of CO previously adsorbed upon a 5 wt% Pt/Al_2O_3 catalyst. Below this the four remaining panels show Ni K-edge XANES and EXAFS (from a net 4 wt% Ni/Al_2O_3 sample) and Pt L_3-edge XANES and EXAFS from a 1 wt% Pt/Al_2O_3 sample. EXAFS data in these cases were collected in 600 s. Reproduced from Marinkovic, N. S., Wang, Q., Frenkel, A. I. (2011) In situ diffuse reflectance IR spectroscopy and X-ray absorption spectroscopy for fast catalytic processes, *J. Synchrotron Radiat.*, 18, 447–455, with permission from the International Union of Crystallography.

14.4 COUPLING TRANSMISSION IR TO SMALL-ANGLE X-RAY SCATTERING (SAXS)

Innocenzi and coworkers [42] demonstrated a method to couple TR with SAXS. This approach elegantly avoids the problem of X-ray TR through generally rather unsuitable IR windows by using a grazing incidence geometry to measure SAXS from mesoporous thin films while they are crystallizing; at the same time, TRIR could be measured through these films (that had been dip-coated onto a silicon wafer). The overall experimental geometry together with exemplary results obtained using this experiment is shown in Figure 14.10.

In this experiment, the SAXS measurements are made using a grazing incidence with the IR applied and collected in TR normal to the surface of the nascent mesoporous thin films. The transmitted IR signal is then bounced from a flat mirror placed behind the sample and eventually to a mercury cadmium telluride (MCT) detector.

To either side (Fig. 14.10) of this novel experimental geometry are given examples of the data obtained from both SAXS and the TRIR as the mesoporous film develops. The central (lower) panel of Figure 14.10 correlates the information arising from the combined use of SAXS and IR. The SAXS shows the appearance (after ca. 10 s) of a diffraction ring (black stars) that identifies the formation of ordered mesoporous domains; the IR tracks the consumption of ethanol (open squares) and H_2O (open circles) and the kinetic differences that exist between them. What this combined experiment clearly shows is exactly how the crystallization process depends on the basic kinetics of ethanol and water evaporation from the dip-coated precursor.

While not a heterogeneous catalysis experiment per se, this elegant methodology has clear and manifold potential applications within fields of study wherein the active element is presented as a model, thin-film catalyst, or a planar device: a wide range of sensing or photoactive materials, all of which are founded upon the use of heterogeneous catalytic events, would seem to be at least one area where this sort of approach might be very useful.

Lastly, there seem to be no a priori reasons why the use of a grazing incidence methodology, to obtain an orthogonal couple with TRIR, should not also work with other X-ray methods such as X-ray absorption and emission spectroscopies (XAS and XES), and where X-ray fluorescence is the required mode of data collection.

14.5 COUPLING TRANSMISSION IR AND TRANSMISSION EXAFS

A second experiment based on TRIR has also recently appeared [44] and used to study the *in-situ* study of Ni_2P systems for desulfurization applications. This approach to IR/X-ray combination is shown in Figure 14.11. Of note is that this is the first approach that makes use of commercially available IR fiber optics. Again it can be seen that a mutually orthogonal IR/X-ray couple is

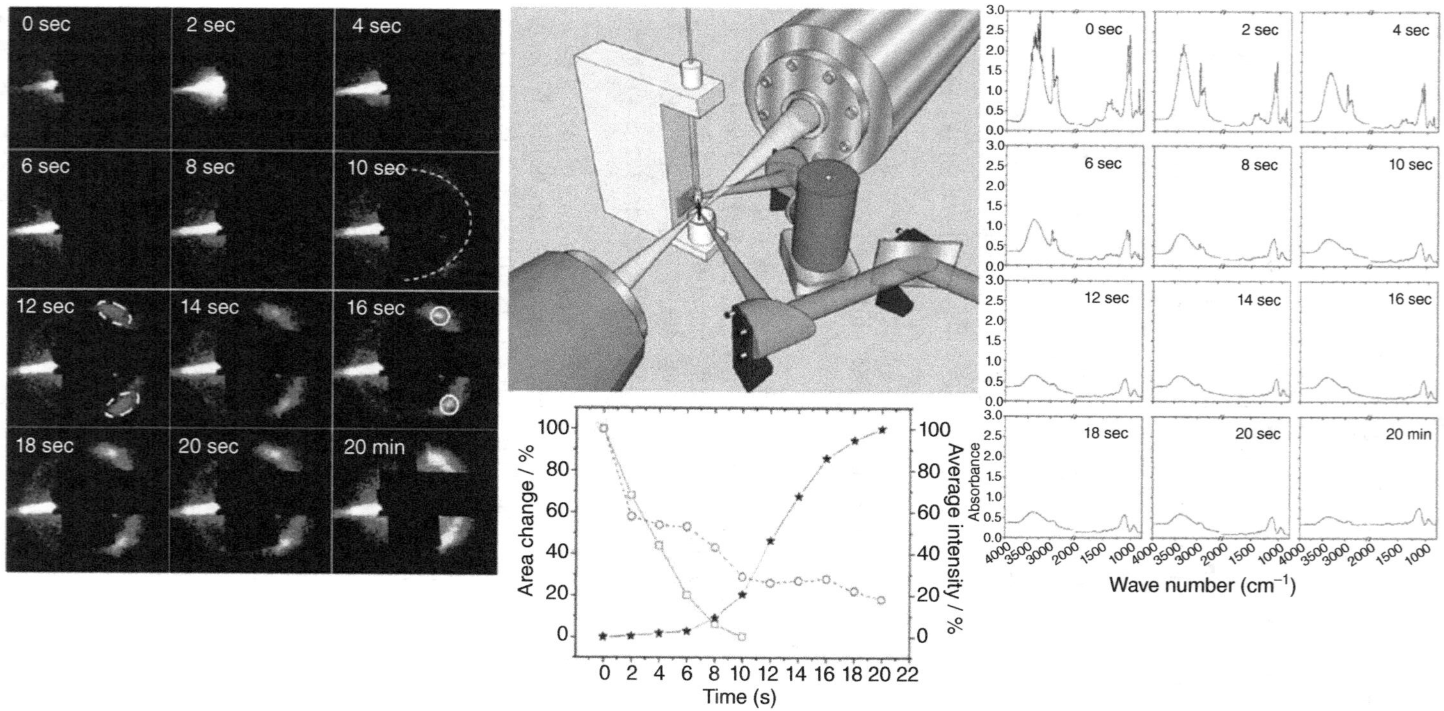

Figure 14.10. The coupling of TRIR and SAXS demonstrated by Innocenzi et al. [42]. Center (top): a schematic representation of the experimental setup. Left: SAXS images taken as a function of time during the crystallization of a mesoporous thin film. Right; the simultaneously obtained TRIR spectra. Center (bottom): the relationship between the appearance of a diffraction ring (filled stars) in the SAXS data with IR bands indicative of water (open circles) and ethanol (open squares). Reproduced from Innocenzi, P., Malfatti, L., Kidchob, T., Costacurta, S., Falcaro, P., Piccinini, M., Marchelli, A., Morini, P., Sali, D., Amenitsch, H. (2007) Time-resolved simultaneous detection of structural and chemical changes during self-assembly of mesostructured films, *J. Phys. Chem. C*, 111, 5345–5350. Copyright (2007) American Chemical Society.

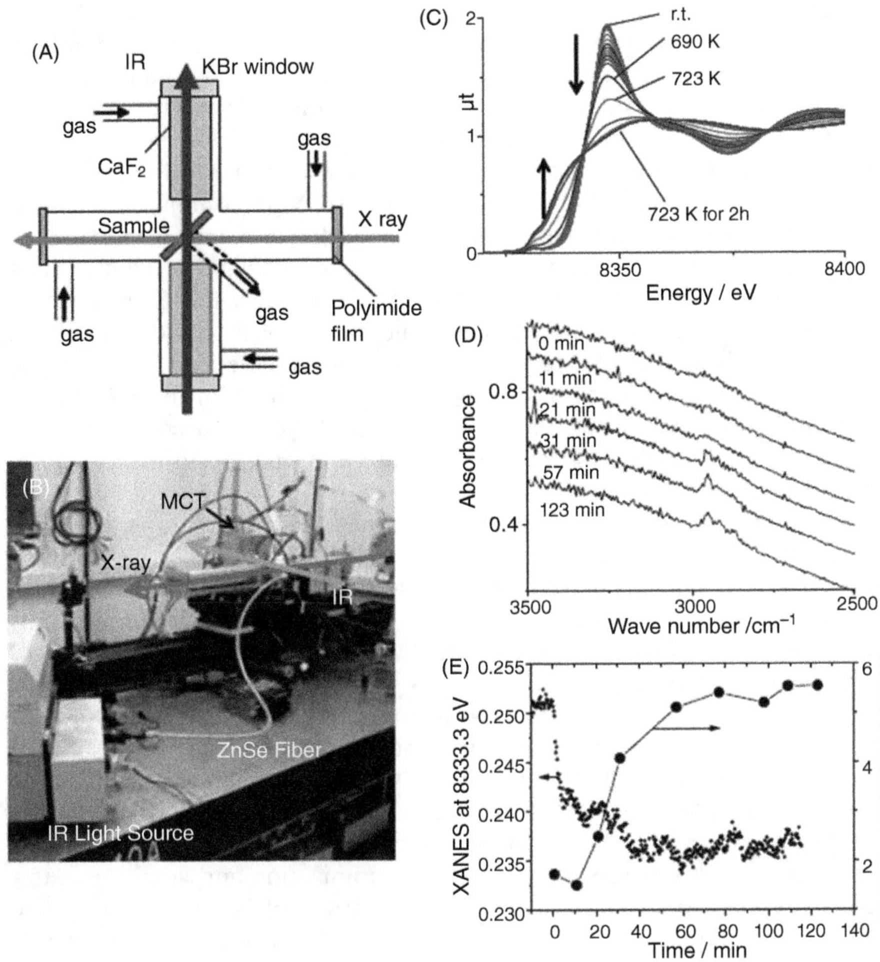

Figure 14.11. (A) Schematic of the TRIR couple to TR-EXAFS due to Bando and coworkers [44]; (B) this setup, mounted on the 9C beamline at the Photon Factory, Japan, showing the use of IR fibers within this experiment. (C) Ni K-edge XANES, from Ni_2P/MCM-41catalysts obtained as a function of temperature; (D) the evolution of IR spectra during isothermal (573 K) HDS catalysis; (E) correlation between a preedge feature observed at 8.333 keV in the Ni K-edge XANES with the area of the C–H stretches due to the formation of tetrahydrothiopene. Reproduced/adapted Bando, K. K., Wada, T., Miymoto, T., Miyazaki, K., Takakusgi, S., Gott, T., Yamaguchi, A., Nomura, M., Oyama, S. T., Asakura, K. (2009) Combined in situ analysis of Ni2P/MCM-41 under hydrodesulfurization conditions—simultaneous observation of QXAFS and FTIR., 14th International Conference on X-Ray Absorption Fine Structure (XAFS14), *J. Phys. Conf. Ser.*, 190, 012158, with permission from the Institute of Physics.

easily achieved within what is, in essence, a TRIR cell, the likes of which have been used for many years in IR spectroscopy. In its first incarnation, therefore, this experiment suffers from a relatively large internal dead volume (though it is easy to see how this might be considerably reduced in the future).

In addition, the sample presentation is in the form of a pressed, self-supporting disk, again an approach typical of that which has often been used over many years within both IR and X-ray spectroscopic communities. This is another solution that neatly avoids any problems due to X-rays having to pass through inconvenient (from an X-ray point of view) IR windowing materials. Though this is a highly practical resolution in many ways, this does however mean that a space velocity/contact time cannot be established and, as has been noted previously, the precise feedstock conditions experienced by the inside of the pressed disk are, a priori, not known [66]. That said, for the TRIR component of the experiment to work, the thickness of the disk used in this study was of the order of only 300 μm and therefore interdisk gas diffusion limitations should be minimized.

The combination of the previously mentioned factors will, however, most likely mean that, while it can be shown (see later section) that this arrangement is highly efficacious in achieving the desired TRIR and EXAFS couple, it will be limited to the study of relatively slow processes, as are studied in its first demonstration. Moreover, the available X-ray path length will be limited by the (sample dependent) requirements of the TRIR to a few hundred micrometers. This means that this experiment is best adapted to relatively soft X-ray absorption edges (especially in combination with very highly absorbing/scattering support materials). In this sense, therefore, the approach of Bando and coworkers [44] constitutes an extremely useful counterpoint to the DRIFTS-based approaches previously discussed.

One can also clearly see that with some optimization and adaptation, this sort of approach could be used to couple a variety of X-ray scattering and diffraction techniques with the TRIR experiment. These are techniques where unduly long TR path lengths through the sample are best avoided and which, therefore, fit well with the overall requirements of the TRIR.

In its first demonstration, this couple is used to follow hydrodesulfurization (HDS) catalysis mediated byNi_2P supported upon mesoporous silica (MCM-41) from the perspective of the Ni K-edge (XANES) and the C–H stretching bands indicative of the hydrogenation of thiophene to tetrahydrothiophene. The latter, in itself, is interesting as such IR bands are generally much weaker than those of CO or NO species that have been generally been studied using the DRIFTS/EXAFS combination: only a very recent investigation of selective oxidation of crotonaldehyde over Pd catalysts [67] has also achieved a similar task, one which is considerably more challenging than those using the somewhat more IR-friendly molecular species.

What emerges from this study is a direct correlation between an evolving Ni K-edge XANES structure and the formation of the tetrahydrothiophene, the hydrogenated product molecule being observed to augment (in IR) with an inverse relation to a preedge feature in the XANES (8333.3 eV). The diminution of this edge feature in the XANES is attributed to the formation of Ni–S coordination subsequent to exposure to thiophene. It is therefore concluded that the formation of this new coordination within the catalyst is the critical element in creating the active (Ni phosphosulfide) HDS catalyst.

14.6 POSSIBILITIES FOR FUTURE DEVELOPMENTS

Having outlined what has been achieved to date in terms of coupling structurally direct X-ray methods with surface and molecular sensitivity of IR, we now consider what may yet be done and how these sorts of combinations might be used in the near future to augment research into heterogeneous catalysis. Within this overall milieu, what we have shown earlier is far from complete in terms of what, should it be wished for, could be achieved.

14.6.1 Further Couples of X-ray Techniques with Infrared Spectroscopy

14.6.1.1 ATR The coupling of ATR with SAXS for the investigation of polymers was demonstrated as early as 1995 [37] and the "combined"—though separate—use of ATR and, for instance EXAFS, has been well utilized [39, 68]. Yet since 1995 nothing further has appeared by way of direct combination of X-ray techniques using this IR variant. A priori there seems no reason why a number of couples based upon ATR cannot be achieved; and ones that may access the increasingly important field of liquid–solid catalysis, ATR having been shown to be extremely well adapted [23–25, 39, 68] for studying such processes.

In addition, the nature of the ATR experiment itself may also permit fluorescence yield detection of X-rays. A conventional geometry for fluorescence detection is not at all easy to achieve within any conventional DRIFTS optics, and therefore the potentially highly attractive couples of fluorescence yield XAFS/IR and combined IR/XES could clearly be realized using an ATR approach.

Lastly, there seems to be much in terms of spatiotemporal studies of working catalyst beds (be they homogeneous, gas–solid, or liquid–solid) that may be achievable by combining X-ray methodologies with a single-bounce ATR/focal plane (FPA) array detector. FPAs are relatively new additions to the canon of IR detectors and offer new ways to image materials from an IR perspective with both high spatial and temporal resolution [69]; again, a combination with one or more X-ray probes seems entirely possible.

14.6.1.2 DRIFTS DRIFTS is clearly best adapted for the interrogation of gas–solid interactions and it is relatively easy, as we have seen, to couple this IR variant with X-ray techniques based upon TR. This does not mean, however, that there is not more that can be done to extend this couple.

Overcoming Limitations due to IR Source Brilliance: DRIFTS/EXAFS the First Row Transition Metals and in Systems Comprising High Absorbing Supports One of the limitations of the basic DRIFTS/EXAFS experiment in terms of application of TR-EXAFS studies has been the following: where the element of interest has a K-edge of ≤10 keV, or indeed the support contains large amounts of very absorbing elements (such as, for instance Ce, La, Zr), the TR path length available for X-ray absorption dwindles to ≤2 mm. These criteria apply to a considerable swathe of catalysts that one would like to be in a position to study using an IR/EXAFS couple.

However, the IR throughput one can apply in situations such as these becomes rapidly diminished as a result of having to reduce the IR focal spot size by reducing the size of an aperture internal to the IR spectrometer. For basic, non-time-resolved experiments at low temperatures, obtaining DRIFTS spectra in tandem with the IR may not be problematic even with this restriction.

Eventually, however, the combination of intrinsically weak levels of diffuse scattering and reduced IR throughput (resulting from the need to progressively stop down the IR to target smaller and smaller sample areas) will severely limit attainable performance. This will be especially so where the chemistry that needs to be followed is neither periodic nor easily reproducible (see later).

In addition, and when using (for instance) MCT detectors, operation at temperatures where IR emission from the sample itself will also eventually make the DRIFTS intractable. In such cases, the IR emission from the hot sample eventually swamps out the diffuse scattering that the experiment needs to measure; as and when this point is arrived at is dependent upon how much diffusely scattered signal one has in the first place.

Figure 14.12 shows the schematic layout of a new DRIFTS experiment commissioned for implementation on the postupgrade dispersive EXAFS line of the ESRF (ID24). This has been designed to overcome some of these limitations, and with the knowledge that sample dilution, in the manner used by Marinkovic et al. [43], is not at all likely to be successful when using an energy-dispersive variant of EXAFS (see for instance Chapter 3).

The central element to this new DRIFTS configuration lies in applying IR and detecting the diffuse scattering along the same optical path. The details of this new configuration will be published elsewhere [70], but the essential result of this new optical setup is the ability to obtain a sub-1 mm focal spot without recourse to an excessive stopping down of the IR source. As such, we should be able to comfortably address sample bed diameters of approximately 1 mm from both dispersive EXAFS and DRIFTS points of view while retaining the highest possible level of IR throughput (from a Varian 680 IR spectrometer, Agilent Technologies, Santa Clara, CA). A number of currently intractable catalytic systems (e.g., the first-row transition metals and more absorbing support materials) should therefore become feasible using this experiment.

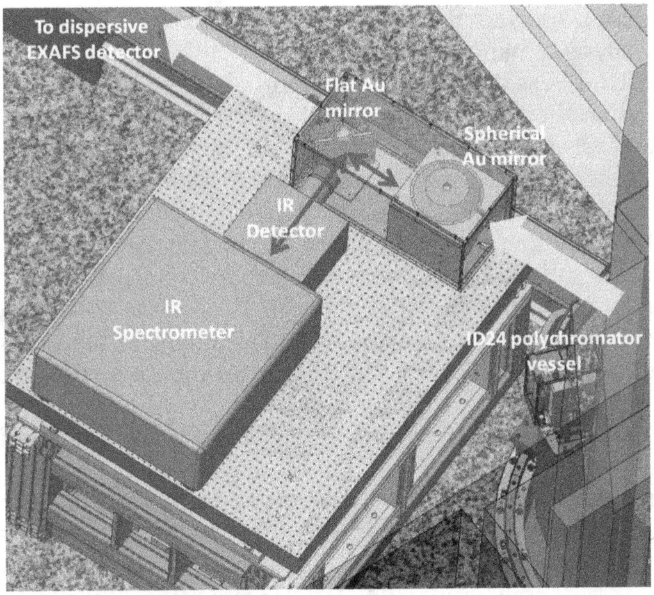

Figure 14.12. Schematic layout of a novel DRIFTS/X-ray experiment to be implemented on one branch (EDXAS_L) of the upgraded dispersive EXAFS line (ID24) of the ESRF during 2012. The large light gray arrows show the path taken by the applied X-rays; the thin darker arrows trace the path taken by the IR light for both illumination of the sample and detection of the diffuse scattering that arise from the catalyst sample.

Coupling DRIFTS to Total X-ray Scattering Measurements and PDF Analysis As we have already seen (Section 14.3.2) the basic DRIFTS/EXAFS couple can be extended to include XRD using very high-energy X-rays. This paves the way for a further logical step: extending this approach to cover total X-ray scattering measurements wherein a subsequent PDF analysis may be undertaken.

Total X-scattering measurements [2, 33, 71, 72] (see also Chapter 5) that permit a PDF analysis for local and midrange structure also strongly favor the use of hard (>50 keV) X-rays. This approach to X-ray scattering provides a further window (complementary to that obtained using EXAFS) into the local order of poorly crystalline/amorphous materials such as many heterogeneous catalysts. Uniquely, however, this method can go beyond the immediate local structure to which EXAFS is physically limited (to ca. 5–6 Å from the excited atom), bridging an important blind spot in the approximately 5–20 Å range, that is, before Bragg diffraction or SAXS start to become viable. Moreover, the PDF technique is extremely suited to the study of amorphous or strongly disordered phases, being less sensitive than XAS techniques to the effect of disorder. The combined application of total X-ray scattering with DRIFTS to study amorphous/disordered solids has the potential to become a key tool for the future analysis of nanocatalytic materials [73].

The underlying reason why the PDF approach favors the use of very high-energy X-rays lies in the high $Q(\text{Å}^{-1})/2\theta$ range (ca. 25–30 Q) required for this method of analysis to be best applied. As with the previous example outlining the combination of DRIFTS with HXRD, the flattening of Ewald's sphere at high X-ray energies means that such Q ranges become feasible to obtain within a solid angle that can be collected on modern flat panel detectors [33, 62] and/or those yielded by specialist sample environment, the latter being a prerequisite for any *in-situ* catalysis experiment. Even at 60 keV, however, 30 Q corresponds to a collection angle of approximately 45° and thus still presents a significant design challenge to realize within a conventional DRIFTS experiment.

Nonetheless, two solutions to enable the DRIFTS/total scattering couple may be suggested.

First, to redesign a commercial DRIFTS system to yield the required solid angles in TR; or, second, to incorporate a tubular microreactor (based upon, for instance, IR transmissive sapphire, rather than quartz) inside the DRIFTS optics instead of the conventional DRIFTS sample cup. At the time of writing, the former possibility is being actively investigated. A prototype reactor that can achieve this couple, and based upon further modifications of the Harrick Praying Mantis DRIFTS accessory, has been tested at Sector II-IDB at the Advanced Photon Source, USA. A second system compatible with Spectra-Tech DRIFTS optics has also been designed but has yet to be tested [74].

We may further note that by enhancing the solid angles for X-ray data collection in either backward- or forward-scattering geometries then it may also become possible to combine DRIFTS with various types of emission spectroscopy (for instance, high-energy resolution XANES), assuming that the X-ray beamlines in question are capable of also collecting X-ray fluorescence in such extreme geometries.

14.6.1.3 Transmission Infrared Of the three major forms of IR spectroscopy, TR is the most easily quantifiable and as such very desirable to be able to use. Following Bando et al. [44] and Innocenzi et al. [42], it is not difficult to see how TRIR coupled to X-ray techniques can be readily augmented and developed. Importantly, TR should be readily combinable with fluorescence yield techniques; as we have seen, it is not a priori easy to see how this might be achieved using conventional geometries for DRIFTS and fluorescence yield EXAFS. Moreover, given the limited sample thicknesses that will have to be used for studying undiluted catalyst samples, it seems clear that this approach is suited to the softer end of the X-ray spectrum and should provide a viable approach to studying materials in the sub-7 keV region: these are X-ray energies where even an augmented and optimized DRIFTS experiment will be very difficult to use effectively. A whole host of combined studies at softer K-edges of, for instance, V, Ti, and Mn, along with a gamut of rare earth and transition metal L-edges, could become possible using a TRIR approach.

14.6.1.4 Model Catalysts and RAIRS For some considerable time, RAIRS has been successfully applied to the study of more fundamental aspects of the behavior of model catalyst systems based upon planar single-crystal metal surfaces [75, 76], thin oxide layers deposited on metal surfaces [77], and (much more rarely) bulk oxide single-crystal surfaces [78]. As such, it is the preeminent IR counterpoint to DRIFTS, ATR, and TR in the world of planar model catalysts.

Surface X-ray diffraction (SXRD) [79] and grazing-incidence small-angle scattering (GISAXS) [80] have evolved immensely in recent years with the maturation of third-generation synchrotron sources. As a result, they have recently begun to contribute much to the bridging of the so-called pressure gap existing between ultrahigh vacuum (UHV) and real-world study of catalysts, just as RAIRS started to do, from a different perspective, a generation previously.

A priori, there is no reason why this well-established IR technique might not be conjoined to these structurally direct X-ray techniques to yield a more complete view of working planar model systems, much in the way that suitable combinations (see earlier) are now being employed to *in-situ* study of high area catalysts. Thus far, however, such experimental couples have yet to be demonstrated.

14.6.1.5 Implementation of Modulation Excitation Methods Alongside developing new experiments based upon coupling available IR variants to a variety of X-ray techniques, the more widespread application of more advanced experimental methodologies and analytical methods is worth mentioning within this overall milieu.

Specifically, in the current context, is the notion of modulation excitation spectroscopy (MES). This provides an analytical framework—based upon principles that, for instance, also underpin lock-in amplification techniques—that can be used to amplify and isolate signals that would be otherwise extremely difficult to track, such as reactive/transient species. For some time now this approach has been applied within the field of IR spectroscopy [46, 81]. Only very recently, however, and as a result of an increasing interaction between vibrational spectroscopic and X-ray based communities interested in how best to study chemical systems *in situ*, has this approach been started to be applied within experiments that couple techniques such as Raman/XRD [82] and DRIFTS/EXAFS [83, 84]. A theoretical assessment of what this might approach might bring to XRD and wherein might lie the limitations of this approach has also recently been published [85].

This analytical approach clearly has within it many potential benefits, not least of which is the greatly increased sensitivity to change that it imbues any given experiment with. That said, it may only be applied in systems wherein a completely reversible response, induced by, for instance, changes in pressure, temperature, or reactant concentration, may be achieved over a sufficient number of modulation events. While this may be a limiting factor in many areas of research, this is exactly the sort of behavior that one would hope to find in any half-decent catalytic system, almost by definition.

As such, and as outlined in much more detail by van Beek et al. in Chapter 15, and also briefly discussed in a previous chapter in relation to dispersive EXAFS (Chapter 3), the continued development, refinement, and more widespread implementation of such approaches within experiments that fuse vibrational spectroscopies with time-resolving X-ray techniques, would seem to have much to offer.

14.6.1.6 Ultrafast Measurements In Section 14.2 we noted that both IR spectroscopy and numerous X-ray based techniques can be operated in manners that access timescales far shorter (nanoseconds or less) than those generally associated with chemical kinetics (microseconds and above). As such, there appears to be no obvious reason why they may not be used synchronously to access more fundamental aspects of catalyst behavior.

Such "pump–probe" experiments have remained, by and large—from a chemical perspective—the preserves of gas-phase or liquid-phase chemistry. However, there are at least two relatively recent examples [28, 29] that show that step-scan IR can be applied in catalytic situations to good effect. Fridell et al. [28] kept to kinetic timescales as they were restricted by the rate of gas switching they could achieve. Thibault-Starzyk and coworkers, however, clearly demonstrated that given the correct setup and "pump"—in their case an IR laser for rapid sample heating—such an experiment can be used to observe rather fundamental, and previously unobserved, aspects of a heterogeneous catalytic function.

Step-scan EXAFS [30], for instance, has also been achievable for some time and recent detector developments (Chapter 3) now mean that dispersive EXAFS too should be far more readily usable in these ultrafast time domains. Highly time-resolved emission spectroscopy within the overall "pump–probe" paradigm has also been recently demonstrated [86].

Technically, therefore, it would seem that in a number of cases, ultrafast dynamical studies that can address the dynamics of physical and/or electronic structure change (using X-ray methods) in the same breath as molecular functionality (using IR) are within reach.

14.7 CONCLUSIONS

Structurally direct X-ray techniques can be combined with surface and molecular sensitivity of IR spectroscopy in a number of different ways, many of which have yet to be demonstrated but lie before us as clear and present opportunities to see further into the behavior of heterogeneous catalysts. Hard

X-ray methods (covering a wide range of length scales) and IR can be applied both over a very wide range of timescales (covering the dynamic, through the kinetic, and to the steady state) and reactive environments. Already, those combinations of IR and X-ray techniques (such as EXAFS, SAXS, and HXRD) that have been demonstrated have shown considerable promise and utility in obtaining new information about catalyst behavior, from more fundamental observations to those of direct applied relevance. It also seems clear that what has been realized thus far is but a fraction of what could be achieved and that a great deal of scope and opportunity is still afforded us in this arena.

ACKNOWLEDGMENTS

We should very much like to acknowledge the support, over a number of years, received from the ESRF (Grenoble, France) that has allowed much of what has been described to be developed and implemented. Marco Di Michiel at ID15 the ESRF is especially thanked for his contributions to the development of the DRIFTS/HXRD combination. M.F.G. thanks the Spanish CICYT (CTQ2010-14872/BQU) for financial support.

M.A.N. should like to gratefully acknowledge the essential role and contributions of Prof. John Evans (University of Southampton, UK) in the development of the first experiment that combined DRIFTS with EXAFS, and the EPSRC (UK), which first funded the development of that system.

M.A.N. should also like to thank the following people/institutions for their collaboration, discussions, advice, and general support: Peter Chupas and Karen Chapman at the Advanced Photon Source (APS), USA; Claudio Marcelli (Frascati, Italy); Anatoly Frenkel (Yeshiva University and Brookhaven National Laboratory [BNL], USA); Andy Dent (Diamond Light Source, UK); and Fred Meunier, (ENSICAEN, France). Pierre van Vaerenbergh (ESRF) is thanked for the schematics of the ESRF DRIFTS/X-ray experiment shown in Figure 14.2B,C; Gary Admans at the ESRF is also thanked for the composition of Figure 14.5.

M.F.G. thanks Drs J.A. Rodríguez and J.C. Hanson of BNL, USA and Prof. J.A. Anderson of Aberdeen University, UK for their long-standing collaboration related to the use of X-ray absorption and diffraction techniques. Also, all coauthors involved in works here described. Apart from people already mentioned, Drs. A. Kubacka, A. Iglesias-Juez, C. Belver, and A. Martínez-Arias from the Instituto de Catálisis y Petrolequímica (CSIC) Spain are thanked for their effort.

REFERENCES

[1] (a) Eischens, R. P., Pliskin, W. A., Francis, S. A. (1952) Infrared spectra of chemisorbed carbon monoxide, *J. Chem. Phys.*, 22, 1786–1788.; (b) Eischens, R. P., Francis,

S. A., Pliskin, W. A. (1956) The effect of surface coverage on the spectra of chemisorbed CO, *J. Phys. Chem.*, 60, 194–201.; (c) Yang, A. C., Garland, C. W. (1957) Infrared studies of carbon monoxide chemisorbed on Rhodium, *J. Phys. Chem.*, 61, 1504–1512.

[2] (a) Gallezot, P. (1984) X-ray techniques in catalysis, in Anderson J. R., Boudart, M. (eds.), *Catalysis, Science and Technology*, Vol. 5, Springer, Berlin, Germany, Chapter 4 and references therein.

[3] (a) Chen, J. S., Sankar, G., Thomas, J. M., Xu, R. R., Greaves, G. N., Waller, D. (1992) Cobalt substituted Aluminophospate molecular sieves—X ray absorption, infrared, spectroscopic, and catalytic studies, *Chem. Mater.*, 4, 1373–1379.; (b) Thomas, J. M., Greaves, N. G., Sankar, G., Wright, P. A., Chen, J., Dent, A. J., Marchese, L. (1994) On the nature of the active site in a CoAPO-18 solid acid catalyst, *Angew. Chem. Int. Ed.*, 33, 11871–11873.; (c) Fernández-García, M., Anderson, J. A., Haller, G. L. (1996) Alloy formation and stability in Pd-Cu bimetallic Catalysts, *J. Phys. Chem.*, 100, 16247–16254.; (d) Fernández-García, M., Martínez-Arias, A., Salamanca, N. L., Coronado, J. M., Anderson, J. A., Conesa, J. C., Soria, J. (1999) Influence of Ceria in Pd for the $CO+O_2$ reaction, *J. Catal.*, 187, 474–485.; (e) Fernández-García, M., Martínez-Arias, A., Rodríguez-Ramos, I., Ferreira-Aparicio, P., Guerrero-Ruiz, A. (1999) Surface and bulk phases of rhodium-copper catalysts, *Langmuir*, 15, 5295–5302.; (f) Fernández-García, M., Martínez-Arias, A., Iglesias-Juez, A., Hungría, A. B., Anderson, J. A., Conesa, J. C., Soria, J. (2001) New Pd/$CexZr_1$-xO_2/Al_2O_3 three-way catalysts. Part 1: catalyst characterization and CO oxidation, *Appl. Catal. B*, 31, 39–50.; (g) Lamberti, C., Prestipino, C., Bonino, F., Capello, L., Bordiga, S., Spoto, G., Zecchina, A., Moreno, S. D., Cremaschi, B., Garilli, M., Marsella, A., Carmello, D., Vidotto, S., Leofanti, G. (2002) The chemistry of the oxychlorination catalyst: an in situ, time-resolved XANES study, *Angew. Chem. Int. Ed.*, 42, 2341–2344.; (h) Newton, M. A., Burnaby, D. G., Dent, A. J., Diaz-Moreno, S., Evans, J., Fiddy, S. G., Neisius, T., Turin, S. (2002) Energy dispersive extended X-ray absorption fine structure, mass spectrometric, and diffuse reflectance infrared studies of the interaction of Al_2O_3 supported $Rh^I(CO)_2Cl$ species with NO and re-formation under CO, *J. Phys. Chem. B*, 106, 4214–4222.; (i) Fiddy, S. G., Newton, M. A., Corker, J. M., Evans, J. (2001) A comparative study of methodologies for the incorporation of Ge into H_1SiO_2-supported, $Pt(Acac)_2$-Derived, Pt-Ge catalysts and the effect of internal oxide structure on particle morphology and CO adsorption, *J. Phys. Chem. B*, 105, 5244–5252.; (j) Fiddy, S. G., Newton, M. A., Campbell, T., Dent, A. J., Harvey, I., Dalvini, G., Turin, S., Evans, J. (2002) Particle development and characterisation in $Pt(acac)_2$ and $Pt(acac)_2/GeBu_4$ derived catalysts supported upon porous and mesoporous SiO_2: effect of reductive environment, and support structure, *Phys. Chem. Chem. Phys.*, 4, 827–834.; (k) Uzun, A., Gates, B. C. (2008) Real-time characterization of formation and breakup of iridium clusters in highly dealuminated zeolite Y, *Angew. Chem. Int. Ed.*, 47, 9245–9248.; (l) Uzun, A., Gates, B. C. (2009) Dynamic structural changes in a molecular zeolite-supported iridium catalyst for ethene hydrogenation, *J. Am. Chem. Soc.*, 131, 15887–15894.; (m) Kulkarni, A., Cates, B. C. (2009) Spectroscopic elucidation of first steps of supported bimetallic cluster formation, *Angew. Chem. Int. Ed.*, 48, 9697–9700.; (n) Agostini, G., Pellegrini, R., Leofanti, G., Bertinetti, L., Bertarione, S., Eroppo, E., Zecchina, A., Lamberti, C. (2009) Determination of the particle size, available surface area, and nature of exposed sites

for Silica-Alumina-Supported Pd nanoparticles: a multitechnical approach, *J. Phys. Chem. C*, 113, 10485–10492.

[4] Couves, J. W., Thomas, J. M., Waller, D., Jones, R. H., Dent, A. J., Derbyshire, G. E., Greaves, G. N. (1991) Tracing the conversion of aurichalcite to a copper catalyst by combined X-ray absorption and diffraction, *Nature*, 354, 465–468.

[5] Sankar, G., Wright, P. A., Natajaran, S., Thomas, J. M., Greaves, G. N., Dent, A. J., Dobson, B. R., Ramsdale, C. A., Jones, R. H. (1993) Combined QuEXAFS-XRD, a new technique in high temperature materials chemistry- An illustrative in-situ study of zinc oxide enhanced solid state production of cordierite from a precursor zeolite, *J. Phys. Chem.*, 97, 9550–9554.

[6] Clausen, B. S., Topsoe, H., Frahm, R. (1998) Application of combined X-ray diffraction and absorption techniques for in situ catalyst characterization, *Adv. Catal.*, 42, 315–344.

[7] Bruckner, A. (2003) Looking on heterogeneous catalytic systems from different perspectives: multitechnique approaches as a new challenge for in situ studies, *Catal. Rev.-Sci. Eng.*, 45, 97–150.

[8] Weckhuysen, B. M. (2003) Determining the active site in a catalytic process: operando spectroscopy is more than a buzzword, *Phys. Chem. Chem. Phys.*, 5, 4351–4360.

[9] Beale, A. M., van der Eerden, A. M. J., Kervinen, K., Newton, M. A., Weckhuysen, B. M. (2005) Adding a third dimension to operando spectroscopy: a combined UV-Vis, Raman and XAFS setup to study heterogeneous catalysts under working conditions, *Chem. Commun. (Camb.)*, 3015–3017.

[10] Tinnemans, S. J., Mesu, J. G., Kervinen, K., Visser, T., Nijhuis, T. A., Beale, A. M., Keller, D. E., van der Eerden, A. M. J., Weckhuysen, B. M. (2006) Operando spectroscopic investigation of supported metal oxide catalysts by combined time-resolved UV-VIS/Raman/on-line mass spectrometry, *Catal. Today*, 113, 3–15.

[11] Newton, M. A., van Beek, W. (2010) Combining synchrotron-based X-ray techniques with vibrational spectroscopies for the in situ study of heterogeneous catalysts: a view from a bridge, *Chem. Soc. Rev.*, 39, 4845–4863.

[12] Bentrup, U. (2010) Combining in situ characterization methods in one set-up: looking with more eyes into the intricate chemistry of the synthesis and working of heterogeneous catalysts, *Chem. Soc. Rev.*, 39(12), 4718–4730.

[13] Korhonen, S. T., Beale, A. M., Newton, M. A., Weckhuysen, B. M. (2011) New insights into the active surface species of silver alumina catalysts in the selective catalytic reduction of NO, *J. Phys. Chem. C*, 115, 885–896.

[14] (a) Griffiths, P. R., de Haseth, J. A. (2007) *Fourier Transform Infrared Spectroscopy*, Wiley, Hoboken, NJ.; (b) Madejova, J. (2003) FTIR techniques in clay mineral studies, *Vib. Spectrosc.*, 31, 1–10.; (c) Ryczkowsky, J. (2001) Application of FTIR to catalytic systems, *Catal. Today*, 68, 263–345.

[15] Roedel, E., Urakawa, A., Kuretiw, S., Baiker, A. (2008) On the local sensitivity of different IR techniques: Ba species relevant in NOx storage-reduction, *Phys. Chem. Chem. Phys.*, 10, 6190–6198.

[16] Urakawa, A., Maeda, N., Baiker, A. (2008) Space- and time-Resolved combined DRIFT and Raman spectroscopy: monitoring dynamic surface and bulk processes during NO_x storage seduction, *Angew. Chem. Int. Ed.*, 47, 9256–9259.

[17] Urakawa, A., Baiker, A. (2009) Space-resolved profiling relevant in heterogeneous catalysis, *Top. Catal.*, 52, 1312–1322.

[18] Maeda, N., Urakawa, A., Baiker, A. (2009) Support effects and chemical gradients along the catalyst bed in NO_x storage-reduction studied by space and time-resolved in situ DRIFTS, *J. Phys. Chem. C*, 113, 16724–16735.

[19] Grunwaldt, J. D., Baiker, A. (2005) Axial variation of the oxidation state of Pt-Rh/Al_2O_3 during partial methane oxidation in a fixed-bed reactor: an in situ X-ray absorption spectroscopy study, *Catal. Lett.*, 99, 5–12.

[20] Newton, M. A., Dent, A. J., Diaz-Moreno, S., Fiddy, S. G., Jyoti, B., Evans, J. (2006) Rapid monitoring of the nature and interconversion of supported catalyst phases and of their Influence upon performance: CO oxidation to CO_2 by γ-Al_2O_3 Supported Rh catalysts, *Chem. Eur. J.*, 12, 1975–1985.

[21] Hannemann, S., Grunwaldt, J. D., van Vegten, N., Baiker, A., Boye, P., Schroer, C. G. (2007) Distinct spatial changes of the catalyst structure inside a fixed-bed microreactor during the partial oxidation of methane over Rh/Al_2O_3, *Catal. Today*, 126, 54–63.

[22] Grunwaldt, J. S., Kimmerle, B., Baiker, A., Boye, P., Schroer, C. G., Glatzel, P., Borca, C. N., Beckmann, F. (2009) Catalysts at work: from integral to spatially resolved X-ray absorption spectroscopy, *Catal. Today*, 145, 267–278.

[23] Bürgi, T., Baiker, A. (2006) Attenuated total reflection infrared spectroscopy of solid catalysts functioning in the presence of liquid-phase reactants, *Adv. Catal.*, 50, 227–283.

[24] Andanson, J.-M., Baiker, A. (2010) Exploring catalytic solid/liquid interfaces by in situ attenuated total reflection infrared spectroscopy, *Chem. Soc. Rev.*, 39, 4571–4584.

[25] Mojet, B. L., Ebbesenz, S. D., Lefferts, L. (2010) Light at the interface: the potential of attenuated total reflection infrared spectroscopy for understanding heterogeneous catalysis in water, *Chem. Soc. Rev.*, 39, 4643–4655.

[26] Mondelli, C., Dal Santo, V., Trovarelli, A., Boaro, M., Fusi, A., Psaro, R., Recchia, A. (2006) An operand DRIFTS-MS study on model $Ce_{0.5}Zr_{0.5}O_2$ redox catalyst: a critical evaluation of DRIFTS and MS data on CO abatement reaction, *Catal. Today*, 113, 81–86.

[27] (a) Mantz, A. W. (1976) Infrared multiplexed studies of transient species, *Appl. Spectrosc.*, 30, 459–461.; (b) Palmer, R. A., Chao, J. L., Dittmar, R. M., Gregoriou, V. G., Plunkett, S. E. (1993) Investigation of time dependent phenomena by use of step-scan FT-IR, *Appl. Spectrosc.*, 47, 1297–1310.

[28] Wille, A., Fridell, E. (2007) Millisecond step-scan FT-IR transmission spectroscopy under transient reaction conditions: CO oxidation over Pt/Al_2O_3, *Appl. Catal. B*, 70, 294–304.

[29] Thibault-Starzyk, F., Seguin, E., Thomas, S., Daturi, M., Arnolds, H., King, D. A. (2009) Real time infrared detection of cyanide flip on silver-alumina NO_x removal catalyst, *Science*, 324, 1048–1051.

[30] (a) Bressler, C., Chergui, M. (2004) Ultrafast X-ray absorption spectroscopy, *Chem. Rev.*, 104, 1781–1812.; (b) Chen, L. X. (2004) Taking snapshots of photoexcited molecules in disordered media by using pulsed synchrotron X-rays, *Angew. Chem. Int. Ed.*, 43, 2886–2905.; (c) Ihee, H., Wulff, M., Kim, J., Adachi, S. (2010)

Ultrafast X-ray scattering; structural dynamics from diatomic to protein molecules, *Int. Rev. Phys. Chem.*, 29, 453–520.

[31] Panine, P., Finet, S., Weiss, T. M., Narayanan, T. (2006) Probing fast kinetics in complex fluids by combined rapid mixing and small angle scattering, *Adv. Colloid Interface Sci.*, 127, 9–18.

[32] Curfs, C., Cano, I. G., Vaughan, G. B. M., Turillas, X., Kvick, A., Rodriguez, M. A. (2002) TiC-NiAl composites obtained by SHS; time resolved XRD study, *J. Eur. Ceram. Soc.*, 22, 1039–1044.

[33] Chupas, P. J., Chapman, K. W., Lee, P. L. (2007) Applications of an amorphous silicon-based area detector for high-resolution, high-sensitivity and fast time-resolved pair distribution function measurements, *J. Appl. Crystallogr.*, 40, 463–470.

[34] Young, N. A., Spicer, M. D. (1990) The 1st combined in situ FTIR and EXAFS study of a matrix-isolated molecule, *J. Mol. Struct.*, 222, 77–86.

[35] Young, N. A. (1996) Is nickel dibromide bent in a nitrogen matrix? A combined, simultaneous, Fourier-transform infrared and X-ray absorption fine structure matrix isolation study, *J. Chem. Soc.-Dalton Trans.*, 249–251.

[36] Wilkin, O. M., Young, N. A. (1999) The development of a combined simultaneous XAFS FTIR facility for the study of matrix species, *J. Synchrotron Radiat.*, 6, 204–206.

[37] Bras, W., Derbyshire, G. E., Bogg, D., Cooke, J., Elwell, M. J., Komanschek, B. U., Naylor, S., Ryan, A. J. (1995) Simultaneous studies of reaction kinetics and structure development in polymer processing, *Science*, 267, 996–999.

[38] Newton, M. A., Jyoti, B., Dent, A. J., Fiddy, S. G., Evans, J. (2004) Synchronous, time resolved, diffuse reflectance FT-IR, energy dispersive EXAFS (EDE) and mass spectrometric investigation of Rh catalysts during NO reduction by CO, *Chem. Commun. (Camb.)*, 2382–2383.

[39] Mondelli, C., Ferri, D., Grunwaldt, J. D., Krumeich, F., Mangold, S., Psaro, R., Baiker, A. (2007) Combined liquid-phase ATR-IR and XAS study of the Bi-promotion in the aerobic oxidation of benzyl alcohol over Pd/Al_2O_3, *J. Catal.*, 252, 77–87.

[40] Grunwaldt, J. D., Baiker, A. (2005) In situ spectroscopic investigation of heterogeneous catalysts and reaction media at high pressure, *Phys . Chem. Chem. Phys*, 7, 3526–3539.

[41] Bentrup, U., Radnik, J., Armbruster, A. M., Leiterer, J., Emmerling, F., Brückner, A. (2009) Linking simultaneous in situ WAXS/SAXS/Raman with Raman/ATR/UV-Vis spectroscopy: comprehensive insight into the synthesis of molybdate catalyst precursors, *Topics in Catal.*, 52, 1350–1359.

[42] Innocenzi, P., Malfatti, L., Kidchob, T., Costacurta, S., Falcaro, P., Piccinini, M., Marchelli, A., Morini, P., Sali, D., Amenitsch, H. (2007) Time-Resolved Simultaneous detection of structural and chemical changes during self-assembly of mesostructured films, *J. Phys. Chem. C*, 111, 5345–5350.

[43] (a) Marinkovic, N. S., Wang, Q., Frenkel, A. I. (2011) In situ diffuse reflectance IR spectroscopy and X-ray absorption spectroscopy for fast catalytic processes, *J. Synchrotron Radiat.*, 18, 447–455.; (b) Marinkovic, N. S., Wang, Q., Barrio, L., Ehrlich, S. N., Khalid, S., Cooper, C., Frenkel, A. I. (2011) Combined in Situ X-ray

absorption and diffuse reflectance infrared spectroscopy: an attractive tool for catalytic investigations, *Nucl. Instrum. Methods Phys. Res. A*, 649, 204–206.

[44] (a) Bando, K. K., Wada, T., Miymoto, T., Miyazaki, K., Takakusgi, S., Gott, T., Yamaguchi, A., Nomura, M., Oyama, S. T., Asakura, K. (2009) Combined in situ analysis of Ni2P/MCM-41 under hydrodesulfurization conditions—simultaneous observation of QXAFS and FTIR., 14th International Conference on X-Ray Absorption Fine Structure (XAFS14), *J. Phys. Conf. Ser.*, 190, 012158.; (b) Bando, K. K., Wada, T., Miymoto, T., Miyazaki, K., Takakusgi, S., Koike, Y., Inada, Y., Nomura, M., Yamaguchi, A., Oyama, S. T., Asakura, A. (2012) Combined in situ QXAFS and FTIR analysis of a Ni phosphide catalyst under hydrodesulfurization conditions, *J. Catal.*, 286, 165–171.

[45] Newton, M. A., Dent, A. J., Fiddy, S. G., Jyoti, B., Evans, J. (2007) Combining diffuse reflectance infrared spectroscopy (DRIFTS), dispersive EXAFS, and mass spectrometry with high time resolution: potential, limitations and application to the study of NO interaction with supported Rh catalysts, *Catal. Today*, 126, 64–72.

[46] Cavers, M., Davidson, J. M., Harkness, I. R., Rees, L. V. C., McDougall, G. S. (1999) Spectroscopic identification of the active site for CO oxidation on Rh/Al_2O_3 by concentration modulation in situ DRIFTS, *J. Catal*, 188, 426–430.

[47] Cavers, M., Davidson, J. M., Harkness, I. R., McDougall, G. S., Rees, L. V. C. (1999) Novel frequency response techniques for the study of kinetics in heterogeneous catalysis, *Stud. Surf. Sci. Catal.*, 122, 65–72.

[48] Newton, M. A. (2009) Synchronous diffuse reflectance infrared absorption spectroscopy (DRIFTS)/EXAFS applied to the structural reactive behaviour of dilute supported Rh/Al_2O_3 catalysts, *Top. Catal.*, 52, 1410–1424.

[49] Newton, M. A., Di Michiel, M., Kubacka, A., Fernández-García, M. (2010) Combining time resolved hard X-ray diffraction (XRD) and diffuse reflectance infrared spectroscopy (DRIFTS) to illuminate CO dissociation and transient carbon storage by supported Pd nanoparticles during CO/NO cycling, *J. Am. Chem. Soc.*, 132, 4540–4541.

[50] Newton, M. A. (2007) Beamsize related phenomena and effective normalisation in energy dispersive EXAFS for the study of heterogeneous catalysts, powder materials, and the processes they mediate: observations, and (some) solutions, *J. Synchrotron Radiat.*, 14, 372–381.

[51] Labiche, J.-C., Mathon, O., Pascarelli, S., Newton, M. A., Ferre, G. G., Curfs, C., Vaughan, G., Homs, A., Carrieras, D. F. (2007) The fast readout low noise camera as a versatile x-ray detector for time resolved dispersive extended x-ray absorption fine structure and diffraction studies of dynamic problems in materials science, chemistry, and catalysis, *Rev. Sci. Instrum.*, 78, 091301.

[52] Newton, M. A., Dent, A. J., Fiddy, S. G., Jyoti, B., Evans, J. (2007) Identification of the surface species responsible for N_2O formation from the chemisorption of NO on Rh/alumina, *Phys. Chem. Chem. Phys.*, 9, 246–249.

[53] Newton, M. A., Dent, A. J., Fiddy, S. G., Jyoti, B., Evans, J. (2007) Particle size effects in Rh/Al_2O_3 catalysts as viewed from a structural, functional, and reactive perspective; the case of the reactive adsorption of NO, *J. Mater. Sci.*, 42, 3288–3298.

[54] Dent, A. J., Evans, J., Fiddy, S. G., Jyoti, B., Newton, M. A., Tromp, M. (2007) Rhodium dispersion during CO/NO conversion, *Angew. Chem. Int. Ed.*, 46, 5356–5358.

[55] Dent, A. J., Evans, J., Fiddy, S. G., Jyoti, B., Newton, M. A., Tromp, M. (2008) Structure–performance relationships of Rh and RhPd alloy supported catalysts using combined EDE/DRIFTS/MS, *Faraday Discuss.*, 138, 287–300.

[56] Newton, M. A., Belver-Coldeira, C., Martinez-Arias, A., Fernández-García, M. (2007) Dynamic *in situ* observation of rapid size and shape change of supported Pd nanoparticles during CO/NO cycling, *Nat. Mater.*, 6, 528–532.

[57] Newton, M. A., Belver-Coldeira, C., Martinez-Arias, A., Fernández-García, M. (2007) "Oxidationless" promotion of rapid Pd redispersion by oxygen during redox CO/(NO + O_2) cycling, *Angew Chem. Int. Ed.*, 46, 8629–8631.

[58] Kubacka, A., Fernandez-Garcia, M., Martinez-Arias, A., Newton, M. A. (2009) Dynamic operando observation of dilute, 1wt% palladium TWC's: simultaneous XAFS/DRIFTS/Mass spectrometry analysis of the effects of CeZr on structure reactivity and performance, *Catal. Today*, 145, 288–293.

[59] Kubacka, A., Martínez-Arias, A., Fernández-García, M., Di Michiel, M. (2010) Multitechnique analysis of supported Pd particles upon dynamic, cycling CO/NO conditions: size-dependence of the structure-activity relationship, *J. Catal.*, 270, 275–284.

[60] Iglesias-Juez, A., Kubacka, A., Fernandez-Garcia, M., Di Michiel, M., Newton, M. A. (2011) Nanoparticulate Pd supported catalysts: size dependent formation of Pd(I)/Pd(0) and their role in CO elimination, *J. Am. Chem. Soc.*, 133, 4484–4489.

[61] Yevick, A., Frenkel, A. I. (2010) Effects of surface disorder on EXAFS modelling of metallic clusters, *Phys. Rev. B*, 81, 115451.

[62] Daniels, J. E., Drakopoulos, M. (2009) High-energy X-ray diffraction using the Pixium 4700 flat-panel detector, *J. Synchrotron Radiat.*, 16, 463–468.

[63] Newton, M. A., Di Michiel, M., Kubacka, A., Iglesias-Juez, A., Fernández-García, M. (2012) Observing oxygen storage and release at work during cycling redox conditions: synergies between noble metal and oxide promoter, *Angew. Chem. Int. Ed.*, 51, 2363–2367.

[64] (a) McCaulley, J. A. (1993) In-situ x-ray absorption spectroscopy studies of hydride and carbide formation in supported palladium catalysts, *J. Phys. Chem.*, 97, 10372–10379.; (b) McCaulley, J. A. (1993) Temperature dependence of the Pd K-edge extended x-ray absorption fine structure of PdC_x (x~0.13), *Phys. Rev. B*, 47, 4873–4879.

[65] Becker, E., Carlsson, P.-A., Kylhammar, L., Newton, M. A., Skoglundh, M. (2011) In situ spectroscopic investigation of low-temperature oxidation of methane over alumina-supported platinum during periodic operation, *J. Phys. Chem. C*, 115, 944–951.

[66] Grunwaldt, J. D., Caravati, M., Hannemenn, S., Baiker, A. (2004) X-ray absorption spectroscopy under reaction conditions: suitability of different reaction cells for combined catalyst characterization and time-resolved studies, *Phys . Chem. Chem. Phys.*, 6, 3037–3047.

[67] Lee, A. F., Ellis, C. V., Naughton, J. N., Newton, M. A., Parlett, C. M., Wilson, K. (2011) Reaction-driven catalyst restructuring and selectivity control in crotyl alcohol aerobic oxidation over palladium, *J. Am. Chem. Soc.*, 133, 5724–5727.

[68] Mondelli, C., Grunwaldt, J. D., Ferri, D., Baiker, A. (2010) Role of Bi promotion and solvent in platinum-catalysed alcohol oxidation probed by in situ X-ray absorption and ATR-IR spectroscopy, *Phys. Chem. Chem. Phys.*, 12, 5307–5316.

[69] (a) Colarusso, P., Kidder, L. H., Levin, I. W., Fraser, J. C., Arens, J. F., Lewis, E. N. (1998) Infrared spectroscopic imaging; from planetary to cellular systems, *Appl. Spectrosc.*, 52, 106A–120A.; (b) Kazarian, S. G., Chen, K. L. A. (2010) Micro- and macro attenuated total reflection Fourier transform infrared spectroscopic imaging, *Appl. Spectrosc.*, 64, 135A–152A.

[70] Newton, M. A., Van Nijhatten, P., Kansiz, M. Manuscript to be submitted.

[71] Petkov, V. (2008) Nanostructure by high-energy X-ray diffraction, *Mater. Today*, 11, 28–38.

[72] Chupas, P. J., Chapman, K. W., Chen, H. L., Grey, C. P. (2009) Application of high-energy X-rays and Pair-Distribution-Function analysis to nano-scale structural studies in catalysis, *Catal. Today*, 145, 213–219.

[73] Fernandez Garcia, M., Belver, C., Hanson, J. C., Wang, X. C., Rodríguez, J. A. (2007) Anatase-TiO_2 nanomaterials: analysis of key parameters controlling crystallization, *J. Am. Chem. Soc.*, 129, 13604–13612.

[74] Beyer, K., Chapman, K. W., Chupas, P. J., Newton, M. A., Zhao, H. Manuscript in preparation.

[75] Hoffmann, F. M. (1983) Infrared reflection-absorption spectroscopy of adsorbed molecules, *Surf. Sci. Rep.*, 3, 107–192.

[76] Chabal, Y. J. (1988) Surface infrared spectroscopy, *Surf. Sci. Rep.*, 8, 211–357.

[77] (a) Xu, X., Vesecky, S. M., Goodman, D. W. (1992) Infrared reflection –absorption spectroscopy at STM studies of model supported Copper catalysts, *Science*, 258, 788–790.; (b) Ozensoy, E., Goodman, D. W. (2004) Vibrational spectroscopic studies on CO adsorption, NO adsorption, CO + NO reaction on Pd model catalysts, *Phys. Chem. Chem. Phys.*, 6, 3765–3778.; (c) Brandt, B., Fischer, J. H., Ludwig, W., Libuda, J., Zaera, F., Schauermann, S., Freund, H. J. (2008) Isomerization and hydrogenation of cis-2-butene on a Pd model catalyst, *J. Phys. Chem. C*, 112, 11408–11420.

[78] (a) Evans, J., Hayden, B. E., Mosselmans, F., Murray, A. (1992) Rhodium germinal dicarbonyl on TiO_2(110), *J. Am. Chem. Soc.*, 114, 6912–6913.; (b) Evans, J., Hayden, B. E., Lu, G. (1996) The adsorption of carbon monoxide on TiO2(110) supported palladium, *Surf. Sci.*, 360, 61–73.; (c) Hayden, B. E., King, A., Newton, M. A. (1997) The alignment of a surface species determined by FT-RAIRS: rhodium gem-dicarbonyl on TiO_2(110), *Chem. Phys. Lett.*, 269, 485–488.

[79] (a) Ackermann, M. D., Pedersen, T. M., Hendriksen, B. L. M, Robach, O., Bobaru, S. C, Quiros, C., Kim, H., Hammer, B., Ferrer, S., Frenken, J. M. W. (2005) Structure and reactivity of surface oxides on Pt(110) during catalytic CO oxidation, *Phys. Rev. Lett.*, 95, 255505.; (b) Ferrer, S., Ackermann, M. D., Lundgren, E. (2007) In situ investigations of chemical reactions on surfaces by X-ray diffraction, *MRS Bull.*, 32, 1010–1014.; (c) van Rijn, R., Ackermann, M. D., Balmes, O., Dufrane, T., Geluk, A., Gonzalez, H., Isern, H., de Kuyper, E., Petit, L., Sole, V. A., Wermeille, D., Felici, R., Frenken, J. M. W. (2010) Ultrahigh vacuum/high pressure flow reactor for surface X-ray diffraction and grazing incidence small angle scattering studies close to conditions for industrial catalysis, *Rev. Sci. Instrum.*, 81, 014101.; (d) Hendriksen, B. L. M., Ackermann, M. D., van Rijn, R., Stoltz, D., Popa, I., Balmes, O., Resta, A., Wermeille, D., Felici, R., Ferrer, S., Frenken, J. M. W. (2010) The roles of steps in surface catalysis and reaction oscillations, *Nat. Chem.*, 2, 730–734.

[80] (a) Saint-Lager, M. C., Bailly, A., Dolle, P., Baudoing-Savois, R., Taunier, P., Garaudee, S., Cucuaro, S., Douillet, S., Geaymond, O., Perroux, G., Tissot, O., Micha, J. S., Ultrich, O., Rietord, F. (2007) New reactor dedicated to in operando studies of catalysts by means of surface X-ray diffraction and grazing incidence small angle scattering, *Rev. Sci. Instrum.*, 78, 083902.; (b) Laoufi, I., Saint-Lager, M. C., Lazzari, R., Jupille, J., Robach, O., Garaudee, S., Cabailh, G., Dolle, P., Cruuel, H., Bailly, A. (2011) Size and catalytic activity of supported gold nanoparticles: an in operando study during CO oxidation, *J. Phys. Chem. C*, 115, 4673–4679.

[81] Baurecht, D., Fringelli, U. P. (2001) Quantitative modulated excitation Fourier transform infrared spectroscopy, *Rev. Sci. Instrum.*, 72, 3782–3792.

[82] Urakawa, A., Van Beek, W., Monrabal-Capilla, M., Galan-Mascaros, J. R., Palin, L., Milanesio, M. (2011) Combined modulation enhanced X-ray powder diffraction and Raman spectroscopic study of structural transitions in the spin crossover materials $[(Fe(Htrz)_2(trz)]BF_4)$, *J. Phys. Chem. C*, 115, 1323–1329.

[83] Ferri, D., Kumar, M. S., Eyssler, A., Korsak, O., Wirz, R., Hug, P., Weidenkaff, A., Newton, M. A. (2010) First steps in combining modulation excitation spectroscopy with synchronous dispersive EXAFS/DRIFTS/mass spectrometry for *in situ* time resolved study of heterogeneous catalysts, *Phys. Chem. Chem. Phys.*, 12, 5634–5646.

[84] Ferri, D., Newton, M. A., Nachtagaal, M. (2011) Modulation excitation X-ray absorption spectroscopy of heterogeneous catalysis, *Top. Catal.*, 54, 1070–1078.

[85] Chernyshov, D., Wouter van Beek, W., Emerich, H., Milanesio, M., Urakawa, A., Viterbo, D., Palin, L., Caliandro, R. (2011) Kinematic diffraction on a structure with periodically varying scattering function, *Acta Crystallogr. A*, 67, 327–335.

[86] Vanko, G., Glatzel, P., Pham, V. T., Abela, R., Grolimund, D., Borca, C. N., Johnson, S. L., Milne, C. J., Bressler, C. (2010) Picosecond time resolved X-ray emission spectroscopy: ultrafast spin-state determination in an iron complex, *Angew. Chem. Int. Ed.*, 49, 5910–5912.

15

XRD–RAMAN AND MODULATION EXCITATION SPECTROSCOPY

WOUTER VAN BEEK, ATSUSHI URAKAWA, AND MARCO MILANESIO

15.1 INTRODUCTION

Chemical and physical processes of both technological and fundamental importance are often very complex. Better understanding of these processes, necessary for their optimization, often come through the use of *in-situ* experimental techniques. This is particularly true for heterogeneous catalytic processes where, on the one hand, firm understanding of the basic transformation or transformation pathways of interest promoted by the catalyst is critical [1]. The fundamental catalytic processes intrinsically take place at the atomic level and proceed extremely fast (from pico- to nanosecond scale). On the other hand, kinetic effects, fundamental for life cycle considerations, are much slower (from millisecond, minutes, to hours) and they play crucial roles in the real world and govern the successful commercialization of a catalyst. In general, *in-situ* techniques now routinely probe the minutes-to-hours timescale while the investigation of the milliseconds-to-seconds timescale is on the way to being routine [2]. When *in-situ* techniques are applied to catalysts, an additional difficulty arises due to the dilution of the active catalytic centers that are often supported or embedded on or in an inert matrix. As a logical consequence, the actual active constituent represents only a minute fraction of the total catalytic body. The need for understanding and controlling the catalysts function, its overall performance, and possibly regeneration are the main driving forces for

In-situ *Characterization of Heterogeneous Catalysts*, First Edition.
Edited by José A. Rodriguez, Jonathan C. Hanson, and Peter J. Chupas.

new experimental techniques. It is a big challenge to develop techniques that are capable of addressing both the fundamental and practical aspects of catalyst design. These techniques need to span over many orders of the time and length scales with sufficient sensitivity to spot the active centers and how these are influenced or not by the surrounding structure [2]. As indicated by the topics of the last four chapters in this book, there is a clear tendency toward combining several analytical tools in a complementary fashion with the ultimate aim of reaching the dreamed rational design of catalysts. The most prominent advantage of such multitechnique approaches obviously lies in the enhanced, correlated information of dynamic chemical and physical processes when (quasi-)simultaneous, synchronized data acquisition is ensured. As a result, one obtains information about the material structure which goes beyond the sum obtained by individual experimental methods. In this chapter we will discuss first complementarities between Raman and X-ray diffraction (XRD), and also including X-ray absorption spectroscopy (XAS). This unique combination has proven its great value in identifying the active sites and species and in understanding factors that influence the long-term performance of catalysts, and several examples will be reviewed. The combination of these techniques can be exploited with an "*in-situ* but separated multitechnique approach" or with an "*in-situ* and simultaneous multitechnique approach." Advantages and drawbacks of the two approaches are discussed through several examples. Second, we will give an introduction to the relatively old concept of *modulation enhancement* that is only recently applied in catalysis. Modulation enhancement can be applied to virtually any analytical technique and boosts enormously the sensitivity by introducing *selectivity* in measurements, and as a consequence the time resolution is also significantly improved. This technique has already demonstrated its great strength unraveling the rate-limiting steps and transformation pathways. Several other examples of modulation enhancement will be reviewed as well as its recent adaptation to diffraction. In the context of catalytic studies, the more specific term of modulation excitation spectroscopy is generally used instead of the more generic term of modulation enhancement. Both terms cover the idea of increasing the sensitivity and introducing selectivity into the measurements. We also propose a potentially very powerful experimental paradigm, currently under development, that combines the two approaches, multiple techniques, and modulation. Finally, we would like to point out that no matter how powerful a technique will be, it will only provide answers on *how* reactions take place. *Why* they come about remains the exclusive property of computational chemistry and human interpretation based on our current knowledge of chemical transformations.

15.2 INSTRUMENTATION

The application of *ex-situ* Raman scattering applied in catalysis has started to grow from the mid-1970s [3], while *ex-situ* diffraction was already present in

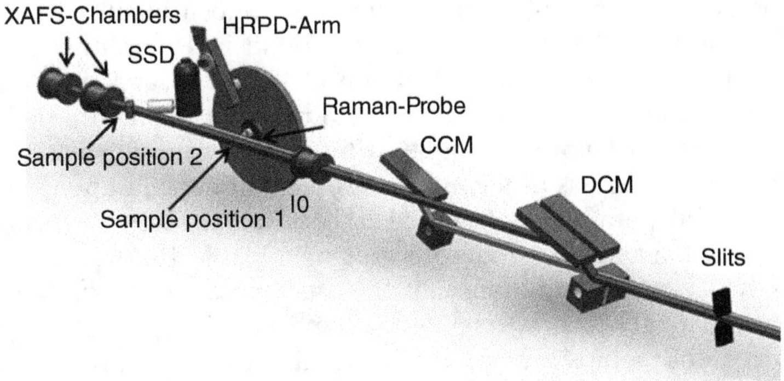

Figure 15.1. SNBL [13] at ESRF for combined XRD–Raman–XAFS.

the 1940s [4]. In the last two decades, continuous technological developments made the individual analytical tools and sampling environments sufficiently advanced to allow fast and sensitive *in-situ* measurements. A logical next step is to combine several techniques *in situ* ideally *without* compromising on data quality. Much of the pioneering developments were made at synchrotrons but many variations on combined setups also exist at home labs [5]. Simultaneous experiments involving Raman scattering and synchrotron-based X-ray techniques were first attempted on polymers in the mid-1990s [6] and from 2003 onward, a relatively large number of papers appeared across different disciplines, reporting simultaneous multitechnique experiments [7]. Several highly specialized beamlines (for microdiffraction, high pressure research, and protein crystallography) have been equipped with online Raman spectrometers. During the last decade, the combination of *in-situ* XRD–Raman for catalysis research was developed by several groups [8–10]. One of these developments [11] resulted into a user facility that provides combined XRD/Raman, and X-ray absorption fine structure (XAFS) as a matter of fact, including all dedicated gas handling and sample environments for catalysis studies (see Fig. 15.1). This beamline started with combining XRD and XAFS before adding Raman scattering. The combination of XRD and XAFS was first reported in 1991 [12]. Despite a general consensus about the usefulness of this combination, it has taken a long time before its full potential was exploited. Especially the quality of the diffraction data has often been severely compromised since the requirements for X-ray optics are very different. In Figure 15.1, one of the solutions to combine high-quality XRD with high-quality XAFS is shown [11].

This beamline is equipped with dedicated monochromators for each technique. A channel-cut (CC) monochromator for diffraction and a double-crystal monochromator (DCM) for XAFS. The monochromators can be automatically and rapidly (within seconds) exchanged, allowing an *optimal* and *stable* wavelength selection for diffraction to be combined with any XAFS edge, avoiding large and slow displacements of the monochromator Bragg angle. More information about this setup can be found in References 11 and 14, examples of its use can be found in References 15 and 16, and more about this combination can be found in Chapter 13. In practice, it is hence not an easy task to combine two or more techniques without compromising on data quality, and ultimately the success of a combined setup will depend upon the careful optimization of all parameters. As noted earlier, Raman scattering is also available on this station and here trade-offs are unavoidable. Before describing the trade-offs, it is worth mentioning that the final success of a combined setup also depends strongly on how reliable and especially how *easy* it is to use and maintain. Portable setups are good to test and learn what is needed but it is hard to exploit the full potential based on temporary installations. Most of the Raman spectrometers on beamlines are hence permanently installed on an optical bench in a dedicated room. The excitation laser and collected Raman signals are transported through fiber optic cables to the experimental stations and returned. Measurement capabilities for offline testing are also often available.

15.2.1 Raman Laser Selection Criteria

Commercial Raman spectrometers are available with excitation wavelengths spanning from the deep ultraviolet (UV) to the near infrared (NIR). NIR and UV lasers are less affected by fluorescence problems. However, NIR systems are strongly limited in *in-situ* applications since black-body radiation impedes its use above 350–400 K, while typical catalyst preparation and *operando* conditions require higher temperatures. Moreover, passing from NIR, from red to green lasers, the penetration depth reduces due to the optical absorption characteristics of many samples. The term surface sensitivity in Raman is deliberately avoided here since a minimum of several hundreds of atomic layers are probed even with a short-wavelength UV laser on a sample strongly absorbing UV radiation [17]. Nevertheless, the difference in probed depth can be exploited and an example of complementarities between the green and red laser is described in a paper not related to catalysis, reporting the kinetic study of a solid-state reaction between fluorene and tetracyanoquinodimethane (TCNQ) [18]. Green lasers are more frequently used in *in-situ* applications and especially in catalysis. Indeed, UV lasers would expand the operation range because of reduced fluorescence, measurement at higher temperatures, and better sensitivity, but come with the disadvantage of being more destructive. In practice, there is no optimal excitation wavelength for all systems, and ideally one would like to have a tunable or multilaser system. Ongoing technological developments (e.g., optical fibers for the UV range) make the implementation of UV spectrometers on beamlines realistic in the near future. Advantages and perspectives of UV Raman in *in-situ* study of catalyst materials are discussed in Section 15.3.4, "Remarks on XRD–Raman in the Present and in the Future."

15.2.2 Raman Remote Probes

A typical Raman objective has a working distance smaller than 2 cm to maximize numerical aperture and the collection of Raman scattered signal. This has major drawbacks in the context of *in-situ* work, the first being the limited freedom in setting up environmental chambers (that are often rather bulky), and the second being the possibility of thermal damage of the probe (objective) due to reactor heating. Extremely long ~10 cm working distance probes have been developed to overcome these two problems, with the side effect of reducing signal-to-noise ratio (S/N) on the Raman side. However, these long focal distances come with the advantage of no extra compromise on the XRD or XAFS side. One can achieve better averaging of signals from a wider spatial range of the sample and the instantaneous laser power is also dramatically reduced. As a consequence, better S/N can be attained since the laser power can be increased without damaging the sample. Some of these remote probes [11] are equipped with a video camera for easy visual alignment and, more importantly, with beam scanners. Raman beam scanners allow better matching of the typical millimeter-sized X-ray beams and micrometer-sized Raman laser spot. Finally, beam scanners introduce spatial resolution as often needed to monitor the reaction front moving from the inlet to the outlet of the reaction cell. See also Chapter 2.

15.2.3 Sample Holder

In 1991, Clausen et al. [19] introduced a plugged flow reactor for *in-situ* XRD. This basic design has been copied, modified, and improved numerous times [11] and it has played a central role in *in-situ* XRD/XAFS and Raman experiments. The sample holder (reactor) itself is a simple glass or quartz capillary connected to gas inlet and outlet lines. Attention must be given to avoid certain glasses that are fluorescent and disturb Raman measurements. Gas atmosphere can be controlled with a mixture of gases down to very low pressure and up to moderate pressure (7–20 bar depending on capillary materials and wall thickness). For higher pressures and/or glass-aggressive solutions, common in catalyst preparation, special chambers with zirconia or mica windows must be employed. *Operando* experiments were performed using the general capillary setup except that reported by O'Brien et al. [20], employing a large reaction chamber to improve fluid dynamics and mimicking laboratory-scale reactors.

The combined XAFS–Raman setup was also developed [21, 22] and also resulted in a user facility. Despite the fact that the performed studies on existing combined instrumentation represent still a rather modest amount in the ocean of catalysis research, it has been employed to all typical catalysis characterization steps: ranging from high-throughput screening to preparation, activation, understanding its activity, and deactivation.

15.3 XRD–RAMAN APPLICATIONS

The whole life cycle of catalysts, from its conception to recycling, is massively studied nowadays as illustrated by the existence of about 20 journals only dedicated to the subject of catalysis. Two typical research themes, catalyst preparation and *operando* experiments, can be easily found in the XRD–Raman work previously carried out.

Preparation and activation of a catalyst are key processes in defining its performance. Catalytic performance, that is, activity, selectivity, and stability, is governed at first by the chemical composition of the employed materials (i.e., the intrinsic activity) and, more importantly, by the details of the preparation and activation procedures, that is, the extrinsic activity acquired from the particular treatments undergone by the catalyst. The treatment history strongly affects crystal size, morphology, and defectivity of catalytic materials, and thus catalytic performance and lifetime. These characteristics are decisive in the performance and can be controlled and/or induced during the preparation and activation phase that usually includes the synthesis of a precursor and then its dispersion onto a support material followed by its activation. A typical example is represented by microporous materials that are in the first step "synthesized" in the presence of organic template molecules that organize and direct the formation of the specific inorganic framework. The "activation" is done in a second step by eliminating the organic component by a well defined procedure (usually thermal treatment in controlled gas atmosphere). Probing the evolution of sample defectivity, morphology, and crystallinity during these two steps allows controlling these elements and more efficient optimization of catalyst synthesis with better catalytic performance and longer catalyst lifetime. Moreover, catalytic activity is strongly affected by the reaction condition and its long-term stability, that is, deactivation and aging. *Operando* experiments simulate actual working conditions, and the interrelations between catalyst characteristics, reaction mixtures, and catalytic performance can be monitored simultaneously.

Regular *in-situ* experiments allow monitoring some of the properties, while combined multitechnique simultaneous experiments tend to the exhaustive characterization of the transformations. Simultaneous multitechnique experiments permit reliable synchronization of the instrumentations with the reaction coordinates, and correlations among the different probes that are hard or impossible to obtain with *in-situ* but separated combined experiments. As a general remark, besides the more efficient use of time, combination of multiple techniques delivers information that is well beyond the linear sum of all techniques. In particular, simultaneous XRD–Raman experiments allow probing structural evolution, ordered-phase composition/evolutions, amorphization processes, and crystal size evolution by XRD and the evolution of chemical bonds, molecular properties, disordered phases, and defectivity by Raman. Concerning the technical difficulty, one might suppose that the "*in-situ* simultaneous XRD–Raman" is far more difficult compared with "*in-situ* but separated XRD–Raman." Naturally, installation of a simultaneous XRD–Raman setup is more demanding than installing only Raman and XRD setups separately in a laboratory, but after this step, performing the "*in-situ* simultaneous XRD–Raman" is less time consuming and yielding even richer information of transformation processes. The step of aligning the different probes is straightforward. These aspects are confirmed by counting the literature related to "*in situ*," "Raman," "diffraction," and "catalysis." Among the found 60 papers, more than half employed an additional technique *in situ*, and about 20 used only one technique, XRD or Raman, *in situ*. Seven papers, that is, more than 10%, applied "*in-situ* simultaneous XRD–Raman" that are the main topic of Section 15.3.1, Section 15.3.2.1, and Section 15.3.2.2, and surprisingly only four are related to "*in-situ* but separated XRD–Raman." It is worth noting that the search was not exhaustive and the numbers are likely underestimated (because of other possible synonyms); however, this brief survey clearly indicates that presently "*in-situ* simultaneous XRD–Raman" is more widespread than "*in-situ* but separated XRD–Raman."

15.3.1 *In-Situ* but Separated XRD–Raman

First, four "*in-situ* but separated XRD–Raman" experiments (in the period of 2004–2009) are briefly summarized to describe the background where "*in-situ* simultaneous XRD–Raman" was developed and to show that combined XRD–Raman laboratory data can give interesting results, in general, on case studies related to simple crystal structure, because of the limited resolution of laboratory diffractometers [23–26]. Finally, a paper by Grandjean et al. [27] combining four separated but "identical" experiments of small-angle X-ray scattering (SAXS)/XRD, XAFS, ultraviolet–visible spectroscopy (UV-Vis) and Raman using synchrotron radiation is described since it connects separated laboratory experiments with synchrotron-based simultaneous experimental development. It is worth noting that these *in-situ* but separated XRD–Raman experiments deal with both preparation and *operando* steps of the investigated catalysts. Studied materials are in general simple oxides and kinetics in the minutes-to-hours timescale.

Mamede et al. [24] studied the thermal activation process of two series of WO_x–CeO_2 catalysts with a W content between 4 and 12 wt% prepared by coprecipitation (PP) and impregnation (IMP). XRD results pointed to the preferential formation of crystalline $Ce_2(WO_4)_3$, which was completed at 1173 K. Crystalline WO_3 was absent in both series of IMP and PP catalysts, while Raman spectra of PP catalysts suggested the formation of amorphous WO_3, not detectable by XRD. Kim et al. [23] studied hydrogen sorption process under *operando* condition using a crystalline titanium nitride (TiN) nanopowder prepared by a mechanochemical reaction by means of high-energy ball milling, supported onto sodium alanate ($NaAlH_4$). During hydrogen sorption cycles, TiN-catalyzed $NaAlH_4$ exhibited a greater hydrogen desorption rate and higher hydrogen capacity than $TiCl_3$-catalyzed $NaAlH_4$. Contradicting thermodynamic predictions, *in-situ* XRD and Raman results revealed that TiN nanopowder remained stable and no by-products (e.g., Ti–Al compounds) were formed in the reaction with $NaAlH_4$ during hydrogen desorption. Conte et al. [25] employed XRD and Raman scattering to investigate the complex polymorph landscape of vanadium phosphorus oxide (VPO) catalysts during its preparation and under *operando* conditions. XRD was used to identify the transformation from $VOHPO_4 \cdot 0.5H_2O$ to $(VO)_2P_2O_7$ and finally to ω-$VOPO_4$ that was supposed to be the active species. ω-$VOPO_4$ stability in reaction condition was studied, because it is a metastable phase, very sensitive to many of the reactants and products. XRD was used to characterize the phase formed during the reactions, and Raman was used to follow the oxidation states and local structure of vanadium. Friedrich et al. [26] studied a number of hydrotalcite-like compounds containing Mg and V, prepared with different Mg:V ratios ranging from 1:1 to 4:1. The effects of variation in preparation procedures, such as preparing the compounds under inert gas, hydrothermal treatments at different temperatures, and different precipitating agents, were examined by different techniques including *in-situ* Raman and *in-situ* XRD. XRD was used to characterize the formation of new phases during reaction condition as a result of catalyst decomposition. Raman was used to follow the growth of pyro- and orthovanadate by identifying the local catalyst structure. In these four studies, XRD and Raman were mainly employed to qualitatively recognize the formed species at long and short range, respectively.

Grandjean et al. [27] studied the crystallization mechanism of CoAPO-5 molecular sieves under hydrothermal conditions. This study, although not simultaneous, can be recognized as a starting point of the *in-situ* and simultaneous setup, since it highlighted the difficulty in aligning separated experiments, further supporting the need of the simultaneous multiprobe setup. The major progression with respect to previous four studies are: the investigation of a complex system (a zeolite that is structurally and chemically more complex than oxides and nitrides studied in previous four case studies), the improvement of time resolution, the use of XRD to obtain also kinetic information and crystallite size analysis, and the combination with other synchrotron techniques (SAXS and XAS). Drawbacks are the time-consuming approach of carrying out four different separated experiments, the effort needed to reproduce the same reaction with similar sampling rate on four different setups, and the difficulty in the time synchronization and reproducibility of the different experiments.

15.3.2 *In-situ* Simultaneous XRD–Raman

Examples of *in-situ* simultaneous XRD–Raman experiments are summarized to highlight the results that can be obtained and the complementarities between the probes.

15.3.2.1 Catalyst Synthesis Experiments Kongmark et al. [28, 29] combined XRD, XAFS, and Raman in one experiment to study the genesis of a

Bi_2MoO_6 catalyst in hydrothermal conditions. The environment of Mo species changed from isolated tetrahedral MoO_4^{2-} in the initial gel to distorted octahedral in γ-Bi_2MoO_6 (Raman, X-ray absorption near-edge structure [XANES]). During the initial stages, spherical $(Bi,Mo)_2O_3$ particles formed a fluorite structure and the growth of anisotropic platelets of γ-Bi_2MoO_6 followed in the second step (XRD). *In-situ* experiments turned out to be crucial since this intermediate compound had not been elucidated in *ex-situ* studies [30] of the crystallization mechanism performed by XRD, transmission electron microscopy (TEM), and Raman spectroscopy. An ad hoc environmental chamber especially dedicated to *in situ* hydrothermal studies was used.

Bentrup et al. [9] studied the preparation of molybdate catalysts. This paper reported on the combined use of SAXS/XRD with Raman scattering to study the genesis of the catalyst from slurry. XRD data were used for phase recognition and for crystallite size analysis. Raman data were exploited to investigate the local structure of molybdate ion but the originality of this study was using Raman spectroscopy as an excellent tool to identify different molybdate species formed in solution and in the solid phase with sizes too small for XRD characterization. Thanks to the synergic use of the three techniques, comprehensive structural information from the atomic level up to the formation of nanoparticles, passing through the liquid phase intermediate, was obtained.

Two papers related to material synthesis are finally mentioned, though not directly related to catalysis, as examples of XRD–Raman application to organic-containing layered and microporous materials that have relevance in catalysis.

Boccaleri et al. [10] reported on the thermal swelling and degradation of stearate–hydrotalcite by XRD–Raman. XRD data gave clear information on the behavior of the inorganic lamellar component of the hybrid materials. Thanks to the Raman data, the behavior of the long alkyl chains in terms of order–disorder or conformational changes (*trans/gauche* ratio) and chain–chain interactions was extracted.

Del Rio et al. [31] characterized the heating process from room temperature (RT) to 573 K of a palygorskite–indigo mixture by *in-situ* XRD–Raman. During this process, the organic dye (indigo) and the clay (palygorskite) interact to form Maya blue (MB), a pigment highly resistant to degradation. It was shown that the formation of a very stable pigment occurs when palygorskite starts to lose zeolitic water, accompanied by a reduction of the crystallographic *a* parameter, and interacting with the clay in such a way that the organic indigo chromophore, which includes C=C and C=O bonds, is altered. Raman was powerful in showing changes at the molecular level, particularly highlighting the changes related to the double bonds C=O and C=C during MB formation, totally inaccessible by diffraction.

15.3.2.2 Operando Experiments Grunwaldt et al. [32] combined XRD/XAFS–Raman and mass spectrometry (MS) in one setup. The structure of palladium during the combustion of methane was studied as a function of temperature up to 800°C. Upon heating in the reaction mixture, a sudden reduction of the palladium oxide particles accompanied by strong sintering occurred (XRD), resulting in less catalytic activity. Raman combined with XAS showed that palladium was reoxidized during cooling but was not as finely dispersed as in its initial, as-prepared state. The strength of XRD–Raman was exploited especially during cooling where the characteristic PdO Raman signal appeared at 608 cm^{-1}. Since this band resulted from a phonon interaction, its presence indicated that the crystallite size had grown large enough in size after sintering to create the collective lattice vibrations and hence the Raman signal. Conversely, no XRD reflections was observed, so authors concluded that PdO crystallite size was sufficiently large to generate a phonon Raman band but too small to allow the detection of a diffraction signal. Raman, in combination with XAS, showed its potential in detecting semiamorphous or very small nanocrystalline phases.

O'Brien et al. [20] have subsequently combined UV-Vis–Raman–XRD and MS with very high energy X-rays (88 keV). Due to the high penetration power of the X-rays, they employed a larger diameter cell (4 mm) mimicking typical laboratory-scale reactors and obtained a better plugged flow behavior without losing XRD data quality. The authors have studied the oxidation of methanol over MoO_3 under *operando* conditions.

Wragg et al. [33] reported on the behavior of a silicoaluminophosphate (SAPO)-34 catalyst during a methanol-to-olefin (MTO) process under *operando* conditions with XRD, Raman, and MS. XRD gave information on the anisotropic changes in the unit cell dimensions in relation to the intermediate species formed in the cages of the SAPO-34 framework. XRD also aided with the estimation of the location and amount of intermediate species by Fourier mapping and Rietveld refinement techniques combined with tapered element oscillating microbalance (TEOM) measurements. Raman was used to study amorphous components formed into the channels and, in particular, indicated that the nature of the coke became increasingly graphitic with time.

Hamon et al. [34] investigated the use of the flexible porous chromium terephthalate $Cr(OH)(O_2C\text{-}C_6H_4\text{-}CO_2)$ for the separation of CO_2 and CH_4 from their mixtures at ambient temperature. The coadsorption of CO_2 and CH_4 was studied by a variety of techniques. XRD allowed studying the breathing motion of the solid upon adsorption of the gas mixtures, while simultaneously measured Raman yielded an estimation of the adsorbed quantities of CO_2 and CH_4, that is, disordered species formed by light atoms that are very hard to locate and impossible to distinguish from one another by XRD.

15.3.3 An Example of XRD/XAFS–Raman during Catalyst Activation and during *Operando* Conditions

De Smit et al. [35] reported on the stability and reactivity of iron carbide phases of a Fischer–Tropsch synthesis (FTS) catalyst, formed during the prepa-

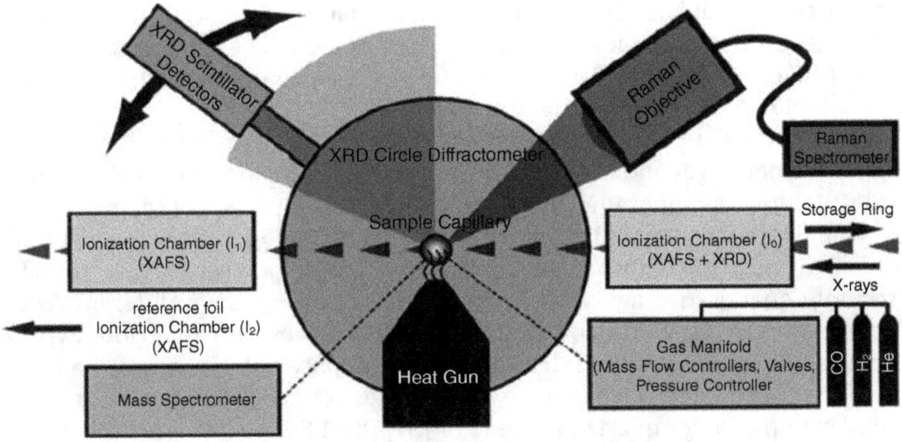

Figure 15.2. Schematic representation of the combined XAFS/XRD–Raman–MS experimental setup for the *in-situ* study of gas–solid reactions. Reprinted from de Smit, E., Cinquini, F., Beale, A. M., Safonova, O. V., van Beek, W., Sautet, P., Weckhuysen, B. M. (2010) Stability and reactivity of ε-χ-θ iron carbide catalyst phases in Fischer-Tropsch synthesis: controlling μ(c), *Journal of the American Chemical Society*, 132, 14928–14941. Copyright (2011) American Chemical Society.

ration and under *operando* conditions. They were affected by both temperature and composition of the gaseous reaction mixture, thus by the change in the carbon chemical potential μ_C. A synergistic combination of experimental and theoretical methods was used. Combined *in-situ* XAFS/XRD–Raman was applied to study the catalysts during pretreatment and under high-pressure FTS *operando* conditions. This paper can be considered as the end point of development and test phase of the simultaneous multitechnique approach at the Swiss–Norwegian beamlines (SNBL) [13] since all probes, together with online effluent gas analysis by MS, were successfully used. Moreover, first-principle theoretical calculations were employed to obtain structures for the prediction of experimental XRD patterns and, more importantly, to depict the energetic landscape of the transformations. The setup scheme is shown in Figure 15.2, illustrating how the various probes are organized and how non-ambient conditions are achieved and controlled.

XRD data were exploited beyond its mere use as a phase recognition tool to determine lattice parameter and crystallite sizes, even if a full use of XRD data by an "*in-situ* structural refinement"[1] was not possible because of small crystal sizes as stated by the authors. XAFS yielded information on the local

[1]"*In-situ* structural refinement" is the full application (lattice determination, structural refinement with atom location, crystallite size analysis) of Rietveld refinement methods to XRD data obtained from an *in-situ* experiment, with the aim of characterizing the evolution of the "structure" during the reaction itself [36].

structure of Fe, highlighting changes in the oxidation state of Fe atoms under different reaction conditions and allowing the detection of amorphous phases. Raman allowed monitoring the evolution of carbonaceous surface species inaccessible by the other techniques.

Concerning the preparation phase, two pretreatments were studied: (1) at higher temperatures and lower CO partial pressures (named "low μ_C" sample) and (2) at lower temperatures and higher CO partial pressures (named "high μ_C" sample). XRD data indicated that the initial α-Fe_2O_3 phase was converted to Fe_3O_4 and FeO before transforming into the respective carbide phases. Accurate quantitative analysis of formed phases was made at different reaction conditions. It is worth noting that some differences with analogous *ex-situ* experiments were attributed due to the higher space velocity applied and to the difference in hydrodynamic flow properties of the capillary reactor compared with the larger quartz tube reactor [37]. XAFS data indicated that there might be a significant contribution of Fe–Fe bonds (similar to the case of the α-Fe phase which is probably not present) that can be attributed to a carbide phase of higher symmetry, with very small crystal sizes and/or amorphous behavior not detected by XRD.

Concerning FTS *operando* experiments, the crystalline phase evolution was fully characterized by XRD, indicating that the two samples behaved quite distinctly. A high μ_C sample was gradually oxidized to Fe_3O_4, while a low μ_C one appeared to be quite stable and was not oxidized to Fe_3O_4. In fact, the increase in atomic displacement parameter (ADP) at isothermal conditions indicated that the total disorder increased during FTS. In accordance with the ADP results, XAFS data clarified that the contribution of carbide phases was underestimated in the XRD analysis because of their amorphous occurrence. Furthermore, the evolution of the local structure around Fe atoms of the high μ_C sample was interpreted similarly to the XRD data but with faster sample degradation, confirming that catalyst degradation started locally and then evolved in the long range. Raman scattering was elegantly employed to characterize carbonaceous species inaccessible by XRD (because they are surface and disordered species) and by XAFS (because they contain no heavy atoms). C=C stretching of polyaromatic rings and conjugated olefins were analyzed [38], with particular attention to G (1595 cm^{-1}) and D bands (1329 cm^{-1}) [39] assigned, respectively, to more graphitic-like carbon and to more disordered carbonaceous species. While the D and G bands initially increased for the high μ_C catalyst, upon reaching the reaction pressure of 10 bar the intensity of the Raman peaks leveled off and was stable for a long time. In sharp contrast, the catalyst pretreated at low μ_C conditions showed a gradual increase in carbon species for a longer time. This information is of paramount importance since graphitic-like carbon formed in μ_C is considered responsible for catalyst deactivation [40].

This case study clearly shows the full potential of the simultaneous approach. All three probes—XRD, XAFS and Raman—are looking at the same spot of the sample during the course of the complete experiment, while the catalytic activity is simultaneously monitored with a mass spectrometer. The variety of rich chemical information, obtained by the three probes combined with the reliable correlation in time, enables the detailed analysis and understanding of the occurring processes. It would have been impossible to reproduce such complex experimental conditions which are dependent on many parameters (gas flux, reaction chamber dimensions, temperature ramp speed, and pressure) on three different, separate instrumentations, to align all the probes concerted with the reaction coordinate and to obtain detailed and comparable results. Not to mention, time saving by avoiding multiple experiments is also a clear advantage.

15.3.4 Remarks on XRD–Raman in the Present and in the Future

In-situ but separated XRD–Raman experiments often exploit only the typical application of XRD, that is, phase recognition. This is due to the limited quality of laboratory XRD data especially under *in-situ* conditions. Nevertheless, detailed studies can be carried out employing such an approach, especially on simple systems when transformations occur slowly (i.e., from minutes to hours). More complex catalytic systems that also need more sophisticated data analysis tools and hence higher quality data, can only be studied on synchrotron-based simultaneous XRD–Raman setups. These setups allow (1) to use XRD Rietveld refinement also at *in-situ* conditions to carry out the sometimes called "*in-situ* structural refinement" [36]; (2) to avoid spatial and time alignment problems between probes; (3) to improve time, peak, and crystallographic resolution, thanks to the synchrotron source; (4) to obtain a clear view of what is a happening at different length scales in the sample by combining XRD/XAS–Raman–MS simultaneously; and (5) to use larger environmental cells with higher pressure ranges mimicking real reactors, where hard X-rays successfully penetrate thicker cell walls. With *in-situ* XRD–Raman, a careful analysis on the long-range (structural refinements and particle size analysis) together with short- and molecular-range length scale investigations can be carried out. Forecasting further development of *in-situ* XRD–Raman setup and applications in catalysis, some remarks can be made. From the technical viewpoint, the use of faster acquisition and higher resolution detectors will make *in-situ* structural refinements routine and straightforward with a time resolution (seconds) similar to that of Raman instruments. Another expected major step forward will be the implementation of *in-situ* UV–Raman–XRD. UV–Raman can avoid fluorescence in many cases, for instance when applied to zeolites. In fact, combined UV–Raman experiments have not yet been reported in conjunction with synchrotron X-ray techniques. This is probably due to the extra experimental complication, the relatively new technology, and thus high costs. If we recall the experiment by Wragg et al. [33], coke formation in zeolites was quantified with diffraction, but the Raman signals were, not surprisingly, severely hampered by the fluorescence induced by the 532 nm laser. Li and Stair [41] pioneered the study of coke formation in zeolites with

UV–Raman intentionally to avoid fluorescence. Furthermore, an additional UV-Vis probe would facilitate the identification and quantification of the coke formation [42, 43]. Such a novel experimental setup, combining high-quality XRD with UV–Raman, UV-Vis, and MS, is expected to be available in the near future. Finally, new ways of exploring details of catalysts by analytical techniques will be the massive application of mathematical tools that allow postexperimental data treatment, with the dual scope of quantitatively correlating the results from different probes and of enhancing the selectivity and sensitivity of the techniques, as described in detail in Section 15.4.

15.4 MODULATION EXCITATION SPECTROSCOPY

The oldest reference to the concept of modulation enhancement (ME) dates from 1953 [44] and in principle ME can be applied to almost any analytical technique [45]. The basic idea of ME is to deliberately expose a sample to a *periodically varying external stimulus*. The stimulus can be any physical variable: temperature, pressure, concentration, electric or magnetic field, and light or any other. Let us suppose that not the whole sample is susceptible to the stimulus but that only one or several parts are. Typically there is some delay between the stimulation and the response of the so-called *active* (the actively responding) part(s). It should be noted that the delay time of different parts of the sample contains information on the response kinetics and transformation pathways. The ME concept performs at best when the sample response is reproducible over time or, in other words, if the sample is not destroyed during the course of the experiment and stays actively responding to the stimulus. In practice, the major active response will take over the frequency of the stimulation. A detection system capable of performing frequency filtering can conveniently extract the signal from only the active part(s), therefore introducing experimental *selectivity* through a proper choice of stimulus. Figure 15.3 illustrates the increase in sensitivity and selectivity.

Let us recall at this moment a typical catalyst where the actual active constituent only represents a minute fraction of the total catalytic body. If one imagines that this catalyst undergoes a periodic switching of gases (e.g., from oxidizing to reducing), it becomes clear that the state of the active species in the catalyst changes. With the help of ME, one is capable of selectively picking up on these changes and in which sequence they occur. This frequency filtering can be performed during data collection by hardware, exploiting a lock-in amplifier, or postexperimentally by software. With a lock-in amplifier, one sets the stimulation frequency and the amplifier searches for the condition when a chosen function shape "locks into" the maxima of the response (in reality, the condition is a multiplied sum of the function and response); this procedure is called phase-sensitive detection (PSD) or demodulation. During the last 20 years, many detection systems in, for example, Raman spectrometers have migrated from slow zero dimension scanning point detectors to fast readout

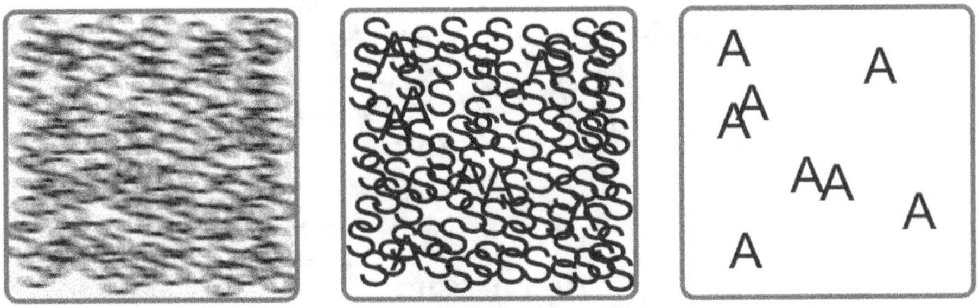

(i) (ii) (iii)

Figure 15.3. Illustration of measurement quality ranging from (i) low sensitivity, (ii) high sensitivity, and (iii) high sensitivity and selectivity (ideal case). "A" and "S" represent the Active responding species and the silent Spectators, respectively. Reprinted from Urakawa, A., Bürgi, T., Baiker, A. (2008) Sensitivity enhancement and dynamic behavior analysis by modulation excitation spectroscopy: principle and application in heterogeneous catalysis, *Chemical Engineering Science*, 63, 4902–4909. Copyright (2008) with permission from Elsevier.

two-dimensional (2-D) charge-coupled devices (CCDs). Thanks to this tremendous increase in data acquisition speed and computational power, it became possible to acquire complete and enough spectra that allow software-based postexperimental demodulation instead of hardware lock-in [45]. These developments have enabled a widespread use of ME and flexible posttreatments of data. Currently, development efforts are made to implement ME for its application in XAFS, diffraction, and even combined with several techniques mentioned later in this chapter and also illustrated already in Chapter 14. The ME section of this chapter is meant to introduce the concept of ME and the mathematics of PSD to bring the necessary understanding for the future users of this technique.

15.4.1 Theory of Modulation Excitation

ME find its strength when it is applied to reactions or processes that can be repeated reproducibly, although in principle it can also be used on irreversible reactions. Generally, a system starts to respond reproducibly after a certain transient period, as illustrated in Figure 15.4.

During one period one acquires typically between 40 and hundreds of spectra. As soon as a reproducible response of the system is achieved, after six or seven periods in Figure 15.4, one can average all the scans acquired at the same moment relative to the start point of each period. When more cycles are added in the summation and averaging, the S/N improves according to Poisson statistics (with $\sqrt{N}$, where N is the number of cycles). In standard *in-situ* experiment design, it is very common that one has to make a trade-off

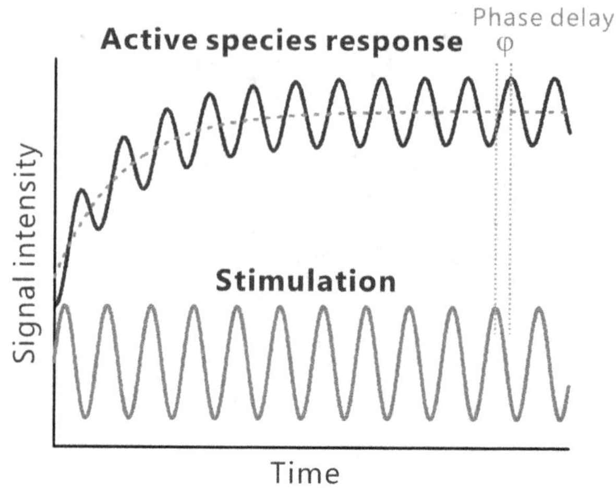

Figure 15.4. Typical profile of active species responding to the stimulation. Note that the delay between stimulation and response as well as the response amplitude contain kinetic information. Reprinted from Urakawa, A., Bürgi, T., Baiker, A. (2008) Sensitivity enhancement and dynamic behavior analysis by modulation excitation spectroscopy: principle and application in heterogeneous catalysis, *Chemical Engineering Science*, 63, 4902–4909. Copyright (2008) with permission from Elsevier.

between S/N and temporal resolution. The advantage of the averaging procedure is that the S/N can be improved *without* compromising temporal resolution. In the end, one period with improved sensitivity is produced. Experimental selectivity, introduced by the choice of a stimulus, becomes enhanced through the mathematical treatment, PSD. Whether the PSD is done with hardware lock-in amplifiers acting during the data acquisition or with postexperimental software demodulation, the underlying theory is the same. The PSD integral is given in Equation 15.1 [45]:

$$A_k(\phi_k^{\mathrm{PSD}}) = \frac{2}{T}\int_0^T A(t)\sin(k\omega t + \phi_k^{\mathrm{PSD}})\,\mathrm{d}t, \tag{15.1}$$

where T represents the length of one period, $A(t)$ is the temporal response of the measured amplitude on the y-axis (dependent on the technique, i.e., absorption for infrared [IR] or intensity for Raman) at a given value on the x-axis (wave number or any other unit also depending on the technique). For clarity, we only consider one x-value here; in reality, all x-values are integrated subsequently. The stimulation frequency is ω, and k is the demodulation index to access higher harmonics of ω that may or may not be present in the signal. The temporal response $A(t)$ multiplied by the sine function shown in Equation 15.1 is integrated over one period with the ϕ_k^{PSD} angle stepwise increasing from

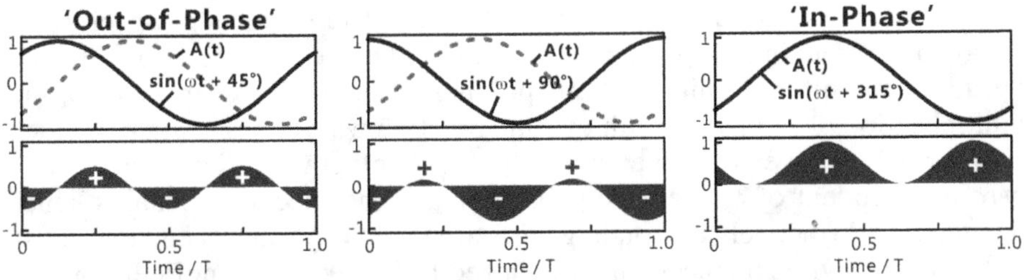

Figure 15.5. Illustration of PSD. Upper panels illustrate experimental response $A(t)$ and the to-be-multiplied sine function used in PSD. Lower panels show the multiplied function and the final value after PSD is given by the integrated area considering the positive and negative sign over the period (T) highlighted by the filled black area. Reprinted from Urakawa, A., Bürgi, T., Baiker, A. (2008) Sensitivity enhancement and dynamic behavior analysis by modulation excitation spectroscopy: principle and application in heterogeneous catalysis, *Chemical Engineering Science*, 63, 4902–4909. Copyright (2008) with permission from Elsevier.

0 to 360°. This sequential integration with increasing ϕ_k^{PSD} transforms the time-domain response $A(t)$ into a so-called phase domain response $A(\phi_k^{PSD})$. In a simple case of sinusoidal response, increasing the ϕ_k^{PSD} angle will shift the to-be-multiplied sine function relative to the temporal response $A(t)$ as illustrated in the three upper panels in Figure 15.5.

The bottom panels show the integrated area after the $A(t)*\sin(k\omega t+\phi_k^{PSD})$ multiplication. The left image and the right image are special cases since the integral values are zero and maximum; in other words, the $A(t)$ and $\sin(k\omega t+\phi_k^{PSD})$ are totally *out-of-phase* and fully *in-phase*, respectively. The middle image is at an intermediate ϕ_k^{PSD} angle. The in-phase condition is of crucial importance for ME analysis. The ϕ_k^{PSD} *in-phase angle*, or lock-in condition, provides, namely, the kinetic information; in other words, when the event occurs within the period. Let us imagine a sample with chemical reaction occurring in two steps, one after the other. The first event reaches the in-phase condition at smaller ϕ_k^{PSD} angles compared with the second. Indeed, there is a simple relation between the ϕ_k^{PSD} angle and the phase delay shown in Figure 15.5 [45, 46]. The in-phase ϕ_k^{PSD} angle is therefore a direct measure of the transformation pathway. If we now consider a complete spectrum, instead of only one $A(t)$, it becomes clear that parts of the spectrum, absorption bands, or peaks with the same in-phase angle happen at the same time and belong to the same species or those appearing at the same time. The correlation between different parts of the spectra, due to identical in-phase angles, can be of great help in understanding what actually happened, since differences in in-phase angles (thus phase delay) provide the information on when and in which sequence they occurred.

We will now also consider in the same sample a species that does not respond to the stimulus and hence does not actively take part in the reaction, such as support material or inactive metal sites. These spectator species will always be present in time-domain data and could make interpretation difficult, because they often dominate the experimental data. However, in PSD, the multiplication of a constant with the sine function will always contain as much positive as negative components and the PSD integral will hence always be *zero* independent of ϕ_k^{PSD}. In the phase domain, the spectator signal is hence deleted and the actively responding parts are *selectively* filtered and extracted. Further sensitivity enhancement is achieved by PSD because noise has a very broad frequency spectrum. In practice, it also turns out that the smallest noise frequency is often still far above a typical experimental stimulation ω, thus allowing also noise elimination or at least major reduction. The PSD integral acts hence in four ways: (1) different in-phase angles directly indicate *when* events occur in a transformation pathway; (2) the same in-phase angles occurring at different parts of the spectra will facilitate the interpretation on *how* the transformation takes place; (3) a proper selection of stimulus changes the concentration of species of interest and spectator signals will go to zero in phase domain, thereby introducing *selectivity*; and, finally, (4) noise is filtered out in phase-domain spectra, enhancing *sensitivity*. The real power of the method is further illustrated with the application examples in the next sections.

15.4.2 Application of ME to Catalysis

The ME technique has great advantages in the investigation of catalytic reactions and mechanisms because of the intrinsic complexity of catalytic processes, particularly heterogeneous ones, involving a wide variety of spectator species such as catalyst support materials and/or solvent. Although the spectators play important roles in catalytic systems as a whole, their signals in analytical measurements often overwhelm the searched information in order to understand reaction mechanisms and key species, for example, catalytic active sites and active species. Therefore, it is desirable to sensitively monitor the species of interests by a selective method. As described earlier, the ME technique utilizes an external perturbation to induce selective detection, and this is very advantageous in *in-situ* studies of catalytic systems because the complexities of the systems increase under common industrial reaction conditions, that is, at high temperature and pressure.

The first application of the ME technique combined with spectroscopy, MES, to study a heterogeneous catalytic system was reported by Bürgi in 2002 for the investigation of the catalytic solid–liquid interface during enantioselective hydrogenation of 4-methoxy-6-methyl-2-pyrone over Pd/TiO_2 [47]. A powerful, almost unique technique able to study solid–liquid interfaces is attenuated total reflection infrared spectroscopy (ATR-IRS). ATR-IRS was combined with MES using the change in the reactant concentration as stimulation in a square-wave form. The reactant concentration perturbation triggered the reaction on and off, thus allowing the detection of key intermediate species and elucidating the main and side transformation pathways during the reaction. The demonstration of sensitivity enhancement by PSD and a convenient way to differentiate the kinetics of responses due to different species, in-phase angle analysis, are shown in Figure 15.6. The time-domain spectra, which were recorded subsequently in one modulation cycle, showed subtle changes induced by the concentration modulation, but the overall signal was dominated by major contributions from invariant silent species. The static signals coming from spectators and most noise were removed in the phase-domain spectra, which showed only temporally changing species. The enhancement of sensitivity can be clearly seen from the level of absorbance in Figure 15.6, identifying confidently the small bands with less than tenths of milliabsorbance. Furthermore, the different gray scale in the in-phase angle plot corresponds to the bands with the same phase delay, that is, from the same species or different

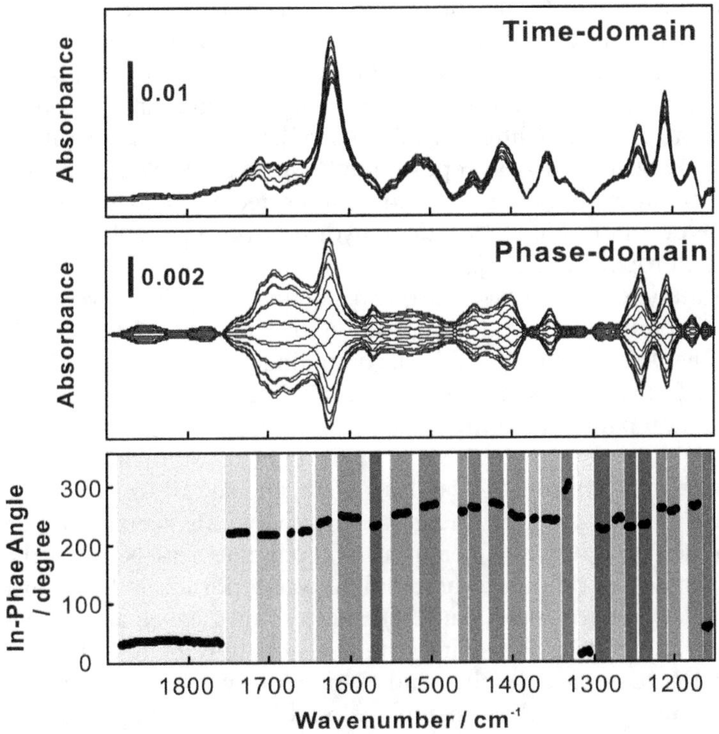

Figure 15.6. Time- and phase-domain ATR-IR spectra and in-phase angle plot of the MES experiment where the concentration of the pyrone was modulated between 0 and 3.6 mmol/L. Reprinted from Urakawa, A., Bürgi, T., Baiker, A. (2008) Sensitivity enhancement and dynamic behavior analysis by modulation excitation spectroscopy: principle and application in heterogeneous catalysis, *Chemical Engineering Science*, 63, 4902–4909. Copyright (2008) with permission from Elsevier.

species appearing at the same time. This in-phase angle analysis is a very effective way to separate contributions from different chemical species while retaining the information about their temporal evolution [46].

Later, the groups of Bürgi, Baiker, and others have investigated various catalytic systems occurring at solid–liquid interfaces by ATR-IR MES. Gisler studied the epoxidation of allylic alcohols over titania–silica aerogel catalysts using concentration MES of reactant and oxidant (*tert*-butyl hydroperoxide, TBHP) to uncover the nature of the interactions between reactants and the catalyst surface hydroxyl groups and the reaction mechanisms [48, 49]. The signals originating from the interactions were very small, and PSD was mandatory to detect the key spectral features of the reaction. Liquid-phase alcohol oxidation was studied by ATR-IR MES by Bürgi and Bieri using Pd/Al_2O_3 [50], capturing a transient, alkoxide intermediate surface species; by Haider et al. using Cu–Mg–Al mixed oxides [51, 52] to elucidate the nature of the basic surface properties; and by Mondelli using Ru–hydroxyapatite [53] to identify the surface Ru sites active in the reaction by the concentration stimulation of the reactant as well as CO as the probing molecule of active sites. Wirz identified the reaction intermediates and pathways of Knoevenagel reaction over silica modified by a covalently bonded basic organic group and also explained the solvent effects based on the distinct molecular interaction modes of reactants with the catalyst in different solvents [54]. Recently, a reactant concentration ATR-IR MES study was reported by Aguirre et al., clearly demonstrating the discrimination between the active, intermediate, and spectator species during the decomposition of acetic anhydride over a phosphotungstic Wells–Dawson heteropoly acid [55].

Photocatalytic systems were also investigated by ATR-IR MES. The investigation of photocatalytic decomposition of malonic acids over TiO_2 was reported by Dolamic and Bürgi [56], combining concentration MES, transient light-on/-off measurements, and isotopic labeling, identifying at least two different reaction pathways in the mineralization of the acid. Recently, Dolamic and Bürgi have also applied MES using light as stimulation, varying the amount of photons hitting the catalysts (TiO_2 and Au/TiO_2), in the mineralization of L-glutamic acid [57]. The great sensitivity enhancement by PSD gave access to the fate of the nitrogen atom of the amino acid. Notably, the formation of cyanide in addition to ammonium was reported and the former promoted the leaching of Au as $Au(CN)_2^-$ species and gave clear guidelines for a better catalyst design. It is also important to note that the mass transfer within the ATR-IR cells used in the studies previously mentioned are diffusion limited because of the close-to-zero velocity of the solution near the ATR crystal (internal reflection element) due to a laminar character of the flow, that is, the concentration of molecules near the ATR crystal (where samples are probed) cannot be altered by convection of the flow but by diffusion driven by concentration difference of solute molecules. The characteristics of mass transport, that is, convection and diffusion, can also be evaluated by concentration MES using molecules with known diffusion constants [58].

Urakawa, Meier, and their colleagues have extended the scope of polarization–modulation infrared reflection–absorption spectroscopy (PM-IRRAS) by combining it with the ME technique for the studies of catalytic gas–solid [59] and solid–liquid interfaces [60]. The sensitivity enhancement by the ME technique enabled the detection with a good time resolution, roughly 1 s, for the technique which was previously not used for time-resolved studies of less than some minutes by increasing the S/N to more than two orders of magnitude. Moreover, the utilization of mutually orthogonal polarizations with respect to the sample surface enabled the detection of both species in the fluid phase (gas/liquid) as well as surface species simultaneously and selectively by a *single* technique. The fundamental process in heterogeneous catalysis, dynamic molecular sorption process, was studied by ME PM-IRRAS, taking sorption processes at the solid–liquid interfaces between alumina and organic acid solution as an example [61]. Based on distinct exchange and adsorption/desorption kinetics, the solute molecule dissolved in liquid phase and two different surface species were identified for salicylic acid due to its rich chemical interactions with the surface of alumina via its two OH groups [61].

So far, the majority of applications of MES in catalysis were reported in combination with IR spectroscopy. Ferri and coworkers recently extended the potential of the technique to XAFS studies of catalysts during the NO reduction by CO [62] and methane oxidation [63]. PSD greatly enhanced the detection sensitivity for the data otherwise too noisy to be analyzed and firmly identified periodically changing spectral features responding to stimulation. The PSD analysis of the XAFS data that are apparently more complex than IR spectroscopy due to the overlapping character of the wiggling signals in the EXAFS region and more efforts in the proper use of PSD analysis for XAFS data are expected to come in the future.

It is worth mentioning that a concentration perturbation with square-wave stimulation is one of the most popular application of MES in catalysis because of the simplicity to generate the stimulation shape just by opening and closing a valve of a desired solution/gas source under a flow condition, as well as the fact that the change in reactant concentration typically influences those of intermediate species and products selectively. Another practical benefit of the shape is given by the multiple odd frequency components (1ω, 3ω, 5ω . . .) contained in square-wave stimulation. The advantage of the PSD is that these components with different frequencies contained in the response can be deconvoluted and separately evaluated [64]. For example, under the assumption of a linear response (i.e., a proportional response to stimulus amplitude), which is valid when stimulation amplitude is small, the PSD with different k (the demodulation index, Eq. 15.1) extracts components oscillating at the frequency $k\omega$. The phase delay determined by higher k PSD is equivalent to that obtained by the experiment carried out at $k\omega$ modulation frequency as the fundamental frequency, 1ω. Also, the amplitude can be easily related by scaling by the factor k [64]. This means that the higher frequency PSD offers fast screening of signal responses at different modulation frequencies and facilitates kinetic analysis.

15.4.3 Further Developments of ME

As already mentioned, development efforts are currently made to adapt ME to XAFS. Two further developments of ME are worth mentioning, the first being the quantitative correlation of signals obtained by different probes [65], the latter being the enhancement obtainable with the application of ME technique to diffraction [66], that is, the newborn technique of modulation enhanced diffraction (MED).

15.4.3.1 Quantitative Correlation Methods The ME technique offers additional advantages when used in the multiprobe detection. For example, the XRD–Raman combination, explained in the first part of this chapter, can gain complementary chemical and structural information on catalyst materials and about short- and long-range ordering. The advantages of multiprobe detection originate from the analysis of correlations in the signal responses observed by each technique. The usefulness of such an analysis has been shown for the study of structural changes of $[Fe(Htrz)_2(trz)]BF_4$ during temperature-induced spin crossover process by ME XRD–Raman [65]. One way to analyze the interrelations between responses in a multiprobe technique is the covariance of, for example, XRD $A(2\theta,t)$ and Raman $B(\nu,t)$ responses. $Cov(2\theta,\nu)$ is calculated as follows:

$$Cov(2\theta,\nu)=\frac{1}{N-1}\sum_{i=1}^{N}A(2\theta)_i'B(\nu)_i',$$

where $A(2\theta)_i'$ and $B(\nu)_i'$ are the XRD pattern or Raman spectrum of the ith data number (in the time axis) subtracted by their mean values, and N is the total number of data points in the time axis in one modulation cycle.

Another way to show such interrelations is to use correlation $Corr(2\theta,\nu)$ which is readily calculated from the covariance as follows,

$$Corr(2\theta,\nu)=\frac{Cov(2\theta,\nu)}{\sigma_{A(2\theta,t)}\sigma_{B(\nu,t)}},$$

where $\sigma_{A(2\theta,t)}$ and $\sigma_{B(\nu,t)}$ are the standard deviations of $A(2\theta,t)$ and $B(\nu,t)$, respectively. The advantage of using correlation is that the quantity shown in covariance is normalized by the standard deviations at chosen 2θ and ν. Therefore, small correlated signals are visible in correlation while they are typically hidden in covariance because the quantity is proportional to the changes in the signal amplitudes. The covariance analysis is yet superior to localize and easily highlight the interrelations between signals.

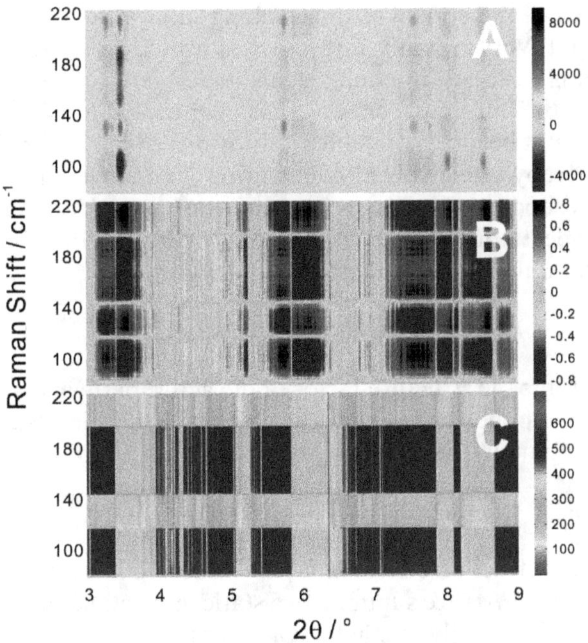

Figure 15.7. 2D XRD-Raman covariance (A), correlation (B), and sum of in-phase angle (C) plots from the spin crossover transition of $[Fe(Htrz)_2(trz)]BF_4$. Reprinted from Urakawa, A., Van Beek, W., Monrabal-Capilla, M., Galán-Mascarós, J. R., Palin, L., Milanesio, M. (2011) Combined, modulation enhanced X-ray powder diffraction and Raman spectroscopic study of structural transitions in the spin crossover material $[Fe(Htrz)_2(trz)](BF_4)$, *Journal of Physical Chemistry C*, 115, 1323–1329. Copyright (2011) American Chemical Society. (See color insert.)

Figure 15.7A,B shows the 2-D plots of covariance and correlation, indicating the localized and sensitive features of the analyses, respectively. This visual way of analysis facilitates the identification of species/crystalline phases with the same kinetic behavior. The two analyses, however, remove the information about the temporal evolution. This important kinetic information can be conveniently retrieved by the sum-of-in-phase-angle plot shown in Figure 15.7C to visualize the evolution of the temporal sequence of the correlated signals because of a one-to-one relationship between the in-phase angle and the phase delay of a signal response.

15.4.3.2 Application of ME to Diffraction One of the basic limitations of XRD originates from the interference nature of the diffraction process itself, therefore limiting detection selectivity and sensitivity. For this reason, XRD does not allow separating contributions from different atomic subsets. These limitations restrain the use of diffraction in catalysis since a catalyst is often diluted, disordered, and defective, thus making catalyst substructure investigation very difficult. Very recently, a theory has been developed, describing the physical meaning of the diffraction signals under external periodic perturbation as in ME [66]. This new technique has been called MED. The developed theory shows that it becomes possible to study directly the subset/substructure of a sample, thereby effectively circumventing the interference nature of diffraction. Let us suppose that a sample consists of an actively responding part (A) and inert spectator (S). The scattered intensity $I(t)$ of this sample when irradiated with monochromatic X-rays can be written as,

$$I(t) = |F_A(t) + F_S|^2 = \left|\left(\overline{F_A} + \delta F_A(t)\right) + F_S\right|^2,$$

where F_A represents the structure factor of the active part varying as a function of time t. F_S corresponds to the structure factor of the silent sublattice. Since F is a complex number, the intensity may be expressed in the following way:

$$I(t) = |F_A(t)|^2 + |F_S|^2 + F_A(t)F_S^* + F_A^*(t)F_S.$$

In a standard diffraction experiment, the structure information by A and S and the four terms of $I(t)$ are summed in a static diffraction pattern and cannot be separated. One can easily see that not all four terms contain time-dependent active parts and the demodulation procedure also used in MED allows probing separately the time-dependent terms of the previous equation. In fact, it can be demonstrated [66] that, thanks to the square present in the $|F_A(t)|^2$ term, the demodulated signal with the doubled frequency of the stimulation contains the crystallographic information on the active substructure only. For clarity, when a periodic varying sinusoidal signal is multiplied by itself, one obtains a periodic signal with the same shape but with the doubled frequency. If one imagines a catalyst with a responding ordered active substructure, only this desired information can be extracted by MED. As described in the MED theory paper [66], a simulated but realistic MED experiment on $LiCoO_2$ (where the Li occupancy is the active species during charging and discharging of the battery and the CoO_2 framework is the spectator) allowed obtaining a 2ω demodulated pattern. This 2ω pattern contains the structural information of lithium only, serving as the proof of principle, and offering the possibility of carrying out a selective diffraction experiment for the investigation of the substructure of the catalyst in *operando* conditions. We foresee a widespread use of the already proven applications of ME in various characterization techniques and the future will tell how successful ME-XAFS and MED developments will be in catalysis research.

REFERENCES

[1] Ertl, G., Knözinger, H., Schüth, F., Weitkamp, J. (eds.) (2008) *Handbook of Heterogeneous Catalysis*, 2nd ed., Wiley-VCH, Weinheim, Germany.

[2] Weckhuysen, B. M. (ed.) (2004) *In-Situ Spectroscopy of Catalysts*, American Scientific Publishers, Stevenson Ranch, CA.

[3] Wachs, I. E. (1999) In situ Raman spectroscopy studies of catalysts, *Topics in Catalysis*, 8, 57–63.

[4] Jellinek, M. H., Fankuchen, I. (1948) The application of X-ray diffraction to the study of solid catalysts, *Advances in Catalysis*, 1, 257–289.

[5] Brückner, A. (2003) Looking on heterogeneous catalytic systems from different perspectives: multitechnique approaches as a new challenge for in situ studies, *Catalysis Reviews-Science and Engineering*, 45, 97–150.

[6] Bryant, G. K., Gleeson, H. F., Ryan, A. J., Fairclough, J. P. A., Bogg, D., Goossens, J. G. P., Bras, W. (1998) Raman spectroscopy combined with small angle X-ray scattering and wide angle X-ray scattering as a tool for the study of phase transitions in polymers, *The Review of Scientific Instruments*, 69, 2114–2117.

[7] Newton, M. A., van Beek, W. (2010) Combining synchrotron-based X-ray techniques with vibrational spectroscopies for the in situ study of heterogeneous catalysts: a view from a bridge, *Chemical Society Reviews*, 39, 4845–4863.

[8] Beale, A. M., van der Eerden, A. M. J., Kervinen, K., Newton, M. A., Weckhuysen, B. M. (2005) Adding a third dimension to operando spectroscopy: a combined UV-Vis, Raman and XAFS setup to study heterogeneous catalysts under working conditions, *Chemical Communications*, 24, 3015–3017.

[9] Bentrup, U., Radnik, J., Armbruster, U., Martin, A., Leiterer, J., Emmerling, F., Brückner, A. (2009) Linking simultaneous in situ WAXS/SAXS/Raman with Raman/ATR/UV-vis spectroscopy: comprehensive insight into the synthesis of molybdate catalyst precursors, *Topics in Catalysis*, 52, 1350–1359.

[10] Boccaleri, E., Carniato, F., Croce, G., Viterbo, D., van Beek, W., Emerich, H., Milanesio, M. (2007) In situ simultaneous Raman/high-resolution X-ray powder diffraction study of transformations occurring in materials at non-ambient conditions, *Journal of Applied Crystallography*, 40, 684–693.

[11] Van Beek, W., Safonova, O. V., Wiker, G., Emerich, H. (2011) SNBL a dedicated beamline for combined in situ X-ray diffraction, X-ray absorption and Raman scattering experiments, *Phase Transitions*, 84, 726–732.

[12] Couves, J. W., Thomas, J. M., Waller, D., Jones, R. H., Dent, A. J., Derbyshire, G. E., Greaves, G. N. (1991) Tracing the conversion of aurichalcite to a copper catalyst by combined X-ray absorption and diffraction, *Nature*, 354, 465–468.

[13] Swiss Norwegian Beamline (BM1, http://www.snbl.eu), ESRF, Grenoble, France. Last accessed December 13, 2012.

[14] Van Beek, W., Carniato, F., Kumar, S., Croce, G., Boccaleri, E., Milanesio, M. (2009) Studying modifications and reactions in materials by simultaneous Raman and X-ray powder diffraction at non-ambient conditions: methods and applications, *Phase Transitions*, 82, 293–302.

[15] Rønning, M., Tsakoumis, N. E., Voronov, A., Johnsen, R. E., Norby, P., van Beek, W., Borg, O., Rytter, E., Holmen, A. (2010) Combined XRD and XANES studies of a Re-promoted Co/gamma-Al_2O_3 catalyst at Fischer-Tropsch synthesis conditions, *Catalysis Today*, 155, 289–295.

[16] Karaca, H., Hong, J. P., Fongarland, P., Roussel, P., Griboval-Constant, A., Lacroix, M., Hortmann, K., Safonova, O. V., Khodakov, A. Y. (2010) In situ XRD investigation

of the evolution of alumina-supported cobalt catalysts under realistic conditions of Fischer-Tropsch synthesis, *Chemical Communications*, 46, 788–790.

[17] Stair, P. C. (2007) The application of UV Raman spectroscopy for the characterization of catalysts and catalytic reactions, *Advances in Catalysis*, 51, 75–98.

[18] Kumar, S., Carniato, F., Arrais, A., Croce, G., Boccaleri, E., Palin, L., van Beek, W., Milanesio, M. (2009) Investigating surface vs bulk kinetics in the formation of a molecular complex via solid-state reaction by simultaneous Raman/X-ray powder diffraction, *Crystal Growth & Design*, 9, 3396–3404.

[19] Clausen, B. S., Steffensen, G., Fabius, B., Villadsen, J., Feidenhansl, R., Topsøe, H. (1991) In situ cell for combined XRD and online catalysis tests—studies of Cu-based water gas shift and methanol catalysts, *Journal of Catalysis*, 132, 524–535.

[20] O'Brien, M. G., Beale, A. M., Jacques, S. D. M., Di Michiel, M., Weckhuysen, B. M. (2009) Spatiotemporal multitechnique imaging of a catalytic solid in action: phase variation and volatilization during molybdenum oxide reduction, *ChemCatChem*, 1, 99–102.

[21] Briois, V., Belin, S., Villain, F., Bouamrane, F., Lucas, H., Lescouezec, R., Julve, M., Verdaguer, M., Tokumoto, M. S., Santilli, C. V., Pulcinelli, S. H., Carrier, X., Krafft, J. M., Jubin, C., Che, M. (2005) New insights for materials science characterisation using different complementary techniques combined with X-ray absorption spectroscopy, *Physica Scripta*, T115, 38–44.

[22] Briois, V., Lutzenkirchen-Hecht, D., Villain, F., Fonda, E., Belin, S., Griesebock, B., Frahm, R. (2005) Time-resolved study of the oxidation of ethanol bycerium(IV) using combined Quick-XANES, UV-Vis, and Raman spectroscopies, *Journal of Physical Chemistry A*, 109, 320–329.

[23] Kim, J. W., Shim, J. H., Kim, S. C., Remhof, A., Borgschulte, A., Friedrichs, O., Gremaud, R., Pendolino, F., Zuttel, A., Cho, Y. W., Oh, K. H. (2009) Catalytic effect of titanium nitride nanopowder on hydrogen desorption properties of $NaAlH_4$ and its stability in $NaAlH_4$, *Journal of Power Sources*, 192, 582–587.

[24] Mamede, A. S., Payen, E., Grange, P., Poncelet, G., Ion, A., Alifanti, M., Pârvulescu, V. I. (2004) Characterization of WO_x/CeO_2 catalysts and their reactivity in the isomerization of hexane, *Journal of Catalysis*, 223, 1–12.

[25] Conte, M., Budroni, G., Bartley, J. K., Taylor, S. H., Carley, A. F., Schmidt, A., Murphy, D. M., Girgsdies, F., Ressler, T., Schlögl, R., Hutchings, G. J. (2006) Chemically induced fast solid-state transitions of omega-$VOPO_4$ in vanadium phosphate catalysts, *Science*, 313, 1270–1273.

[26] Friedrich, H. B., Mahomed, A. S. (2008) The oxidative dehydrogenation of n-octane to styrene using catalysts derived from hydrotalcite-like precursors, *Applied Catalysis A: General*, 347, 11–22.

[27] Grandjean, D., Beale, A. M., Petukhov, A. V., Weckhuysen, B. M. (2005) Unraveling the crystallization mechanism of CoAPO-5 molecular sieves under hydrothermal conditions, *Journal of the American Chemical Society*, 127, 14454–14465.

[28] Kongmark, C., Martis, V., Rubbens, A., Pirovano, C., Löfberg, A., Sankar, G., Bordes-Richard, E., Vannier, R. N., Van Beek, W. (2009) Elucidating the genesis of Bi_2MoO_6 catalyst by combination of synchrotron radiation experiments and Raman scattering, *Chemical Communications*, 32, 4850–4852.

[29] Kongmark, C., Martis, V., Pirovano, C., Löfberg, A., Van Beek, W., Sankar, G., Rubbens, A., Cristol, S., Vannier, R. N., Bordes-Richard, E. (2011) Synthesis of γ-Bi_2MoO_6 catalyst studied by combined high-resolution powder diffraction, XANES and Raman spectroscopy, *Catalysis Today*, 157, 257–262.

[30] Li, H. H., Liu, C. Y., Li, K. W., Wang, H. (2008) Preparation, characterization and photocatalytic properties of nanoplate Bi_2MoO_6 catalysts, *Journal of Materials Science*, 43, 7026–7034.

[31] Del Rio, M. S., Boccaleri, E., Milanesio, M., Croce, G., van Beek, W., Tsiantos, C., Chyssikos, G. D., Gionis, V., Kacandes, G. H., Suarez, M., Garcia-Romero, E. (2009) A combined synchrotron powder diffraction and vibrational study of the thermal treatment of palygorskite-indigo to produce Maya blue, *Journal of Materials Science*, 44, 5524–5536.

[32] Grunwaldt, J. D., van Vegten, N., Baiker, A., van Beek, W. (2009) Insight into the structure of Pd/ZrO_2 during the total oxidation of methane using combined in situ XRD, X-ray absorption and Raman spectroscopy, *Journal of Physics: Conference Series*, 190, 012160.

[33] Wragg, D. S., Johnsen, R. E., Balasundaram, M., Norby, P., Fjellvåg, H., Grønvold, A., Fuglerud, T., Hafizovic, J., Vistad, Ø. B., Akporiaye, D. (2009) SAPO-34 methanol-to-olefin catalysts under working conditions: a combined in situ powder X-ray diffraction, mass spectrometry and Raman study, *Journal of Catalysis*, 268, 290–296.

[34] Hamon, L., Llewellyn, P. L., Devic, T., Ghoufi, A., Clet, G., Guillerm, V., Pirngruber, G. D., Maurin, G., Serre, C., Driver, G., van Beek, W., Jolimaitre, E., Vimont, A., Daturi, M., Ferey, G. (2009) Co-adsorption and separation of CO_2-CH_4 mixtures in the highly flexible MIL-53(Cr) MOF, *Journal of the American Chemical Society*, 131, 17490–17499.

[35] de Smit, E., Cinquini, F., Beale, A. M., Safonova, O. V., van Beek, W., Sautet, P., Weckhuysen, B. M. (2010) Stability and reactivity of ε-χ-θ iron carbide catalyst phases in Fischer-Tropsch synthesis: controlling mu(c), *Journal of the American Chemical Society*, 132, 14928–14941.

[36] Agostini, G., Lamberti, C., Palin, L., Milanesio, M., Danilina, N., Xu, B., Janousch, M., van Bokhoven, J. A. (2010) In situ XAS and XRPD parametric Rietveld refinement to understand dealumination of Y zeolite catalyst, *Journal of the American Chemical Society*, 132, 667–678.

[37] Grunwaldt, J. D., Caravati, M., Hannemann, S., Baiker, A. (2004) X-ray absorption spectroscopy under reaction conditions: suitability of different reaction cells for combined catalyst characterization and time-resolved studies, *Physical Chemistry Chemical Physics*, 6, 3037–3047.

[38] Chua, Y. T., Stair, P. C. (2003) An ultraviolet Raman spectroscopic study of coke formation in methanol to hydrocarbons conversion over zeolite H-MFI, *Journal of Catalysis*, 213, 39–46.

[39] Katagiri, G., Ishida, H., Ishitani, A. (1988) Raman-spectra of graphite edge planes, *Carbon*, 26, 565–571.

[40] Xu, J., Bartholomew, C. R. (2005) Temperature-programmed hydrogenation (TPH) and in situ Mossbauer spectroscopy studies of carbonaceous species on silica-supported iron Fischer-Tropsch catalysts, *Journal of Physical Chemistry B*, 109, 2392–2403.

[41] Li, C., Stair, P. C. (1997) Ultraviolet Raman spectroscopy characterization of coke formation in zeolites, *Catalysis Today*, 33, 353–360.

[42] Tinnemans, S. J., Kox, M. H. F., Nijhuis, T. A., Visser, T., Weckhuysen, B. M. (2005) Real time quantitative Raman spectroscopy of supported metal oxide catalysts without the need of an internal standard, *Physical Chemistry Chemical Physics*, 7, 211–216.

[43] Wu, Z. L., Zhang, C., Stair, P. C. (2006) Influence of absorption on quantitative analysis in Raman spectroscopy, *Catalysis Today*, 113, 40–47.

[44] Bair, E. J., Lund, J. T., Cross, P. C. (1953) Chemical modulation—a method for investigating transient molecules, *The Journal of Chemical Physics*, 21, 759–759.

[45] Baurecht, D., Fringeli, U. P. (2001) Quantitative modulated excitation Fourier transform infrared spectroscopy, *The Review of Scientific Instruments*, 72, 3782–3792.

[46] Urakawa, A., Bürgi, T., Baiker, A. (2008) Sensitivity enhancement and dynamic behavior analysis by modulation excitation spectroscopy: principle and application in heterogeneous catalysis, *Chemical Engineering Science*, 63, 4902–4909.

[47] Bürgi, T., Baiker, A. (2002) In situ infrared spectroscopy of catalytic solid-liquid interfaces using phase-sensitive detection: enantioselective hydrogenation of a pyrone over Pd/TiO_2, *Journal of Physical Chemistry B*, 106, 10649–10658.

[48] Gisler, A., Bürgi, T., Baiker, A. (2003) Epoxidation on titania-silica aerogel catalysts studied by attenuated total reflection Fourier transform infrared and modulation spectroscopy, *Physical Chemistry Chemical Physics*, 5, 3539–3548.

[49] Gisler, A., Bürgi, T., Baiker, A. (2004) Epoxidation of cyclic allylic alcohols on titania-silica aerogels studied by attenuated total reflection infrared and modulation spectroscopy, *Journal of Catalysis*, 222, 461–469.

[50] Bürgi, T., Bieri, M. (2004) Time-resolved in situ ATR Spectroscopy of 2-propanol oxidation over $Pd/A12O_3$: evidence for 2-propoxide intermediate, *Journal of Physical Chemistry B*, 108, 13364–13369.

[51] Haider, P., Baiker, A. (2007) Gold supported on Cu-Mg-Al-mixed oxides: strong enhancement of activity in aerobic alcohol oxidation by concerted effect of copper and magnesium, *Journal of Catalysis*, 248, 175–187.

[52] Haider, P., Grunwaldt, J. D., Baiker, A. (2009) Gold supported on Mg, Al and Cu containing mixed oxides: relation between surface properties and behavior in catalytic aerobic oxidation of 1-phenylethanol, *Catalysis Today*, 141, 349–354.

[53] Mondelli, C., Ferri, D., Baiker, A. (2008) Ruthenium at work in Ru-hydroxyapatite during the aerobic oxidation of benzyl alcohol: an in situ ATR-IR spectroscopy study, *Journal of Catalysis*, 258, 170–176.

[54] Wirz, R., Ferri, D., Baiker, A. (2006) ATR-IR spectroscopy of pendant NH_2 groups on silica involved in the Knoevenagel condensation, *Langmuir*, 22, 3698–3706.

[55] Aguirre, A., Bonivardi, A. L., Matkovic, S. R., Briand, L. E., Collins, S. E. (2011) ATR-FTIR study of the decomposition of acetic anhydride on fosfotungstic Wells-Dawson heteropoly acid using concentration-modulation excitation spectroscopy, *Topics in Catalysis*, 54, 229–235.

[56] Dolamic, I., Bürgi, T. (2007) Photocatalysis of dicarboxylic acids over TiO_2: an in situ ATR-IR study, *Journal of Catalysis*, 248, 268–276.

[57] Dolamic, I., Bürgi, T. (2011) In situ ATR-IR study on the photocatalytic decomposition of amino acids over Au/TiO_2 and TiO_2, *Journal of Physical Chemistry C*, 115, 2228–2234.

[58] Urakawa, A., Wirz, R., Bürgi, T., Baiker, A. (2003) ATR-IR flow-through cell for concentration modulation excitation spectroscopy: diffusion experiments and simulations, *Journal of Physical Chemistry B*, 107, 13061–13068.

[59] Urakawa, A., Bürgi, T., Schlapfer, H. P., Baiker, A. (2006) Simultaneous in situ monitoring of surface and gas species and surface properties by modulation excitation polarization-modulation infrared reflection-absorption spectroscopy: CO oxidation over Pt film, *The Journal of Chemical Physics*, 124, 054717.

[60] Meier, D. M., Urakawa, A., Mader, R., Baiker, A. (2009) Design and performance of a flow-through polarization-modulation infrared reflection-absorption spectroscopy cell for time-resolved simultaneous surface and liquid phase detection under concentration and temperature perturbations, *The Review of Scientific Instruments*, 80, 094101-1–094101-11.

[61] Meier, D. M., Urakawa, A., Baiker, A. (2009) Adsorption behavior of salicylic, benzoic, and 2-methyl-2-hexenoic acid on alumina: an in situ modulation excitation PM-IRRAS study, *Physical Chemistry Chemical Physics*, 11, 10132–10139.

[62] Ferri, D., Kumar, M. S., Wirz, R., Eyssler, A., Korsak, O., Hug, P., Weidenkaff, A., Newton, M. A. (2010) First steps in combining modulation excitation spectroscopy with synchronous dispersive EXAFS/DRIFTS/mass spectrometry for in situ time resolved study of heterogeneous catalysts, *Physical Chemistry Chemical Physics*, 12, 5634–5646.

[63] Eyssler, A., Kleymenov, E., Kupferschmid, A., Nachtegaal, M., Kumar, M. S., Hug, P., Weidenkaff, A., Ferri, D. (2011) Improvement of catalytic activity of $LaFe_{0.95}Pd_{0.05}O_3$ for methane oxidation under transient conditions, *Journal of Physical Chemistry C*, 115, 1231–1239.

[64] Urakawa, A., Bürgi, T., Baiker, A. (2006) Kinetic analysis using square-wave stimulation in modulation excitation spectroscopy: mixing property of a flow-through PM-IRRAS cell, *Chemical Physics*, 324, 653–658.

[65] Urakawa, A., Van Beek, W., Monrabal-Capilla, M., Galán-Mascarós, J. R., Palin, L., Milanesio, M. (2011) Combined, modulation enhanced X-ray powder diffraction and Raman spectroscopic study of structural transitions in the spin crossover material $[Fe(Htrz)_2(trz)](BF_4)$, *Journal of Physical Chemistry C*, 115, 1323–1329.

[66] Chernyshov, D., Van Beek, W., Emerich, H., Milanesio, M., Urakawa, A., Viterbo, D., Palin, L., Caliandro, R. Kinematic diffraction on a structure with periodically varying scattering function, *Acta Crystallographica. Section A, Foundations of Crystallography*, A67, 327–335.

16

CATALYST IMAGING USING SYNCHROTRON-BASED MULTITECHNIQUE APPROACHES

ANDREW M. BEALE, JAVIER RUIZ-MARTINEZ, AND
BERT M. WECKHUYSEN

16.1 INTRODUCTION

As the earlier book chapters testify, synchrotron radiation (SR)-based methods have proven a powerful asset in the characterizing of catalytic samples. The properties of SR such as its brightness (>10^{10} s/mrad^2/mm^2), its coherency covering a large swathe of the electromagnetic spectrum (>0.1–140,000 keV), the uniformity of the beam at the sample, coupled with the availability of high-performance detectors enabling the recording of high-quality data (with regard to resolution and signal-to-noise ratio) as a function of time (<ms) or space (<μm), are what makes this possible. As Figure 16.1 illustrates, currently for many characterization techniques the ultimate resolution that can be obtained is already at the nanoscale in terms of spatial resolution, approaching the nanoscale with respect to time, while close to the microelectronvolt scale with regard to energy. For the most part, it is the time parameter for which SR has traditionally been exploited for catalysis research purposes, enabling a

In-situ *Characterization of Heterogeneous Catalysts*, First Edition.
Edited by José A. Rodriguez, Jonathan C. Hanson, and Peter J. Chupas.

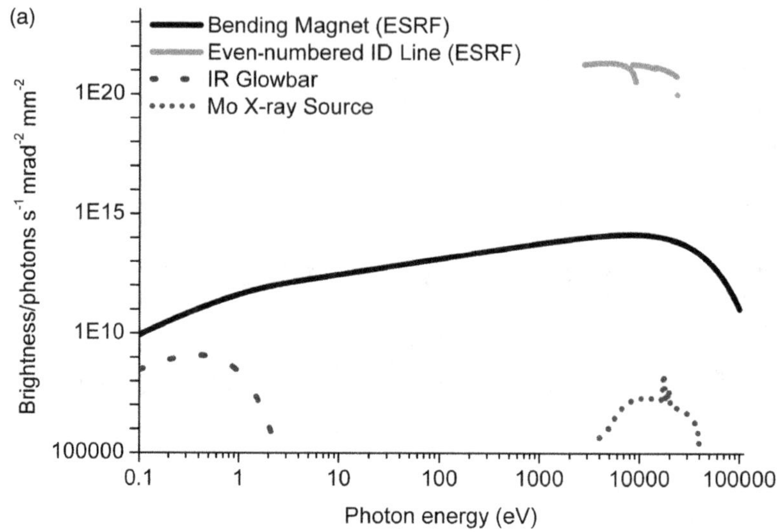

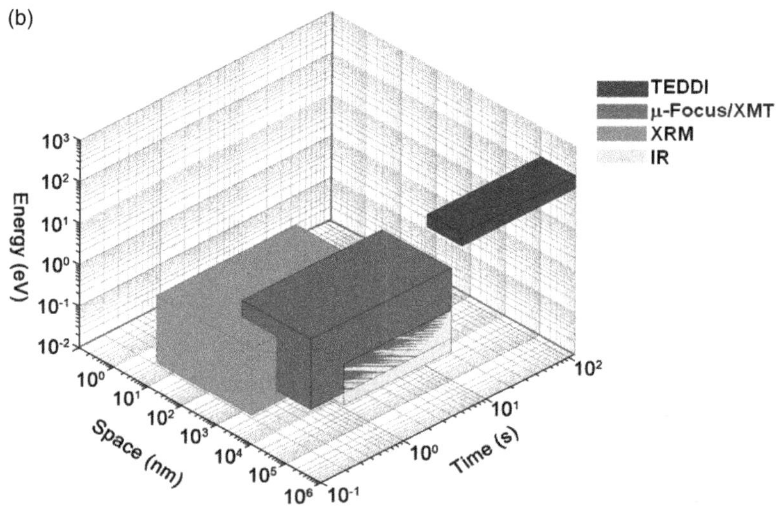

Figure 16.1. (a) Typical spectra (brightness against photon energy) comparing an IR glowbar source, an X-ray tube (40 kV) with a Mo source and synchrotron radiation emitted by a bending magnet and by typical even-numbered insertion device lines at the European Synchrotron Radiation Facility (ESRF) (6 GeV source, 200 mA current). (b) 3-D representation of the spatial, temporal, and resolution regimes available for synchrotron-based imaging techniques: XRM (X-ray microscopy), XMT (X-ray microtomography), IR (infrared), and TEDDI (tomographic energy dispersive diffraction imaging).

great number of detailed *in-situ* and *operando* studies of heterogeneous and homogeneous catalysts under reaction conditions [1–5]. In comparison, the high spatial resolution available at SR facilities is somewhat less exploited. Catalytic solids, however, possess a three-dimensional (3-D) structure which is rarely uniform. The problem is that such deviations in uniformity result in a heterogeneous surface or composition and potentially a number of different "chemical sites" which may or may not possess catalytic activity. Therefore, there is a risk that by focusing on one part (alternatively termed a zero-dimensional [0-D] study) of a catalyst system when performing an *in-situ* study, one could end up drawing conclusions that might not be entirely representative of the process/sample under study. In order, then, to effectively evaluate and characterize catalytic solids in their entirety, relevant information should be obtained from the atomic scale and up to, where appropriate, the millimeter scale over the entire 3-D structure of the solid. Furthermore, in order to generate the necessary structure–activity relationships, these multidimensional data should ideally be acquired under reaction conditions. This is of course no mean feat considering the unavoidable trade-off between obtaining interpretable data as a function of time and space. Clearly undeterred by this challenge, a number of SR-based spatiotemporal spectroscopic and scattering studies have been performed in recent times, in which detailed insight has been obtained into both the behavior of the catalyst sample and the reactants/products at the reaction interface [6–9]. Most of these imaging studies have focused on the use of X-rays to measure the sample's proclivity for absorbance/emission or scattering of a beam >100 eV in energy which have been used to study single-catalyst particles up to entire reactor beds in order to obtain, for example, elemental composition, local atomic coordination/oxidation state, crystallinity, and density; the highly penetrating nature of X-rays (high mean-free path) means often such data are obtained under real reaction conditions with a sufficient time resolution to reveal chemical and structural change. In numerous cases, however, it has been demonstrated that the behavior of the sample under investigation is nonuniform [7]. This chapter, which builds upon a tutorial review we published recently, provides an overview of this burgeoning field and gives a flavor that illustrates the possibilities for future catalyst characterization, primarily from a technique-based perspective [6]. We begin by showcasing the power of SR methods to study individual particles before moving on to consider examples which have employed higher energy beams to examine packed beds and larger samples in general, thereby illustrating the scope of the techniques currently available and the insight they can give. Where possible, we have tried to showcase data that have been obtained under "steady-state" reaction conditions although we do not completely exclude static pre-/postmortem studies where a study nicely illustrates a thus far untapped potential for future studies. Finally, we end by considering longer-term possibilities for catalyst imaging in terms of recent and protracted developments which could enable new and better measurement possibilities.

16.2 IMAGING CATALYTIC SOLIDS WITH SYNCHROTRON RADIATION

16.2.1 Imaging the Effects of Acidity on the Catalytic Activity in Zeolite Crystals

16.2.1.1 Infrared Microspectroscopy Reveals the Distribution of Products and Reactants in Single Crystals From a catalytic perspective, perhaps the most important property exhibited by microporous zeolitic materials is acidity [10]. The origin of this acidity are the Al^{3+} ions, which isomorphously substitute some of the Si^{4+} within the microporous framework, generating an excess charge which, when charge compensated for by protons, form Brønsted acid sites ((Si–O(H)–Al)). At the same time, a capacity to expand their coordination sphere renders Al^{3+} species capable of exhibiting Lewis acidity. The bulk distribution of these sites within zeolite crystals can therefore have a significant influence on the catalytic properties of the industrially applied zeolite. Stavitski et al. recently demonstrated in the little literature there is on the subject of SR infrared (IR) imaging insight into the strength of these Brønsted acid sites and into the nature and location of the reaction product within polycrystalline H-ZSM-5 [11, 12]. In these studies a 5×5 μm beam was utilized to study $100 \times 20 \times 20$ μm, coffin-shaped H-ZSM-5 crystals during acid catalyzed oligomerization of 4-fluorostyrene and 4-chlorostyrene at 373 K as a function of time in both zero and two dimensions. Initially they utilized SR-IR to follow the conversion of the 4-fluorostyrene monomer (by tracking the disappearance of a band at 1510 cm^{-1}) to an unknown oligomer product characterized by a new band at 1534 cm^{-1}. Density functional theory (DFT) calculations later revealed the most likely origin of the 1534 cm^{-1} band to be a dimeric carbocation 1,3-bis(4-fluoro/chlorophenyl)-2-buten-1-ylium, and furthermore, that such species, when specifically oriented along the length of the straight channels within H-ZSM-5, would exhibit a polarization dependency. Since H-ZSM-5 crystals possess a complex intergrowth structure comprising a number of subunits around a common [001] crystallographic axis, a nonuniform two-dimensional (2-D) channel network results, in which the edges of the crystal contain straight pores open at the surface of the crystal, while the "center region" possesses straight channels which run perpendicular to the two ends of the crystal [13]. Hence what is seen in a 2-D raster map shown in Figure 16.2b is a signal intensity distribution of the dimeric carbocation (1534 cm^{-1} band) that directly reflects this internal structure with light polarized along the direction of the straight channels in the main body of the crystal, "illuminating" the oligomers polymerized along the lengths of the straight channel, resulting in a more intense signal in the center of the crystal rather than at the edges.

As with styrenic compounds, so the presence of a number of IR active modes in thiophenic compounds also renders them suitable for study with IR. Furthermore, such compounds comprise the majority of sulfur species within

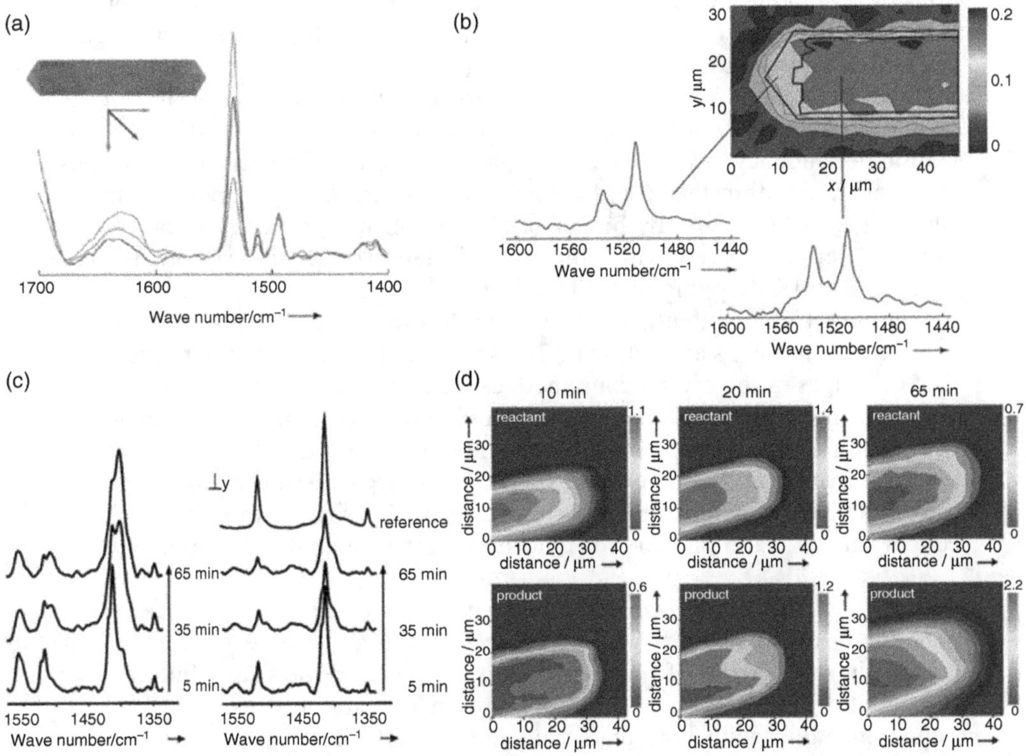

Figure 16.2. (a) Synchrotron-based IR spectra of a large H-ZSM-5 crystal in contact with 4-chlorostyrene taken with radiation polarized in different directions as indicated. The figure illustrates the polarization dependency of the band for the product signal at approximately 1540 cm^{-1}. (b) Intensity of the IR band at 1534 cm^{-1} for the 1,3-bis (4-fluorophenyl)-2-buten-1-ylium cation mapped over H-ZSM-5 after reaction. To the left and below the intensity map are example IR spectra taken from the edge and the body of the crystal, demonstrating differences in the intensity ratio of the bands. (c) IR spectra of the ring stretching region measured in the center of the crystal during reaction of 2-chlorothiophene over a HZSM-5 crystal after 5, 35, and 65 min of reaction. The incident polarization was set to parallel to the *y*-direction and perpendicular to the *y*-direction, respectively. (d) Corresponding 2-D IR intensity maps for the 1412 cm^{-1} reactant band (top) and 1401 cm^{-1} product band (bottom). Panel a from Reference 12, copyright John Wiley & Sons 2010; panel b from Reference 11, copyright John Wiley & Sons 2008. Panels c and d from Reference 14, copyright John Wiley & Sons 2009.

automotive fuels, and while predominantly are removed via the process of hydrodesulfurization (HDS), they can also be selectively trapped by Brønsted acid-bearing zeolites such as H-ZSM-5. With this in mind, a third showcase study by Kox et al. examined the uptake and distribution of 2-chlorothiophene and its reaction products within large H-ZSM-5 crystals using a combination of SR-IR and coherent anti-Stokes Raman scattering (CARS), the latter non-SR technique providing additional lateral resolution so as to allow for 3-D imaging of the crystal [14]. As with the styrene oligomerization study, a gradual decrease in the 2-chlorothiophene (C–C stretch at 1412 cm^{-1}) monomer and the subsequent formation of oligomeric (determined by a red shift of the C–C stretch to 1401 cm^{-1}) reaction products were observed. The strong polarization dependency of the signal was again ascribed to the polymer molecules being aligned within the straight channels of the zeolite micropores. Impressively due to the inherently higher brightness of the synchrotron (ca. 50–100 times) when compared with conventional lab sources, it was possible to perform 2-D IR mapping of a 30 × 20 μm segment of the crystal with a time resolution of 10 min; sufficiently fast to follow reactant consumption and product formation as a function of 2-D space (Fig. 16.2c,d) and demonstrated that the edges of the crystal contained less reactant and product, regardless of reaction time. Unfortunately corresponding attempts to map the distribution of reaction products using CARS were not successful due to excessive fluorescence from the reaction products, although CARS measurements were able to reveal again the importance of the intergrowth structure of the crystal and concurrent diffusion barriers for determining the signal distribution.

16.2.1.2 μ-X-ray Absorption Fine Structure to Examine the Interaction between Framework and Adsorbed Species An alternative method to following the conversion of reactants into products and their distribution over the zeolite crystal is the μ-X-ray absorption fine structure (μ-XAFS) technique, since this yields both elemental and chemical insight and has also been used to study large (100 × 20 × 20 μm) H-ZSM-5 zeolite crystals during their exposure to thiophenic compounds at 493 K [15]. For such purposes, a soft–medium X-ray beam is focused on the sample using Fresnel zone plates to deliver a rectangular beam (0.21 × 0.8 μm^2). The setup also utilizes a scanning double-crystal monochromator to enable full X-ray absorption near edge structure (XANES) spectra to be obtained and mapped onto a pixelated Ge detector so as to be able to probe and understand the nature of the framework–thiophene interaction. Fluorescence maps of the various components from an individual zeolite H-ZSM-5 crystal revealed an even distribution of the framework components (Si/Al) as well as the "sorbed" thiophenic compound(s). Despite being unable to determine precisely a structure of the thiophene–Brønsted acid site complex, DFT modeling of the XANES data was able to reveal that the S atom was likely to be in close proximity (ca. 2.5 Å) to a framework O atom. Ultraviolet–visible spectroscopy (UV-Vis) and confocal fluorescence microscopy were then used to identify the presence of two types of thiophenic species within the zeolite crystal; dimeric carbocations (a band at 410 nm in the UV-Vis data) and polymeric carbocations (590 nm) aligned along the straight channels in the body of the crystal. The presence, then, of at least two different S-containing species within the crystal may well explain the difficulty in obtaining a more detailed view of the S interaction with the zeolite framework—a common problem for XAFS measurements in general.

Polymeric thiophenic species were only present within the zeolite crystal and not on a physical mix of silica/boehmite and was indicative of the importance of the "harder" zeolite Brønsted acid sites for catalytic decomposition of thiophenic compounds. While this study focused on obtaining comparatively soft X-ray data, it should be noted that many microfocus beamlines are also able to map the harder K-edges, typically up to and including the end of the first-row transition-metal elements.

16.2.2 Supported Catalysts

Supported catalysts play a vital role in heterogeneous catalysts since they are used in a number of industrial processes such as oil refining, emission control, and the manufacturing of bulk and fine chemicals [16]. They typically comprise metal/metal oxides that are well dispersed on a porous support and owe their activity to the formation of species with particular structural and electronic properties. However, they provide a real challenge when it comes to understanding the nature of these active sites since it is seldom seen that such catalytically active species form exclusively. Nanoimaging techniques able to distinguish between active/inactive species by virtue of their spectroscopic signatures thus provide a possible solution to this quandary [17]. This is especially so if the technique is able to probe both the solid catalyst and the reactant species at the catalyst surface.

16.2.2.1 Scanning Transmission X-ray Microscopy for Imaging Catalytic Solids at the Nanoscale Soft X-ray transmission microscopy (STXM) is a technique that uses soft X-rays (between 100 and 2000 eV) and probes the 1s core (K-edges) electrons of light elements or the 2p core ($L_{2,3}$ edges) of heavier elements [17, 18]. High spatial resolution (30 nm spot size) is achieved by focusing a monochromatized beam via a Fresnel zone plate lens before the sample is then raster-scanned (1–5 ms per pixel) with synchronized detection of transmitted X-rays to generate an image (Fig. 16.3a). Two-dimensional *chemical* images are obtained as a function of beam energy ($\Delta E \sim 25$ eV/edge of interest), resulting in an image sequence (stack) scans which when reconstructed results in an energy spectrum (XANES scan) at each pixel. Thus the technique yields information on valence state/bonding state of the elements of interest at the nanoscale, although due to its soft nature, signal attenuation is a problem (the penetration distance in air being only 0.4–4 mm) and limits the sample thickness to approximately 20 μm. Furthermore, in order to perform catalytic experiments a specially designed nanoreactor is required, central to which are two 10-nm-thick SiN_x windows to enable a sufficient number of photons to enter/exit the cell. The reactor is then mounted on to a piezoelectric stage with nanometer precision, thus allowing for the recording of STXM images with a 35 × 35 nm step size, and the setup has recently been used in order to follow key phase changes and the evolution and location of carbon species on Fe_2O_3 supported on SiO_2 catalysts during both prereduction and

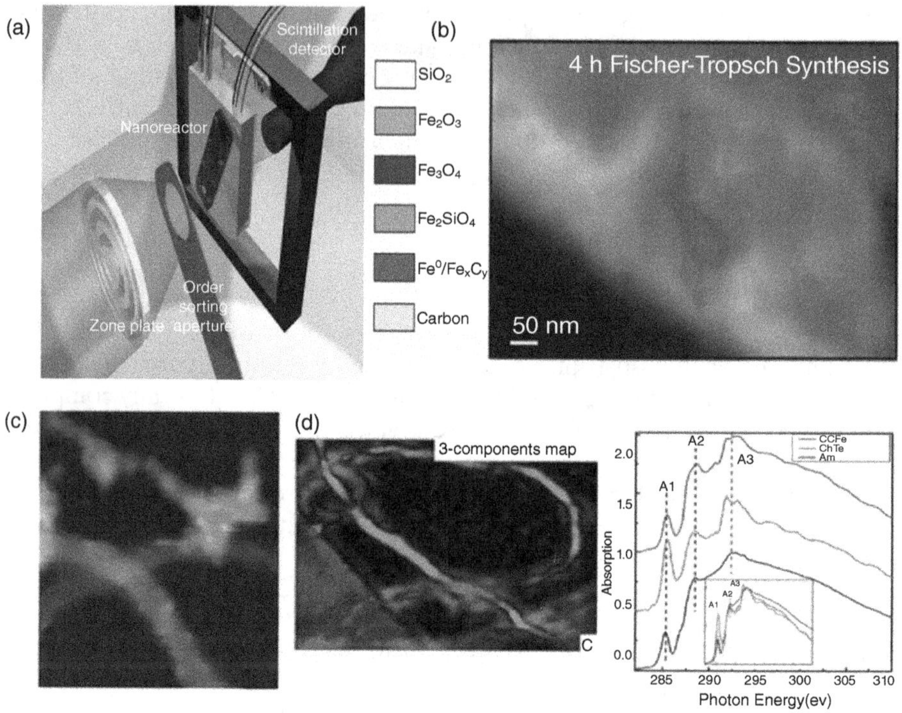

Figure 16.3. (a) Setup of an STXM microscope with, from left to right, the zone plate lens, order-sorting aperture, the nanoreactor, and the scintillation detector. (b) Example nanoscale X-ray images which detail the state of the iron (initially as Fe_2O_3 supported on SiO_2) as it evolves under the Fischer–Tropsch reaction conditions. (c) Three-component STXM chemical map of RuO_2/MWCNTs. Green: RuO_2; blue: MWCNT; and red: O-containing groups at the externs of the MWCNTs. (d) Three-component STXM map of the TiO_2-impregnated carbon nanofibers comprising of the following: blue: amorphous carbon; red: CCFs; and green: carbon nanotubes. On the right-hand side are the corresponding spectra for the individual components with a list of spectral regions used to identify the unique components. Panels c and d adapted with permission from Reference 17, copyright Royal Society of Chemistry 2010. (See color insert.)

the Fischer–Tropsch synthesis (FTS) at 1.2 bar and 250°C [18]. Two-dimensional images were created using the C K-edge (284.2 eV), O K-edge (543.1 eV), and Fe L_2- and L_3-edges (706.8 and 719.9 eV, respectively) to distinguish between the species present in the sample in order to probe the relationship between the various elemental components. The starting phase comprises essentially a mixture of α-Fe_2O_3 and SiO_2. After undergoing "activation" in H_2 at 350°C, the catalyst becomes partially reduced; the α-Fe_2O_3 phase, for example, transforms into a mixture of Fe_3O_4, Fe_2SiO_4, and FeO [19]. During FTS reaction, (Fig. 16.3b), the majority of Fe phases become metallic Fe (20%) and Fe_2SiO_4

(75%), respectively. Analysis of the C K-edge data via a deconvolution of the spectra by linear combination fitting shows that C is colocated in Fe-rich regions but perhaps more importantly, from the C K-edge spectra, it was possible to identify these species as sp^2-based hybridized species, typical of FeC-type phases, whereas in contrast in Fe-poor regions, more sp^3-like FTS reaction products were found. It was proposed that the presence of reactant carbon species in iron-deficient areas indicates that the support might have some role in the spillover of (hydro)carbon species from the metal to the support, thereby preventing blocking of the active sites of the catalyst.

Although not per se a catalytic study, recent articles by both Zhou et al. and Zhong et al. demonstrated quite elegantly the importance of coupling nanoresolution with that of the chemical insight [20, 21]. Here STXM was used to chemically image both RuO_2- and TiO_2-impregnated carbon nanotube bundles in order to reveal insight into the chemical process of oxide loading onto such "supports." In the study by Zhou et al., 2-D STXM maps were created from the C and O K-edge signals, respectively, to produce chemical maps detailing the distribution of the impregnated RuO_2 and the carbon nanotubes (Fig. 16.3c). O K-edge measurements revealed that RuO_2 is not uniformly coated on the tubes, while simultaneously showing that terminal O groups cover nearly the whole surface of the tubes. C K-edge measurements, however, showed that the graphitic framework of the tubes remained largely intact upon coating but that in places where RuO_2 was present, a 0.3 eV (0.05 eV resolution is in fact achievable) red shift for the s–π^* transition and a reduction in its intensity was concluded by the authors to be due to charge transfer from RuO_2 to the C 2p-derived π^* states at the RuO_2–nanotube interface. An additional feature at 289.3 eV indicated a more strongly oxidized C atom than that found for the $-CO_2H$ functional group and was suggestive of a strong RuO_2–multiwall carbon nanotube (MWCNT) interaction. A largely similar study by Zhong and coworkers, working now with TiO_2-impregnated nanotubes, revealed (using 2-D C and O K-edge STXM maps) first that TiO_2 was present inside the carbon nanotube and the presence of a number of carbon environments within the bundle (Fig. 16.3d). Much of the obtained C K-edge signal could be ascribed to the nanotubes but oxygen functionalization of the surface using nitric acid had also introduced two new C environments: amorphous carbon species caused by nanotube collapse and "carboxylated carbonaceous fragments" or CCFs at the peripheries of the nanotubes. Confirmatory evidence of such CCFs was also found in the O K-edge data and was also speculated to be the interface point for the TiO_2 within the nanotubes, although due to the data having been collected in transmission mode, no clear evidence for such a signal could be determined. Such studies, however, demonstrate unequivocally the importance for such a technique in the designing of functional materials and the understanding of their chemical properties.

16.2.2.2 Imaging Individual Catalyst Particles with μ-XAFS STXM has proved to be a very suitable technique for imaging catalyst particles with nanoscopic resolution as well as extraordinary chemical information. Conversely, this technique falls short on giving insight into detailed local coordination structure of active and inactive phases in a single catalyst particle. This can only be achieved by measuring the EXAFS region of the XAFS spectrum. Recently, a work of Tada et al. demonstrated the potential of μ-XANES and μ-EXAFS for "chemical imaging" individual catalyst particles, with $1000 \times 800\ nm^2$ resolution [22]. They studied a single particle of a $NiO_x/Ce_2Zr_2O_y$ ($0 \leq x \leq 1.7 \leq y \leq 8$) catalyst, supported on an SiO_2 membrane. This catalyst is used for CH_4 steam reforming, and it is well known that the catalytic performance of the NiO_x active phase depends on the oxidation state and local structure of the aforementioned species. Thus, Ni species on the reduced $Ce_2Zr_2O_7$ are active for CH_4 steam reforming, while those on the oxidized $Ce_2Zr_2O_8$ support are entirely inert. The sample was mounted in an *in-situ* XAFS cell, and subsequently 2-D μ-X-ray fluorescence (μ-XRF) images were collected in order to map the Ni and Ce species at the single particle level. Ni-edge μ-XANES spectra of individual particles reveal that NiO species are present in the oxidized $NiO_x/Ce_2Zr_2O_8$ catalyst, while metallic Ni was found in the reduced $NiO_x/Ce_2Zr_2O_7$ one. From the Ni K-edge μ-EXAFS spectrum of the oxidized catalysts, it was possible to reveal the local structure of the inactive NiO species. A comparison of the μ-XAFS with conventional XAFS data of the bulk catalyst demonstrates that single particle measurements represent the overall behavior of the bulk catalyst.

16.2.3 Imaging of Packed Bed Reactors

An industrial reactor normally operates under conditions whereby the entire catalyst volume is used to carry out a reaction. This ensures longevity in reactor performance as well as avoiding further reaction of the desired product. Where possible, most *in-situ* studies on catalyst samples aim to mimic these "plug-flow" conditions so as be able to attach a degree of transferability to the data obtained. Typically then the industrial process is mimicked by using samples that have been sieve-fractioned in packed bed reactors which allow sufficient X-ray transmission. Recently a number of reports have appeared in the literature in which a section of a catalytic bed has been imaged in real time during a catalytic reaction in order to examine the "uniformity" of the species present as a function of time. Both supported and bulk oxide catalysts have been interrogated, revealing both nonuniformity of the degree of reduction and volatilization of the active phase.

16.2.3.1 Real-Time Analysis of an Rh/Al_2O_3 Packed Bed Plug-Flow Reactor Methane partial oxidation is regarded as one of the most promising technologies for hydrogen production in plants of medium size [23]. The modus operandi of the catalyst is, however, not known, although it was proposed from previous studies that the nature of the catalyst changes along the length of the bed. Hence, 2-D mapping of a 2.5% Rh/Al_2O_3 catalyst during reaction was performed in order to shed insight into the reaction mechanism. The catalyst was loaded into a 1 mm thick capillary (ca. 10 mm bed length) and imaged using a square beam 1 × 1 mm (1 × *h*) and using a charge-coupled device (CCD) camera/area detector to map along the axial coordinate [23, 24]. Data were recorded as a function of 1–2 eV steps from 23,190 to 23,375 eV (around the Rh K-edge), resulting in a total exposure time of 30 s/point. The resultant images are derived per pixel as a function of the intensity $I\ (E, x, y)$ or $I_0\ (E, x, y)$ for images taken with or without sample, respectively. The data are then "stacked" before the integral of the absorption is then calculated for each pixel so that for any given pixel, a full absorption spectrum is available. Additional least squares fitting of these spectra was used to produce quantitative 2-D maps of the Rh oxidation state distribution in the catalytic reactor.

Figure 16.4 contains the resultant 2-D X-ray transmission images of the reactor during the partial oxidation of methane at 331–378°C (6 h) during the early stages of ignition. From the Rh K-edge XANES data, it is possible to distinguish between Rh in its oxidized (3+) or reduced state (0) from features in the XANES. X-ray absorption is in general stronger on the left-hand side of the reactor, although interestingly a sharp (ca. 100–200 μm) diagonal gradient of Rh (0) can be seen at what appears to be an interface which remains in place for several hours afterward. This, the authors argued, was evidence for a two-stage process occurring within the reactor comprising low-temperature methane combustion in the first part (Rh^{3+}-containing part) and methane reforming in the second part (where Rh (0) dominates). The position of the Rh^{3+}/Rh^0 front is dependent upon space velocity and reaction temperature as shown in Figure 16.4: an increase in both results in movement of the Rh^{3+}/Rh^0 front toward the inlet. Importantly such a correlation of structure and function was made possible by virtue of coupling an online mass spectrometer to characterize the product stream at the reactor outlet.

16.2.3.2 Two-Dimensional Cross-Sectional Imaging of a CuO/ZnO-Containing Packed Bed Reactor The previous study utilized a letterbox beam for the full-field 2-D imaging of a reactor bed, however the approach does not provide sufficient depth resolution so as to be able to observe variation from within the reactor. Such depth information allows for evaluating whether bed bypassing is a problem in order to ensure that the reactor is working under plug-flow conditions. This can only be achieved by performing microtomography-type experiments as has been done in order to study a 30% CuO/ZnO catalyst undergoing reduction/oxidation (redox) cycling in 4% H_2/He and 2% O_2/He at 200–300°C gas mixtures [25]. Computed tomography (CT) scans were performed whereby a full XANES spectrum was collected at each location in a virtual section (x,y) through the catalyst bed. The approach utilized a quick-scanning monochromator and data acquisition system with a scanning pencil beam microprobe to record absorption spectra as a function of a number of translations (r) and rotations (ϕ); ideally, the number of translations ≈ rotations so that 180° is covered. In this work, XANES spectra per projection were

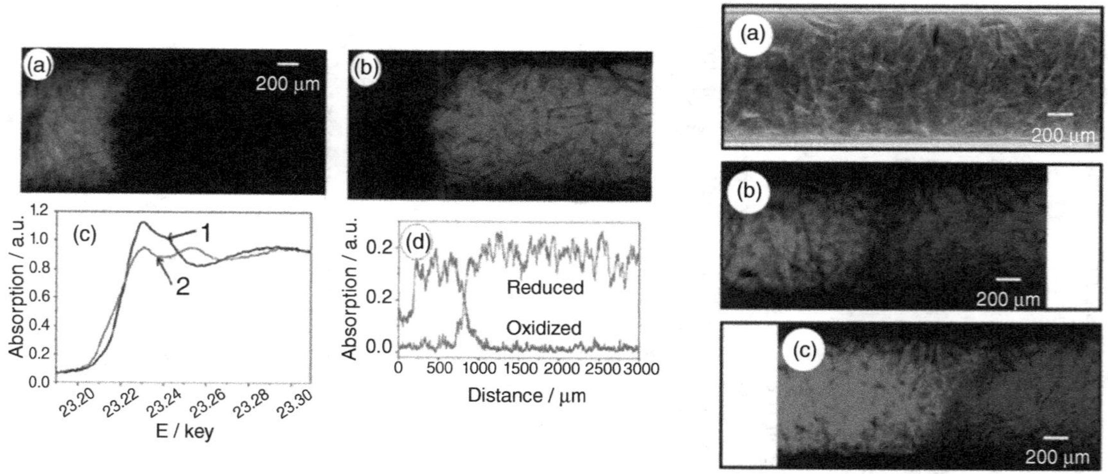

Figure 16.4. Left: Extracted components from the analysis of 160 dark- and flat-field corrected transmission images. (a) Oxidized Rh species; (b) reduced Rh species; (c) corresponding XANES spectra form oxidized (1, black line) and spectra reduced (2, gray line); (d) relative concentration of the oxidized (dark) and reduced (gray) Rh particles in the axis of the fixed bed reactor (conditions: 362°C, space velocity 1.9×10^5 h^{-1}). Right: Structural changes over the catalyst bed due to variation of the space velocity at 338°C. (a) Flat- and dark-field corrected image of catalyst bed; (b) amount of oxidized (light gray) and reduced (dark gray) at 338°C (space velocity 1.9×10^5 h^{-1}); (c) 338°C (space velocity 3.8×10^5 h^{-1}). Reprinted with permission from Reference 24, copyright Elsevier 2007.

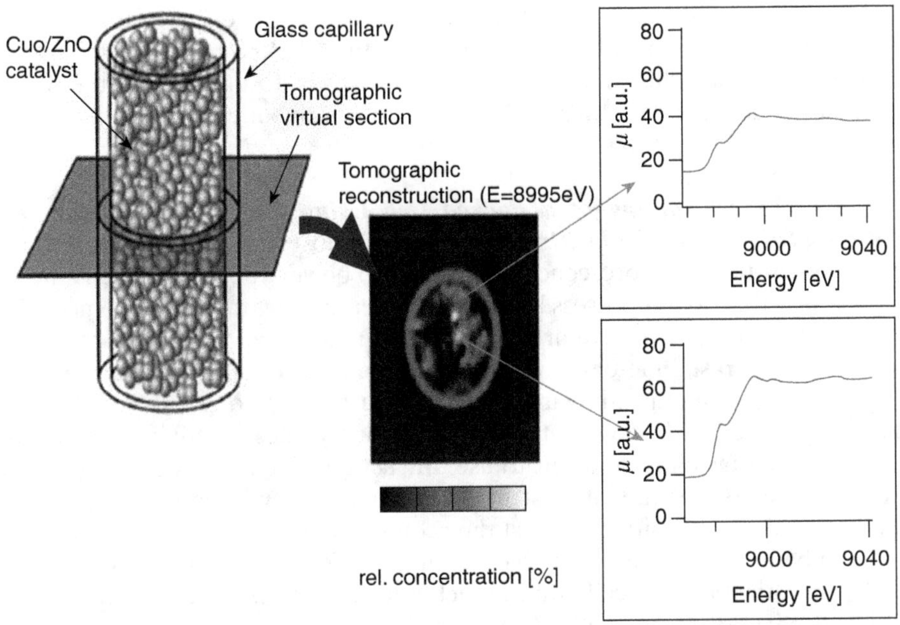

Figure 16.5. Scanning microtomography combined with X-ray absorption spectroscopy at the Cu K-edge. In a virtual slice of a capillary (here the catalyst bed contains Cu/ZnO particles in BN), the X-ray absorption of the sample as function of energy is measured by scanning the sample in translation (90 steps of 10 μm each) and rotation (101 steps over 360 s), acquiring at each position a full absorption spectrum (see gray arrows). The absorption coefficient can be reconstructed for each energy at each location on the virtual slice. Thus, in each location on the reconstructed slice, a full XANES (in principle, it could also be possible to obtain EXAFS data, although to date this has yet to be reported) spectrum is obtained. From these data, the concentration of the different Cu species can be extracted by fitting a linear combination of reference spectra to the spectrum at each location on the virtual slice. Adapted with permission from Reference 25, copyright American Institute of Physics 2003.

acquired within 1 s and could be acquired in either fluorescence or transmission mode, although in the former case, the signal is strongly attenuated.

A reconstructed 2-D slice through the catalyst bed at an energy of 8995 eV is given in Figure 16.5 and to illustrate the chemical information available, XANES spectra from two pixels (grey arrows) in the reconstructed 2-D image are also given. While the 2-D slice revealed density differences across the sample, analysis of the XANES spectra by linear combination fitting for each pixel using reference spectra of Cu^0, Cu_2O, and CuO revealed differences in the degree of Cu reduction. This compositional analysis, when expressed as a 2-D slice, resulted in a series of images pertaining to the amounts of Cu(I) and Cu(0), respectively. These results demonstrate clearly what is possible with such a technique although if, as the authors propose, the energy range of the monochromator can be adjusted to cover the full EXAFS regime, it might be possible to extend this method in order to attempt to understand the local structure in more detail, thus making this approach suited to tackling more complex systems.

16.2.3.3 Line Scanning of a Packed Bed Using Multiple Techniques to Follow Molybdenum Volatilization For the most part, imaging based on X-ray absorption or fluorescence has proven to be very powerful when studying samples that do not possess long-range order. However, there are a notable number of catalysts, particularly those used in selective oxidation, that are comprised of crystalline oxide phases—the oxide material providing a reservoir of oxygen anions, allowing the catalytic surface to perform selective oxygen insertion/hydrogen abstraction accordingly. When studying such catalysts it is then far more relevant to use diffraction-based methods in order to examine the role of the catalyst in a particular catalytic process. However, diffraction data acquisition, at least the acquisition of data of sufficient quality so as to be able to extract valuable "chemical information" from the process under investigation, typically takes much longer than XAFS data acquisition. This is particularly so if standard "bending line magnets" are used, which offer optimal brightness at energies lower than 15 keV. However, these problems can be overcome by using very high energy beamlines where, by virtue of an undulator insertion device, optimal brightness can be found at very high energies (i.e., 88.85 keV/brightness (10^{12} p/s/0.1 BW at 200 mA). The use of the more penetrating hard X-rays significantly reduces the problem of absorption by the sample and allows for the use of much larger, "more realistic" reactors as well as causing a flattening of the Ewald sphere, enabling rapid collection (this can be on the millisecond timescale for digital flat-panel X-ray detectors) of a large d range over a small angle in transmission geometry [26]. Furthermore, a larger Compton cross section results at these energies, allowing for the collection of the inelastically scattered signal. A particular advantage of X-ray imaging is that it is often possible to simultaneously collect more than one signal from the incident beam, that is, transmission and fluorescence, transmitted and scattered beam, so as to allow for obtaining complimentary and highly congruent information on the sample. Data from both the elastic and inelastic signals were recently utilized in a study by O'Brien et al. who looked at the problem of molybdenum loss from methanol selective oxidation catalysts [27]. Up until this study, there had been some debate in the literature as to how this occurs exactly, whether, for example, the formation of volatile Mo-methanol species were responsible or else this was the way in which water in the feed was known to lead to deactivation. Thus methanol and water were fed separately in the gas stream onto a packed bed within a large 5 mm internal diameter (i.d.) capillary reactor (chosen to improve sampling statistics as well as to work under more plug-flow-like conditions). In order to determine whether Mo volatilization occurred from MoO_3 at reaction temperature more

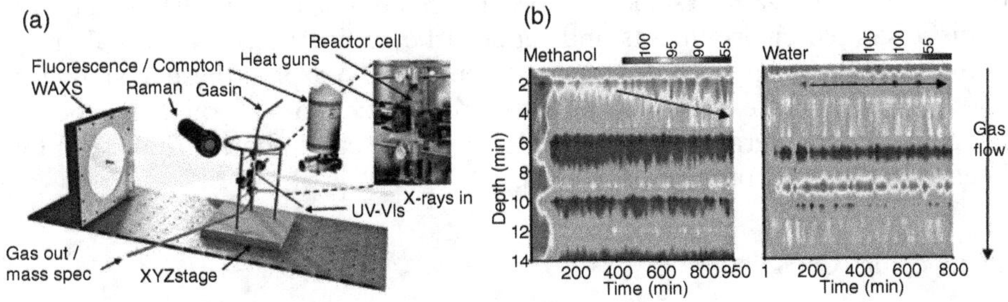

Figure 16.6. (a) A schematic of the novel multitechnique high-energy setup for spatiotemporal characterization studies on catalytic solids. The SR techniques WAXS, fluorescence, and Compton scattering were utilized, while in this experiment complementary Raman, UV-Vis, and online mass spectrometry were also used. The reactor consists of a large (5.5 mm ∅, 0.55 g) quartz reactor heated by air guns in an "open" architecture. (b) Spatiotemporal Compton scattering intensity plots through the depth of the reactor under both methanol and water vapor conditions over extended time periods. Reprinted from Reference 27, copyright John Wiley & Sons 2009. (See color insert.)

readily in a methanol or water atmosphere, transmission X-ray diffraction (XRD) data, the inelastic and fluorescence signals, UV-Vis, and Raman data were recorded at the same time down the length of the reactor bed (~1.1 cm) as a function of time. Since methanol is a reductant, MoO_3 was shown by all techniques to reduce to form MoO_2, although this was via the intermediate phases $Mo_xO_yC_z$ and MoO_xH_y. The rate at which MoO_2 formed varied down the length of the bed with the outlet of the reactor containing more of the phase than the inlet. Little evidence for Mo volatilization could be seen from the diffraction data while the sample was too dark to obtain a reliable spectroscopic signal; however, on closer inspection of the inelastic scattering/fluorescence data (Fig. 16.6), clear evidence could be seen of a loss of Mo due to an increase in the measured inelastic signal at the top of the reactor bed. It was proposed that this was caused by a density decrease and so could not be due to the formation of more MoO_2, since this is denser than the MoO_3, $Mo_xO_yC_z$, and MoO_xH_y phases and therefore in all likelihood was due to Mo loss. It was not possible to tell which of the observed phases was the principal source of the volatilized Mo, however. In contrast, no such changes were observed during treatment in water or during similar experiments performed on the Fe_2MoO_4 phase (the principle component of methanol selective oxidation catalysts) [26].

Of late, there has been something of a realization from within the catalysis community of the benefits of using very hard X-rays (here we define very hard X-rays as being >30 keV) for *in-situ* studies on catalysts. As remarked by O'Brien et al. [26] in a recent article, there are many advantages to be found in using such hard X-rays, including the possibility to obtain higher quality data (both with regard to signal-to-noise ratio and time resolution) using more realistic sample environments and reduced beam damage, and a number of publications have appeared which have been able to exploit these advantages [28, 29]. Furthermore, such hard X-rays can be utilized to carry out pencil beam imaging and tomography studies under reaction conditions, as has recently been performed by our group [30, 31].

16.2.4 Fuel Cells

16.2.4.1 Imaging Water Formation during Fuel Cell Operation In polymer electrolyte membrane (PEM) fuel cells, hydrogen and oxygen react to form water and electrical energy. For this to happen, protons migrate through the membrane and recombine with oxygen on the surface of the cathodic catalyst to form water. It has been proposed, however, that for optimal operation, rapid removal of water is required from the catalyst surface and the adjacent porous gas diffusion layer (GDL) [32]. Failure to remove this water can hinder the transport of reactant and product gases and limit the maximum achievable power density. In order to investigate this process, Manke et al. performed X-ray microcomputed tomography (micro-CT) (3 μm beam, 5 s acquisition time) to gain a realistic image of the transport and production of liquid water [32, 33]. The experiment was performed at 13 keV since this energy represents a suitable trade-off between obtaining sufficient transmission yet sensitivity to water formation. A 2048 × 2048 pixel camera setup (Princeton VersArray 2048B [Princeton Instruments, Trenton, NJ] with a Gadox scintillator screen) was used to capture image area sizes up to 7 × 7 mm with pixel sizes between 1.5 and 3.5 μm. Imaging was performed on a single cell through an 8 mm hole (sealed) in the metal end plates and focused on the 1 mm flow-field channels within the blank graphite composite plates. A typical raw image is shown in Figure 16.7, containing the flow-field channels on the cathode and anode side. The white spots observable in the image are the sought-after water clusters, which are displayed with such a high accuracy that it is possible to determine their contact angle and orientation in space. The amount of liquid water produced in the cell was demonstrated to depend strongly upon the electric output current density i_0 and the fraction of total gas input which is consumed in the cell at the electrodes. This can be seen in Figure 16.7, which contains an image of the distribution of water for two different operating conditions. Under comparatively "dry conditions," only small water clusters of about 10–20 μm in diameter were visible within the GDL pores, indicating that the majority of water is in the gas phase. Under wetter conditions the water clusters appear both more numerous and larger, reaching sizes up to 300 μm in diameter. Interestingly, the mode of water transport was observed to differ depending on the position in the cell; an "eruptive transport process" (where many droplets of liquid water form in the cathodic flow-field channel after ejection from the GDL) was observed at some spots while at others a continuous flow was observed. Eruptive transport is characterized by

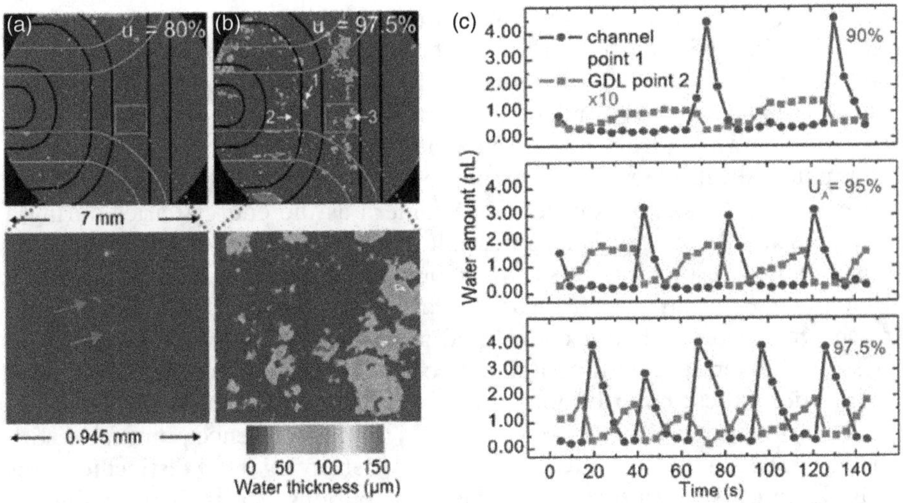

Figure 16.7. (a) and (b) Normalized radiographs of the operating fuel cell displaying the water distribution for two different operating conditions ("dry" [a] and "wet" [b]) with corresponding enlargements of the areas marked in red in the second row. Flow-field channels are marked by dark lines. Arrows point to examples of water clusters, whereas the white arrows labeled 1, 2, and 3 refer to points where water formation was profiled in the study. (c) Time dependence of the water amount at the three locations marked in (b) for three different utilization rates U_A during operation. Adapted with permission from Reference 32, copyright American Institute of Physics 2007.

the filling and emptying (through evaporation) of specific pores in the GDL over a period of 100 s and occurred with an almost constant frequency over the duration of the experiment (45 min), with the frequency also showing a dependency on the amount of water present in the vapor (U_A)—see Figure 16.7b. The authors concluded that the reason for water buildup occurring in only some pores was due to inhomogeneities in the GDL fiber network. Critically, then, it was concluded that such water probably plays a non-negligible role in water transport and particularly so at high current densities. Furthermore, these observations were very much in agreement with previously proposed theoretical models and the recent findings from fluorescent imaging studies.

16.2.4.2 Hard X-ray Nanoimaging of a Prospective Fuel Cell Although essentially a proof-of-principle study, Nelson and coworkers demonstrated recently the potential of hard X-ray microscopy (where $E > 2$ keV) to perform "chemical imaging" (XANES measurements) experiments with a spatial resolution comparable to that of STXM (~30 nm) and used it to image a fuel cell/

test object. A beam of energy ~9 keV was used to investigate the oxidation state of Ni in a NiO-coated Ni foil material [34]. The setup, shown in Figure 16.8, differs from the STXM setup in that the monochromatic X-rays are focused using a capillary condenser on to a sample located on a stage capable of rotation and *x*,*y*,*z* translation. The zone plate is located behind the sample and is able to translate with the sample so as to still be able to obtain a magnified image of the sample on the CCD camera as the energy varies during a XANES scan. As such, it is also possible to perform 3-D tomographic scans with "chemical resolution," as shown in Figure 16.8. Data from this initial study in Figure 16.8 "clean" spectra from regions where Ni and NiO predominate are shown in addition to a convoluted spectrum taken, where the physical states of Ni overlap. By fitting a linear combination of reference Ni and NiO contributions, these contributions could be easily deconvoluted. In particular, the strong "white line" at 8350 eV served as a clear reference point for distinguishing between the NiO and the Ni foil. By identifying these distinct features in the XANES spectra for the various contributions, it was then possible to perform "targeted" tomographic scans at the energy levels of interest, resulting in a reconstructed image of the total volume fraction of the phase in question using ray tracing methods. The final digitized form of the region of interest is shown in Figure 16.8b for Ni foil and NiO regions. The authors propose that ultimately such a technique could yield useful information into the 3-D behavior of Ni oxidation and the structural modifications that occur during lithium intercalation/deintercalation phenomena during battery charge and discharge cycles. However, and perhaps more importantly, the highly penetrative nature of the harder X-ray source allows for the acquisition of data through more "typical" reactor cell media (i.e., quartz) than that which can be used by its brother technique, STXM.

16.2.5 Imaging of a Single Catalyst Body

For maximum catalytic performance, the active component of a supported catalyst is normally dispersed upon structures known as catalyst bodies comprising the support phase shaped and sized according to the end application [35]. There are at least five different bodies commonly used: extrudates, pellets, monoliths, foams, and spheres. Crucial to the functioning of these catalyst "bodies" is the preparation method by which active and body are brought together [35, 36]. The efficiency of these catalysts depends on the speciation and microdistribution of the active phases and their stability during the catalytic reaction. It is therefore of importance to understand the physical and chemical processes involved in the preparation of active catalysts. The preparation can be manipulated to obtain uniform or nonuniform microdistributions, depending on which is required for optimal reaction conditions. Recently, X-ray imaging methods have been used to obtain a better understanding of the preparation process of transition metal containing catalyst extrudates. These studies have used, namely, a spectroscopic technique (absorption

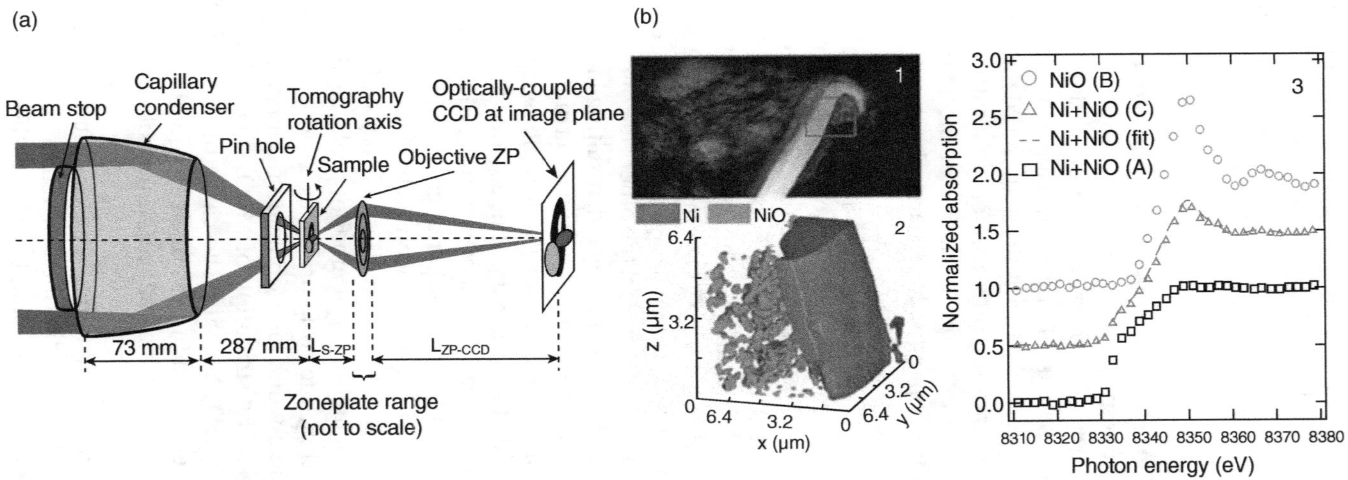

Figure 16.8. (a) Schematic of the hard X-ray STXM setup applied in the full field. Monochromatic X-rays are focused onto the sample using a capillary condenser. A combination of beamstop and pinhole block out the unfocused X-rays. For tomography measurement, the sample is rotated in the beam path and a series of absorption images are taken. (b) A single transmission image (1) taken from the series of images that comprise the tomographic scan of an Ni foil/NiO powder sample with the region of interest highlighted. The image is shown in absorption mode. The sample is reconstructed from the transmission images and converted to a digitized form (2) for further analysis. The distinction between oxidation states is enabled by image subtraction at the energy levels associated with the primary features in the spectra. Transmission images taken at X-ray energies of 8326, 8334, 8350, and 8370 eV illustrate the capability to distinguish Ni oxidation states. A linear combination of the Ni and NiO spectra, shown as the dashed line, reproduces the spectrum for the region of overlapping Ni–NiO and corroborates the spectra obtained. Adapted with permission from Reference 34, copyright American Institute of Physics 2011.

microtomography) as well as a novel scattering-based technique (tomographic energy dispersive diffraction imaging [TEDDI]) to study and obtain spatial information in both 2-D and 3-D, on the distribution of chemical species within catalyst bodies during preparation both under static and under dynamic conditions.

16.2.5.1 Dynamic Imaging of an Extrudate during Impregnation and Calcination Absorption microtomography has been used, for example, by Grunwaldt et al. to study the effect of immersion time upon the ingress of $CuCl_2$ in solution within γ-Al_2O_3 pellets of the same shape and size [37]. Immersion times of 1 and 10 min were used with the acquired data from the study and enabled for the clear observation of movement of a diffusion front toward the center of the pellet. However, the thickness of the impregnated part exhibited radial and axial variation, most likely caused by a variation in the pellet structure as a result of the extrusion process.

Beale et al., however, employed TEDDI to profile supported Mo/γ-Al_2O_3 and CoMo/γ-Al_2O_3 catalysts [38]. In truth, TEDDI is not limited to obtaining just scattering information since it utilizes a white beam and energy-dispersive detector, and thus also yields element composition information for elements where $Z > 23$. The technique works by pushing a volume element of the X-ray beam (for these experiments $x,y,z = 0.1 \times 0.1 \times 0.1$ mm) through the sample, with the diffracted beam/fluorescence signal then being collected on an energy-dispersive Ge detector, via postsample collimation, the other side of the source. Many such volume element locations are then collected, resulting in a full 2-D scan where each "point" on the 2-D dataset contains a full spectrum containing information on both the diffracting properties of the sample as well as the elemental composition [39]. Provided at least one unique "diagnostic" peak can be identified for each crystalline phase/element present, TEDDI offers a means of obtaining concentration maps for a number of crystalline phases and elements within a catalyst body in a time- and space-resolved manner. Figure 16.9 shows a typical detector response from a "pixel" of a Mo/γ-Al_2O_3 catalyst body (1.5 mm $\phi \times 10$ mm l) after calcination. A number of peaks can be identified in this pattern, which are either due to fluorescence or diffraction. The distribution of these signals in the z,y-direction obtained from peak profiling point to the formation of an eggshell distribution of MoO_3 (Fig. 16.9b) and a catalyst that was not effective for HDS activity. The cause of this "eggshell" distribution was proposed to be either due to insufficient equilibration time during impregnation step or due to a too high concentration of molybdenum used.

When coimpregnating with Co, no crystalline phases result, as the 2-D cross-sectional scans *revealed* broadly even distribution in these elements over the sample. However, down the length of the sample, various "hot spots" were discovered, implying a buildup of the active components in specific parts of the extrudate, although it was not clear why the buildup occurred specifically at these positions. This, however, may be again related to variation in the

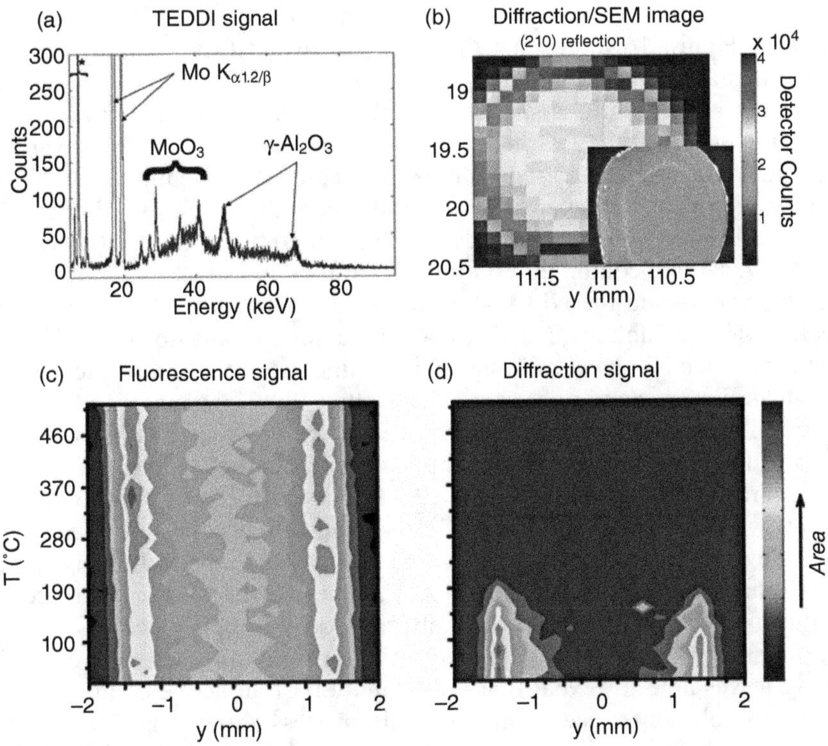

Figure 16.9. (a) TEDDI detector signal together with a typical (b) Y–Z diffraction map created from the (210) Bragg peak from the MoO_3 phase from a Mo/Al_2O_3 catalyst body after calcination at 500°C. Confirmatory evidence for the preferential location of Mo at the periphery of the pellet is also given in the SEM image (inset). (c) $Ni_{K\alpha}$ fluorescence signal at 7.5 keV recorded during the *in-situ* calcination of $[Ni(en)_3](NO_3)_2/\gamma\text{-}Al_2O_3$ pellet and (d) YT maps for the diffraction peak at 39.3 keV of the $[Ni(en)_3](NO_3)_2$ crystalline phase. The $y = 0$ point corresponds to the center of the pellet. Reprinted from Reference 40, copyright American Chemical Society 2009. (See color insert.)

internal structure of the pellet [38]. Through the development of an environmental cell located on a movable (x,y,z) stage with temperature and gas control, the application of this technique was extended to the performing of time-resolved 2-D TEDDI measurements. In this work, Espinosa-Alonso et al. studied *in situ* the formation of two $Ni/\gamma\text{-}Al_2O_3$ hydrogenation catalysts prepared via the impregnation of two different precursor complexes ($[Ni(en)_3](NO_3)_2$ and $[Ni(en)(H_2O)_4]Cl_2$) onto 3×3 μm catalyst bodies [40]. The motivation for studying these two complexes was that previously it had been shown that after impregnation and drying the precursors were distributed differently, the nitrate precursor being more evenly distributed than the chloride. However, it was not known whether this would remain the case after calcination. Shown in Figure 16.9c,d, are the results from the calcination of the $[Ni(en)_3](NO_3)_2$ precursor. Figure 16.9d reveals that the $[Ni(en)_3](NO_3)_2$ sample contained crystalline material at the beginning pertaining to a $[Ni(en)_3](NO_3)_2$ precursor phase, with a preferential distribution at the peripheries of the extrudate. On calcination in He, the precursor breaks down on reaching approximately 200°C to yield 5-nm-sized crystallites of face-centered cubic (fcc) Ni, the majority of which was again located predominantly at the periphery ("egg-shell" distribution) of the 2-D cross section. In contrast, the $[Ni(en)(H_2O)_4]Cl_2/\gamma\text{-}Al_2O_3$ sample contained no initial crystalline phase although Ni fluorescence maps revealed the Ni to be more evenly distributed. At the beginning of calcination an unknown crystalline precursor initially forms, which also yields a final fcc Ni phase, although this occurs via a number of intermediate steps, including a second unknown phase, followed by Ni_3C and finally crystalline fcc Ni. Interestingly, both 2-D *yz* and the dynamic *xy* scans revealed significant differences in the phase composition across the sample; for example, in the core of the pellet, Ni_3C species accounted for 40% of the total Ni present, while at the edge this amounts to less than 20%—the remainder comprising fcc Ni. As the dynamic *xy* scans revealed, fcc Ni is the only crystalline Ni phase present on reaching 460°C, although the average particle size was much larger 24–34 nm, most likely caused by the presence of Cl. While then this study could not provide mechanistic insight into the phase evolution process, it was able to reveal that the formation of an active component is a multistep evolution process and that this process shows significant spatial variation. Improving the acquisition time and data quality will be key to obtaining more insight into the evolutionary process, leading to the formation of a final crystalline phase. To a large extent, this could be realized by combining a brighter beam with a more sensitive detector, although considering the gain in signal quality when using high energy monochromatic beams, performing TADDI-type, X-ray diffraction microtomography (XRD-CT) experiments are an altogether better option [41–43].

16.2.5.2 Imaging the Effects of the Deposition Method on the Catalytic Activity of Ni/γ-Al_2O_3-Coated Metal Foams Structured metal catalysts, formed by an active catalyst layer deposited on a structured support such as a honeycomb, plate, foil, or foam, possess a number of advantages over pelletized catalysts including lower pressure drop across the bed, a higher mechanical strength, and a large geometric surface area. Furthermore, their greater heat transfer capacity minimizes thermal gradients and hot spots. A key challenge, however, to their utilization is the coating of the active on the support so as to ensure good performance and stability. Conventional wash-coat methods have been compared with electrochemical deposition (ECD) methods for the production of a steam-reforming catalyst (Ni/Al_2O_3) on a metallic FeCrAlY support in a recent publication by Basile et al., and the resultant coatings were characterized in terms of both structure and chemical composition using a combination of combined XRF, X-ray powder diffraction

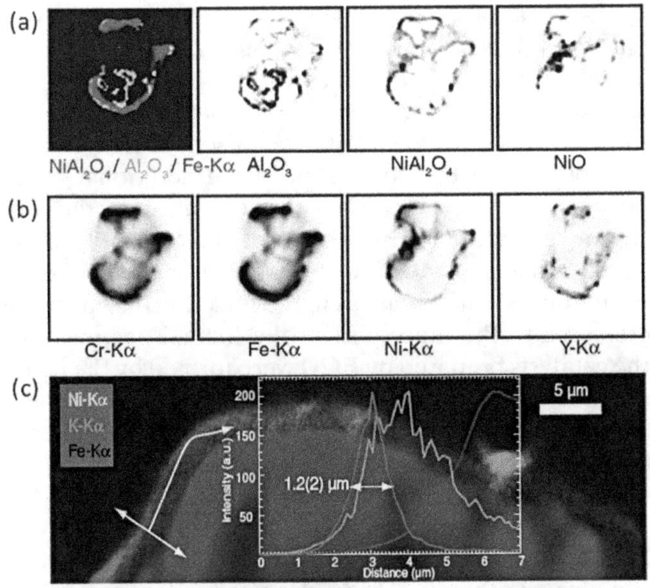

Figure 16.10. (a) Diffraction phase maps created from diagnostic diffraction peaks obtained by XRF/XRPD tomography measurements. (b) Corresponding XRF maps. (c) RGB plot of elemental maps of Ni, K and Fe [49 × 14.5 μm] of an ECD sample prior to calcination obtained by XRF tomography. The thickness of the hydrotalcite layer represented by its nickel content is 1.2 ± 0.2 μm, averaged from 34 one-dimensional cross sections. Adapted with permission from Reference 44, copyright Wiley-VCH 2010. (See color insert.)

tomography measurement, and separate absorption tomography [44]. The combined μ-XRF and μ-XRD were performed at 25 keV with a beam 3.4 × 1.8 μm in size, with the μ-XRF data being collected using two Vortex silicon drift detectors (Hitachi High-Technologies Science America, Inc., Northridge, CA) at 90°.

Initial X-ray absorption tomography data for the ECD sample revealed the foams to be both hollow yet containing a thin layer of Al_2O_3 at the surface, which provided the anchoring points for the active phase. Corresponding μ-XRF data shown in Figure 16.10b revealed that the metallic FeCrAlY support comprised of predominantly well-dispersed Cr and Fe, although with Y localized in spots. However, the μ-XRF also revealed that in some cases the active (Ni) coverage was not uniform for both ECD samples. μ-XRD tomography data showed that regions containing large amounts of "active" comprised of predominantly NiO in addition to α-Al_2O_3 and $NiAl_2O_4$; the source of this excess NiO was thought to be due to an excessive buildup of the hydrotalcite precursor. Analysis of the XRD peak full widths yielded an average active coating thickness of 2 μm; this was also confirmed by μ-XRF tomography data as shown in Figure 16.10c. In contrast, the conventionally prepared sample contained on average a thicker active coating, although this was somewhat more uneven; some large zones with no coating were observed next to areas with thicker coating. More α-Al_2O_3 was also found on the sample which was thought to be due to the application of an additional boehmite wash, although it was not possible to rule out that oxidation of the foam had also taken place. However, unlike the ECD samples, the NiO and $NiAl_2O_4$ distributions were very similar and much more uniform and therefore showed the same molar ratios on the microscale as in the bulk. Furthermore, the washcoat had also ingressed into the inner part of the foam. Thus overall it was concluded that the catalysts prepared by ECD were formed by the phases expected from the decomposition of HT compounds.

The catalysts were then compared in performance for endothermic methane activation. It was observed that while one of the ECD samples deactivated gradually with time on stream, the other showed a good and stable performance with a performance to rival that of a commercial catalyst, despite its lower nickel loading. It was proposed by μ-XRD tomography that the cause of deactivation for one of the ECD samples was the loss of the coating during reaction; the resultant exposed parts lead to carbon deposition over the uncovered support, or partial sintering of the catalyst. In contrast, the steady-state performance of the other ECD sample pointed toward a good stability/adhesion/uniformity of the catalytic layer. It should be noted, however, that in the active ECD sample, both the fcc Ni phase and the alumina-containing layers appeared to be more intermixed after reaction, suggesting that the severe reaction conditions (hydrothermal conditions and high temperature) cause migration of Al from the metallic foam into the catalyst washcoat, and nickel diffusion into the alumina takes place.

16.3 CURRENT AND FUTURE DEVELOPMENTS IN THE FIELD AND THE POSSIBILITIES FOR CATALYST IMAGING

For the sake of brevity, the examples given earlier serve to illustrate what is possible but really only represent a tip of the iceberg. There are a number of possibilities to image catalysts with a variety of SR-based techniques and in the succeeding sections we detail some additional techniques which have yet to be employed for such purposes.

16.3.1 Nanodiffraction Imaging

To date, STXM of the hard and soft varieties have delivered spectroscopic information with spatial resolutions in the order of 10 s of nm. Recently however, data were reported on the diffraction equivalent of STXM, known as scanning X-ray diffraction microscopy (SXDM) [45]. With a pixel size to rival (18 nm), to date this technique has been employed in both transmission and, more recently, in reflection mode [46]. In some of the first experiments performed, the technique was used to image a Fresnel zone plate. Diffraction patterns (201×201, 300-nm illumination spot) were collected from the sample, each one of the patterns being collected for only 50 ms—this fast collection time being possible in no small part due to the use of the recently developed fast single-photon-counting Pilatus detector (SLS Detector Group, Paul Scherrer Institute, Villigen, Switzerland) with no readout noise [47]. While clearly only a proof-of-principle experiment, it is clear that there is tremendous potential for studying catalyst samples in order to see, for example, the behavior/role of crystalline nanoparticles or supports in a catalytic process or, for example, the presence of nanodomains of crystalline material in zeolite samples [48]. It is predicted that a combination of higher coherent flux and improved focusing optics should eventually provide conditions for sub-10 nm resolution. Should it be possible, for example, to obtain data of sufficient quality so as to be able to perform Rietveld analysis on a pixel-wise basis, using, for example, parametric methods, it would be possible to characterize the sample beyond a simple phase composition analysis [49]. Such an approach could yield new insight into metal–support interactions, potentially leading to a better understanding as to why some catalysts are more active than others despite possessing the same composition [50].

16.3.2 Broadening the Scope of Scanning X-ray Microscopy

Ultimately researchers would like to examine the role/behavior in a single particle for a catalytic process. Hence the momentous drive to better the currently available resolution of microscopy techniques offers some exciting possibilities in the future. Meanwhile, there is an alternative approach that has been developed which can enable the possibility to obtain better spatial resolution without the need to develop a stable nanobeam. In a recent development, nano-/μ-XRD has been coupled with atomic force microscopy (AFM) [51, 52]. The benefit of coupling XRD with AFM is that the subnanometer spatial resolution of the latter technique can be used to provide more detailed structural insight than the conventional nano-/μ-XRD measurements, yielding, for example, information regarding the uniformity of form and shape of a number of nanoobjects across a sample. This combination allows for alignment of the sample with respect to the AFM tip and the incident X-ray beam to a very high accuracy, and therefore conceivably the “lack” of resolution from the XRD perspective could be overcome by intelligent sample preparation such as the dispersal of nanoparticles on a sample grid. In the first reported study using this setup, the elastic response in nanoobjects was quantified during the application of a well-controlled uniaxial pressure. In these first conceptual experiments, SiGe islands grown on Si(001) were investigated and when pressure was applied, correspondence could be observed between variations in the object lattice parameters with shifts of the resonance frequency of the AFM tuning fork, thereby allowing for the inference of the Young's modulus from an individual nanoobject. Although not sharing the same spatial resolution, the idea of coupling a non-SR technique to obtain better spatial resolution has also been developed elsewhere with, for example, the coupling of a micro-Raman probe head to which delivers a focused laser spot (ca. 1 μm in diameter) to the same position on the sample as the X-ray beam (for μ-XRD). The microprobe is coupled to the beamline control software for automatic triggering during X-ray data collection and, therefore, the two techniques can be used to simultaneously map an area of the sample [53]. Furthermore, an absorption-based setup has been developed, whereby the optical fiber tip of a scanning near-field optical microscope (SNOM), which is used to detect the X-ray excited optical luminescence (XEOL) of a sample, was irradiated with an SR beam [54, 55]. The "XAS-SNOM" microscope prototype works by collecting the XEOL in near field using a tapered optical fiber probe glued to an oscillating quartz tuning fork. A shear force feedback is employed both to regulate the probe–sample distance and to record the topographical image of the surface. The absolute positioning of the tip and of the sample with respect to the X-ray microbeam is performed using a six-independent-axis system. In order to collect the XEOL signal using the optical fiber tip, the imaging is carried out by scanning the sample under the probe that is kept fixed with respect to the radiation beam. With such a setup, a lateral resolution of about 100–200 nm and vertical resolution of 3 nm has been possible and has, to date, been used to study pure ZnO and mixed $ZnWO_4$–ZnO nanostructured thin films, obtaining atomic maps of both Zn and W distributions. In addition, it is also possible to record high-quality XANES spectra for chemical imaging purposes; thus, with improved lateral resolution, this technique combination could rival/complement TXM as the ideal nanoimaging technique for characterizing catalyst solids.

16.3.3 Multiple Technique Data

As was demonstrated in a number of examples previously discussed, it is possible to simultaneously collect multiple signals from the same incident source beam so as to yield different types of information on a sample, often at no additional experimental cost. In some cases, two different signals are collected on the same detector (i.e., the fluorescence and diffraction signals in TEDDI) or else a separate detector is used for each technique (XRF and X-ray microtomography [XMT]) [38, 56]. An additional advantage of such setups, particularly so in the former case, is that the signal comes from the same spot on the sample. Other such technique combinations that have yet been used to image catalysts could be, for example, simultaneous small-angle and wide-angle X-ray scattering (SAXS/WAXS) signals or pair distribution function (PDF)/SAXS/WAXS in either a radiographic or tomographic format, provided a 2-D detector covering a sufficiently large *d* range were available [57]. By utilizing a photodiode as a beamstop, it would also be possible to collect XMT data, thus allowing for the simultaneous acquisition of XMT/SAXS/WAXS data, but perhaps more importantly the collection of the transmitted X-ray signal would enable facile "correction" of signal attenuation when reconstructing the scattering data [58]. Somewhat more interesting are the recent endeavors to develop synchrotron beamlines in which two separate beams can be focused on the same part of the sample at the same time [59]. This allows, for example, for simultaneous imaging with two beams utilizing, for example, a combination of fixed monochromator/scanning monochromator setup for combined diffraction/XAFS scanning. Alternatively, and as has been recently proposed, a portion of the storage ring beam could be taken off at two points of a bending magnet to yield both IR and X-ray radiation which could be focused to the same part of the sample, resulting in a very powerful technique combination for catalyst imaging purposes [60].

16.3.4 Obtaining More Chemical Information

For spectroscopy-based techniques, one of the major problems with collecting imaging data is the limited chemical information that can be extracted from it. This is primarily due to the current apparatus being limited in the energy range over which a spectra can be acquired (typically this is limited to the XANES region of at best a few 100 eV) while maintaining the desired nanometer spatial resolution. While for the very soft elements (i.e., C, N, O) this is not a problem, for harder elements more chemical information (i.e., bond lengths, coordination numbers, disorder) is available from the EXAFS region of an XAFS spectrum and, as such, should it be possible to obtain such data from a single particle, would prove very powerful. One could imagine that if it were not possible to maintain the spatial resolution during a monochromator sweep, instead the spectra could be divided up into several regions, whereby a full spectrum per pixel could be obtained by stitching the regions together. To some extent at least, it has been shown that it is possible to acquire at least μ-EXAFS data at both soft and hard energies albeit with significant trade-off in time resolution since a single spectrum takes 10 s of minutes to acquire [22, 25, 61].

Perhaps, then, an altogether more promising approach to improving the chemical resolution is through the acquisition of data using a recently developed hard X-ray spectroscopic tomography with a unique sensitivity to light elements [62–64]. The technique measures the inelastic X-ray scattered, or X-ray Raman scattered signal from a single shot of a square ($30 \times 30\ \mu m^2$) X-ray beam; thus, the technique is a form of direct tomography (DT) as opposed to angle-scan reconstruction-based CT. The inelastic X-ray scattering (IXS) signal comes from core electron excitations and gives similar information to XAFS since the energy transfer corresponds to the absorbed photon energy. The inelestically scattered X-ray signal is recovered from the energy loss spectrum via a curved analyzer crystal measured with a very high energy resolution (~1 eV) and is therefore not only element specific but can be sensitive to the chemical nature and type of species present. This is particularly significant for soft edges since it enables their measurement with hard X-rays from within bulk samples up to the millimeter or even centimeter scale. Clearly, then, such an approach could prove extremely revealing when imaging catalyst bodies, for example, providing insight into the behavior of the oxide support or else the complexing agent during impregnation. Perhaps more pertinently, should it be possible to perform imaging experiments at the nanoscale with such an apparatus, then the problem of specialized nanoreactors that are so necessary for STXM experiments would be circumvented.

16.3.5 Looking beyond the Storage Ring

Finally, it is worth considering some recent developments in source technology "beyond the ring," as it were, and the influence that these could have on future imaging studies. Perhaps first and foremost, one cannot ignore the significance and potential of the free-electron laser (FEL). Recent projects such as the soft X-ray FLASH facility in Germany and the Linac Coherent Light Source at SLAC National Accelerator Laboratory are poised to take center stage as the premier source of tunable, intense, coherent photons [65, 66]. FELs, for example, provide an unprecedented enhancement anywhere from 10^6 to 10^{11} in the peak brightness when compared to an SR storage ring, enabling the delivery of intense ultrashort pulses allowing for the acquisition of femtosecond time resolution and microelectronvolt spectral resolution. Such a facility is therefore able to provide "single-shot" X-ray imaging experiments of single particles or else could be used to explore the temporal evolution of phenomena such as atomic motion, phase transitions, and chemical reactions in real time [67, 68]. However, one should be aware that there can be a downside to the increased brightness of FELs in that a more intense beam leads to an increased thermal load and possibility for photoionization. This of course needs to be taken into consideration for time-resolved imaging purposes and that sacrificial trade-offs might need to be made between the spatial, time, and chemical resolution as a function of the overall radiation dose to the sample in an experiment. Ways exist in which the beam exposure can be effectively managed, that is, continuous movement of the beam or else by controlling the exposure by shuttering, or else maybe via a judicious choice of wavelength or edge energy (i.e., K vs. L_3) for sampling [69]. However, one could also consider redesigning the experiment so as to obtain an average image of the structure from a series of instantaneous "single-shot" experiments on a number of samples as has been done for the imaging of protein structures [68].

Despite the major advantages of SR light, the development of new laboratory X-ray sources continues. Recent developments have concentrated on making light sources which are small enough to fit into a laboratory yet possess fluxes that are comparable to those of a third-generation source. Examples include high-intensity microfocus X-ray sources with dramatically higher flux and ultrashort duration pulse, enabling dynamic measurements, and the highly versatile and commercially available tabletop synchrotrons which have already been successfully employed for, among other things, phase contrast and projection XMT, XAFS, XRF analysis, and IR absorption spectroscopy [70, 71]. The expected developments here in both source spectral brightness and detector sensitivity may well eventually allow for SR imaging in the home laboratory.

ACKNOWLEDGMENTS

The authors would like to thank the Nederlandse Organisatie voor Wetenschappelijk Onderzoek (NWO) and the National Research School Catalysis Controlled by Chemical Design (NRSCC) for funding. We would also like to thank Dr. Simon Jacques, Dr. Marco Di Michiel, Dr. Matthew G. O'Brien, and Prof. Paul Barnes for their insight and technical know-how which has led to the publication of some of the research work featured in this chapter.

REFERENCES

[1] Foster, A. J., Lobo, R. F. (2010) *Chemical Society Reviews*, 39, 4783.
[2] Thomas, J. M. (1997) *Chemistry—A European Journal*, 3, 1557.
[3] Bruckner, A. (2003) *Catalysis Reviews: Science and Engineering*, 45, 97.
[4] Tinnemans, S. J., Mesu, J. G., Kervinen, K., Visser, T., Nijhuis, T. A., Beale, A. M., Keller, D. E., van der Eerden, A. M. J., Weckhuysen, B. M. (2006) *Catalysis Today*, 113, 3.
[5] Sankar, G., Thomas, J. M. (1999) *Topics in Catalysis*, 8, 1.
[6] Beale, A. M., Jacques, S. D. M., Weckhuysen, B. M. (2010) *Chemical Society Reviews*, 39, 4656.
[7] Weckhuysen, B. M. (2009) *Angewandte Chemie International Edition*, 48, 4910.
[8] Urakawa, A., Baiker, A. (2009) *Topics in Catalysis*, 52, 1312.
[9] Grunwaldt, J. D., Schroer, C. G. (2010) *Chemical Society Reviews*, 39, 4741.
[10] Barthomeuf, D. (1987) *Materials Chemistry and Physics*, 17, 49.
[11] Stavitski, E., Kox, M. H. F., Swart, I., de Groot, F. M. F., Weckhuysen, B. M. (2008) *Angewandte Chemie International Edition*, 47, 3543.
[12] Stavitski, E., Pidko, E., Kox, M., Hensen, E., van Santen, R., Weckhuysen, B. (2010) *Chemistry—A European Journal*, 16, 9340.
[13] Karwacki, L., Kox, M. H. F., de Winter, D. A. M., Drury, M. R., Meeldijk, J. D., Stavitski, E., Schmidt, W., Mertens, M., Cubillas, P., John, N., Chan, A., Kahn, N., Bare, S. R., Anderson, M., Kornatowski, J., Weckhuysen, B. M. (2009) *Nature Materials*, 8, 959.
[14] Kox, M. H. F., Domke, K. F., Day, J. P. R., Rago, G., Stavitski, E., Bonn, M., Weckhuysen, B. M. (2009) *Angewandte Chemie International Edition*, 48, 8990.
[15] Kox, M. H. F., Mijovilovich, A., Sattler, J., Stavitski, E., Weckhuysen, B. M. (2010) *ChemCatChem*, 2, 564.

[16] Regalbuto, J. (ed.) (2007) *Catalyst Preparation: Science and Engineering*, CRC Press, Boca Raton, FL.

[17] de Groot, F. M. F., de Smit, E., van Schooneveld, M. M., Aramburo, L. R., Weckhuysen, B. M. (2010) *ChemPhysChem*, 11, 951.

[18] de Smit, E., Swart, I., Creemer, J. F., Hoveling, G. H., Gilles, M. K., Tyliszczak, T., Kooyman, P. J., Zandbergen, H. W., Morin, C., Weckhuysen, B. M., de Groot, F. M. F. (2008) *Nature*, 456, 222.

[19] de Smit, E., Swart, I., Creemer, J. F., Karunakaran, C., Bertwistle, D., Zandbergen, H. W., de Groot, F. M. F., Weckhuysen, B. M. (2009) *Angewandte Chemie International Edition*, 48, 3632.

[20] Zhong, J., Xie, T., Deng, J., Sun, X., Pan, X., Bao, X., Wu, Z. (2011) *Chemical Communications*, 47, 8373.

[21] Zhou, J. G., Wang, J., Fang, H. T., Wu, C. X., Cutler, J. N., Sham, T. K. (2010) *Chemical Communications*, 46, 2778.

[22] Tada, M., Ishiguro, N., Uruga, T., Tanida, H., Terada, Y., Nagamatsu, S.-I., Iwasawa, Y., Ohkoshi, S.-I. (2011) *Physical Chemistry Chemical Physics*, 13, 14910.

[23] Grunwaldt, J. D., Hannemann, S., Schroer, C. G., Baiker, A. (2006) *Journal of Physical Chemistry B*, 110, 8674.

[24] Hannemann, S., Grunwaldt, J. D., van Vegten, N., Baiker, A., Boye, P., Schroer, C. G. (2007) *Catalysis Today*, 126, 54.

[25] Schroer, C. G., Kuhlmann, M., Gunzler, T. F., Lengeler, B., Richwin, M., Griesebock, B., Lutzenkirchen-Hecht, D., Frahm, R., Ziegler, E., Mashayekhi, A., Haeffner, D. R., Grunwaldt, J. D., Baiker, A. (2003) *Applied Physics Letters*, 82, 3360.

[26] O'Brien, M. G., Beale, A. M., Jacques, S. D. M., Di Michiel, M., Weckhuysen, B. M. (2011) *Applied Catalysis A: General*, 391, 468.

[27] O'Brien, M. G., Beale, A. M., Jacques, S. D. M., Di Michiel, M., Weckhuysen, B. M. (2009) *ChemCatChem*, 1, 99.

[28] O'Brien, M. G., Beale, A. M., Jacques, S. D. M., Buslaps, T., Honkimaki, V., Weckhuysen, B. M. (2009) *Journal of Physical Chemistry C*, 113, 4890.

[29] Newton, M. A., Di Michiel, M., Kubacka, A., Fernandez-Garcia, M. (2010) *Journal of the American Chemical Society*, 132, 4540.

[30] O'Brien, M. G., Jacques, S. D., Di Michiel, M., Weckhuysen, B. M., Beale, A. M. (2012) *Chemical Science*, 3, 509.

[31] Jacques, S. D. M., Di Michiel, M. D., Beale, A. M., Sochi, T., O'Brien, M. G., Espinosa-Alonso, L., Weckhuysen, B. M., Barnes, P. (2011) *Angewandte Chemie International Edition*, 50, 10148.

[32] Manke, I., Hartnig, C., Grunerbel, M., Lehnert, W., Kardjilov, N., Haibel, A., Hilger, A., Banhart, J., Riesemeier, H. (2007) *Applied Physics Letters*, 90, 174105.

[33] Manke, I., Hartnig, C., Kardjilov, N., Riesemeier, H., Goebbels, J., Kuhn, R., Kruger, P., Banhart, J. (2010) *Fuel Cells*, 10, 26.

[34] Nelson, G. J., Harris, W. M., Izzo, J. R., Grew, K. N., Chiu, W. K. S., Chu, Y. S., Yi, J., Andrews, J. C., Liu, Y. J., Pianetta, P. (2011) *Applied Physics Letters*, 98, 173109.

[35] Neimark, A. V., Kheifez, L. I., Fenelonov, V. B. (1981) *Industrial & Engineering Chemistry Product Research and Development*, 20, 439.

[36] Espinosa-Alonso, L., Beale, A. M., Weckhuysen, B. M. (2010) *Accounts of Chemical Research*, 43, 1279.

[37] Grunwaldt, J. D., Kimmerle, B., Baiker, A., Boye, P., Schroer, C. G., Glatzel, P., Borca, C. N., Beckmann, F. (2009) *Catalysis Today*, 145, 267.

[38] Beale, A. M., Jacques, S. D. M., Bergwerff, J. A., Barnes, P., Weckhuysen, B. M. (2007) *Angewandte Chemie International Edition*, 46, 8832.

[39] Hall, C., Barnes, P., Cockcroft, J. K., Jacques, S. D. M., Jupe, A. C., Turrillas, X., Hanfland, M., Hausermann, D. (1996) *Analytical Communications*, 33, 245.

[40] Espinosa-Alonso, L., O'Brien, M. G., Jacques, S. D. M., Beale, A. M., de Jong, K. P., Barnes, P., Weckhuysen, B. M. (2009) *Journal of the American Chemical Society*, 131, 16932.

[41] Bleuet, P., Welcomme, E., Dooryhee, E., Susini, J., Hodeau, J. L., Walter, P. (2008) *Nature Materials*, 7, 468.

[42] Cernik, R. J., Khor, K. H., Hansson, C. (2008) *Journal of the Royal Society Interface*, 5, 477.

[43] Middelkoop, V., Boldrin, P., Peel, M., Buslaps, T., Barnes, P., Darr, J. A., Jacques, S. D. M. (2009) *Chemistry of Materials*, 21, 2430.

[44] Basile, F., Benito, P., Bugani, S., De Nolf, W., Fornasari, G., Janssens, K., Morselli, L., Scavetta, E., Tonelli, D., Vaccari, A. (2010) *Advanced Functional Materials*, 20, 4117.

[45] Thibault, P., Dierolf, M., Menzel, A., Bunk, O., David, C., Pfeiffer, F. (2008) *Science*, 321, 379.

[46] Roy, S., Parks, D., Seu, K. A., Su, R., Turner, J. J., Chao, W., Anderson, E. H., Cabrini, S., Kevan, S. D. (2011) *Nature Photonics*, 5, 243.

[47] Broennimann, C., Eikenberry, E. F., Henrich, B., Horisberger, R., Huelsen, G., Pohl, E., Schmitt, B., Schulze-Briese, C., Suzuki, M., Tomizaki, T., Toyokawa, H., Wagner, A. (2006) *Journal of Synchrotron Radiation*, 13, 120.

[48] Davis, T. M., Drews, T. O., Ramanan, H., He, C., Dong, J. S., Schnablegger, H., Katsoulakis, M. A., Kokkoli, E., McCormick, A. V., Penn, R. L., Tsapatsis, M. (2006) *Nature Materials*, 5, 400.

[49] Stinton, G. W., Evans, J. S. O. (2007) *Journal of Applied Crystallography*, 40, 87.

[50] Kasatkin, I., Kurr, P., Kniep, B., Trunschke, A., Schlogl, R. (2007) *Angewandte Chemie International Edition*, 46, 7324.

[51] Rodrigues, M. S., Cornelius, T. W., Scheler, T., Mocuta, C., Malachias, A., Magalhaes-Paniago, R., Dhez, O., Comin, F., Metzger, T. H., Chevrier, J. (2009) *Journal of Applied Physics*, 106, 103525.

[52] Scheler, T., Rodrigues, M., Cornelius, T. W., Mocuta, C., Malachias, A., Magalhaes-Paniago, R., Comin, F., Chevrier, J., Metzger, T. H. (2009) *Applied Physics Letters*, 94, 023109.

[53] Davies, R. J., Burghammer, M., Riekel, C. (2009) *Journal of Synchrotron Radiation*, 16, 22.

[54] Larcheri, S., Rocca, F., Jandard, F., Pailharey, D., Graziola, R., Kuzmin, A., Purans, J. (2008) *The Review of Scientific Instruments*, 79, 013702.

[55] Larcheri, S., Rocca, F., Pailharey, D., Jandard, F., Graziola, R., Kuzmin, A., Kalendarev, R., Purans, J. (2009) *Micron*, 40, 61.

[56] Jones, K. W., Feng, H., Lanzirotti, A., Mahajan, D. (2005) *Topics in Catalysis*, 32, 263.

[57] Becker, J., Bremholm, M., Tyrsted, C., Pauw, B., Jensen, K. M. O., Eltzholt, J., Christensen, M., Iversen, B. B. (2010) *Journal of Applied Crystallography*, 43, 729.

[58] Nikitenko, S., Beale, A. M., van der Eerden, A. M. J., Jacques, S. D. M., Leynaud, O., O'Brien, M. G., Detollenaere, D., Kaptein, R., Weckhuysen, B. M., Bras, W. (2008) *Journal of Synchrotron Radiation*, 15, 632.

[59] Simmance, K., Sankar, G., Bell, R. G., Prestipino, C., van Beek, W. (2010) *Physical Chemistry Chemical Physics*, 12, 559.

[60] Marcelli, A., Hampai, D., Xu, W., Malfatti, L., Innocenzi, P. (2009) *Acta Physica Polonica A*, 115, 489.

[61] Schroer, C. G., Kuhlmann, M., Gunzler, T. F., Lengeler, B., Richwin, M., Griesebock, B., Lutzenkirchen-Hecht, D., Frahm, R., Ziegler, E., Mashayekhi, A., Haeffner, D. R., Grunwaldt, J. D., Baiker, A. (2004) *Proceedings of SPIE 5535*, Developments in X-Ray Tomography IV, Bellingham, WA, p. 715.

[62] Glatzel, P., Bergmann, U. (2005) *Coordination Chemistry Reviews*, 249, 65.

[63] de Groot, F. (2005) *Coordination Chemistry Reviews*, 249, 31.

[64] Huotari, S., Pylkkanen, T., Verbeni, R., Monaco, G., Hamalainen, K. (2011) *Nature Materials*, 10, 489.

[65] Barletta, W. A., Bisognano, J., Corlett, J. N., Emma, P., Huang, Z., Kim, K. J., Lindberg, R., Murphy, J. B., Neil, G. R., Nguyen, D. C., Pellegrini, C., Rimmer, R. A., Sannibale, F., Stupakov, G., Walker, R. P., Zholents, A. A. (2010) *Nuclear Instruments & Methods in Physics Research. Section A, Accelerators, Spectrometers, Detectors and Associated Equipment*, 618, 69.

[66] Tiedtke, K., Azima, A., von Bargen, N., Bittner, L., Bonfigt, S., Dusterer, S., Faatz, B., Fruhling, U., Gensch, M., Gerth, C., Guerassimova, N., Hahn, U., Hans, T., Hesse, M., Honkavaar, K., Jastrow, U., Juranic, P., Kapitzki, S., Keitel, B., Kracht, T., Kuhlmann, M., Li, W. B., Martins, M., Nunez, T., Plonjes, E., Redlin, H., Saldin, E. L., Schneidmiller, E. A., Schneider, J. R., Schreiber, S., Stojanovic, N., Tavella, F., Toleikis, S., Treusch, R., Weigelt, H., Wellhofer, M., Wabnitz, H., Yurkov, M. V., Feldhaus, J. (2009) *New Journal of Physics*, 11, 111104.

[67] Seibert, M. M., Ekeberg, T., Maia, F., Svenda, M., Andreasson, J., Jonsson, O., Odic, D., Iwan, B., Rocker, A., Westphal, D., Hantke, M., DePonte, D. P., Barty, A., Schulz, J., Gumprecht, L., Coppola, N., Aquila, A., Liang, M. N., White, T. A., Martin, A., Caleman, C., Stern, S., Abergel, C., Seltzer, V., Claverie, J. M., Bostedt, C., Bozek, J. D., Boutet, S., Miahnahri, A. A., Messerschmidt, M., Krzywinski, J., Williams, G., Hodgson, K. O., Bogan, M. J., Hampton, C. Y., Sierra, R. G., Starodub, D., Andersson, I., Bajt, S., Barthelmess, M., Spence, J. C. H., Fromme, P., Weierstall, U., Kirian, R., Hunter, M., Doak, R. B., Marchesini, S., Hau-Riege, S., Frank, M., Shoeman, R. L., Lomb, L., Epp, S. W., Hartmann, R., Rolles, D., Rudenko, A., Schmidt, C., Foucar, L., Kimmel, N., Holl, P., Rudek, B., Erk, B., Homke, A., Reich, C., Pietschner, D., Weidenspointner, G., Struder, L., Hauser, G., Gorke, H., Ullrich, J., Schlichting, I., Herrmann, S., Schaller, G., Schopper, F., Soltau, H., Kuhnel, K. U., Andritschke, R., Schroter, C. D., Krasniqi, F., Bott, M., Schorb, S., Rupp, D., Adolph, M., Gorkhover, T., Hirsemann, H., Potdevin, G., Graafsma, H., Nilsson, B., Chapman, H. N., Hajdu, J. (2011) *Nature*, 470, 78.

[68] Caleman, C. C. C., Huldt, G., Maia, F., Ortiz, C., Parak, F. G., Hajdu, J., van der Spoel, D., Chapman, H. N., Timneanu, N. (2011) *ACS Nano*, 5, 139.

[69] Owen, R. L., Holton, J. M., Schulze-Briese, C., Garman, E. F. (2009) *Journal of Synchrotron Radiation*, 16, 143.

[70] Fouras, A., Kitchen, M. J., Dubsky, S., Lewis, R. A., Hooper, S. B., Hourigan, K. (2009) *Journal of Applied Physics*, 105, 102009.

[71] Haque, M. M., Yamada, H., Moon, A., Yamada, M. (2009) *Journal of Synchrotron Radiation*, 16, 299.

INDEX

In-situ *Characterization of Heterogeneous Catalysts*, First Edition.
Edited by José A. Rodriguez, Jonathan C. Hanson, and Peter J. Chupas.

www.ingramcontent.com/pod-product-compliance
Lightning Source LLC
LaVergne TN
LVHW011552200926
848484LV00001B/305